S.B.Lm

Lecture Notes in Artificial Intelligence 2117

Subseries of Lecture Notes in Computer Science
Edited by J. G. Carbonell and J. Siekmann

Lecture Notes in Computer Science

Edited by G. Goos, J. Hartmanis and J. van Leeuwen

Springer
Berlin
Heidelberg
New York
Barcelona
Hong Kong
London
Milan
Paris
Singapore
Tokyo

Meurig Beynon Chrystopher L. Nehaniv
Kerstin Dautenhahn (Eds.)

Cognitive Technology: Instruments of Mind

4th International Conference, CT 2001
Coventry, UK, August 6-9, 2001
Proceedings

Springer

Series Editors

Jaime G. Carbonell,Carnegie Mellon University, Pittsburgh, PA, USA
Jörg Siekmann, University of Saarland, Saarbrücken, Germany

Volume Editors

Meurig Beynon
University of Warwick
The Empirical Modeling Research Group, Department of Computer Science
Coventry, CV4 7AL, U.K.
E-mail: wmb@dcs.warwick.ac.uk

Chrystopher L. Nehaniv
Kerstin Dautenhahn
University of Hertfordshire
Adaptive Systems Research Group,
Faculty of Engineering and Information Sciences
College Lane, Hatfield Herts AL10 9AB., U.K.
E-mail: {C.L.Nehaniv/K.Dautenhahn}@herts.ac.uk

Cataloging-in-Publication Data applied for

Die Deutsche Bibliothek - CIP-Einheitsaufnahme

Cognitive technology: instruments of mind : 4th international conference ;
proceedings / CT 2001, Warwick, UK, August 6 - 9, 2001. Meurig Beynon .. (ed.). - Berlin ; Heidelberg ; New York ; Barcelona ; Hong Kong ; London ; Milan ; Paris ; Singapore ; Tokyo : Springer, 2001
(Lecture notes in computer science ; Vol. 2117 : Lecture notes in artificial intelligence)
ISBN 3-540-42406-7

CR Subject Classification (1998): I.2, I.3.7, K.3.1, K.4.3, H.5.3

ISBN 3-540-42406-7 Springer-Verlag Berlin Heidelberg New York

Springer-Verlag Berlin Heidelberg New York
a member of BertelsmannSpringer Science+Business Media GmbH

http://www.springer.de

Printed in Germany

Typesetting: Camera-ready by author, data conversion by PTP Berlin, Stefan Sossna
Printed on acid-free paper SPIN 10839867 06/3142 5 4 3 2 1 0

Preface
Cognitive Technology: Instruments of Mind

Cognitive Technology is the study of the impact of technology on human cognition, the externalization of technology from the human mind, and the pragmatics of tools. It promotes the view that human beings should develop methods to predict, analyse, and optimize aspects of human-tool relationship in a manner that respects human wholeness. In particular the development of new tools such as virtual environments, new computer devices, and software tools has been too little concerned with the impacts these technologies will have on human cognitive and social capacities. Our tools change what we are and how we relate to the world around us. They need to be developed in a manner that both extends human capabilities while ensuring an appropriate cognitive fit between organism and instrument. The principal theme of the CT 2001 conference and volume is declared in its title: Instruments of Mind.

Cognitive Technology is concerned with the interaction between two worlds: that of the mind and that of the machine. In science and engineering, this interaction is often explored by posing the question: how can technology be best tailored to human cognition? But as the history of technological developments has consistently shown, cognition is also fashioned by technology. Technologies as diverse as writing, electricity generation, and the silicon chip all illustrate the profound and dynamic impact of technology upon ourselves and our conceptions of the world. The instruments afforded by these technologies continue to evolve and to shape the minds that first conceived them.

The technologies of the third millennium promise mind-machine interactions of unprecedented intimacy and subtlety. These interactions embrace radically new kinds of experience that force us to re-examine fundamental concepts of embodiment and consciousness which frame our understanding of the relationship between minds and machines. The implications of these interactions will hinge on the ways in which humans make meanings out of these new experiences. The conference and these proceedings address this issue using the diverse perspectives afforded by a wide range of disciplines, and evidence drawn from both contemporary developments and the history of technology. Its aim is to deepen our insight into the potential influence of current and future technologies over people and society.

1 The Making of Meaning

The CT 2001 conference focuses on the core question of how technology contributes to the making of meaning. 'The making of meaning' is to be broadly interpreted as referring to all the activities by which significance is attached to the actions of people and machines engaging with a technology. For a new technology, meaning is in the first instance associated with intended and preconceived applications. The pioneers of the motor car were first preoccupied with refining the car engine, supplying the primary driver controls, building basic roads. As a technology matures, new meanings typically emerge, as skills are acquired, and unforeseen functionality is identified. Driving skills and protocols evolve, the car becomes a status symbol, the drivers are subject to road rage. A new technology typically establishes a pattern of usage, and an associated social organization. Driving regulations are introduced, and the organization of families, industries, and cities comes to reflect greater mobility and autonomy. This in turn spawns languages and conventions that are universally understood by proficient users of the technology. New features and classifications of road are created, and resources to provide services, information, and training about cars and driving are developed. Established technologies supply the metaphors that influence the ways in which we interpret and communicate our experience. Access to autonomous travel is perceived as a norm, neighbouring cities converge, metaphors such as "giving a proposal the green light" and "stepping on the gas" invade our language.

The contribution of technology to the making of meaning through these processes has been analyzed in many ways: in the design and creation of technologies and artifacts themselves; in the psychological, sociological, and historical analysis of their individual and corporate use; and in the philosophical implications for our modes of thought and ways of communicating. A proper understanding of the processes of mutual co-evolution and adaptation which shape our interaction with the technology of the computer age will ultimately require a holistic rather than a reductionist approach. Given our current understanding of these matters, an integrative and holistic account is inevitably a long term ambition, but it is an ambition which must not be forgotten. With this in mind, CT 2001 addresses the core question of how technology affects the making of meaning from the following perspectives taking both empirical and more analytical or philosophical approaches.

2 The Personal and Experiential

The impact of technology upon individuals is central to our understanding of the making of meaning. Technologies such as writing and number systems have provided us for a long time with the ability to extend our cognitive and conceptual operations, and various new technologies take this further by offering enhanced representational and perceptual capacities which change the nature of human experience as an embodied condition. This raises very difficult questions about the role of embodiment, affect, and consciousness in the making of meaning, as individuals begin to operate with altered or novel perceptual capacities in virtual or real environments which are seemingly unconstrained in the relationships they permit between self and world, and self and other. It also has implications relating to ethics and aesthetics, and thus to psychological well-being.

3 The Social

The persons who are affected by a technology will not only change their role in the constitution of their social world, but are also affected by how that technology is embedded in and changes their social order. Consequently, any proper understanding of the conference theme must turn to macrosociological accounts of the impact of technology. We are already witnessing how new technology is rapidly changing the temporal and spatial dimensions of communication and decision making, and how this is having a differential impact on sections of society. It can isolate those who do not have access to it, but it can also bring together those who were previously separated by custom, prejudice, or geography. These changes are potentially of great significance for the structuring of society and the access to political power and economic resources of different persons and groups. This raises important questions concerning the access to and organization and regulation of these technologies.

4 The History of Technology

Whilst we live in times of great technological change, technologies which have a major impact are not novel. Studies of Cognitive Technology have, for the most part, been focused upon contemporary and emerging computer-based technology, but there is no reason why studies of earlier technologies cannot yield important lessons. Indeed it would be foolish to ignore what can be learned from an analysis, comparative or otherwise, of technologies which have gone the full cycle from invention and introduction, to acceptance and maturity, to the point where they become a seemingly natural part of the world for all. This analysis would necessarily focus on the co-evolution of technologies, societies, and persons as each adapts to the changing circumstances.

5 Education and Individual Development

Any newly born child faces the challenges provided by the technologies of the society into which he or she is born, and must develop in some appropriate fashion if he or she is to prosper. The sense and meaning which they find in a technology may differ from that which their parents found in it at an earlier stage of its introduction or development. This has consequences for both the individual and social perspectives mentioned above, and it is important to understand how each new generation comes to understand and respond to the meanings of a technology for itself. Technologies are also significant in individual development in the ways in which they offer differing kinds of educational engagement and experience. Constructivist approaches to learning highlight a potentially key role for technology in education. Understanding the current role and future scope of educational technology is intimately bound up with understanding how it is implicated in the making of meaning. This motivates a re-evaluation of traditional theories of knowledge representation and of educational development in the light of, for example, new advances in web-based learning and mind-computer interfaces.

6 Creating, Designing, and Engineering

Ultimately, each of these perspectives is only of more than academic interest if it can be translated into understandings which can affect the processes of invention and design. Consequently, CT 2001 considers such translations in the light of particular engineering practices, both successful and unsuccessful. The contemporary context for design highlights the need for a more holistic approach to design such as Cognitive Technology commends. Key issues include: the need to take account of requirements that cannot be preconceived, but evolve through feedback and adaptation in use; the problems of devising abstract models of mind and machine to support the design of applications that use new technologies (such as virtual reality, robotics, and brain-mediated interaction); and the paradoxical way in which the social and technical infrastructures that enfranchize particular technologies can obstruct alternative creative developments.

June 2001

Meurig Beynon
Chrystopher L. Nehaniv
Kerstin Dautenhahn
David Good
Barbara Gorayska
Jacob Mey

Cognitive Technology: Instruments of Mind

The Fourth International Conference on Cognitive Technology: Instruments of Mind held Monday 6th - Thursday 9th August, 2001 at the University of Warwick, United Kingdom, was hosted by the Empirical Modelling Laboratory, Department of Computer Science, University of Warwick.
CT 2001 was supported by the Computer Science Department of the University of Warwick, U.K.; the Adaptive Systems Research Group of the University of Hertfordshire, U.K.; the Cognitive Technology Society (CTS); the Media Interface and Network Design (MIND) Labs of Michigan State University, U.S.A., host of CT'99; the University of Aizu, Japan, host of CT'97; the City University of Hong Kong, host of CT'95; and also by Springer-Verlag, publisher of these proceedings, and by John Benjamins Publishing, publisher of CTS's *International Journal of Cognition and Technology.*

Conference Chair

Meurig Beynon	University of Warwick, U.K.

Scientific Program Chairs

Kerstin Dautenhahn	University of Hertfordshire, U.K.
Chrystopher L. Nehaniv	University of Hertfordshire, U.K.

Conference Committee

(the above and)

David Good	University of Cambridge, U.K.
Barbara Gorayska	City University of Hong Kong
Jacob Mey	Odense University, Denmark

Invited Plenary Speakers

Steve Benford, Tom Rodden	University of Nottingham, U.K.
Martin Campbell-Kelly	University of Warwick, U.K.
Andy Clark	University of Sussex, U.K.
Judith Donath	MIT Media Lab, U.S.A.
David Gooding	University of Bath, U.K.
Steve Talbott	The Nature Institute, U.S.A.

International Program Committee

Table of Contents

Presence in Virtual Environments

Human Activity & Human Computing

Implications for Technology

Computing and People

Education & Cognition

Learning

Narrative and Story-Telling

Interfaces

Cognitive Dimensions

Society & Technology

Human Work and Communities

Human-Technology Relationships

Cognitive Technology: Tool or Instrument?

Barbara Gorayska[1], Jonathon P. Marsh[2], and Jacob L. Mey[3]

[1]City University of Hong Kong
csgoray@cityu.edu.hk
[2]Higher Colleges of Technology, United Arab Emirates
jonathan.marsh@hct.ac.ae
[3]University of Southern Denmark, Odense
jam@language.sdu.dk

Motto:
"Toiling to live, that we may live to toil"
(Wm. Morris, 19th century Utopian thinker,
quoted in Ehn 1988:1, 371)

Abstract. This paper discusses the tool aspect of the cognitive artifacts often referred to as 'instruments of mind'. Having established the basic distinction between tool and instrument, the authors then go on to review the notion of artifact itself, and discuss the potential for mind change that is inherent in the use of 'mental' instruments such as the computer. It is pointed out that the relationship between the mind and its instruments is a dialectic one, and that the 'reflexivity' inherent in this relationship constitutes the very nature of our interaction with cognitive instruments, such as it is studied in Cognitive Technology.

1 Phenomenology of Terminology

Why, in our everyday use of language, do we make a distinction between the terms 'tool' and 'instrument'? Closely related, they don't seem to quite acquire the status of the synonymous. We talk about instruments for making music; we have surgical instruments; we are familiar with the instruments on the dashboard of a car or in the cockpit of a plane, and would never think to refer to them as tools. Alternately we talk of a tool box, carpenter's tools, bicycle tools, gardening tools, etc. and would become disoriented should someone speak of them as instruments. Clearly we make a distinction but upon what basis? It is possible that an examination of this distinction may lead to insights into the ways humans construct and orchestrate environmental interactions and events. This paper explores that possibility.

1.1 So What's the Real Difference?

One clear (but maybe superficial) example of how we differentiate between tools and instruments emerges when we compare their representation in a car: i.e. the instruments are found on the dashboard while the tools are in the trunk. Can we say

M. Beynon, C.L. Nehaniv, and K. Dautenhahn (Eds.): CT 2001, LNAI 2117, pp. 1-16, 2001.

then that distinction is simply a matter of relative location and importance of function? We think not (Is a cigarette lighter a tool? An instrument? Or a gadget?). However the example is useful in that is suggests that the distinctions we make between the various objects we purposefully use are in some critical sense socially constructed.

Take a hammer. A hammer as such is just a hammer; following Marx; we can say that it only becomes a *tool* by becoming 'socialized', that is, by entering the production process.[1] This socialization is critical to the determination of its status. What is interesting is that the determination process is governed by an emphasis on user need and user skill, not on the object as a physical entity. Consider that the same hammer, in the hands of a physician, becomes an *instrument* rather than a tool (e.g. when testing for reflexes).

1.2 Emphasis on the User

Pelle Ehn has remarked that "the tool perspective takes the labor process as its origin rather than data or information flow" (1988:375). The importance of this perspective becomes evident if we consider instruments to be a particular class of highly specialized tools, designed to help the worker perform an operation in a skilled manner, or vice versa, when instruments are 'created' out of tools by the skill of the operator (e.g. it is the skill of the surgeon that converts the humble kitchen knife into a surgical instrument).

Still, it is not simply the measure of technical sophistication which transforms a tool into an instrument; tools can be highly sophisticated, but still not qualify as instruments. For instance, if I upgrade the common goad-stick (a long rod with a nail attached to its end) to an electric or electronic device for goading on my cattle, it remains a tool, even though I am employing relatively complex technology. But if I govern the animals by remote control and am to finely steer their movements from a distance, in the fashion of a boy flying his miniature airplane, then I'm using an instrument. Here, it is my skills in manipulating the device to a purposeful end that constitutes the *instrument* as such.

1.3 Emphasis on the Nature of the Operation

We further differentiate between tools and instruments based on the operations they perform. Tools are essentially for constructing, repairing, or modifying. Instruments are for 'instructing' or 'guiding' a process[2]. When my car breaks down, I don't start banging on the instruments on the dashboard; I get out my wrench and jack and other tools and try to fix the problem.

---

1 In the capitalist mode of production, the tool furthermore becomes capital. Cf.: "A negro is a negro. Only under certain conditions [that is, in a slave economy] does he becomes a slave". (Marx 1971:155; quoted in Ehn 1988:97)

2 (in accordance with the term's etymology: Latin *instruere* ("to guide") plus the suffix *-mentum*, as in *tegu-mentum* 'that by which I cover' (e.g. a blanket, a roof), *indu-mentum* 'that which I put on' (a garment)

Whereas a tool is for running or 'debugging' an operation, an instrument 'instructs', 'in-forms' it. While the tool directly extends and amplifies user skills, the instrument serves to govern and help manage those skills. Both tools and instruments have to do with the use of skills and the performing of an operation. However, in the case of the tool, emphasis is more on the operation and the kind of work being performed, while in the case of the instrument, emphasis is more on user skills and the way in which the work is being done.

Given this distinction, is it then possible to define any specific object as belonging exclusively to either class? Again we think not. Instead it may be reasonable to define a class of objects, which we will call artifacts, which may be functionally defined as having the characteristic of being able to be purposefully used by a conscious reasoning agent. Within this class there are a number of subclasses such as "tools", "utensils" "instruments", "gadgets", "machines", 'prosthetics", "organisms", et cetera. Each of these subclasses has determinant characteristics but are not mutually exclusive. Hence subclass membership of any particular artifact, on the occasion of its use, may be determined by the context and purpose of use as well as the measure of attention applied by the user. A smooth continuum between the cognitive characteristics of the user and the functional potential of the artifact is implied. Unpacking the parameters of that continuum may help us to better formulate more environmentally sensitive and humane instrument design principles.

1.4 On Artifacts, Cognitive, and Others

The notion of artifact was originally coined in physical anthropology and archaeology. It is used to indicate the presence of a human agent in a piece of nature, as for example a fetish or a tool. If we find such an 'artificial' object in nature, our first thought is that there has been somebody out there who made it, or put it there. By extension and extrapolation the artifact can be used in various other ways: e.g. to support a claim for the existence of God.[3] To further extend the notion of human agency as critically important to the defining of an artifact, one could consider finding to be an act by an agent that transforms the object or piece of nature into an artifact by the mere fact of its having been found. Such is the case of the so-called *objet trouvé*, the odd item (possibly itself already an artifact) that enters the artist's conception of nature, as expressed in the work of art.

As the name suggests, cognitive artifacts may be considered to be of a different, special kind: they have something to do with cognition, with the way humans 'cognizingly' enter and represent the world. A good example is the artifact commonly known as the book. Norman gives the following description.

"Cognitive artifacts are tools, cognitive tools. But how they interact with the mind and what results they deliver depend upon how they are used. A book is a cognitive tool only for those who know how to read, but even then, what kind of tool it is depends upon how the reader employs it." (1993:47)

---

[3] This is the first of the four 'canonical' proofs, often called 'the watch in the desert proof'; among the remaining three, the best known is the so-called 'ontological one', of dubious fame.

Considering the tool as an artifact of cognition means that we somehow must be able to use it in our cognitive operations. At the low end of toolness we find such devices as the cairn or other stone artifacts, possibly used in measuring the time or seasons, or a system of arrows designed to point the way to food, or the primitive stone cutter's chisel, when he uses it to produce an inscription. In the case of the book, the feedback that various readers get from the book may be quite different, depending on their world orientation and cognitive characteristics. If we throw the book at someone (in the non-metaphoric, extra-judiciary sense), we use the book as a weapon. Children use books to tear them apart. Older children look at pictures. Adults (and younger proficient readers) spurn pictures and want to go directly to the text itself. Mature readers take in all these 'bookish aspects' and synthesize them into one smooth, well-adapted reader behavior.

Obviously the complexity of the tool is one important factor to consider (see more on this below, section 3). But complexity in itself is not definitive. As we have said, a tool can be extremely complex, yet not transcend its 'toolness'. This is where representation enters the picture.

Cognitive artifacts represent the world to us. Charles S. Peirce distinguished between three ways of representing: by *indexes* (the arrow pointing to some location), by *icons* (the artifact having a certain resemblance to the object represented, such as a common pictogram for 'No Smoking', and by *symbols* (tools such as words, that represent through a mental operation of recognition that has nothing to do with their shape).

But representation is not just a state: it involves a process, an activity. Humans are representing animals: they attach meaning to the things they do, meaning that is very often far removed from the simple representation of the activity itself. For instance, walking as such is pure locomotion and has no meaning other than to move from one place to another; but if a member of the Roman Senate picks up his toga and walks from one end of the Senate Chamber to the other, in order to take up his place there along with his colleagues taking similar action, he performs an act of *voting*: 'going with your feet to a decision', *pedibus eundo in sententiam*. The movement becomes a 'motion', as we nowadays call it, seconded and approved by the body, the feet[4].

But what of cognitive artifacts and their characteristics? How do they represent meaning and action? As Norman points out, "to understand cognitive artifacts we must begin with an understanding of representation" (1993:49). But this is not enough. Far from it: the best representation only comes alive on the condition that we have a representer who actively interprets the represented. In other words, the artifact (of whatever kind) must be such that it not only exhibits enough complexity to offer a complete (or at least passable) representation, but also represents in such a way that the people using it have no trouble identifying what is represented and how it works.

Moreover, whenever an artifact represents, it does so only on the condition that its way of representing is adapted to, and adaptable by, both the represented and the representer. This adaptation, however, shouldn't be seen as a quality given 'once and for all'; adaptation is a process of give and take, of mutual conditioning, in short a *dialectical* communicative process. We would like to explore in greater detail how this process affects our view of the mental instruments we use. However, first we

---

[4] Ernst Cassirer called humans 'symbolic animals', which is of course the same idea, but applied more narrowly to language as the symbolic representation *par excellence*

would like to situate that exploration within the context of several propositions. We propose that a) all artifacts are in some measure cognitive artifacts; b) the mind itself can be viewed as a goal driven cognitive artifact that brains develop as an aid to interacting with the environment through the organization and integration of perceptual data; and c) all artifacts can be situated on a continuum of purposeful use between the extremes of raw material and the mind (see Fig. 1).

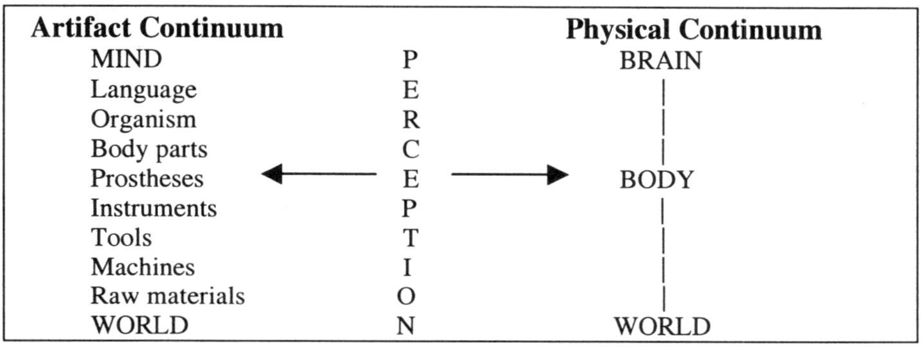

Fig. 1. Relationship between a continuum of artifact subclasses and the physical continuum between a perceiving brain and the environment.

The computer is a cognitive artifact *par excellence*. But what do we mean when we call it an 'instrument' of mind? To answer this question, we need to establish the basis upon which we distinguish between subclasses on this continuum of artifacts. Such distinction is important to informing the design of complex artifacts such as computers and for understanding the way we interact with them. This will be discussed further in section 4, below. But before we do that, let's turn to another question, viz.: whose instruments and whose minds are we talking about when we discuss 'instruments of mind'?

2 Whose Instruments, Whose Mind?

2.1 The Computer Worker and His/Her Tools

Normally, we consider the worker as the one who is in control of his or her tools. Other people are kindly but strongly reminded to stay out of the work place and not touch the tools unless allowed to do so. Here is a warning that appeared on the door of the main computer room at Yale University's Department of Computer Science in the early eighties (Fig. 2):

ACHTUNG!!!

Alles touristen and Non-technischen Lookens Peepers!

Das Machine control is nicht fur Gerfingerpoken und Mittengrabben. Oderwise is easy Schnappen der Spriggenwerk, Blownfuse, und Poppencorken mit Spitzensparken. Der Machine is Diggen by Experten only. Is nicht fur Gerverken by das Dumnkopfen. Das Rubberneken Sightseenen Keepen das Cottenpicken Hands in das Pockets. So Relaxen und Watchen das Blinkenlight.

Fig. 2. Warning at Yale University's Department of Computer Science.

The message is clear, despite (or maybe even thanks to?) its baroque Germanesque 'command language': "Hands Off, All Non-Expert Users"! The persons behind the message, the experts themselves, did not want any interference with their instruments from people who didn't know how to use them. In other words, the experts exercised propriety rights over the computer: the computer was properly theirs.

In yet another sense, the computer was the experts': the design that was prevalent in the early years centered around the people who knew. Thus, the whole operation of the computer room had something of the arcane atmosphere to it that normally is associated with secret societies and exotic priesthoods. In one sense the computers were instruments designed to be manipulated by experts only, not by common users. However, the operations that these experts performed had mostly to do with maintenance and repair. Hence in another sense the central computer was a massive, central tool that served a whole community of 'end-users' (notice the term 'server', which seems appropriate in this context).

All of this changed with the advent of workstations and later, personal computers. Gradually, design, too, shifted: from being oriented towards 'experts only' to a concern for the 'dummies' (cf. the name of a family of very successful computer instruction books, whose general heading reads 'X For Dummies' (where 'X' could be 'UNIX', 'MS Word', 'Mac OS' or any other computer-related subject). End-users (the ones using the computer in their daily work) were now the focus of design, and the word 'user-friendly' became a shibboleth for the implementation of programs that could assist the user in his or her daily work at the terminal. What earlier had been a tool for performing specific operations (scientific calculation 'number crunching', strict 'word processing', or at most some primitive computer game such as 'Adventure') now started to live a life of its own as an instrument of the mind. Users started to be creative with their texts in ways that had not been possible before and as a result, the instrument not only facilitated the workings of the mind, but in addition, and more importantly, influenced the very way the mind functioned.

2.2 Who Owns Our Minds?

The question could be raised as to whether, as a matter of principle, minds can be owned, period. Hasn't Schiller told us that 'Thoughts are free' (*Die Gedanken sind frei*)? Even if we speak of a 'captive mind', or a 'captive audience', or of being 'captivated' by some presentation, we are using metaphoric terminology; we don't really envision our minds or our thoughts as belonging to someone else than ourselves.

Yet, there may be more to the metaphor than meets the eye (or the mind, for that matter). If we think of the mind in a modular way, as suggested by Marvin Minsky (1986) in his famous model of 'the society of mind', where several semi-independent 'operators' join to produce what we see as a unitary operation, then we can also take the next step, in which we reproduce the modules of the brain on an artificial basis.

To turn to a more recent development, given that, in the terminology of Artificial Life (AL), we can define (biological) life itself as a function (Helmreich 1998), we certainly can define certain parts of the human 'wetware' as programmable units, to be incorporated in physically independent, replaceable parts that can operate in the fashion of a computer component.[5] And once we have taken that step, it is easy to imagine that a market for such replaceable 'wetware' spare parts comes into being, where electronic brain simulation components are bought and sold. If you can't afford to have your brain (partially) replaced, you can lease an artificial brain, or part of it; that part then essentially belongs to, and is under the control of, the company or person who purchased it initially. It can be remote programmed and steered, and people will truly 'be like machines' (Mey 1982).

As we see, the viewpoints commonly advocated by Cognitive Technology in these matters, especially with regard to the way the computer impacts on our minds, imply a number of grave consequences that we ought to consider carefully in the light of our assumptions about the mind as an artifact. Is it the case that the artifact will itself become the main player in this game? How can we vouchsafe for the freedom and independence of the individual, while keeping the beneficial effects of 'computerizing' the mind? Without giving up our view of CT as 'mind-boggling', we still don't want to end up in what one could call the extreme and perverse end of CT, where everything's just computational (as in the illustrious 'computational tar-pit', to use Alan Perlis' expression; cf. Mey 2001:178), and where the 'soul of language' (Gorayska 1994), the mind itself, is forever gone.

3 Mind Changes Instrument, Instrument Changes Mind

Don Norman has remarked that "artifacts [including such things as tools and instruments] change the tasks we do" (1993:78); but it is equally true that they change

---

[5] The idea, originally due to researchers like Christopher Langton and his associates (see the articles in Langton 1989, 1992, 1994), that AL is all about essentially programmable functions replicating "the logical form of life", is expressed as follows by Stefan Helmreich: "formal and material properties of entities can be usefully separated, and what really matters is form" (1998:211). For some fascinating visionary illustrations from the sci-fi literature, compare the works of William Gibson, e.g. *Neuromancer* (1984).

our minds, such that the tasks we seem to be performing are not only not the same tasks any longer, but in addition, we consider ourselves as changed in relation to the tasks.

A housewife owning a vacuum cleaner is changed by this fact of ownership, as many have remarked when this household gadget (supposed to relieve the lives of countless women around the world) turned out to be a mighty tyrant in its own right, adjusting and raising the standards for housework that had been prevalent until then[6].

Among the things that are said to distinguish tools from instruments are the amount of 'guidance' that the artifact allows and respectively expects us to exert. Instruments provide feedback on a process; tools usually just make the process happen, and do not adjust themselves to change the product as it emerges. A glass blower's pipe is a tool that permits the blower to shape the product at the end of the pipe, and currently adjust its form until the desired shape has been obtained. But if we want to do the same, using a mechanized artifact such as an 'automatic glassblowing machine', we will need to have controls steering the blowing process: instruments that allow us to keep track of the ongoing process, and adjust it so that the machine continues to operate satisfactorily. But at the same time, the instruments act on our minds; they are an intermediate point on the continuum between mind and tool, more 'reflexive' than tools and coming closer to what Norman calls a 'compositional medium', where 'compositionality' is understood as the quality of mind that "allows [affords] adding new representations, modifying and manipulating old ones, and then performing comparisons" (1993:247).

There is, of course, always some reflexive activity involved in even the simplest mechanical tools and their operation. We hear the difference in sound when the saw is half way through a log, and when it approaches the bottom side of the log; should we be so unfortunate as to graze a hidden nail that someone has hammered into the tree many years ago, again, the sound will stop us from sawing on. However, there is a considerable distance from, and difference between, the reflexivity that is happening in the mind when we think and compare the process with its outcome (all the time adjusting and evaluating, then adjusting again, along different dimensions) and the reflexivity involved in tool use, where the reflexive cycle is unidimensional and short-lived (as in the saw example).

We want to suggest that it is a lesser or greater distance that allows us to project the various artifacts that join mind and environment along its scale, and that this scale is matched by another one, going in the opposite direction, that of feedback. Feedback is inversely proportionate to distance, such that the artifacts closest to the human brain have the most extensive and most 'compositional' feedback (i.e. they feed information back into the brain along different dimensions, and allow for comparisons between the various effects of the process). At the bottom of the scale, we find the world, as it presents its raw materials to us, offering itself for development by the means of intentional action. This relationship is graphically depicted in Fig. 3.

Since a defining characteristic of membership in any of the arbitrary artifact subclasses on the continuum is that of the relative distance they reflect between an aritfactual construction and the brain of the user, individual members are determined by the degree of alignment between the user's cognitive characteristics and the

---

[6] Another more extensively examined example is that of the leaf-blower discussed in J. Mey, 1996.

representation(s) of the artifact. The less distance between them, the more direct the interconnection between their inherent operations.

A good example is an artificial limb as compared with its natural equivalent. In the case of an artificial limb, say a hand, the natural nerve connections between the brain and the prosthetic hand have been severed and only certain of them have been artificially replaced; consequently the functionality of the artificial hand is greatly reduced. We can still perform basic functions such as grasping or moving objects, but we can no longer feel any impact force upon striking, or respond to variations of temperature or texture upon touch. A clear distinction between prostheses and body parts, both conceived of as instances of artifacts, therefore, falls naturally out from the discrepancy in the degree of functional connectivity. The same considerations apply in discriminating between prostheses and instruments, instruments and machines, machines and tools, or tools and the raw materials available in the perceivable environment.

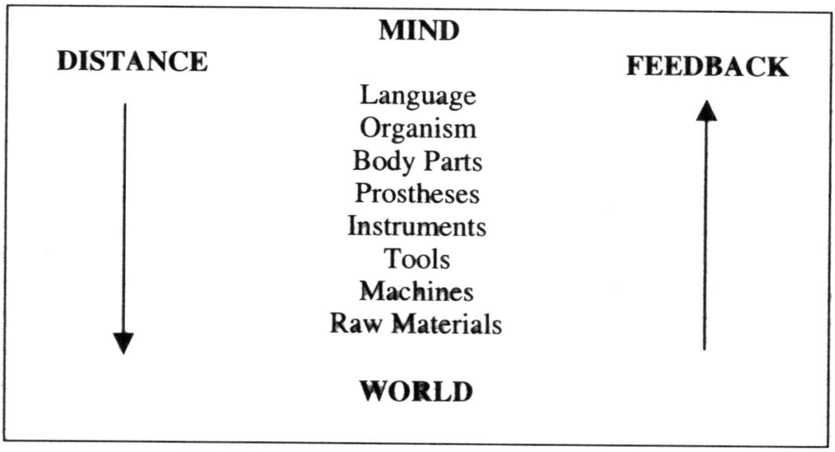

Fig. 3. Distance and feedback as inversely proportional determinants of artifact subclasses. The broad class of artifacts maps onto the continuum in ways determined by the ability of a given artifact to be used for a specific purpose by an external agent. Once again, this allows any member of the class of artifacts to be used in a variety of subclasses depending on the context, intention, and measure of attention of the user, as explained in greater detail below.

Generally speaking, the degree to which an artifact is connected to the functions of the brain determines the full measure in which the brain is able to attend to the feedback it receives from the interface. The more externalized or detached an artifact is from our brain and the physical body, the lower the measure of brain-artifact connectivity and alignment, and consequently the lower the ability to attend to feedback. The more the ability to attend to feedback is constrained, the lower the ability of the user to adapt to and or beneficially modify the artifact.

The significance of feedback from an artifact may further be amplified or dampened by contextual variables both external and internal to the user. Essential factors here are: 1) an external environmental potential for or constraint on action (partially determined by the variations in connectivity discussed above); 2) our internal motivational states; and 3) linguistic or cultural mediation in our interpretation of perceptual data.

Feedback mechanisms register changes in degrees of satisfaction with respect to currently detected needs (motivational states). Since the attention of the user as a perceiving agent is governed by what matters most to him or her at the precise moment of perception, the degree of attention paid to intervening variables in generating behaviors is directly linked to our receptivity to feedback. When we consciously attend to feedback, linguistic and cultural variables come into play and constitute an additional set of constraints that act as filters on perception and limit the potential for action.

Perception and adaptation are dialectically interdependent. How we as users of artifacts adapt to the environment that we inhabit, depends on how that environment constrains our perception of our needs. How we subsequently adapt the environment such that it satisfies the newly perceived needs depends in turn on the available resources that such needs allow us to perceive. Cognitive techniques emerge to aid this process first as mental artifacts (*vide* categorical perception or generating effective goal-action schemata) which may or may not be later embodied as sets of socially shared artifacts. Embodied artifacts become a part of the environment and can be further adapted by others. They may also cause us all to further adapt to their use. The process is highly recursive and all variable values are subject to constant change.

Socially transmitted language and cultural variation also influence the subclass determination of artifacts. This process appears arbitrary but may in fact also be strongly determined by the goals we pursue and the perceivable environmental characteristics within a cultural community. Each subclass binds together a number of cognitive/perceptual filters that constrain the perception of needs, as well as of the objects and states within the environment that can satisfy those needs. Consequently, effective action sequences can only be generated within the delineated frames of reference. In other words, the determination of relevance is seen as a function of the relationship between perceived needs and actions on objects in the environment that satisfy them. The perceptual filters associated with each subclass of artifacts constitute culturally specific constraints on our ability to assign relevance and consequently pay attention to the purpose of a particular artifact.

As we have said earlier, the nature of user attention is critical in determining the subclass of an artifact, because it triggers the process of adaptation and conditions our capacity to perceive purpose in an artifact by mediating the relationship between feedback and distance. Attention can be measured in terms of longevity (the length of time a set of perceptions remains in focus) and in terms of intensity (the degree to which cognitive processing capabilities are brought to bear on the object of perception, e.g. absence or presence of response across a varying number of processing nodes). The significance of feedback from an artifact may be amplified or dampened by contextual variables, both external and internal to the receiver. It is by nature volatile. In the example of a man sawing wood, despite the consistency of information, the feedback taken from the saw cannot be seen as constant across all similar instances. The variance is in the receiver. The man sawing his four hundredth log will receive feedback from the saw very differently than the man sawing his first. Adaptation has occurred, changing the required parameters of attention.

The notion of measurable adaptability is key to the establishment of useful design principles. Their achievement is highly dependent on the measure of user attention they entail. Elsewhere, we have discussed the notion of relevance as the anvil of attention (Gorayska and Marsh, 1996). It follows that relevance governs the extent to which adaptability is established in an artifact. It also governs the way in which we are able to overlay "tools" with "instrumental" characteristics through application. The more we attend to the feedback provided by an artifact, the more we are able to attenuate the purpose to which we are applying it and the higher it climbs on the continuum. Hence a scalpel can become an instrument and a computer can become a tool. Relevance parameters that govern the distinction between subclasses on the artifact continuum are shown in Fig. 4.

PROCESSES
Adaptation: initiation, control
Feedback: recipient, provider
Attention: scope, intensity, longevity

SYSTEMIC ORGANIZATION
Complexity: degrees of freedom
Intensity: simple, complex

AGENCY
Free or dependent
External or internal

Fig. 4. Relevance parameters governing the distinction between subclasses on the artifact continuum

Set in this framework, Cognitive Technology as a discipline explores the processes by which cognitive techniques (schemata) come into being and how they mediate human perception such that they govern user-environment adaptation. Its primary goal is the development of methods and practices for designing artifacts that maximize human benefit. By seeking to further unpack the relationship between human agents and artifacts, we may gain some insight into how recursive changes in perception about desired outcomes condition and determine the generation and design of technology.

4 Design vs. Use

Given the considerations above, what are principles of good design? In particular, what principles should guide us in designing artifacts that embody the principles of Cognitive Technology? We will consider these questions first from the point of the user, then from that of the designer.

4.1 The Use(r)

In questions of design, the user is traditionally thought of as an 'end-user', that is the person at the keyboard, not the persons dealing with software or hardware at higher levels of use (such as software support managers, systems analysts and programmers, and so on).

These users are the ones that have to live 'with' the software, not 'off' it; hence the demand that instruments of mind must be such that the user is able to live with them. This claim for 'convivial' tools (originally formulated by Ivan Illich, 1973) centers around the notion of a dynamic, changing context, in which a living organism adapts itself to constantly new environments, new challenges, new instruments, in short, a new life. With the reservations formulated elsewhere (Mey 1998), the user, in order "to survive as a skillful tool-user of computer artifacts, [has] to adapt to changes, not just get more experienced with the tools [he/she] already knows." (Ehn 1988:394)

4.2 The Design(er)

The first and most important question to raise in the context of CT design is: Who is the designer working for? (A variation on the theme of: 'Whose minds, whose instruments?'). The designer is a Janus-like figure, who at the same time must keep an eye on the market and his or her employer (and more indirectly, on his or her own status in the market place or the company), and on the user, as the final arbiter in matters of what sells and what doesn't.

As Norman once remarked, even the very notion of 'user-friendliness' can be read in different ways, and not all of them are to the benefit of the user. What the designer thinks the user needs, and consequently incorporates into his or her idea of 'being friendly', may well turn out to be a highly irritating feature of an overblown information content (often repeated to infinity) or a set of instructions that, once internalized, are not 'helpful' any longer (a good example is the 'Help' function on many personal computers: one has to wade through oceans of 'friendly' advice before one gets to the 'meat'; shortcuts are possible, but only practical if you know how to formulate your request for help in the exactly right manner). "Start with the needs of the user" (Norman 1986:59-61) is a good, general recipe for design, but it has to be made more concrete to be useful.

Anticipating the needs and wants of a user can be a tricky business. The traditional profile of a user as one who just wants to have a tool to assist him or her in a particular activity (e.g., replacing the typewriter in the production of written texts) is no longer valid. Users want *instruments*, not just tools, and they want instruments of the mind, that is to say, devices that can help them change their mind behavior, literally 'blow their minds'.

Between replication and expansion of the mind, we have to steer a precarious course. It is all right to opt for a 'user-centered design', in Norman's (1986) felicitous phraseology; but the users' minds and their instruments should not be centered solely on their personal goals, or on the limited goals set by the producer. As we have seen above, in section 3, the mind needs to change in the direction of greater flexibility and true adaptation, not mindless repetition and mechanistic adaptivity (Mey 1998).

4.3 Use to Design, Design to Use

Ehn has drawn our attention to the fact that there is an apparent contradiction built into the very notion of design. Just as "the division of labor is not only social but also technical, [the] design and use of artifacts [read: instruments] is not only technical but also social." (1988:101) What this means is that the technical aspects of shaping an instrument to the specifications of a user inevitably will have an effect on the user him-/herself; and conversely, the influence that the user exerts in choosing between diverse instruments (say, computer software or applications) will have its influence on the way design is carried out.

But not only that: the relationship is one of true *dialectics*, as best illustrated by the general case of tool-use. Improving my tools will not only enhance the quality of my product; as a result, my relationship to the work process will change, and so will my mental attitude, my understanding of what I am doing. Changing the tool thus eventually changes the worker, who then in turn will change the use of the tool and the work process, and eventually the product itself.

In all this, we recognize the fundamental processes that are seen as characteristic for CT. The mind creates its instrument (in our case, subsumed under the general label of 'computer technology'; but using the instrument does not leave the users (and their minds) untouched. This dialectic of mind and instrument has to reflect itself in the design process, designers and users interacting in the creation of mental instruments. Only in this way can the 'hidden dimensions' of design be brought out into the open: to wit, the social relations that are encapsulated in the design, as well as the social labor that is 'congealed' in it, to use Marx's expression. The computer as an instrument of mind is "neither natural nor given" (Ehn 1988:100); the same goes for its use, which has to be negotiated in a context of social responsibility and human adaptability.

5 Conclusion

When talking about instruments of mind, we have to be clear as to what our terms are. In the preceding, we have mostly talked about the various interpretations of the terms 'aritfact', 'tool', and 'instrument', and how they fit our 'mental' picture of the world and ourselves, including our relationship to the work we are doing. The question of what is meant by 'the mind' is a much more vast and tricky one.

As Lindsay has remarked, talking of minds and possible 'mindchanges' as a result of using the computer, "[t]he model of mind which has been most influential in western thought is that associated with Cartesian dualism" (1999:50). In this view, the mind is considered some kind of unit separate from the body, but in contrast to the latter, immaterial and dimensionless. And after the behaviorists' 'mindless dreams' of the late nineteenth and twentieth centuries, the mind seems to have had some kind of revival, a 'ghost' reappearing in the 'machine', to borrow Ryle's expression (quoted by Lindsay, *ibid.*).

But rather than go into the debates around the existence or status of this most elusive of 'aritfacts', the mind (see Lindsay 1999 for a fruitful exposition-cum-discussion as regards the possibilities of 'mindchanges'), we would rather, in the guise of a conclusion, tell you a story about the ways in which the mind operates in relation to one of its instruments, the computer. Or maybe it would be better to start the story at the other end, looking at the computer with our mental eyes, and ask ourselves: What kind of instrument do we want this 'aritfact' to be?

The answer, of course, depends on the way we look at the instrument. To vary Keats, the instrument's beauty, or usefulness, is in the eyes of the beholder. Briefly: the instrument is only as good as the mind that is beholding it, or that it is beholden to. And this has some serious repercussions on the way we see the future of our mind-computer cooperation and interaction.

In the early days of the computer's entry on the mental scene, it was thought of as a practical device for calculating and numbering. It evolved into a tool for replacing drudgery (in mental operations such as dictionary look-up and making concordances). A little bit later (this was in the early sixties), some smart educators found that the computer could be programmed to take over certain tasks that they did not feel motivated to perform, such as the drilling of students in grammar, or the grading of exam papers. The outcomes of such thoughts are well-known: we got the phenomenon called CAI (computer assisted instruction), in which the laboratory drills that used to be performed individually by the students using a tape recorder and a microphone, or just pencil and paper, now were introduced in the form of programmed functions on the screen, where instant feedback (and a possible symbolic remuneration) became the propelling force in the learning process.

In this case, we see how the computer, reproducing a mindless program, became mindless itself. The computer as instrument of learning was no better than the lackluster routines that it replaced; but while the latter always could be repaired by the presence of a live teacher, computer-aided instruction had no such options and as a result, this use of the computer instrument quickly faded into the background.

Grading exam questions in computerized mode led to another atrocity: the 'multiple choice' test. The early computers operated with a programming input derived from the original Hollerith punched cards (Hollerith's patent was acquired by IBM in the late eighteen-nineties, and laid the foundations for its later success and prospering). The thought naturally arose: If one can feed information about programming into the computer in this way, to be executed purely mechanically, then why not try and feed information about students' learning progress into the machine in the same way, and have it 'executed' (read: tested) in a similar fashion?

In order to realize this project, one had to have something that was analogous to the Hollerith system: a punched card with a certain number of options. The five part multiple choice question is therefore a direct derivate of the original 32-slot IBM punch card (we have no idea where the number five came from), and limiting the options for answers in this way, to be executed by punching holes in a card (or by blackening a box on a form) became the standard for both questionnaires and examinations. (We refrain from commenting on recent catastrophic consequences of this punching technique in national elections).

Rather than checking on understanding and learning, the multiple choice test was a check on the presence or absence of certain elements of knowledge, to be separated out and identified as discrete amounts of factual 'stuff'. The only right answer of the five revealed the presence of this knowledge, whereas all the other options, no matter how ingenious, and perhaps more interesting as answers, were discarded as 'wrong'. Again, a mindless procedure incorporating itself in a mindless instrument.

The conclusion is that our instruments, the computers, never will be better than our minds. Rather than worrying too much about the facilities and enhancements (of a mere technical nature) that one may want to introduce into one's computer programming or software 'packages', we should ask ourselves what the mental options are that we want the computer to 'instrumentalize', or help us carry out. To stay in our educational example: the whole difference between grading a paper in the classical way and testing a student's knowledge with the aid of the computer is in the mind and in the mental operations it presupposes.

When the two young students in Roger Schank's latest book, *Scrooge meets Dick and Jane* (2001), finally get to talk to Old Joe, the only remaining teacher of the 'old school', they are astonished to learn from him that in his time, he used to give grades for class and home work, rather than merely reporting test results.

"Do you mean they really gave grades for papers in those days?" the girl asked. 'They certainly did'", Joe responded. (Schank 2001:96)

In the book, Scrooge (reincarnated as the director of a mega-testing company) finally repents, and reverts to the teachings of his old professor John Dewey, vowing to put an end to the mindless instrumentalization of learning, abandoning all his test equipment and mindless testing operations, in order to go back to an old fashioned 'sharing of the minds'.

We put the horse before the cart; so, too, we must put the instrument before the mind. In order to improve the quality of our instrument of mind, the computer, we have to learn how to improve the mind itself.

References

Ehn, Pelle. 1988. *Work-oriented design of computer artifacts.* Stockholm: Arbetslivscentrum.

Gibson, William. 1984. *Neuromancer.* London: Collins.

Gorayska, Barbara. 1994. 'How not to lose the soul of language'. *Journal of Pragmatics* 22(5): 536-547.

Gorayska, Barbara and Jonathon P. Marsh, 1996. Epistemic Technology and Relevance Analysis: Rethinking Cognitive Technology. In Gorayska and Mey, eds. *Cognitive Technology: In search of a humane interface*. Amsterdam: North-Holland.

Gorayska, Barbara & Jonathon P. Marsh. 1999. 'Investigations in Cognitive Technology: Questioning perspective'. In: Marsh, Gorayska & Mey, eds. *Humane Interfaces: Questions of method and practice in Cognitive Technology*, 17-43. Amsterdam: North-Holland.

Helmreich, Stefan. 1998. 'Replicating reproduction in Artificial Life: Or, the essence of life in the age of virtual electronic reproduction'. In: Sarah Franklin & Heléna Ragoné, eds., *Reproducing reproduction: Kinship, power and technological innovation*. Philadelphia: University of Pennsylvania Press. pp. 207-234.

Illich, Ivan. 1973. *Tools for conviviality*. London: Calders & Boyar.

Langton, Christopher (ed.) 1989. *Artificial Life*. Redwood City, Calif.: Addison-Wesley.

Langton, Christopher, Charles Taylor, J. Doyne Farmer & Steen Rasmussen (eds.) 1992. *Artificial Life II*. Redwood City, Calif.: Addison-Wesley.

Langton, Christopher (ed.) 1994. *Artificial Life III*. Redwood City, Calif.: Addison-Wesley.

Lindsay, Roger O. 1999. 'Can we change our minds?' In: Marsh, Gorayska & Mey, eds. 1999:45-69.

Marsh, Jonathon P., Barbara Gorayska & Jacob L. Mey, eds. 1999. *Humane interfaces: Questions of method and practice in Cognitive Technology*. Amsterdam: North-Holland.

Marx, Karl. 1971. *Wage labour and capital*. In: T.B. Bottomore, *Karl Marx, Selected writings in sociology and social philosophy*. Harmondsworth: Penguin. [1848]

Mey, Jacob L. 1982. 'And ye shall be as machines'. *Journal of Pragmatics* 8(5/6):757-797.

Mey, Jacob L. 1996. 'Cognitive Technology — Technological Cognition'. *AI & Society* 10:226-232.

Mey, Jacob L. 1998. 'Adaptability'. In: J.L. Mey, ed., *Concise Encyclopedia of Pragmatics*. Oxford: ElsevierScience. pp. 5-7.

Mey, Jacob L. 2001. *Pragmatics: An introduction*. Malden, Mass. & Oxford, England: Blackwell Publishers. (Second, greatly enlarged and entirely revised edition). [1993]

Minsky, Marvin. 1986. *The society of mind*. London: Heinemann.

Norman, Donald A. 1986. 'Cognitive engineering'. In: D.A. Norman & S.W. Draper (eds.), *User-centered system design*. Hillsdale, N.J. & London, England: Lawrence Erlbaum. pp. 31-61.

Norman, Donald A. 1993. *Things that make us smart: Defending human attributes in the age of the machine*. Reading, Mass. etc.: Addison-Wesley.

Ortony, Andrew. 1993. *Metaphor and thought*. Cambridge: Cambridge University Press. (Second edition) [1979]

Schank, Roger C. 2001. *Scrooge meets Dick and Jane*. Hillsdale, , N.J.: Erlbaum.

Natural-Born Cyborgs?

Andy Clark

School of Cognitive and Computing Sciences
University of Sussex
Brighton BN1 9QH
U.K.
andycl@cogs.susx.ac.uk

'Soon, perhaps, it will be impossible to tell where human ends and machines begins'.
Maureen McHugh, *China Mountain Zhang*, p. 214

'The machine is us, our processes, an aspect of our embodiment ... We are responsible for boundaries. We are they ... I would rather be a cyborg than a goddess'.
Donna Haraway, "A Cyborg Manifesto", in *Simians, Cyborgs, and Women*, pp. 180-181

Abstract. Cognitive technologies, ancient and modern, are best understood (I suggest) as deep and integral parts of the problem-solving systems we identify as human intelligence. They are best seen as proper parts of the computational apparatus that constitutes our minds. Understanding what is distinctive about human reason thus involves understanding the complementary contributions of both biology and (broadly speaking) technology, as well as the dense, reciprocal patterns of causal and co-evolutionary influence that run between them.

My body is an electronic virgin. I incorporate no silicon chips, no retinal or cochlear implants, no pacemaker. I don't even wear glasses (though I do wear clothes). But I am slowly becoming more and more a Cyborg. So are you. Pretty soon, and still without the need for wires, surgery or bodily alterations, we shall be kin to the Terminator, to Eve 8, to Cable...just fill in your favorite fictional Cyborg. Perhaps we already are. For we shall be Cyborgs not in the merely superficial sense of combining flesh and wires, but in the more profound sense of being human-technology symbionts: thinking and reasoning systems whose minds and selves are spread across biological brain and non-biological circuitry.

This may sound like futuristic mumbo-jumbo, and I happily confess that I wrote the preceding paragraph with an eye to catching your attention, even if only by the dangerous route of courting your disapproval! But I do believe that it is the plain and literal truth. I believe, to be clear, that it is above all a SCIENTIFIC truth, a reflection of some deep and important facts about (a whiff of paradox here?) our special, and distinctively HUMAN nature. And certainly, I don't think this tendency towards cognitive hybridization is a modern development. Rather, it is an aspect of our huanity which is as basic and ancient as the use of speech, and which has been extending

M. Beynon, C.L. Nehaniv, and K. Dautenhahn (Eds.): CT 2001, LNAI 2117, pp. 17-24, 2001.

its territory ever since. We see some of the 'cognitive fossil trail' of the Cyborg trait in the historical procession of potent Cognitive Technologies that begins with speech and counting, morphs first into written text and numerals, then into early printing (without moveable typefaces), on to the revolutions of moveable typefaces and the printing press, and most recently to the digital encodings that bring text, sound and image into a uniform and widely transmissible format. Such technologies, once up-and-running in the various appliances and institutions that surround us, do far more than merely allow for the external storage and transmission of ideas.

What's more, their use, reach and transformative powers are escalating. New waves of user-sensitive technology will bring this age-old process to a climax, as our minds and identities become ever more deeply enmeshed in a non-biological matrix of machines, tools, props, codes and semi-intelligent daily objects. We humans have always been adept at dovetailing our minds and skills to the shape of our current tools and aids. But when those tools and aids start dovetailing back- when our technologies actively, automatically, and continually tailor themselves to us, just as we do to them-then the line between tool and user becomes flimsy indeed. Such technologies will be less like tools and more like part of the mental apparatus of the person. They will remain tools in only the thin and ultimately paradoxical sense in which my own unconsciously operating neural structures (my hippocampus, my posterior parietal cortex) are tools. I do not really 'use' my brain. There is no user quite so ephemeral. Rather, the operation of the brain makes me who and what I am. So too with these new waves of sensitive, interactive technologies. As our worlds become smarter, and get to know us better and better, it becomes harder and harder to say where the world stops and the person begins

What are these technologies? They are many, and various. They include potent, portable machinery linking the user to an increasingly responsive world-wide-web. But they include also, and perhaps ultimately more importantly, the gradual smartenng-up and interconnection of the many everyday objects which populate our homes and offices. This brief note, however, is not going to be about new techology. Rather, it is about us, about our sense of self, and about the nature of the human mind. The goal is not to guess at what we might soon become, but to better appreciate what we already are: creatures whose minds are special precisely because they are tailor-made for multiple mergers and coalitions.

Cognitive technologies, ancient and modern, are best understood (I suggest) as deep and integral parts of the problem-solving systems we identify as human intel-igence. They are best seen as proper parts of the computational apparatus that constitutes our minds. If we do not always see this, or if the idea seems outlandish or absurd, that is because we are in the grip of a simple prejudice: the prejudice that whatever matters about MY mind must depend solely on what goes on inside my own biological skin-bag, inside the ancient fortress of skin and skull. But this fortress has been built to be breached. It is a structure whose virtue lies in part in it's capacity to delicately gear its activities to collaborate with external, non-biological sources of order so as (originally) to better solve the problems of survival and reproduction.

Thus consider two brief examples: one old (see the Epilogue to Clark (1997)) and one new. The old one first. Take the familiar process of writing an academic paper. Confronted, at last, with the shiny finished product the good materialist may find herself congratulating her brain on its good work. But this is misleading. It is misleading not simply because (as usual) most of the ideas were not our own anyway, but because the structure, form and flow of the final product often depends heavily on the complex ways the brain co-operates with, and depends on, various special features of the media and technologies with which it continually interacts. We tend to think of our biological brains as the point source of the whole final content. But if we look a little more closely what we may often find is that the biological brain participated in some potent and iterated loops through the cognitive technological environment. We began, perhaps, by looking over some old notes, then turned to some original sources. As we read, our brain generated a few fragmentary, on-the-spot responses which were duly stored as marks on the page, or in the margins. This cycle repeats, pausing to loop back to the original plans and sketches, amending them in the same fragmentary, on-the-spot fashion. This whole process of critiquing, re-arranging, streamlining and linking is deeply informed by quite specific properties of the external media, which allow the sequence of simple reactions to become organized and grow (hopefully) into something like an argument. The brain's role is crucial and special. But it is not the whole story. In fact, the true power and beauty of the brain's role is that it acts as a mediating factor in a variety of complex and iterated processes which continually loop between brain, body and technological environment. And it is this larger system which solves the problem. We thus confront the cognitive equivalent of Dawkins' (1982) vision of the extended phenotype. The intelligent process just *is* the spatially and temporally extended one which zigzags between brain, body and world.

Or consider, to take a superficially very different kind of case, the role of sketching in certain processes of artistic creation. Van Leeuwen, Verstijnen and Hekkert (1999) offer a careful account of the creation of certain forms of abstract art, depicting such creation as heavily dependent upon "an interactive process of imagining, sketching and evaluating [then re-sketching, re-evaluating, etc.]" (op cit p. 180). The question the authors pursue is: why the need to sketch? Why not simply imagine the final artwork "in the mind's eye" and then execute it directly on the canvas? The answer they develop, in great detail and using multiple real case-studies, is that human thought is constrained, in mental imagery, in some very specific ways in which it is *not* constrained during on-line perception. In particular, our mental images seem to be more interpretatively fixed: less able to reveal novel forms and components. Suggestive evidence for such constraints includes the intriguing demonstration (Chambers and Reisberg (1989)) that it is much harder to discover (for the first time) the second interpretation of an ambiguous figure (such as the duck/rabbit) in recall and imagination than when confronted with a real drawing. Good imagers, who proved unable to discover a second interpretation in the mind's eye, were able nonetheless to draw what they had seen from memory and, by then perceptually inspecting their own unaided drawing, to find the second interpretation. Certain forms of abstract art, Van Leeuwen et al go on to argue, likewise, depend heavily on the deliberate creation of "multi-layered meanings" – cases where a visual form, on continued inspection, supports multiple different structural interpretations. Given the postulated constraints on mental imagery, it is likely that the discovery of such multiply interpretable forms will depend heavily on the kind of trial and error process in which we first sketch and then perceptually (not merely imaginatively) re-encounter visual forms, which we can then tweak and re-sketch so as to create a product that supports an increasingly multi-layered set of structural interpretations. This description of artistic creativity is strikingly similar, it seems to me, to our story about academic creativity. The sketch-pad is not just a convenience for the artist, nor simply a kind of external memory or durable medium for the storage of particular ideas. Instead, the iterated process of externalizing and re-perceiving is integral to the process of artistic cognition itself.

One useful way to understand the cognitive role of many of our self-created cognitive technologies is thus as affording *complementary* operations to those that come most naturally to biological brains. Consider here the connectionist image (McClelland, Rumelhart and the PDP Research Group 1986, Clark 1989) of bio-ogical brains as pattern-completing engines. Such devices are adept at linking patterns of current sensory input with associated information: you hear the first bars of the song and recall the rest, you see the rat's tail and conjure the image of the rat. Computational engines of that broad class prove extremely good at tasks such as sensori-motor co-ordination, face recognition, voice recognition, etc. But they are not well-suited to deductive logic, planning, and the typical tasks of sequential reason. They are, roughly speaking, "Good at Frisbee, Bad at Logic" – a cognitive profile that is at once familiar and alien. Familiar, because human intelligence clearly has something of that flavor. Yet alien, because we repeatedly transcend these limits, planning family vacations, running economies, solving complex sequential problems, etc., etc. A powerful hypothesis, which I first encountered in Rumelhart, Smolensky, McClelland and Hinton (1986), is that we transcend these limits, in large part, by combining the internal operation of a connectionist, pattern-completing device with a variety of external operations and tools which serve to reduce various complex, sequential problems to an ordered set of simpler pattern-completing operations of the kind our brains are most comfortable with. Thus, to borrow the classic illustration, we may tackle the problem of long multiplication by using pen, paper and numerical symbols. We then engage in a process of external symbol manipulations and storage so as to reduce the complex problem to a sequence of simple pattern-completing steps that we already command, first multiplying 9 by 7 and storing the result on paper, then 9 by 6, and so on. The value of the use of pen, paper, and number symbols is thus that – in the words of Ed Hutchins;

> "[Such tools] permit the [users] to do the tasks that need to be done while doing the kinds of things people are good at: recognizing patterns, modeling simple dynamics of the world, and manipulating objects in the environment." Hutchins (1995) p. 155

This description nicely captures what is best about *good* examples of cognitive technology: recent word-processing packages, web browsers, mouse and icon sysems, etc. (It also suggests, of course, what is wrong with many of our first attempts at creating such tools – the skills needed to use those environments (early VCR's, word-processors, etc.) were *precisely* those that biological brains find hardest to support, such as the recall and execution of long, essentially arbitrary, sequences of operations. See Norman (1999) for further discussion.

The conjecture, then, is that one large jump or discontinuity in human cognitive evolution involves the distinctive way human brains repeatedly create and exploit various species of cognitive technology so as to expand and re-shape the space of human reason. We – more than any other creature on the planet – deploy non-bio-logical elements (instruments, media, notations) to *complement* our basic biological modes of processing, creating extended cognitive systems whose computational and problem-solving profiles are quire different from those of the naked brain.

The true significance of recent work on "embodied, embedded" problem-solving (see Clark 1997 for a review) may thus lie not in the endless debates over the use or abuse of notions like internal representation, but in the careful depiction of complex, looping, multi-layered interactions between the brain, the body and reliable features of the local problem-solving environment. Internal representations will, almost certainly, feature in this story. But so will external representations, and artifacts, and problem-transforming tricks. The right way to "scale-up" the lessons of connectionist research (and simple robotics- see e.g. Brooks 1991, Beer 1995) so as to illuminate human thought and reason is to recognize that human brains maintain an intricate cognitive dance with an ecologically novel, and immensely empowering, environment: the world of symbols, media, formalisms, texts, speech, instruments and culture. The computational circuitry of human cognition flows both within and beyond the head, through this extended network in ways which radically transform the space of human thought and reason.

Such a point is not new, and has been well-made by a variety of theorists working in many different traditions. This brief and impressionistic sketch is not the place to delve deeply into the provenance of the idea, but some names to conjure with include Vygotsky, Bruner, Dennett, Hutchins, Norman and (to a greater or lesser extent) all those currently working on so-called 'situated cognition'. My own work on the idea (see Clark 1997,1998, 1999) also owes much to a brief collaboration with David Chalmers (see our paper, 'The Extended Mind' in ANALYSIS 58: 1: 1998 p.7-19). I believe, however, that the idea of human cognition as subsisting in a hybrid, extended architecture (one which includes aspects of the brain and of the cognitive technological envelope in which our brains develop and operate) remains vastly under-appreciated. We cannot understand what is special and distinctively powerful about human thought and reason by simply paying lip-service to the importance of the web of surrounding Cognitive Technologies. Instead, we need to understand in detail how our brains dovetail their problem-solving activities to these additional resources, and how the larger systems thus created operate, change and evolve. In addition, and perhaps more philosophically, we need to understand that the very ideas of minds and persons are not limited to the biological skin-bag, and that our sense of self, place and potential are all malleable constructs ready to expand, change or contract at surprisingly short notice.

A natural question to press, of course, is this: since no other species on the planet builds as varied, complex and open-ended designer environments as we do (the claim, after all, is that this is why we are *special*), what is it that allowed this process to get off the ground in our species in such a spectacular way? And isn't *that*, whatever it is, what really matters? Otherwise put, even if it's the designer environments that makes us so intelligent, what biological difference lets us build/discover/use them in the first place?

This is a serious, important and largely unresolved question. Clearly, there must be some (perhaps quite small) biological difference that lets us get our collective foot in the designer environment door - what can it be? The story I currently favor located the difference in a biological innovation for greater neural plasticity combined with the extended period of protected learning called "childhood" Thus Quartz (1999) and Quartz and Sejnowski (1997) present strong evidence for a vision of human cortex (especially the most evolutionarily recent structures such as neocortex and prefrontal cortex) as an "organ of plasticity" whose role is to dovetail the learner to encountered structures and regularities, and to allow the brain to make the most of reliable external problem-solving resources. This "neural constructivist" vision depicts neural (especially cortical) growth as experience - dependent, and as involving the actual construction of new neural circuitry (synapses, axons, dendrites) rather than just the fine-tuning of circuitry whose basic shape and form is already determined. One upshot is that the learning device *itself* changes as a result of organism-environmental interactions - learning does not just alter the knowledge base for a fixed computational engine, it alters the internal computational architecture itself. Evidence for this neural constructivist view comes primarily from recent neuroscientific studies (especially work in developmental cognitive neuroscience). Key studies here include work involving cortical transplants, in which chunks of visual cortex were grafted into other cortical locations (such as somatosensory or auditory cortex) and proved plastic enough to develop the response characteristics appropriate to the new location (see Schlagger and O'Leary (1991)), work showing the deep dependence of specific cortical response characteristics on developmental interactions between parts of cortex and specific kinds of input signal (Chenn, (1997)) and a growing body of constructivist work in Artificial Neural Networks: connectionist networks in which the architecture (number of units and layers, etc.) itself alters as learning progresses - see e.g. Quartz and Sejnowski (1997). The take home message is that immature cortex is surprisingly homogeneous, and that it 'requires afferent input, both intrinsically generated and environmentally determined, for its regional specialization' (Quartz (1999) p.49).

So great, in fact, is the plasticity of immature cortex (and especially, according to Quartz and Sejnowski, that of prefrontal cortex) that O'Leary dubs it 'proto-cortex'. The linguistic and technological environment in which the brain grows and develops is thus poised to function as the anchor point around which such flexible neural resources adapt and fit. Such neural plasticity is, of course, not restricted to the human species (in fact, some of the early work on cortical transplants was performed on rats) , though our brains do look to be far and away the most plastic of them all. Combined with this plasticity, however, we benefit from a unique kind of developmental space- the unusually protracted human childhood.

In a recent evolutionary account which comports perfectly with the neural constructivist vision, Griffiths and Stotz (2000) argue that the long human childhood provides a unique window of opportunity in which "cultural scaffolding [can] change the dynamics of the cognitive system in a way that opens up new cognitive possibilities" (op cit p.11) These authors argue against what they nicely describe as the "dualist account of human biology and human culture" according to which biological evolution must first create the "anatomically modern human" and is then followed by the long and ongoing process of cultural evolution. Such a picture, they suggest, invites us to believe in something like a basic biological human nature, gradually co-opted and obscured by the trappings and effects of culture and society. But this vision (which is perhaps not so far removed from that found in some of the more excessive versions of evolutionary psychology) is akin, they argue, to looking for the true nature of the ant by "removing the distorting influence of the nest" (op cit p.10). Instead we humans are, by nature, products of a complex and heterogeneous developmental matrix in which culture, technology and biology are pretty well inextricably intermingled. The upshot, in their own words, is that:

> "The individual representational system is part of a larger representational environment which extends far beyond the skin. Cognitive

processes actually involve as components what are more traditionally conceived as the expressions of thought and the objects of thought. Situated cognition takes place within complex social structures which 'scaffold' the individual by means of artifactual, linguistic and institutional devices...[and]..culture makes humans as much as the reverse." (Griffiths and Stotz (2000)).

In short it is a mistake to posit a biologically fixed "human nature" with a simple "wrap-around" of tools and culture. For the tools and culture are indeed as much determiners of our nature as products of it. Ours are (by nature) unusually plastic brains whose biologically proper functioning has always involved the recruitment and exploitation of non-biological props and scaffolds. More so than any other creature on the planet, we humans are indeed *natural-born cyborgs*, factory tweaked and primed so as to be ready to participate in cognitive and computational architectures whose bounds far exceed those of skin and skull.

All this adds interesting complexity to recent evolutionary psychological accounts (see e.g. Pinker (1997)) which emphasize our ancestral environments. For we must now take into account a plastic evolutionary overlay which yields a constantly moving target, an extended cognitive architecture whose constancy lies mainly in its continual openness to change. Even granting that the biological innovations which got this ball rolling may have consisted only in some small tweaks to an ancestral repertoire, the upshot of this subtle alteration is now a sudden, massive leap in cognitive-architectural space. For the cognitive machinery is now intrinsically geared to self-transformation, artifact-based expansion, and a snowballing/bootstrapping process of computational and representational growth. The machinery of human reason (the environmentally extended apparatus of our distinctively human intelligence) thus turns out to be rooted in a biologically incremental progression while simultaneously existing on the far side of a precipitous cliff in cognitive-architectural space.

Conclusions

The project of understanding human thought and reason is easily misconstrued. It is misconstrued as the project of understanding what is special about the human brain. No doubt there *is* something special about our brains. But understanding our peculiar profiles as reasoners, thinkers and knowers of our worlds requires an even broader perspective: one that targets multiple brains and bodies operating in specially constructed environments replete with artifacts, external symbols, and all the variegated scaffoldings of science, art and culture. Understanding what is distinctive about human reason thus involves understanding the complementary contributions of both biology and (broadly speaking) technology, as well as the dense, reciprocal patterns of causal and co-evolutionary influence that run between them.

For us humans there is nothing quite so natural as to be bio-technological hybrids: cyborgs of an unassuming stripe. For we benefit from extended cognitive architectures comprising biological and non-biological elements, delicately intertwined. We are cognitive hybrids who occupy a region of design space radically different from those of our biological forbears. Taking this idea on board, and transforming it into a balanced scientific account of mind, should be a prime objective for the Cognitive Sciences of the next few hundred years.

References

Beer, R. (1995). "A Dynamical Systems Perspective Perspective on Agent-Environment Interaction." *Artificial Intelligence* 72: 173-215.

Brooks, R. (1991). "Intelligence without representation." *Artificial Intelligence* 47: 139-159.

Chambers D, and Reisberg,D (1989)" Can Mental Images Be Ambiguous?" *Journal of Experimental Psychology: Human Perception and Performance* II(3) 317-328.

Chenn, A. (1997). Development of the Cerebral Cortex in W. Cowan, T. Jessel and S. Ziputsky (eds) *Molecular and Cellular Approaches to Neural Development* Oxford, England, Oxford University Press 440-473.

Clark, A. (1989). *Microcognition: Philosophy, Cognitive Science and Parallel Distributed Processing*. Cambridge, MIT Press.

Clark, A. (1997). *Being There: Putting Brain, Body and World Together Again*. Cambridge, MA, MIT Press.

Clark, A. (1998). Magic Words: How Language Augments Human Computation. *Language and Thought*. J. Boucher and P. Carruthers. Cambridge, Cambridge University Press.

Clark, A (1999). "An Embodied Cognitive Science?" *Trends In Cognitive Sciences* 3:9:1999: 345-351.

Clark, A. and Chalmers, D. (1998). "The Extended Mind." *Analysis* 58: 7-19.

Dawkins, R. (1982). *The Extended Phenotype* (New York: Oxford University Press).

Dennett, D. (1996). *Kinds of Minds*. New York, Basic Books.

Griffiths, P. E. and K. Stotz (2000). How the mind grows: A developmental perspective on the biology of cognition. *Synthese* 122(1-2): 29-51.

Hutchins , E. (1995). *Cognition In The Wild* Cambridge, MA, MIT Press.

Norman, D. (1999). *The Invisible Computer* Cambridge, MA, MIT Press.

Pinker, S. (1997). *How the Mind Works* New York, Norton.

Quartz, S. (1999). The Constructivist Brain *Trends In Cognitive Science* 3:2: 48-57.

Quartz, S. and Sejnowski, T (1997). The Neural Basis of Cognitive Development: A Constructivist Manifesto *Behavioral and Brain Sciences* 20:537-596.

Rumelhart, D. Smolensky, P. McClelland, D. and Hinton, G. (1986). Schemata and Sequential Thought Processes in PDP Models, in *Parallel Distributed Processing: Explorations in the Microstructure of Cognition, vol 2*, MIT Press, Cambridge, MA p.7-57.

Schlagger, B. and O'Leary, D. (1991). Potential of Visual Cortex to Develop an Array of Functional Units Unique to Somatosensory Cortex *Science* 252 1556-1560.

Van Leeuwen, C., Verstijnen, I. and Hekkert, P. (1999). Common unconscious dynamics underlie common conscious effects: a case study in the interactive nature of perception and creation. In S. Jordan (ed) *Modeling Consciousness Across the Disciplines* Lanhan, MD, University Press of America.

Fact and Artifact: Reification and Drift in the History and Growth of Interactive Software Systems

Martin Loomes and Chrystopher L. Nehaniv

Department of Computer Science
University of Hertfordshire
Hatfield, Herts AL10 9AB
United Kingdom

"Caught in a net of language of our own invention, we overestimate the language's impartiality. Each concept, at the time of its invention no more than a concise way of grasping many issues, quickly becomes a precept. We take the step from description to criterion too easily, so that what is at first a useful tool becomes a bigoted preoccupation."

– *C. Alexander (pp. 69–70, 1964)*

Abstract. We discuss the processes and forces informing artifact design and the subsequent drift in requirements and interests in the long-term growth of reified systems. We describe, following Latour, the strategies of technoscience in making artifacts into "facts" and consider their impact on human life and activity. Drawing from the history of word-processing systems in particular and interactive software systems in general, we illustrate the drift in requirements and context of use that create new needs (including possibly inappropriate ones). We draw attention to the dynamics creating such needs and raise questions regarding the appropriateness of technology-driven drift that shapes the interactive systems around us. The viewpoint is toward software design and evolution in the long-term and we promote the critical recircumscription of problem spaces in order to use technology to improve human life rather than to merely integrate and increase the functionality of existing technologies.

1 Artifacts in Human Contexts

Cognitive technology seeks to optimize the relationship of humans and their tools (Gorayska & Mey 1996). Other fields such as Human-Computer Interaction (HCI) and Software Engineering play important roles in such an endeavour, but their roles and the assumptions implicit in their current practice merit critical examination. One of the major problems facing those of us who wish to talk about complex issues regarding "technology" and "people" is that inevitably we seek simple models, analogies and terminology to help our discourse, but equally inevitably there is a tendency to allow these artifacts to cross over from the

M. Beynon, C.L. Nehaniv, and K. Dautenhahn (Eds.): CT 2001, LNAI 2117, pp. 25–39, 2001.

artificial to the factual. For example, what starts out as a convenient shorthand, well understood to have severe limitations and dangers in use by its originators, can rapidly cross over into being, and become a valid object for investigation in its own right.

1.1 From Artifacts to Reality

> *"It is surely true of all tools, that by making some things easier they direct activity and thinking from things that are more difficult; but what is easy and what is difficult are partly set by the available tools, and so we are carried along by a sequence of largely arbitrary and sometimes unfortunate features of our technology, including our language. Human intelligence is very largely Artificial Intelligence, and even our hopes and fears (and our moral commitments, for they are set by possibililities of achievement) are largely set by exisiting technology."*
>
> – *R. L. Gregory (p. 51, 1984)*

As Bruno Latour has noted, once this process starts, it is all too easy for a complex web of literature and activity to arise which, being too complex and powerful to challenge, leads to the *artifact* being *reified* as a *fact* – an artifiact, or candidate for becoming something real, is of only local importance and with a local and possibly transient existence. Becoming a "fact" means becoming a real thing in the world, of which there are many instances or examples of usages.[1] Moreover, once such a status has been achieved, anyone wishing to challenge this change of status is usually forced to do so by criticizing aspects of what has "become", thus acknowledging existence on route to the challenge. Reification of a representation of an artifact – the acceptance of "it" as a persistent existing entity – can carry inertia that diverts resources toward and gives rise to versions of the artifact when it is accepted as valid topic of discourse. This inertial mechanism plays a critical role in the externalization of technology, as well as in its persistence, and in its willing or unwilling acceptance by a community of practice. The reified artifact is subject to forces that result in "it" being employed in slowly changing and *drifting* contexts of usage in the struggle for resources and unforeseen change in requirements. This is illustrated below by examining in some detail the strategies of technoscience (following (Latour 1988)) and applying this analysis to the particular example of word-processing software systems and then more briefly interactive systems in general.

If the Cognitive Technology community is serious in its desire to improve the interaction between the people and technology, it must be prepared to challenge some of the fundamental notions that have become institutionalized as "truths" within the wider Interface Design and Software Engineering communities. This

[1] An exegesis of this is to consider "fact" as the thing made, and thus "artifact" as an artificial thing made or discussed, whence to "become fact" is for an artificial made or discussed thing to be withstand trials of strength, have the support of recruited resources in networks, and be accepted as real.

will lead to a deeper understanding of the epistemic loop from cognition to technology (Gorayska & Marsh 1996), and an awareness of why technologies may develop in an inhumane manner, or fail to develop at all. It may also lead to insights into the particular dangers, and possibly remedies, to current modes of practice in interactive systems and the humane evolution of these technologies. In this paper we will start the process by calling into question the status of a few terms as frequently used in the literature, and consequently challenge a few of the widely held beliefs, or (more likely) unquestioned assumptions, upon which debates and design activities are often founded.

1.2 A Pathology of Current Software Engineering Practices – "The System" as Artifact

Let us start by questioning the assumption that it makes sense to discuss "the interface between a system and the people who interact with it", and that such discussions should take place during the design of the system. This sentiment is found in many textbooks on software design and HCI and underpins the very notion of "user-centered design". But what does it mean? One implicit assumption here is that there exists a single interface between "a system" and some generic concept of "people who interact with it". Moreover, this single interface exists, in some sense, throughout the design of the system. (Of course, more sophisticated versions of this acknowledge the need for several views or modes of interface –or, in effect, a set of interfaces – for the system, for a variety of diverse users of the system. Nevertheless, the design problems are still formulated with respect to a single, reified, pre-supposed system, referred to, already before the fact, as "the system".) One of the reasons for the blind acceptance of such an "institutional truth" is that Software Engineering has become dominated by the life-cycle model of design, which dictates the way in which system "development" is rationalized, discussed and criticized. Within this model, there is the assumption that "the system" exists in some embryonic form from the very start of the process, and that the task of the designer is to nurture development and allow maturity to be reached in good health. Typically, the system equivalent of "DNA" is encoded into a "requirements document", which presumably captures the notion of the "systemness" that will be evident in the mature adult of the species. Occasionally, for practical reasons, the immature system will be cloned, allowing development to continue to version 2, whilst exploiting the current behaviour of version 1. We are drawing attention to fact that "the system" as discussed by designers, the representations or prototypes of "the system" in the course of its development, and "the system" as a physical, deployed entity, and maintained or modified versions of "the system", on which perhaps other systems will be based are not a single entity with any uniform mode ontological status. They are treated as an entity that has a single identity whose existence is tacitly accepted and validated in unspoken assumptions implicit when speaking of "the system".[2]

[2] Note that "the system" may, to the unintiated, still seem to have an identity – a name – albeit with version numbers added. Users will still refer to "it" as though "it" has some existence as an entity.

1.3 Requirements and System Drift

Of course, we might point out that "systems are not built like this", but note the reaction of the model-defenders to attacks of this type. Rather than saying "of course not, this is a model and it has limitations" they hack the model to accommodate surface similarities with the challenging scenario. Thus multiple feedback loops abound. This adds the complexity of just what is circulating within the model? It is usually presented that each level in the life cycle focuses attention on a different representation of the system, with more detail being added. Thus the design process is partially one of translation between representations, and partly one of refinement. Exactly what is fed back in the model is unclear, but that need not concern us here. The important point to note is that the concept of "the interface between a system and the people who interact with it" is no longer quite so simple. We must either accept that the system does not exist until its final "real" version is released, or we must accept that the concept of interaction with a system includes interactions with representations of the system. These two choices provide for very different scope of analysis. If we take the former view, then considering this interaction prior to the release of the system becomes rather difficult. If we accept that a modality of possibility is quite acceptable, we can sensibly discuss interactions between a possible system and its possible users.

If we accept the (somewhat far-fetched) notion that the "systemness" of the system is encoded from the earliest representation, then we have still managed to keep our focus on a single object of interaction (provided we accept the notion of a generic, unchanging user). If, however, we take the view that each representation or realization of the system is effectively a different object for purposes of interaction, then our interface problem becomes one of responding to the evaluation of a series of interactions between people and representations (or prototypes or realizations) of systems. Given the feedback in the model, there is no necessary ordering on the requirements, implementation, and design of these representations. Moreover, if we accept that people change over time, and the situations in which their responsive actions occur also change and have a material influence on the behaviour of the people, then this series of interactions involves a constantly changing set of changing agents. The inertia of talking about the one "system" may lead us to apply assumptions or analyses valid for one context, for one representation and user type, or from one setting for a realization or system prototype to others. Of course, recognizing that there are series of systems and interactors allows one to avoid such fallacies.

Against this background, the seemingly rational, and simple, suggestion that we should consider such interactions as part of the design process becomes a platitude: we simply do not know how to do it. As designers, we do something anyway, but we may not have enough experience, understanding, or foresight to do it well. However there are glimmerings of hope in methodologies such as *situated design* (Greenbaum & Kyng 1991) that put the user's work practices and interests at the center of software development. But so far these methods, and other related laudable ones of participatory design and user-centered design, cannot adequately address long-term software evolution issues in the context of changing requirements (cf. (Goguen & Jirotka 1994, Nehaniv 2000)). Of course, for a very simple system, where we restrict attention to a small subset of representations, consider interactions related only to a small subset of tasks, and do not question too deeply the value judgements that we make regarding these, a plausible rationalized reconstruction may be possible. The danger of so doing, however, is that we add credence to the precepts that such action is what should be done, and that failure to do so in other situations is simply a reflection of poor practice. Current work in HCI, cognitive modelling, and user-centered design often serves to reinforce these prejudices, and by accepting a reified view of "the system" allows valid findings from one limited context to be overgeneralized to others where they might no longer apply. Thus, once again, reification occurs and "analysis of the user interface" takes on the status of a "thing" which has meaning outside of the limited number of cases where we know how such meaning can be ascribed.

2 The Growth of Networks in Technoscience

One way in which such issues can be addressed is to acknowledge that the design of most systems is not a process of simple construction, but one of research, where outcomes, methods and directions are not known in advance, but emerge in a variety of forms, and under a variety of pressures. This has been referred to elsewhere as the "theory-building view" (e.g. (Loomes & Jones 1998)), and is also captured by Latour in his sociological view of the processes of science and technology. In particular, Latour makes the distinction between a *diffusion* model of the spread of technology, where a concept that is well-understood by some driving force is disseminated through society, and the *network* model, where ideas undergo translations (in the strict sense of the term) as they are passed around a network of actors (which may themselves be human or technological in nature) (Latour 1988, Latour 1996).

2.1 Diffusion

The diffusion model sees a few scientists or inventors as the sources of the impact of technology. It allows them to see themselves as the driving force behind technoscience and glamorizes their role. We note that companies marketing a new product use this type of glamor to introduce it into the market:

"HyperCard is a new kind of application – a unique information environment for your Apple MacInctosh computer. Use it to look for and store information – words, charts, pictures, digitized photographs – about any subject that suits you. Any piece of information in HyperCard can connect to any other piece of information, so you can find out what you want to know in as much or as little detail as you need."

- HyperCard User's Guide (p. xvi, 1987) (Apple Computer Inc. 1987)

Here HyperCard is sold as an innovative new technology, with almost magical powers. This approach recruits the interest of enthusiasts, who then make the technology take off. (Actually to "find out what you want to know in as much or as little detail as you need" it is actually necessary – for someone – to put what will want to know in the HyperCard system in as much detail as you might like, in advance of your being able to find it. Similar phenomena can be observed in the history and growth of the world wide web.) The marketing hype served to *seed* the growth of a network of enthusiasts committed to building a community of use that is necessary for the product to succeed at base level. In fact, the initial oversell aids in the struggle for identity and the attempt to engage allies. The HyperCard community lasted a long time, and might still have been been able to persist to the present day, had it been able to develop beyond the MacIntosh.

2.2 Networks

Under the network model, it is not the case that some truth, encapsulated in early manifestations of the system, is gradually reflected in some artifact. Rather, it is through the adoption of some artifact that a set of ideas gains truth. Recognizing the combined emergence of the series of situated users and communities together with the series of representations, prototypes, models and versions of artifacts in which processes of reification and inertia play important roles gives a deeper view (and a less mystical one) of the social and cognitive context in which the human-tool relationship evolves. Extending networks of influence, recruiting allies and resources, including huge numbers of people who do development, production, marketing, support the work, and use the artifact are what is behind the impact of technoscience.

In the evolution of interactive software systems, it is often possible to identify a small group of designers who identify requirements and develop them to introduce the initial artifact, but with time, if the network of influence and users grows, control becomes distributed and spreads throughout it. As a result of many pressures on and within the network, the technology may then drift, although no one is likely to be in a controlling, strong position to re-examine the embedded assumptions nor whether they still serve the requirements of users in now changed contexts anymore. We illustrate how this happens below with some concrete examples from the history of word-processing systems, after we examine the dynamics of networks in technoscience.

2.3 Network Dynamics

We review Latour's identification of strategies on the recruitment and translation of resources and interests faced by a fact-builder (Latour 1988). Influences such as these can be seen in "systems" such as word-processing software which now have a substantial impact on much daily human activity. Latour identifies five general mechanisms for the technoscientist who is working to extend his/her network of influence in building new facts – i.e. making artifacts real – which we illustrate by successful arguments for word-processing (and office automation) software:

1. *"I want what you want"* -
   The designer says: You want to produce typed documents, business letters, etc. You need an electronic typewriter to run on a general purpose computer. You couldn't have known this, as you scarely know of the existence of general purpose computers, let alone their capabilities. This will let you edit the document and also save it for re-use.
2. *"I want it why don't you?"* (Your usual way is cut off, resources can be co-opted/enlisted) -
   companies says: We must move on, there will be no more typewriter or dedicated word-processing machine. Everyone *will* move to a word-processing software package on their machine.
3. *"If you make a short detour through our methods..."* -
   marketing and training say: I know using our word-processing package is more complicated than your old typewriter (which we've now thrown away), but once you've mastered "it", you're set for life.
4. *Reshuffling interests and goals* -
   a) *displacing goals* -
      Now the rhetoric moves further from marketing/training into the companies adopting technology: All of our major competitors use word-processing systems far superior to ours, we must move on...
   b) *inventing new goals* -
      We must be able to *communicate* electronically, so you must use *this* word-processing package. To do this we need to *standardize* the format of our communicated documents to use *this* package.
   c) *inventing new groups* -
      We must move into the *desktop publishing* community. Simply producing typewritten sheets is not enough, we must have production quality documents.
   d) *rendering the detour invisible – drift, translating away from the original intent as the number of interests increase.*
      We must continue to upgrade our word-processing software in line with the latest office automation suite.
   e) *winning trials of attribution* -
      Is there any *real* alternative? Everyone in the *real* world uses it!
5. *"Whatever you want, you want this as well"* – (becoming indispensible, your facts become obligatory passage points for all interests to pass through on the way to wherever they are going.)
   This package is in line with our office suite [compatible] therefore we should choose it. It is back-compatible, so you will be able to access your old stuff. Besides, it converts any alien standard (if anyone would want to use one for some reason) into our standard. More than that, although we don't know why anyone should want to, it is possible to save documents in an alien standard with the loss of some functionality, since we haven't bothered to conform exactly to those alien standards (anyway of decreasing importance) ...

In technoscience artifacts become real (i.e. facts), in part, by applying these strategies and winning trials of strength.

3 Examples from the History of Interactive Software Systems

In this section, we discuss these ideas through examples drawn from the history of interactive software systems, focusing on word-processing technologies and their descendents in changing contexts now involving such technologies as e-mail and agents. We suggest that the Cognitive Technology community needs to promote the critical evaluation of the very basis of Software Design as currently formulated and its integration into society, before real progress can be made in realizing Cognitive Technology's objectives for humane interactive systems. The phenomena of drift illustrated below serve to make clear that a *long-term* view is necessary for understanding the manner in which a persistent technology (such as "the system" of a word-processing software package) impacts the human realm.

In studying the history of interactive software systems, we identified dimensions of *content*, *presentation*, *communication*, *embeddability*, and *security* as relevant. We focus here on the evolution of word-processing software.[3]

3.1 Drift of "the User" and "the System" in Word-Processing

Originally the users for which electronic typewriters and then the first word-processing software packages were developed were typists. A small group of designers could write down requirements for producing hardcopies of typed text. *Presentation* was that of a single typeface on single sheets of US letter or A4 paper. The typewriter had no memory and what was typed was incribed only on the paper, but now *content* was recorded as data stored in computer memory, and that could be retrieved. Content could be stored in various character sets and encodings.

On a general purpose computer, the typewriter can begin second-guessing the author. Content could eventually be monitored, with spelling checks, punctuation checking and correction becoming automatable – "e.g. have" is 'corrected' to "e.g. Have" since – the software reasons – following a period, "have" must begin a new sentence and so should be capitalized. Similarly, when she types a line starting with the "1." or "A.", the software guesses that the author is making a list and inserts "2." or "B." etc. on the next lines. An author may really be making a list and welcome the saved effort, but she may not (e.g. typing, "A. Jones wishes to purchase product X" (carriage return; the software inserts "B."). The author asks, Can this *second-guessing* of my key strokes be turned off? Not easily, but probably in the next version... With content monitoring and 'correction'

[3] We recall that, for software systems – just as in biology – *evolution* does NOT imply *progress*!

issues of *delegation* and *autonomy* arise that require careful balancing to find appropriate levels for particular tasks and users (Dautenhahn & Nehaniv 2000). The user can write macros, use "grammar checking", and thesaurus facilities.

With internal markup of content, elements of presentation began to be supported, fonts of variable sizes, lists, tables, and hypertext. *Embeddability* and *linking* permit the incorporaton of graphics, movies, sound, web-access, and executable code (including e.g., computer viruses, that may be activated when the document is opened).

With more than a single typist now recruited by the technological network, a possibility of – and hence later a need for – *communication* between multiple authors arises. Supported by *storage* and *retrieval* of documents, and by *transmission* of data within and between office suites and across computer networks, issues of *portablity*, *compatability*, and *interoperability* begin to arise. One author working on a persistent document over time and multiple authors working on a single document concurrently find they would like or need automated version management, annotation markup, consistency checking, and merging of documents (e.g., Lotus Notes, unix sccs, rcs, merge). Forms and embedded spreadsheets allow *autonomous data transfer*; e.g. a table in a document changed on one machine is linked to a database that updates the table for all copies of the document. *Activity* by applets / javascript / cgi-bin scripts and other self-launching bits of code is embeded and incorporated into something still called a 'document'.

Security becomes an issue as active embedded content arises, and as it is unknown just who is accessing your database server through software clients of unknown location and origin activated by the document.

Functionality bloat loads successive versions of the software suite with additional capabilities. These, together with 'bug fixes' serve as fodder for marketing, and justification for new releases. Note, however, these are not new products, but "the system" in a new guise. Interoperability, portability, composability of software components must be supported. Remote accessiblity ("my tools anywhere, anytime") lets the user connect to his data via mobile phones, hand-held devices while rock climbing or commuting. Anything that can be done, must be done to stay ahead. *Integration* and *increased bandwidth*, linking the bits of software so that they can talk to each other on the 'desktop', in the home, or via the internet to allow the user maximal access to information, become goals in and of themselves.

Today some quarter of a century down the line, who is the user of the system? The user of the release of say, Word 2003, is much less likely to be a typist, and more likely to be an author or a reader of the document. The "systemness" of the word-processing software package has followed "it" throughout its history, loading "it" with the embedded assumptions of requirements built in. The requirements of this user are not the same as those of the typist in 1975. But we still speak of "the word-processing system".

3.2 A Tab Margins Example

The act of inserting a tab or setting right- and left- margins on a page produced at a typewriter could be done by moving left- and right- tabs along a ruler-like scale on the typewriter. Word-processing software often includes such a visible tab at the top of the display under a tool bar of other functions. The old ruler scale of the mechanical typewriter is still there, persisting from the early days of when designers identified it as a requirement for their electronic software running on general purpose computer as a replacement for this typewriter. Sections of text within a stored document are internally marked to indicated the tabbing information set on the tab bar when the text was entered. The tab bar is always visible, but its effect is not global on the text of the document. Since little blocks of text (possibly only one line long) are tagged with tabbing information, why does it still make sense to display a single tab ruler? The tabbing information on the many different blocks of text within the same document can differ wildly. It is unclear – as it is not visible to the user – to which portions of the text the tabs seen on the ruler scale actually apply. The system still retains some of its roots – the ruler with pseudo-mechanical tabs that can be move, for example, – but now the ruler is rarely used by the user to control the software – rather the software changes the settings on the ruler.

The tab bar has persisted – apparently without any reanalysis of why it is there and whether or not its apparently quite sensible design for the original application some decades ago is still appropriate. Since the tabbing information is actually local to blocks of text, why not use for instance a pop-up box showing these details when a cursor lingers over the text? As far as we know, no one has asked such questions. The tab bar and its integration into internal data representation in early word-processing software has simply been carried along. Does it still make sense to think of using a word-processing system as a kind of typewriter?

People used to producing all documents with word-processing packages tend to generalize this behavior and carry along the old metaphors such packages were motivated by. For example, in preparing a conference poster our natural reaction would be to use such packages. But we found the others motivated in another way can made quite a different choice: Some of our colleagues went through quite different reasoning in the choice of tools for the same task: MS-Powerpoint is for presentations, so to make a poster presentation, use it, set the pagesize to A1 and print out a huge poster. This shows how the names we give our tools can inform the scope of applicability we ascribe to them as well as our manner of employing them, and they can carry much of their history with them despite deep changes in the contexts of use.

3.3 From Wants to Needs

The character of the word-processing software is now not determined by examining user requirements, other than the requirements of the current community of users working now in a very different context. The designers, marketing, training, managers, users, companies, and their competitors in the course of a long and distributed history of activity have made the word-processing package real and indispensible for the user's survival in the organization. This network exerts tremendous pressures on the directions of future drift and development of the word-processing "system". The original requirements that were there at the outset no longer define "the system". User-testing of the word-processing system is not driven by what the user might want if the requirements were recircumscribed today. *What the user wants* is now determined by many pressures on the network of organizations that rely on word-processing software, and what the user now *needs* to survive in the organization. (See also Sect. 2.3 above.) This is how "*wants*" has become "*needs*".

4 Conclusions

4.1 Directions for Technological Change

What persists in the history of interactive software systems? Many things, file formats, communities of users, legacy computer code. The goals and interests of the designers of a particular piece of interactive software technology may not persist. The purposes for which particular technology was designed may be forgotten, lost, co-opted, lose significance, drift in scope of applicability, and so on, as the context of use and work pratices change and as the pressures of the networks that recruit it vary. Instead of blindly integrating the obvious extant technologies, we argue that it makes better sense to re-evaluate software against the needs of its users. Moreover, the needs that have been created by the pressures and history of the technology should be examined. Can any of them be removed? What can we do without? (Note that it may not be possible for the next version of word-processing package XYZ to *remove* functionality, even if this would be an improvement!) Can new technological networks be grown that will supersede the existing technologies in a more humane manner? Are we missing whole important areas of possibility for developing humane technologies? For example, affective and narrative grounding characteristics of human users have for example been completely ignored by the drive for technological intergration (Nehaniv 1999).

Historically software producers continue with the push for integration, more broadband channels and interactivity in software other than but linking to the word-processing system descendants of the humble typewriter. Virtual presence in business meetings and interactive sports viewing facilities and TV touch-screens that us sell products and on-line software trading packages do "commerce at light speed", while a searchable and integrated web is touted as the "classroom of the future"– education with no teachers. Access to vast amounts information (of often dubious quality) is growing at an unprecedented rate. The ability to navigate through it, distill and relate it to other information, and find and use meaningful content to manipulate and act on the world through it or to have software agents act on our behalf (cf. (Dautenhahn & Nehaniv 2000, Goguen 2001)) present us with psychological, social and ethical challenges that we ignore at our peril. Many of the proposed technologies and integrations will prove only to be artifacts, fading into the past; but others will become reified facts that are part of the human environment.

4.2 Choosing Problem and Design Spaces

The circumscription of a problem space constrains and informs the partitioning of this problem space with requirements, abstractions, functionality, and tasks by designers. In turn a system designed to address the problem in the constrained problem space has been insulated from the circumscription of that problem space. Each time someone refers to "the system", he or she reifies and reinforces this circumscription and partitioning of the problem space. This is a tacit consequence from the outset when one begins or resumes "system analysis".

Under the pressures of various interests impinging on a network that has reified an artifact, the context of its use, the requirements and created needs of the users are all subject to change and drift from the initial set of requirements circumscribed by the system's initial designers. It is well-recognized that the system may fail to meet and address many needs of its users as the result of such requirements change. But more is true, the problems of the problem spaces that the artifact was designed to solve may no longer exist. The reified artifact now drifts subject to pressures which have nothing to do with a humane way of life for its users. Solving the problems of integrating the artifact with other tools and actors and of developing products and technology become primary interests of the network. The design spaces that are candidates for dealing with the problem space are thus constrained, and not critically re-evaluted.

Reflecting on human interests, possibly involving a complete recircumscription of other overlapping problem spaces, can lead us to new, more humane solutions to the questions of other problem spaces we should be asking rather than the ones that we are currently asking constrained by the limitations of our current problem space.

4.3 Letting Human Interests Lead (Rather Than Technology)

As Gorayska, Marsh and Mey (Gorayska, Marsh & Mey 1997) point out, it makes humane sense to put the metaphorical "horse before the cart". It makes humane sense to consider impact on the human mind of technologies and tools rather than to be led by goals of developing technologies for their own sake. The motto is that "*better is better*" rather than "*more is better*". Rather than, *How can we improve technology?*, we need to ask, *How can we use technology to improve the mind?* (Gorayska, Marsh & Mey 2001).

Much research and industry is being led by technology. Driving questions have been: *How can we improve this technology? How can we increase the bandwidth? How can we integrate the existing technologies? How can mobile phones be integrated with the internet? How can we provide more access anywhere for anyone at anytime? How can we attract more buyers to our product?* Such questions serve to put the cart of technology before the horse of human interests, which they tacitly ignore.

Instead, by asking, *What do we want the mind to be?*, one begins to turn the situation around. *How can we enhance human cognitive and social capabilities in a humane manner that respects human wholeness?* In these questions technology becomes a means to serve humane goals rather than the driver of a run-away coach that drags human beings along.

4.4 Technology Integration vs. Re-circumscription of Problem Spaces

A simple technology-driven mindset, deprecated above, asks *"How can we integrate or extend existing technologies?"* Integration of technologies may or may not serve human needs. Should we support an integration of telecommunications technologies that enables automobile drivers to watch television on their mobile phones while navigating a motorway? Safety concerns suggest that we should not. The social and cognitive impact of technological integration are ignored at great peril. Rather than integration, recircumscription of problem spaces (and therefore design spaces and therefore of "systems") may be more appropriate for more humane technology that serves real human needs.

Software engineering practice is often geared to supporting or replacing human work practices by the capture of requirements for and the automation of existing work practices. It is often argued that this is cost-effective, since it seeks to reduce human effort. The introduction of such systems also serves interests of redistributing political power. However, with poor automation human workers often have to work around the inflexible constraints of such systems. Without maintenance of the software and without adapting it to the changing requirements of the organization, such systems can become more of a burden than the paper systems and human-human interactions they were intended to automate. The automated software system intended to support previous practices can become an *obligatory* passage point for workers in the organization although it does not adequately meet their needs and loses relevance to these needs as they change over time.

The use of word-processing software to compose e-mail messages to which are attached documents and that are then sent to others on other platforms creates the problem of incompatability of file formats. Integrators therefore have motivation to produce software converting between various system type, version, and file formats, and variously supported standards. This integration helps makes the locus of integration an obligatory passage point for its users (we see this in particular web-browsers, editors (e.g. emacs, MS-Word, etc.)), with numerous consequences for the users, both immediate and for their long-term use of interactive system technologies.

Looking instead toward the evolvability of such systems, we can ask questions on the circumscription of the problem spaces by stepping outside current practice. Why use different file formats to begin with? How can we design systems which do not require such conversion? How can we build information systems that will be used in ways not yet predictable? What if this system is to be part of something else? How can we make it able to handle that without knowing in advance? What will promote robustnes to requirements change? (cf. (Berners-Lee 1998, Berners-Lee, Hendler & Lassila 2001, Nehaniv 2000, Conrad 1983, Conrad 1990).) Who or what becomes indispensible if a technology is adopted? The software developers? the person with the combination to the safe at the bank? Microsoft? Monsanto? Telephone and telecommunications companies?

By understanding the dynamics of technology change and drift, we hope that designers and users of artifacts will become aware of the networks in which their activity has been shaped by their chosen tools. Letting technology-driven drift carry us blindly into strange and inhumane worlds is the cost of continued inaction or misdirected action. Designers and managers can have an impact by asking whether and how to recircumscribe problem spaces and how to redefine the directions of technology appropriately. Some tools for analysis and some first questions to ask have been given throughout this paper. The choices in answering them should be guided by reflection on developing human interests and humane lifestyles in our relationship to technology.

References

Apple Computer Inc. (1987), *Hypercard User's Guide: Apple MacIntosh Hypercard.*

Berners-Lee, T. (1998), Evolvability, *in* 'Seventh International World Wide Web Conference WWW7 – 14-18 April 1998, Brisbane Australia'. on-line at `http://www.w3.org/Talks/1998/0415-Evolvability/overview.htm`.

Berners-Lee, T., Hendler, J. & Lassila, O. (2001), 'The Semantic Web', *Scientific American* **294**(5), 28–37.

Conrad, M. (1983), *Adaptability: The Significance of Variability from Molecule to Ecosystem*, Plenum.

Conrad, M. (1990), 'The Geometry of Evolution', *Biosystems* **24**, 61–81.

Dautenhahn, K. & Nehaniv, C. L. (2000), Living with Socially Intelligent Agents: A Cognitive Technology View, *in* K. Dautenhahn, ed., 'Human Cognition and Agent Technology', John Benjamins Publishing Company, pp. 415–426.

Dehnert, E. (1986), The Dialectic of Technology and Culture, *in* M. Amsler, ed., 'The Languages of Creativity', Univ. Delaware Press, pp. 109–141.

Goguen, J. (2001), Are Agents an Answer or a Question?, *in* 'JSAI-Synsophy International Workshop on Social Intelligence Design, 21-22 May 2001, Matsue, Japan'. on-line at: `http://www-cse.ucsd.edu/users/goguen/pubs/ps/agents.ps.gz`.

Goguen, J. & Jirotka, M. (1994), Requirements Engineering as the Reconciliation of Technical and Social Issues, *in* J. Goguen & M. Jirotka, eds, 'Requirements Engineering: Social and Technical Issues', Academic Press, pp. 165–199.

Gorayska, B. & Marsh, J. P. (1996), Epistemic Tehcnology and Relevance Analsysis: Rethinking Cognitive Technology, *in* B. Gorayska & J. L. Mey, eds, 'Cognitive Technology: In Search of a Humane Interface', Elseiver North-Holland.

Gorayska, B. & Mey, J. L. (1996), Of Minds and Men, *in* B. Gorayska & J. L. Mey, eds, 'Cognitive Technology: In Search of a Humane Interface', Elseiver North-Holland.

Gorayska, B., Marsh, J. P. & Mey, J. L. (1997), Putting the Horse Before the Cart: Formulating and exploring methods for studying Cognitive Technology, *in* J. P. Marsh, C. L. Nehaniv & B. Gorayska, eds, 'Proceedings of the Second International Conference on Cognitive Technology: Humanizing the Information Age (CT'97)', IEEE Computer Society Press, pp. 1–8.

Gorayska, B., Marsh, J. P. & Mey, J. L. (2001), Cognitive Technology: Tool or Instrument?, *in* 'Cognitive Technology: Instruments of Mind', Vol. 2117 (this volume), Springer Lecture Notes in Computer Science.

Greenbaum, J. & Kyng, M. (1991), Introduction to situated design, *in* J. Greenbaum & M. Kyng, eds, 'Design at Work: Cooperative Design of Computer Systems', Lawrence Erlbaum Associates.

Latour, B. (1988), *Science in Action: How to Follow Scientists and Engineers through Society*, Harvard University Press.

Latour, B. (1996), *ARAMIS or the Love of Technology*, Harvard University Press.

Loomes, M. & Jones, S. (1998), Requirements Engineering: A Perspective through Theory-Building, *in* 'Proc. Third International IEEE Conference on Requirements Engineering', IEEE Computer Society Press, pp. 100–107.

Nehaniv, C. L. (1999), Story-Telling and Emotion: Cognitive Technology Considerations in Networking Temporally and Affectively Grounded Minds, *in* 'Third International Conference on Cognitive Technology: Networked Minds (CT'99), Aug. 11-14', San Francisco/Silicon Valley, USA, pp. 313–322.

Nehaniv, C. L. (2000), Evolvability in Biological, Artifacts, and Software Systems, *in* C. C. Maley & E. Boudreau, eds, 'Artificial Life 7 Workshop Proceedings', Reed College, pp. 17–21.

Rapp, F. (1981), *Analytical Philosophy of Technology*, D. Reidel Publishing.

Suchman, L. A. (1987), *Situated Plans and Actions*, Cambridge.

Turski, W. (1981), Software Stability, *in* 'Systems Architecture: Proc 6th ACM European Regional Conference', Westbury House, pp. 107–116.

Thinking Together in Concept Design for Future Products — Emergent Features for Computer Support

Tuomo Tuikka and Kari Kuutti

Department of Information Processing Science, University of Oulu, P.O.Box 3000, 90014 Oulu, Finland
{Tuomo.Tuikka,Kari.Kuutti}@oulu.fi

Abstract. This paper points out how we can design systems in order to support concept designers who collaborate in creating new product concepts, small hand held electronics devices. An understanding of concept design is presented using concepts from Activity Theory to analyse design practice. The three levels of Activity: Activity, Action and Operation are discussed in order to reveal how Hypothetical User Activity is embedded as a mediator in Design Activity. This finding is used as an informing concept to design computer systems for synchronous collaboration of geographically distributed concept designers. Two exemplary systems are used to instantiate how the conceptual understanding has been useful for our systems design efforts. It is concluded that this treatment has lead us into consider CSCW applications as systems which mediate meaning.

1 Introduction

Concept design of small hand held electronics devices is an activity where new electronics gadgets, e.g. mobile telephones or heart rate monitors, are collaboratively created by a group of interdisciplinary people. Among these disciplines are software, electronic, mechanical, and industrial engineering, and marketing. Concept design is the early phase of product development process, which provides frames for further development of the product. Uncertainty is characteristical to this phase, in the beginning of the work no one of the designers knows what kind of product concept the group will end up with (Tuikka 1997). Consequently, the design situations are dynamic, and rich communication between participants is necessary. Many means for conveying the designers understanding of the emerging concept are used during the design situation. Examples of such means are gestures, drawings or listings (Tang 1989).

Design is situated action in the sense that the designers do improvise when they convey their ideas to others and respond to contingency during the improvisation (Suchman 1987). Besides talking, sketching and gesturing, we have observed that a number of artifacts available in a situation are used to augment designer's gestures by artifacts thus conveying what has come into their mind during (Tuikka and Kuutti 2000).

M. Beynon, C.L. Nehaniv, and K. Dautenhahn (Eds.): CT 2001, LNAI 2117, pp. 40-54, 2001.

Because of the intensity of the needed communication, a typical concept design session has happened in one place. Increasing globalization, networking, subcontracting etc. has had an influence, however, that a design team more and more consists of participants located in different cities, countries or even continents, and collecting them in the same place is slow and expensive. Thus there is increasing interest in to conduct also concept design sessions in a distributed way over a communication network.

There are several alternative venues how this might be supported by different communication channels like emerging collaboration features in CAD software, using a 3D environment (Büscher, Mogensen et al. 1999), or real-time video (Scrivener, Harris et al. 1993), for example. Plain increase of the capability of communication channels between cooperation partners does not, according to (Karsenty 1997) automatically increase the efficiency of communication. He suggests that rather than looking for a mix of technologies ensuring a 'maximum shared environment' one should provide participants in remote collaboration with a limited set of tools simply ensuring an 'optimal shared environment'

We believe that it is possible to proceed towards this optimal shared environment by studying the nature of cooperation process, and the information that must be communicated during it. In this paper we analyze a concept design process we have been able to observe at a detailed level. The paper has three purposes:

First, to show that although the end result of the process is an idea of a technical device, something else was also constructed during the process, something that was absolutely vital for the end result. A set of hypothetical user actions generated during the process served as the central mediating instrument: the features of the device idea were created and shaped in thought-experiments with these actions. The set of actions served as a scaffolding in constructing the concept itself. In our example process, this scaffolding was lost after the idea solidified, but that does not necessarily need to be the case. In this paper we call the set of actions a *hypothetical activity*.

Second, to show that the actions cooperatively constructed and used in the process have several layers that must be taken into account. There are purposeful actions themselves – doing something with the device-to-be. But besides this, there is contextual information about the actions (where it happens, who else was there, what did happen before action) to make them meaningful and grounded. And the actions discussed were mostly *combined, fluent actions*. This means that it was tacitly assumed, that the hypothetical user had already developed certain skills in manipulating the physical device. For example, in scrolling a phone number list by repeatedly pushing a button it was assumed, that the focus of the user is in identifying the right number, not in pushing the button. This means, that the former action of finding a button and pushing it had already become an automated operation, subsumed into a fluent, combined action of number-finding.

Finally, the paper discusses about a prototype system that has been developed to study, how to support a design discourse between distributed designers on issues like skilled use of a non-existing device.

2 The Role of Hypothetical Activity in Concept Design

2.1 The Case

The material for the following analysis comes from a fieldwork in an experimental concept design project simulation. Due to several reasons it is very difficult to gain access to real concept design projects of this type, and thus it was decided in a research project (VIRPI) to do a simulation instead. Although such simulation naturally lacks some constraints of real life (see (Tuikka and Kuutti 2000) for a more accurate discussion about the nature of the process), we believe that the process is a plausible representation of concept design experiments actually done in companies. This is also supported by the fact that after the end result was made public, the project was contacted by companies that were interested in developing the concept further into a commercial product.

The concept development team consisted of people involved in actual product development or research on product development. As normal in concept design, the people involved had different backgrounds: industrial design, mechanical engineering, electronics engineering and software engineering. 6-8 persons participated in different workshops. The experience of the participants varied from very experienced designers to novices (researchers). The team had four successive concept design workshops which were observed and three of them videotaped. These workshops lasted for altogether 12 hours.

The team was given a very open task to create a concept for a novel telecommunications device. Figure 1 shows a visualisation of final product concept produced in the workshops. It is a mobile communication device, a'pen-shaped user interface for a mobile telephone', placed in the picture in a sales box.

Fig. 1. Pen-shaped user interface for a mobile telephone (Photo courtesy of Metsävainio Design).

The material collected from the workshops has been analysed in detail and various findings have been reported for example in (Tuikka & Kuutti 2000, Tuikka 2001). This paper concentrates only one aspect of the material: the relation between the idea of the technical device and how it was thought to be used.

2.1.1 Hypotethical User Activity in Developing a Concept

It is no surprise that there is dicussion about actions user might do with the concept thing to shaped, but it is a bit surprising how important role these user actions have: there was quite little discussion about concentrating on the device itself, without a reference to a possible use situation or a way to use it. Moreover, the actions discussed were not simple and atomistic, but there was a certain structure in them.

We will use activity, action, operation hierarchy of Activity Theory (Leontjev 1978) as a useful conceptual framework. Activity can be described as a hierarchy with three levels (cf. Figure 2): *activities* realised through chains of *actions*, which are carried out through *operations*. Activities are larger formations that give actions the context where they can be seen sensible and purposeful. Activities typically involve other persons than a single actor.

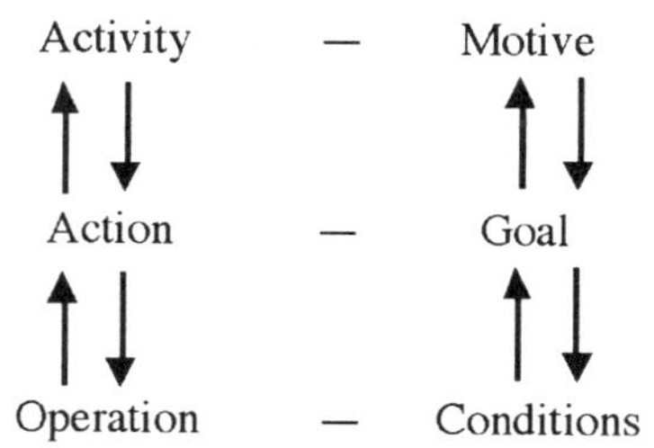

Fig. 2. Three levels of activity (Leontjev 1978)

Actions are directed towards a goal. These goals are consciously aspired, thus requiring attention of the subject of action. While actions are conscious, operations are not. Operations are former actions that are transformed into automatic routines, triggered by suitable conditions in the course of action. Operations are always subordinate parts of a larger action, making it smooth and fluent. It can be characterized that actions describes a task, activity describes a meaningful context of tasks, and when discussing about operations we address fluent and skilled performing of a task.

Our observation is that the dicussion about user actions in concept design contains elements from all these three levels of hierarchy. In the dicussions during the workshops, a set of hypothetical but contextualized, fluent actions with the device-to-be were constructed parallel to the development of the device idea itself; this set is vital for the emergence of the concept. We call this set a *hypothetical user activity*, the possible future user activity which links the future use of the device with the current, ongoing concept design situation. The reason why we call it hypothetical is that it is not based on any particular real activity, it is neither a complete activity in activity theory sense, but a collection of plausible short scenarios that 'might be' a part of a meaningful larger whole. In the sessions a common understanding of hypothetical user activity was collaboratively generated by the designers. Characteristical to this was that the use situations were addressed through the activity.

We believe that the using of the three activity levels in the set of hypothetical actions is necessary for the development of the concept, and therefore we must maintain a possibility to communicate about them also in a distributed design situation. Next we will show how the levels are present in the discussion, and the rest of the paper will explore how they can be utilized in developing requirements for a design support system.

2.1.2 Examples of the Existence of Hierarchical Levels

The middle level, goal-directed actions, is obvious: most of the discussion was about them. We will here concentrate only on the upper and lower levels.

The activity level refers to a larger whole, that makes individual actions meaningful. Also references to something outside the actions that could influence some aspect of the concept, like references to culture, division of labour etc. belong to activity level.

Example one: one of the designers' suggested that the group could design a telephone which would be inserted into a tooth cavity filling. The argument was that: "...I have heard that in Hollywood, someone has inserted a positioning device into tooth of a child." Thus, in case of kidnapping the phone would always be available. Suggestion was soon discarded, but it is an example where only referring to an (extraordinary) activity makes sense to an idea.

Another example of general activity-level references, was the discussion on 'a woman communicating' in a mobile context. Designers (all male) had a vivid and long discussion about female cultural acceptance factors, reasons for communicating, people contacting to, etc. that were assumed to differ from the "mainstream" mobile phone design ideas of that time. (As a curiosity it can be mentioned that this discussion was one of the major "shapers" of some central aspects of the final product concept.)

The third example is a suggestion to design a very simple device with only one button, or with no buttons at all, "like the command badge in Star Trek". Touching this would contact the caller to a secretary or to the company exchange, who then would connect the call further. Again, without the explanation of the larger context the action itself would be useless.

2.1.3 Existence of the Operation Level in Discussions

(Bardram 1998) argues that the design of computer support should view cooperative breakdowns (Bødker 1990) as an important resource in design. In the examples we can identify breakdowns in order to illustrate how designers argue for their perspectives in design situation.

The first example was called as the handedness problem of a pen-shaped phone. One of the designers pointed out, using an occasional pen as an aid, that there would be a problem with the concept if a left handed person would start making a phone call: "Look at the display, it is on the right hand side of the device, if we roll the pen to the left hand, the display will be upside down". This also leads to the problem that the button numbers are upside down. The user has another option, which is — not to roll — but to move it to the left hand. Then the microphone and loudspeaker do not match with ear and mouth, however.

The example shows how hypothetical user action was referred at since there the hypothetical user has an actual conscious goal which she wants to accomplish to make a phonecall. The handedness problem creates a potential breakdown in the hypothetical users behavior. A left-handed user cannot make a fluent combined action of making a call but he or she must concentrate on device itself. This way the discussion addressed the level of operations.

Interestingly, the designer in this case and many other cases acts out the user action. This acting out helps designers' to convey and furnish their arguments. Acting out the user action, with the problems or benefits of use of certain approach, was usually conducted in ways which required kind of 're-living' the hypothetical future situation. Users' gestures were therefore mimicked beforehand. User activity, action and operation brings the intention, and meaning to the gestures and any use of the situational artifacts.

Another example can be found in the discussion about the number of buttons in a pen-shaped form; should there be a full mobile telephone set or only a few special buttons. A designer argued that if the user happens to have big fingers, it is quite difficult to press the small buttons which had been designed on the device. Thus, the user interface approach would be better. Here again, the problem in the design makes the hypothetical user to concentrate on the operations, forcing the user from action of making a phonecall to action of pressing the number buttons. In this occasion a normal ballpoint pen was taken from pocket, and used to illustrate the dimensions of the concept. Also, the pressing was acted out, leading to a convincing presentation.

2.2 Mediation

Concept design can be characterized as work practice where designers engage in situated action (Suchman 1987). An important resource for designers in their situated action is their history and their personal development in the culture they belong to. Everything they do in their actions is related to their own culture, whether they create a combination of new techniques or a 'new innovation'. Situationality is manifested especially well in the way concept designers' use artifacts to *mediate* their understanding.

As we discussed earlier, designers' use various means to share their vision of the concept idea. The product concept is a collaboratively constructed design object in the level of design activity; it is in the interest of the designers is to create a new product concept in a certain period of time. Furthermore, in the beginning of the concept design process there is no 'shared' object of design. Thus, designers must construct the object through collaboration. Various representations and situational artifacts are used to convey designers tentative ideas. These artifacts can be anything, a pen, eye-glasses or a quick sketch, i.e. situationally used artifact to discuss the design options and tentative ideas.

Mediation by artifacts is another important concept of Activity Theory in addition to the three levels of activity:

"A key idea in Activity Theory is the notion of mediation by artifacts. Artifacts, broadly defined include instruments, signs, language, and machines to mediate activity and are created by people to control their own behavior. Artifacts carry with them particular culture and history and are persistent structures that stretch across activities through time and space."(Nardi 1996)

Figure 3 illustrates what kind of design artifacts can be used in order to represent and construct the shared design object. The shared design object is transformed during the process. According to the discussion above this design object is not only the idea of the device, but the idea of the device and hypothetical user activity that evolve together (Fig. 3).

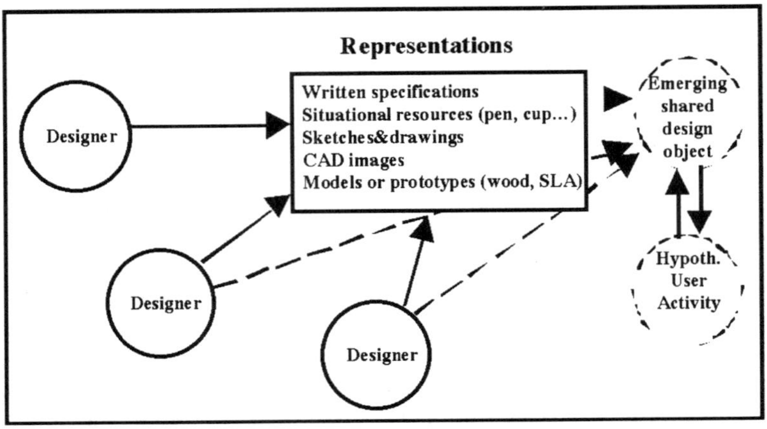

Fig. 3. Designers, various representations and the shared design object.

In contrast to any representation which is physical, the hypothetical user activity can be considered as a psychological artifact (Vygotsky 1978). Other examples of such psychological artifacts are heuristics, individual and collective experiences, scientific results and methods (Bardram 1998). In our concept design case the psychological artifacts, such as hypothetical user actions, are continuously modified and shaped to meet the evolving needs. On one hand, the designers' vision of hypothetical user activity forms limits of what kind of users' actions and operations will there be in the future. On the other hand, its' purpose is to mediate multidisciplinary designers' understanding of the future product concept.

In conclusion, it is our understanding that in actual concept design situation, the existence of a hypothetical user activity gives meaning to the use of representations and gestures. In the next section we will discuss how the understanding of hypothetical user activity as a mediating artifact can inform the identification of requirements and design ideas for a computer system to support distributed concept design.

3 Designing Systems for Concept Designers

Contemporary computer systems used by product designers are usually Computer-Aided Design (CAD) systems, which support individuals' work on a design task. While these systems have been successfully used for this purpose, the vendors have recently been adding features to support cooperative design of geographically distributed product designers. This trend in CAD technology has been enabled by the World Wide Web technology. The most common solution is that the CAD data can be viewed with proprietary 'viewers'. Such features are used for product data distribution and are helpful in, for instance, communicating the product concept status via a company intranet.

Virtual prototypes for concept design communication have been used in such domains as aircraft industry, and automotive industry. Electronics and telecommunications industry these applications are CAD for industrial and mechanical design or various simulation tools for user interface design.

The aim in our research project has been to provide photo-quality, functional virtual prototypes to support geographically distributed cooperative design of electronics devices (Tuikka and Salmela 1998; Salmela and Tuikka 1999; Salmela and Kyllönen 2000). In our terms, a virtual prototype simulates product features with a degree of functional realism comparable at least to a physical prototype. A shared virtual prototype allows multiple designers to collaborate on the product concept over time and distance.

3.1 Mediation through Computer Systems

Much of our work in shared virtual prototyping has been studying feasibility of virtual prototyping through the Internet, and providing solutions to distributed simulation (Salmela and Tuikka 1999). We will not go here into the software architectures, or such WWW 'standards' as Virtual Reality Modeling Language (VRML), Java or Java3D, but only note that these are a mean to introduce some ideas and concepts in order to achieve goals on higher abstraction level.

We have found out that computer supporting geographically distributed creative interdisciplinary designers in synchronous collaboration is difficult for two reasons, first there is no clear shared object to deal with, which we can use as a mediator, second, since the design situation is dynamic, the designers' use any kind of situational means in order to convey their design ideas. The difficulty of the area can be illustrated with an example from the field. In the start of our work we were given a requirement of building a system which would be like digital wax, which the designer should be able to mold with their hands. This is, however, practically impossible, and in any case this 'wax' would be in use of one designer. The second problem is, we think, one of the most challenging for computer systems designers: the problem is to find a well-designed computer environment for fluent communication of designers. Certainly, we could use existing direct video, audio and shared web pages etc. to computer support designers' collaboration. But due to the nature of concept design, these means provide only partial support for concept designers'. For instance, physical dimensions and dynamics of the shared objects are not explicitly addressed.

The idea with using hypothetical user activity as a concept to inform computer systems design is based on the assumption that this user activity, action and operation is a common denominator which the interdisciplinary designers can all relate to. But what is the link to systems design, i.e. how can we use this conceptual understanding to inform design. The purpose of the finding has been to map new ways of computer support, i.e. helping us to find a new view on what should be in the system, and also giving us new perspective into the use of computer systems. Thus, supporting us going beyond preconceived design of collaboration systems.

Table 1 presents the hypothetical user (HU) activity as a delineator and issues which can be linked from the designers activity to computer systems requirements.

Table 1. Forming computer support requirements for the designers' activity in general level.

Levels of activity	Requirements: computer system should support esigners' activity in:
Hypothetical user (HU) activity	•mediation of the designers' understanding of hypothetical users' (HU) activity
HU action	•mediation of designers' understanding of the HU actions. •designers' acting out the HU actions. •designers' argumentation on the contradictions of the object of design
HU operation	•designers argumentation on the contradictions of the object of design

This presentation gives direction for the innovation of features in the future CSCW systems and systems to support collaborative design. The benefit of this perspective comes apparent when we start asking what the specific requirement means, and how should we design and construct these features into a computer system.

3.2 Bringing Design Ideas Forward through Application Demonstrations

The purpose of our research demonstrations is not to implement all the ideas that can be derived from the field work, but these applications have rather served as means for discussion with our research and industrial partners. The first application is called WebShaman (Shared manufacturing), a web-based application and the second DigiLoop, due to its approach in implementation.

3.2.1 WebShaman

We started implementing WebShaman in 1997 and have developed several tools and techniques to make creating virtual prototypes easy, supporting collaboration, and simulation of functionality. This development has taken place with VTT Electronics at Oulu (Salmela and Kyllönen 2000).

WebShaman is based on the VRML 2.0, controlled with a set of Java tools, implemented in a dynamic framework, allowing loading a selected set of tools. The user interface has a VRML window and a Java applet to control it (Figure 4). This dichotomy is due to technical reasons and leads to an implementation where the virtual prototype is on the left in the figure, while the tools used to control it are on the right.

WebShaman introduces a three-dimensional functional virtual prototype as a design object representation for collaborating designers. Since World Wide Web is the platform these designers can be geographically distributed. In this implementation we assumed that the object of design has one or several forms, and the virtual prototype is an outcome from a CAD application, where all the exact measures can be dealt with.

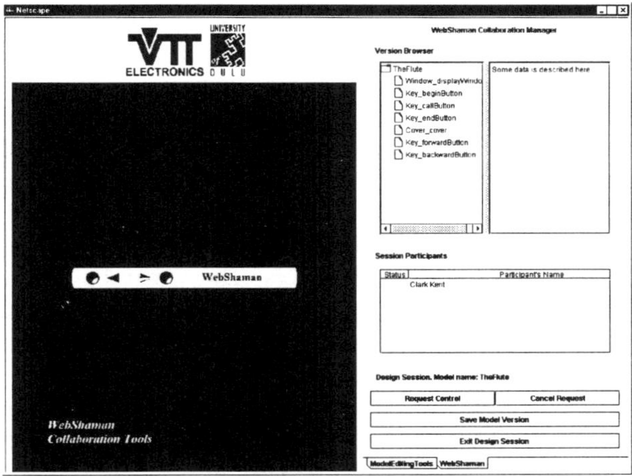

Fig. 5. WebShaman, virtual prototype on the left, and Java tools on the right. The Java tools in the figure are for collaboration support, other tools can be selected as well.

The designer can quickly access a web page and try out the virtual prototype. It is possible to view the virtual prototype in 3D, press the buttons, and see the effect of the user interactions on the display, i.e. try out the functionality. This is due to the simulation application which resides in the net. Thus the application is aware of the virtual prototype components, which allows their selection and manipulation.

Some of the basic tools in collaboration systems are shared cursor and views (Prakash and Shim 1994; Anderson, Jasnoch et al. 1996). Collaboration tools under the virtual prototype in Fig. 5 are a 3D shared cursor (arrow), and a virtual trackball for synchronous object manipulation. Awareness and control mechanisms allow the designers to go into synchronous session where observation or object manipulation are possible.

In design session one of the users is in control of the virtual prototype shared by the designers. Other designers can take turns manipulating the prototype, and everyone will see the list of designers updated with a number indicating a request for a turn. All the designers participating in the session can see the interactions of the one who is in control. Virtual prototype rotation and functionality, i.e. button presses and their feedback are synchronously shown on every web page which is linked in collaboration. When the color or size of a component is changed, the others will see the new color as well.

Of course the concept design is still in the designers' responsibility, WebShaman only mediates some issues of user activity among designers. We assume that there is a conference telephone connection for the group, which is used for discussion, argumentation and coordination of tasks. Agreements such as decision on a focus group and use activity can be done using the telephone. Besides visualisation, virtual prototyping, however, is aimed to support the designers in expressing what kind of functionality the user can or cannot do. Now we come to the design ideas for the application, and I will present them in form of scenarios.

A scenario could be that one of the designers creates visualisations of one concept. In the end there will be 8 or 9 versions which solve the problem given, this is a usual case in concept design. These visualisations are converted to virtual prototypes which accomplish those features which the designer wants to discuss on. The designer can create a set of cases which supports his argumentation on doing these specific visualisations, and also may present some contradictions in the selected design. Thus, the user actions for the session are prerecorded and argued for discussion. In the session, the user activity is known, i.e. a woman using a mobile telephone in order to call her secretary for arranging flights. Hence, the user actions can be reflected against the concept under scrutiny with the virtual prototypes, which gives a shared object representation for the designers. There may be common notes pages open, in order to collect comments on the suggestions and to do meeting minutes.

Hypothetical user action is reflected in prerecorded series of actions which the designer can use in his argumentation. Every operation is step by step executed on the virtual prototype, which provides a 3D user interface to the design object. The gathered information can be organized in form of user activities, actions, and operations, which makes the argumentation easy to revert to. There may be several virtual prototypes in use at the same time, thus allowing designers to juxtapose their options.

Digital videocameras allow designers to take snapshots on their use of situational artifacts in their environment and distribute the digital data. In a design session this kind of information will be large, and a user interface which allows delineating information according to activity and action would be quite useful. These are examples of ideas, and some of them have implemented in WebShaman. Table 2 instantiates some of the design ideas which can be derived from presented conceptual understanding.

Table 2. Forming computer support requirements for the designers' activity.

Levels of activity	Design ideas
HU activity	•Use telephone or netphone to mediate designers' understanding of hypothetical users' (HU) activity •Build or use shared text application
HU action	•Create user interface for recording use of virtual prototype step by step (operations) •Create user interface for delineating data in form of activity, and action •Build an information tree where situational videodata can be attached •Make information artifact based, virtual prototype can be used as a mean to open up data on argumentation (cf.(Reeves and Shipman 1992))
HU operation	•Step by step playback of the user operations •Saving argumentation referring to the operation-action dynamics, i.e. breakdowns which are caused by design contradictions.

Table 2 shows only an extract of the many ideas which our treatment of data has produced. Thus, we can say that the experiment has been successful. The ideas, however, do not point out exactly how the user interfaces should be done, so something is left for the user interface designer to do. Drawing conclusions how user interface should be done should be based on participative design with the future users. We have done this iteratively, and found out certain solutions that fit our purpose.

One major problem which was referred earlier also is that since the design object is co-constructed, there is no shared object to deal with. In the case of virtual prototyping, it is the task of industrial designer to create the first version of the shared design object. But a product concept is not only a visualisation, also functionality, i.e. what happens when you press a virtual prototype button, and mechanical properties, such as dimensions must be addressed. In WebShaman the functionality is built using a set of simulation tools by user interface engineers. Since the user interface does not need to be totally finished at this point of time there may be only partial implementations. Although WebShaman supports simulation, it is meant only for communication between geographically distributed designers, i.e. there is no exact measurement data.

3.2.2 Some Critique on WebShaman Approach

The first critique we received was that this approach does not support the critical part of concept design, i.e. the quick creation of ideas, where representations change quickly. Second, one should be able to take hold of the object, in order to really understand its dimensions, and third, one cannot actually use the surrounding physical environment in order argue for a concept. Neither of these three arguments, however, conflicts with the basic understanding we presented of how designers do their work. As we see this criticism, these claims manifest the need of designers for an approach that introduces computer support for user actions, but we should also try to find other possible facilities for them to express what they mean.

3.2.3 DigiLoop

What does it mean when we say that we should support quick creation of ideas? To someone in the design group this would mean sketching (Gross and Do 1996). To someone else who is not that able to draw, it would mean innovative discussion. It has been argued here that the ideas are generated during a co-constructive design session, situational artifacts, and that acting out the uses of the future device are important for idea generation. Altogether, the environment where designers are is important, that is, where they do the acting out. Acting out refers here to all the pointing and gesturing in 'physical' space and use of situational artifacts in order to convey ones' understanding of a future device.

Figure 6 depicts a use situation of the (previously also WebShaman) DigiLoop system which was a result of this consideration. The flat screen in DigiLoop is a window to the virtual design world and its virtual objects. While watching a 3D virtual prototype through the window, the user holds a physical object in his hand. The physical object is anything which can be used to represent the size of the future device, a pen, an existing mobile telephone, a watch. In product design industry widely used stereolithography or wooden model could be used as such physical object as well.

Bringing the physical object to the virtual world requires some technical devices. The box on the right is Polhemus Fastrack, which tracks where the hand and the physical mock-up are. The smaller box is the sender, and the two receivers are placed on the dataglove and the mock-up. The (Virtual Technologies) dataglove can be used to recognize finger movements. With all this technology combined, we have an application where the user can interact with the virtual prototype while touching the mock-up. The user sees the virtual prototype wrapped on the physical mock-up. The distance between the mock-up and the virtual prototype can be determined by the system, and the users' mind connects these two. The virtual prototype simulates mechanical movements, such as button presses, and functionality, i.e. response to button presses on display.

Fig. 6. DigiLoop in use. Virtual prototype and hand representation seen on the flat screen. The artifact in the users' hand is a whiteboard pen.

The mock-up, i.e. the design object representation, can be brought into the designers' physical environment. It can be used to evaluate concept usability in the designer/user environment and to test its use with other artifacts, such as a purse or a pocket, or the users' physical properties, including the hand or the ear (a pen can be behind the ear).

DigiLoop is built with Java and Java 3D, and we have developed a driver for Polhemus. We have not yet designed or implemented tools for sharing of a virtual prototype in DigiLoop, but designers should naturally be able to share these use situations. A next step, is obviously to make an interface with collaborative tools we have been developing. This application has not been yet tested with users, since our aim is to generate means for collaboration. There are also usability issues which must be addressed before the application is suitable for actual use. Besides that the idea has been generated through a conceptual understanding from a field study of concept design, there are also other benefits: the idea is fairly simple, there is not too much cognitive load with it (no helmet), designers' environment can be used for evaluation of user actions, the virtual prototype actually exists in the physical space, if the artifact is moved, the virtual prototype will follow it, thus DigiLoop, it shows only part of the virtual reality.

4 Conclusion

Hypothetical user activity as a mediator among designers is a finding from studying concept design, which would have been difficult to point out without Activity

Theory. This has lead us to a path, which has helped us to create application design ideas and argue on their validity in real use. The development of these applications has convinced us that in our application instead of exploiting What You See Is What I See (WYSIWIS) paradigm, we use an approach which could be phrased as: What You See Is What I Mean (WYSIWIM). Thus, we give appreciation to the meaning of what the designer is doing, and conveying this meaning to other designers.

Acknowledgments. This work has been done in the Finnish national VIRVE project (1999-2001). The author wishes to acknowledge Finnish Technology Development Center, TEKES, Nokia Mobile Phones, Cybelius Software, Polar Electro, Metsävainio Design, and C3 Suunnittelu for funding this research and VTT Electronics of research collaboration. Thanks are due to Pertti Repo and Virtu Halttunen who have done most of the software.

References

Anderson, B. G., U. Jasnoch, et al. (1996): Coconut—A virtual prototyping environment. *Computer Graphics Topics* **8**(Mar. 1996): 20-22.

Bardram, J. (1998): Designing for the Dynamics of Cooperative Work Activities. ACM 1998 *Conference on Computer Supported Cooperative Work CSCW '98*, Seattle, USA, ACM Press 89-98

Bardram, J. E. (1997): Plans as Situated Action: An Activity Theory Approach to Workflow Systems, *European Conference on Computer Supported Cooperative Work ECSCW '97*, Lancaster, UK, Kluwer Academic Publishers 17-32

Bødker, S. (1990). *Through the Interface — A human activity approach to user interface design*. Hillsdale, New Jersey, Lawrence Erlbaum.

Büscher, M., P. Mogensen, et al. (1999): The Manufaktur: Supporting work practice in (landscape) architecture. *European Conference on Computer Supported Cooperative Work ECSCW '99*, Copenhagen, Denmark, Kluwer Academic Publishers 21-40

Engeström, Y. (1987): *Learning by Expanding*. Helsinki, Finland, Orienta-Konsultit Oy.

Engeström, Y. (1993): Developmental studies of work as a testbench of activity theory. *Understanding Practice: Perspectives on Activity and Context.* S. Chaiklin and J. Lave. Cambridge: 64-103

Gross, M. D. and E. Y.-L. Do (1996): Ambiguous Intentions: A Paper-Like Interface for Creative Design. *ACM Symposium on User Interface Software and Technology*, ACM Press 183-192

Karsenty, L. (1997): Effects of the amount of shared information on communication efficiency in side by side and remote help dialogues. *European Conference on Computer Supported Cooperative Work ECSCW '97*, Lancaster, UK, Kluwer Academic Publishers 49-64

Kuutti, K. (1991): Activity Theory and its applications to information systems research and development. *Information Systems Research Arena of the 90's.* H.-E. Nissen, H. K. Klein and R. Hirschheim (ed.). Amsterdam, North-Holland: 529-549

Kuutti, K. (1991): The Concept of Activity as a Basic Unit for CSCW Research. *European Conference on Computer Supported Cooperative Work ECSCW '91*, Amsterdam, Kluwer 249-264

Kuutti, K. (1996): Activity Theory as a potential framework for human-computer interaction research. *Context and Consciousness: Activity Theory and Human Computer Interaction.* B. A. Nardi (ed.). Cambridge, MIT Press 17-44

Leontjev, A. N. (1978): *Activity, Consciousness and Personality*. Englewood Cliffs, NJ, USA, Prentice-Hall.

Nardi, B. A., (ed.) (1996): *Context and Consciousness*. Cambridge, Massachusetts, USA, The MIT Press.

Nardi, B. A. (1996): Studying Context: A Comparison of Activity Theory, Situated Action Models, and Distributed Cognition. *Context and Consciousness: Activity Theory and Human Computer Interaction*. B. A. Nardi (ed.). Cambridge, Massachusetts, USA, The MIT Press 69-102

Prakash, A. and H. S. Shim (1994): DistView: Support for Building Efficient Collaborative Applications using Replicated Objects. *Conference on Computer Supported Cooperative Work CSCW '94*, Chapel Hill, NC, USA, ACM Press 153-164

Reeves, B. and F. Shipman (1992): Supporting Communication between Designers with Artifact-Centered Evolving Information Spaces. *Conference on Computer Supported Cooperative Work CSCW '92*, Toronto, Canada.

Salmela, M. and H. Kyllönen (2000): Smart Virtual Prototypes: Distributed 3D Product Simulations for Web Based Environments. *Web3D/VRML Symposium 2000*, Monterey, CA, USA, ACM Press 87-94

Salmela, M. and T. Tuikka (1999): Smart Virtual Prototypes for Web-Based Development Environments. *International Conference on Web-based Modeling&Simulation WEBSIM '99*, San Fransisco, CA, USA 127-133

Scrivener, S. A. R., D. Harris, et al. (1993): Designing at a distance via real-time designer-to-designer interaction. *Design Studies* **14**(3): 261-282

Suchman, L. A. (1987): *Plans and situated actions*. Cambridge, UK, Cambridge University Press.

Tang, J. C. (1989): *Toward an Understanding of the Use of Shared Workspaces by Design Teams*. Dept. of Mechanical Engineering, Stanford University: 173.

Tuikka, T. (1997): Searching Requirements for a System to Support Cooperative Concept Design in Product Development. *Designing Interactive Systems DIS '97*, Amsterdam, The Netherlands, ACM Press 395-404

Tuikka, T. (2001): User Actions as a Mediator for Concept Designers. *Proceedings of Hawaii International Conference on Systems Sciences HICSS '34 (on CD)*, Hawaii, USA, IEEE.

Tuikka, T. and K. Kuutti (2000): Making New Design Ideas More Concrete. *Knowledge-Based Systems* **13**(6): 387-394

Tuikka, T. and M. Salmela (1998): Facilitating Designer — Customer Communication in the World Wide Web. *Internet Research: Electronic Networking Applications and Policy* **8**(5): 442-451

Vygotsky, L. (1978). *Mind in Society: The development of higher mental processes*. Cambridge, MA, USA.

Extended Abstract

The Space of Cognitive Technology: The Design Medium and Cognitive Properties of Virtual Space

Frank Biocca

Media Interface and Network Design (M.I.N.D.) Labs
Dept. of Telecommunication
Michigan State University, U.S.A.
Biocca@msu.edu

By its name, cyberspace is a spatial medium. Metaphorically, all information is presented as locations in an abstract representation space. As in pictures and television, cyberspace inherits the properties of pictorial and screen space. But with 3D interactive media such as immersive virtual reality, augmented reality, and room computing, cyberspace adds dramatically new ways to use space to represent information around the body of the user, around the virtual environment, and as distributed representations around the physical environment. One of the more visible ways in which spatial properties of virtual and augmented reality systems are used is in the area of scientific visualizations. But this is only one among a set of emerging applications that emphasize spatial representation and make use of spatial abilities.

Humans are spatial animals. A great deal of the cortex, especially in areas such as the superior colliculus, is dedicated to processing spatial egocentric and allocentric spatial relations. Humans use space in many different ways including for computation.

Spatial representation in advanced virtual environments is probably one of the most powerful system for spatial representation and manipulation ever developed. At one level, computer graphics, the basis for virtual environments, owes its origin to our understanding of the rules of spatial perception and cognition.

But cyberspace as a design, communication, and cognitive environment remains largely unexplored. While the cognitive processing of cyberspace shares properties with the processing of physical space, the two remain different because (1) distortions in mediated spatial representations have no equivalent in physical space, (2) mediated Environments such as those created by virtual reality allow ways to represent, use, and manipulate space in manners that have no equivalent in physical space, and (3) virtual environments represent space with only a subset of cues found in physical environments.

Much remains to be understood and explored about the perception, manipulation, and design of the space in cyberspace. In this paper we intend to begin to map the perceptual, behavioral, and semantic properties of cyberspace. We would like to further explore how the design of virtual environments can make use of the properties of spatial cognition. Specifically, we would like to examine the cognitive properties of spatial position in virtual environments. We want to treat virtual space as a cognitive technology.

M. Beynon, C.L. Nehaniv, and K. Dautenhahn (Eds.): CT 2001, LNAI 2117, pp. 55-56, 2001.

Our journey into the topic of spatial position is motivated by a scientific instinct that there are powerful properties of spatial representation that can be harnessed to further improve the representation and organization, human learning, cognitive performance, and human collaboration and communication in virtual environments.

We hope to support and elaborate the cognitive implications of the following fundamental theoretical hypotheses:

- Positions in space may come "precoded" with useful cognitive properties that can be used in the design of information systems.
- Positions in virtual space are cognitively anisotropic, that is, different locations and directions exhibit different cognitive properties.
- Positions in virtual space have different perceptual, behavioral, semantic, and memory properties.
- There are design advantages and disadvantages of placing information in different spatial positions.

If most of the above are true, the designers of information systems need to know how the user's mind makes use of spatial information. For example, it would be valuable to know how users think, use, and assign possible meaning to a location in space around the body, the screen, objects in the virtual environment, and the 3D virtual environment. In this paper, we will concentrate on how information objects and tools might be mapped to spatial positions in virtual environments to optimize the meaning or utility of information. Along the way we will explore the possible cognitive properties of virtual space.

Acknowledgements. This work was supported by the National Science Foundation Grants IIS 00-82743ITR,and CISE #9911123, the MSU Foundation, and the Ameritech Endowment.

Can Social Interaction Skills Be Taught by a Social Agent? The Role of a Robotic Mediator in Autism Therapy

Iain Werry[2], Kerstin Dautenhahn[1], Bernard Ogden[1], and William Harwin[2]

[1] Adaptive Systems Research Group, University of Hertfordshire, Department of Computer Science, College Lane, Hatfield, Hertfordshire, AL10 9AB, UK
{kerstin,bernard}@aurora-project.com
[2] Department of Cybernetics, University of Reading, Whiteknights, PO Box 225, Reading, RG6 6AY, UK
iain@aurora-project.com, w.s.harwin@reading.ac.uk

Abstract. Increasingly socially intelligent agents (software or robotic) are used in education, rehabilitation and therapy. This paper discusses the role of interactive, mobile robots as social mediators in the particular domain of autism therapy. This research is part of the project AURORA that studies how mobile robots can be used to teach children with autism basic interaction skills that are important in social interactions among humans. Results from a particular series of trials involving pairs of two children and a mobile robot are described. The results show that the scenario with pairs of children and a robot creates a very interesting social context which gives rise to a variety of different social and non-social interaction patterns, demonstrating the specific problems but also abilities of children with autism in social interactions. Future work will include a closer analysis of interactional structure in human-human and robot-human interaction. We outline a particular framework that we are investigating.

1 Introduction

Increasingly new technologies are used in educational and therapeutic contexts, among them *socially intelligent agents*, i.e. agents that show aspects of human-style intelligence [Dautenhahn 98, 99b]. New classrooms are designed, involving computer technology in order to facilitate learning, creativity and collaboration among children [Bobick *et al.* 99]. The NIMIS project gives an example where virtual worlds are designed as places of imagination, virtual and interactive theatres where children can create and explore their own stories [Machado & Paiva 00]. *Carmen's Bright Ideas* is an interactive animation simulating a counseling session for mothers whose children undergo cancer treatment [Marsella *et al.* 00]. It is hoped, that by getting engaged in the scenario, users empathize with its characters and can reflect on their own situation by assisting the virtual characters in their decisions. For many years LOGO/LEGO systems with physical or simulated robots have been used in education,

M. Beynon, C.L. Nehaniv, and K. Dautenhahn (Eds.): CT 2001, LNAI 2117, pp. 57-74, 2001.

supporting an exploratory, constructionist approach towards learning [Papert 80]. In recent years a new generation of interactive *social robots* has been designed as research platforms [Breazeal & Scassellati 99], toys (e.g. Sony's Aibo, the Robota dolls [Billard 00]) or educational tools [Druin & Hendler 00; Cooper *et al.* 99]. Such robots are specifically designed for interactions with humans, and often interactivity is the primary purpose. Social robots are therefore clearly different from service robots that may or may not interact with people, but which are primarily designed for a particular purpose such as a tour guide [Schulte *et al.* 99].

For many years computer and (more recently) virtual environment technology is used in autism therapy, see e.g. [Colby & Smith 71; Colby 73; Russo *et al.* 78; Panyan 84; Chen & Bernard-Opitz 93; Strickland 96; Powell 96; Murray 97; Moore 98; Blocher 99; Charitos *et al.* 00; Parsons *et al.* 00]. Such work shows that many people with autism interact very 'naturally' with computer technology, and that such technology provides a safe and predictable environment that can be used in an exploratory and creative manner. A few projects also investigate robotic devices. Weir and Emanuel (1976) studied one child with autism who used a keyboard to control a remote-controlled robotic device. Plaisant *et al.* (2000) are studying how a remote controlled robot can teach children about emotions. Michaud *et al.* (2000) developed different robotic designs and did initial studies with children with autism.

This article addresses the potential of interactive, social robots as therapeutic tools or 'toys' in autism therapy, research that is done in the project AURORA. The AURORA project uses a single interactive robot and studies different scenarios in order to characterize and evaluate qualitatively and quantitatively the children's behavior when playing with the robot. The one-child-one-robot scenario is our basic approach, with the robot in the role of a therapeutic teaching device, a tool that can be used to teach children with autism basic social interaction skills. However, a second very important role of the robot that we investigate in this paper, is its role of a *social mediator*, a tool that mediates (encourages and facilitates) social behavior among children, and among children and adults. In designing such a social mediator it is important to study carefully how children with autism interact with the robot, how their particular difficulties and abilities influence the interactions, and how the robot can be used so as to encourage particular 'desirable' types of interactions, i.e. interactions that children with autism might generalize, and that then could help in daily interactions with humans outside the classroom context.

2 The AURORA Project

The AURORA project was initiated in 1998, with the goal of developing a robotic agent which would aid in the therapy of children with autism.

According to the National Autistic Society, children with autism show impairments in three key areas: social interaction, social communication and imaginative play. Additionally, these children appear to find comfort in repetitive and monotonous activities, and avoid the complex interactions involved with people, whether they are teachers or relatives. This leads to an inhibited ability in social situations, which provokes fear and further avoidance.

The aim of the AURORA project is to provide both a mediator and teaching aid for the children, giving them an opportunity to practice and explore their behavior in social situations, as well as to provide a focus of interest and attention for the children in order to stimulate interpersonal relations and interactions. Also, a robotic agent has the properties of repeatability and stability, while presenting communication along a limited number of channels, and robots have become familiar to children through television and the media. These factors allow the children to relate to the robot on a simple level, without the fear and complexity of human interaction, and allow the children to learn simple skills such as role-playing and turn-taking games. The robot should be used by the teachers and be flexible enough to reach a number of children's levels, not replacing the personal interaction common in classes and traditional therapy methods, but providing an additional tool. More details about the project's background, objectives, the robot used in the trials, and autism therapy are given in [Dautenhahn 99; Werry & Dautenhahn 99, 00, 01; Werry *et al.* 01].

3 The Two-Children-One-Robot Case Study

A number of trials have been performed to date. The trials involve children with autism between 8 and 12 years of age, mostly male (all children reported on in this paper are male). Trials at an early stage in the project were intended to gauge the level of interaction to be expected from the children and to provide a baseline for further development work. The first positive result was that the children were not afraid to interact and approach the robot. This was an important result as, if the children had been cautious of the robot, then the interaction would not have been comfortable and natural for them.

Later trials were performed to evaluate the type and nature of the interactions involved in order to direct further development. Trials were structured to allow the children to interact with the robot as they would in a natural environment, with a minimum of guidance. The robot that is currently being used in the project is small mobile robot with 8 infrared sensors positioned around its body and 1 positional heat sensor mounted at the front. Its dimensions are 38cm by 30cm by 21 cm. It is very robust and relatively lightweight (6.5 kg). A speech box has been added to the robot so that it is able to producing speech, i.e. certain phrases such as 'I can't see you', triggered by the robot's own state and the way children interact with it.

One series of trials studied how children with autism interact with the mobile robot as opposed to a non-robotic toy [Werry *et al.* 01]. Here, a quantitative evaluation technique based on micro-behaviors was developed. By applying the quantitative evaluation technique, interaction patterns of individual children could be characterized, data necessary for further studies. This article focuses on a particular series of trials with pairs of children, investigating the role of a mobile robot as a social mediator. For this particular series of trials it seemed important to go beyond a micro-level of description and study patterns of interaction that focus on the social level. Results are therefore presented qualitatively in the format of descriptive narratives.

3.1 Setup

The trial sessions took place in a room at the children's school (Radlett Lodge School, Hertfordshire, United Kingdom). This has the advantages that the children are not required to move out of the building, the disruption to their normal classes is minimal, and the environment is known to them. These factors reduce anxiety and allow the children to behave and interact without additional distractions and also allow the staff of the school to be present as advisors.

The room used was approximately three meters by four meters in size, and had a single door and a single window, overlooking the school car park. The only features present in the room were chairs for the experimenters, and the robot. Also present was a member of staff from the school, who arrived with the children and was familiar with them. They were on hand to advise on the mood of the children and when to terminate the trial, as well as to restrain and provide occasional interaction and prompts with the children. Two experimenters were also present in the room, although they restricted their actions to documenting the trials with a video camera and to assist the robot when it inevitably became overturned or switched off.

Six children participated in the trials, in three pairs. These children had been selected prior to the trial session by the teachers of the school. The children consisted of four who had had a previous experience with the robot, and two who had not seen or interacted with the robot before. The children were chosen on the basis that they would provide an interesting response with the robot, either because they enjoyed technology or because they were more outgoing. Additionally, the pairing of the children was also performed by the teachers, based partially on which children would provide 'interesting' pairings and which were available at the same times and could be brought out of their normal class without disruption. It should be noted that this selection method is preferential to children who are of a higher functioning level, as these children are able to tolerate a disrupted schedule and are automatically selected by the teachers for their interaction abilities.

The robot is programmed using a behavior-based architecture (e.g. [Arkin 98]). It performs a number of actions, primary among these are avoidance of obstacles, including people, following of a heat source and the generation of simple speech and phrases. These behaviors allow the robot to follow and to be chased by the children, while at the same time producing brief utterances for those children who are able to respond to speech. The result of this is a robot which provokes responses from the children and is pro-active in its interaction by moving towards the children and prompting by the use of speech.

Within these confines, the children are able to perform any action or behavior which would not damage either the robot, themselves or the environment. Other members of staff at the school were seated outside the room and so the children were even able to leave the room if they wished. No pressure was placed on the children to perform a specific type of interaction or behavior, or even to interact with the robot at all. This is due to the fact that any type of interaction was welcomed as informative on the state of the robot and the development and ability level of the children. The trials lasted until either the teacher or one of the experimenters requested an end, due to the state of the robot or the emotional state of the children. On average, the trials lasted nine minutes.

3.2 Results

3.2.1 First Pair – Child A, Child B

While child A only started at the school only 2 months before the trial, this pairing was the strongest socially of the trial pairs. The trial started with child B, who had encountered the robot previously, entering the room and commenting on its ability and speed compared to an earlier trial, without directly interacting with the robot. Child A then entered the room and watched the robot until child B pointed and drew attention to it. Both children then watched the robot as it moved around them, and child B questioned the experimenters about the robot's designers. Child A then questioned the experimenters, who he had not previously met, and child B intercepted the robot as it headed towards the door, and picked up the robot to turn it around. The robot then uttered a phrase, which child A heard and became excited with. He then questioned the experimenters about the robot's speech, and directly interacted with the robot to bring it into the center of the room. Both children then watched the robot. Child A then knelt down to look at the robot more closely while child B asked questions about the sensors on the robot and picked it up briefly. Child A then tentatively touched the robot. Both children became puzzled when the robot stopped moving and moved in front of it at the experimenters' suggestion.

Child A seemed continually fascinated by the robot's speech and crawled around to look at it. He then backed away from the robot in an effort to avoid it and continued to move away when it followed him. When the robot lost him, he moved back into range and then away again in order to allow the robot to continue to follow him. At this point, child A noted that one of the wheels of the robot was loose and falling off and an experimenter intervened to fix the robot. The children seemed to lose interest in the robot as it was being repaired, and focused their attention on one of the, now unattended, video cameras.

When the robot was again ready, both children transferred their attention back to it, and child A asked what the name of the robot was. Child B then tried to get the attention of the robot by tapping on the top of it. He then asked about other programs for the robot and child A asked about the buttons at the rear of the robot, child B answered that they were for re-programming. Child B continued to wonder about other programs for the robot and showed one of the experimenters his 'Robot Wars' hat.

Child A appeared to grow tired with the robot and jumped over it before leaving the room to talk to the teaching staff who were observing outside the room. He came back into the room and the children discussed what else the robot should do. Both children then attempted to make the robot follow them simultaneously.

The children then talked about what they would like the robot to be able to do and one of the experimenters showed the children how they were able to chase the robot by moving close to it, which child A then tried to do. Both children talked about robots that they had seen and things that they would like the robot to do and child A then showed child B how to chase the robot. At this point, the wheel again fell off the robot and the trial session was ended.

Fig. 1. Child A and child B playing with the robot

The trial showed a number of positive results and the children were reasonably social and interactive, both with the robot and with each other. Both children were vocal and talked both about and to the robot throughout the session, in particular about what the robot 'should' do, and both children became excited with the robot. The children were uninhibited with the robot and were able to interact naturally, by touching and operating it. In particular, the robot provided a platform for interaction in the case of the children chasing it. This was demonstrated to child A who then instructed child B, suggesting learning by imitation and social interaction. The fact that these children were naturally quite sociable with each other was demonstrated in the session and they were obviously comfortable together.

The robot was a strong focus for child-child interaction and child-robot interaction and it worked well in promoting interaction generally. However, it should be noted that the children were of a higher functioning level than the other pairs in the trial and this may influence the number and types of interactions observed.

3.2.2 Second Pair – Child C, Child D

The second pair of children also consisted of one child who had encountered the robot before (child C), and one who had not (child D). Child C had recently become interested in the car park and the cars outside of the building, while we were told that child D had little interest in people generally unless they functioned as an audience for him. Additionally, child D arrived at the school only one month previously, and both the children were in the same class and so were familiar with each other, although they were not as strong socially as the previous pair.

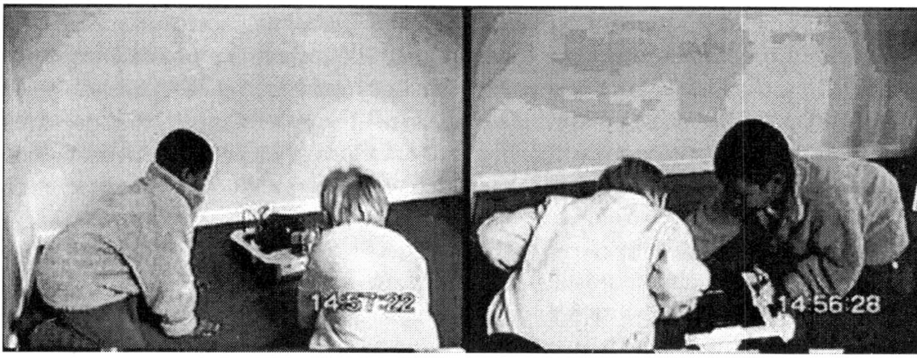

Fig. 2. Child B and child C playing with the robot

Child C was instantly interested in the robot and immediately started to push and touch it – interactions which had been prevalent in the previous trials. When child D entered the room, he circled the robot and stood watching as child C pushed it. When it began to speak, child D bent down to tell the robot to 'rip up the carpet'. Both children interacted with the robot simultaneously, but without interacting with each other. As the robot became stuck against the wall both children pushed it away, and watched as it moved around a little. Child D was aware that the robot was able to locate him and asked how this was possible. He then became agitated about the robot's speech as child C pushed the robot again.

The teacher present calmed child D a little and drew attention again to the robot. Child D again told the robot to 'rip up the carpet' and both children touched and interacted with the robot but without acknowledging each other, however while child C pushed the robot, child D was forced to move in order to stay in front of the robot.

The teacher again drew the attention of child D to the robot's speech, while child C sat to the side. Child D then circled the robot and child C pushed it a little, which continued as child D talked to the teacher about the robot and its abilities.

Both children then appeared to grow bored and their attention shifted to other parts of the room. Child D then picked up the robot and the teacher took it off him, and child C walked around the room. With neither child in front of it, the robot stopped moving and child D asked why this had happened. The robot then moved towards the teacher and child D asked it if he could go home. Child C moved to the window and spent the majority of the remainder of the session watching the car park, and child D got closer to the robot and interacted with it by touching it, forcing it to become stuck by obstructing it.

When the robot was again able to move, child D chased and started to push it around, as child C had done at the beginning of the trial. He then became interested in the buttons on the rear of the robot and turned it off. The teacher was able to restart the robot, and child D then shouted at it in an attempt to get it to move. He then became hyperactive and began shouting. He was then instructed by the teacher to 'be nice' to the robot and ignored or watched it without any direct interaction.

This session demonstrated the common phenomenon of non-social play, both children attempted to interact with the robot without acknowledging the presence of the other. They were able to do this only to a certain degree; however, with one robot and two children, situations occurred where this was not completely possible, as one child moved the robot the second was forced to move in turn to be able to interact with it. This is an area which demands further research as it may provide a means of encouraging social interaction in children with autism.

This pair of children interacted with the robot at a lower level (compared to the previous pair), for example treating it as a toy and pushing it around the floor. It is important that the robot is robust enough to withstand this type of operation. Child C became interested in the car park towards the end of the session and this restricted the interactions with the robot. Previous trials have taken place in a room without an external window and so this type of distraction has been limited.

3.2.3 Third Pair – Child E, Child F

This pairing was the least social, although both children are in the same class as the first pair and both had encountered the robot in previous sessions. The session began with child E interested in the robot and manipulating the robot's front sensor to point in different directions, while child F seemed uninterested in the robot, and interacted with the teachers outside the room. The robot then backed away from child E, who extended his arms as it moved and then withdrew from the robot so that it moved towards him again, as he did this he told the robot to 'leave me alone'. He then repeated this interaction but moved his whole body position towards and away from the robot. Child F then re-entered the room and put a foot on the robot, without acknowledging the presence of child E. Child E continued to dominate the interaction with the robot while child F walked around the robot and ignored the play in the center, before again leaving the room. All this time, child E operated the robot through his relative position with it and gave it vocal commands and directions of movement. As child F again entered the room, child E became interested in the door and this allowed child F to become focused on the robot a little more, crouching down and looking at it more closely. Child E then resumed his interaction, giving the robot vocal commands while moving towards it, and child F diverted his attention away from the robot and again moved around the room.

When the robot spoke, child F seemed pleased and approached the robot once again, even though child E was still interacting with it, and then moved away again. He approached the robot again when it said the same phrase and examined it and touched it. Child E then obstructed the hands of child F and took control of the robot and pushed it. Child F again touched the robot and child E again pushed his hand away, as the robot moved between them. Child E again obstructed child F from touching the robot and told child F to 'leave him alone', even mentioning the name of child F to instruct him not to touch the robot.

Child F then backed away from the robot and observed the interaction between it and child E, before again touching the robot and again being told to 'leave that alone' and 'leave him alone' by child E who again mentioned the name of child F. Child E also used his body to obstruct child F and pushed him away bodily. Child E also pointed out the robot's sensor as its 'eyes'.

Child E then backed away from the robot, and walked around it, and asked it for its name. Child F again lost interest in the robot and started to ask the teacher about the window and to walk around the room. Child E then started to overturn the robot, and one of the experimenters was forced to intervene to avoid the robot becoming damaged, turning it right way up again.

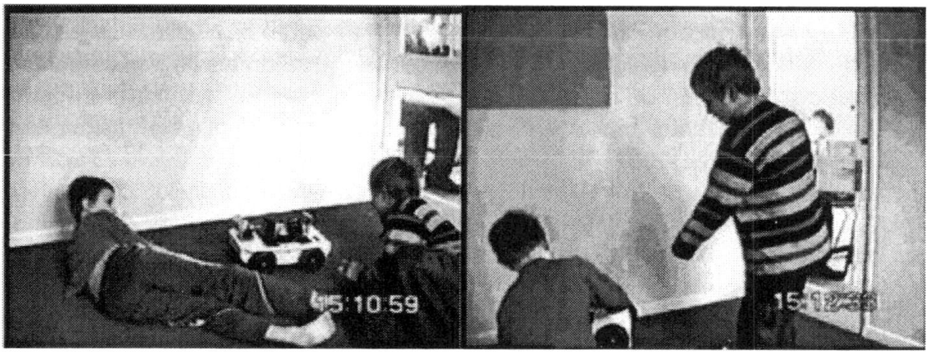

Fig. 3. Child E and child F playing with the robot

Child F continued to circle the robot as child E watched and waited for the experimenter to replace the heat sensor which had become dislodged. When this was done, child E interacted with the robot and again obstructed child F, again began to overturn the robot and again the experimenter was forced to intervene. This pattern was repeated a third time and a fourth before the session was ended.

This session shows a repeating pattern of interaction: child E is dominant in the interaction and does not allow child F to interact. However, he waits for the experimenter to finish with the robot when the robot becomes overturned. He is also relatively vocal and in his obstruction of child F becomes social. He also displays a variety of interactions with the robot and understands how to operate it and that it will follow and avoid him. Child F became interested in the robot initially after it provided verbal stimulus, and takes opportunities to interact with the robot when child E is otherwise occupied or distracted.

3.3 Discussion of Results

These trial sessions resulted in a number of important points. Something which is perhaps not initially obvious is that the social relationships between the children were preserved. While the children did not show a marked increase in their sociability, the fact that they behaved in the same way as they do in other situations and that their social interactions, and lack of them, was preserved shows that the robot scenario did not present an entirely different environment for them. This result is particularly important given the autistic child's inhibited ability to generalize, with any ability learned in a certain environment not being carried to a second, different environment. This means that there is an increased possibility of any developments made with the robot to be used in other situations.

Secondly, the ability of the robot to provide a focus of attention and shared attention was demonstrated in the first pair, as an experimenter showed child A how to interact with the robot by demonstrating that it would back away from objects. Later in the session, child A then explicitly demonstrated this in order to instruct child B, showing social interaction, social learning (possibly imitation) and demonstration of a new skill. All of these areas are known to be difficult for children with autism [Jordan 99, Nadel *et al.* 99].

The role of the robot as a mediator also became apparent in a number of situations where the children interacted with the teachers about the robot. A number of the children asked the teacher present about the robot and it became the focus of attention for child-teacher interaction, child-experimenter interaction and child-child interaction. For example, both the children in the first pair discussed the robot with the experimenters, talking about what it is and is not able to do, as well as what they thought that it should do.

Additionally, the fact that children with autism are generally seen to play non-socially was observed in the trial sessions. Both children in each pair only had a single robot and no other items to play with. This lead to the situation where both children attempted to use the robot individually and in isolation of each other, inevitably leading to conflicts. Future developments of the robot could take advantage of this situation, providing a method of interaction which would allow two children to operate the robot simultaneously. This would mean that the children would not directly interact with each other, but would do so through the robot as a mediator. Such scenarios could support the transition from non-social play (a child playing alone with the robot), to non-interactive play (both children playing with the robot simultaneously but without interaction between them), to social and interactive play (the children playing with each other, with or without the robot as a toy).

The preconceptions about robots carried by the children also influenced their interaction, with the first pair discussing the television program 'Robot Wars' with the experimenter and with child D instructing the robot to become destructive. This can be an important factor in exactly what the children expect from the robot and the ways in which they expect to be able to interact with it. However, the fact that the robot was not destructive did not seem to inhibit the interaction greatly, except in a few cases where the children asked the robot to perform some specific actions.

In general, the behaviors implemented on the robot performed well. A few of the children were aware that the robot is able to follow them and that they are also able to chase it. Also, a few children were able to learn this, both through instruction and through learning by observation. An important fact is that the interactions with children who did not make use of the follow and chase behaviors were not obstructed by these behaviors. For example, child C was able to push the robot around without interference from the robot's behaviors. The robot's ability to produce speech had an impact on the interactions. While the children did not seem to use the speech for their direct interactions, a number of children became drawn to the robot after they learned that it produced verbal comments. This initiated further interactions and prompted a closer inspection by the children. In this respect, the actual phrases produced by the robot seem to be less important than the fact that the robot is able to speak. A few of the children attempted to operate the robot through the use of vocal commands, which were ineffectual and if implemented the robot would unlikely be able to obey them as they were relatively complex. In this way, the use of verbal commands for the robot seems to have a limited potential. It should also be remembered that children with autism can have a large range of abilities and reactions. To simply develop the ability of the robot to obey speech commands would neglect those children who are non-verbal. Similarly, to expand the speech produced by the robot is to risk children of a lower ability becoming afraid and confused with the robot, and the speech becoming too distracting.

4 A Closer Look: Interactional Structure and Implications for Robot-Human Interaction

The qualitative and rather macro-level descriptions given in section 3 reveal that during interactions of the children with the robot very interesting types and structures of interaction can be observed (such as instruction, cooperation, possibly imitation etc.). In the long-term we hope that a) knowledge about such structures will help us in the development of a vision system that can automatically identify and track certain interaction patterns with therapeutic relevance which will aid the evaluation of trials tremendously, and b) that knowledge about interactional structures will allow us to make the robot more 'socially responsive', going beyond reactive, behavior-based interaction. Such issues are addressed in the interactive vision project [Ogden & Dautenhahn 01a, 01b] which is a sub-part of the AURORA project and explores the interactional structure of human-human and human-robot interaction in more detail, strongly inspired by literature from sociology, psychology and anthropology. This section will outline the main issues and concepts that we identified as important ingredients of a framework of interactional structure that we will investigate in the context of interactions of children with autism with robots.

Generally, humans as one kind of interactive system are very advanced and it is certainly unrealistic at the present time to develop robotic systems based on many aspects of human interactional structure: for example, human interactional 'rules' must be applied according to the current context [Button 90], where context has a number of specific meanings including the personality and history of the interactants, the relationship between interactants, the physical nature of the place in which the interaction is occurring and the type of institution and culture in which the interaction is taking place. In order to apply knowledge on interactional structures to scenarios studied in the AURORA project the complexity of human interactional behavior needs to be reduced so that we can realistically deal with it. We propose achieving this in two ways: by viewing interaction as locally structured and by dealing with interaction only at a metacommunicative level. We believe that this allows us to develop a system that is simple enough that it can realistically be implemented at the present time but that is sufficiently advanced to both provide the basis for a system that will extend the interactive capabilities of robots and provide a useful framework for understanding robot-human interaction. Also, on a conceptual level, a focus on locally structured interaction and a subsemantic level better fits a bottom-up view on human social behavior, and human intelligence in general [Dautenhahn 97, 99]. Such a view means that we are particularly interested in *emergent* properties in human-human and human-robot interaction, with a focus on *interaction dynamics* and changes in *interaction structures* over time.

4.1 Globally Structured Interactions

Viewing interaction as locally structured contrasts with the globally-structured view of interaction, which is perhaps best exemplified by the concept of scripts [Schank & Abelson 77]. While scripts were not specifically designed to explain human interaction, they are an ideal way to view globally structured human interaction. In this view a given type of interaction has an associated script specifying the actions that should occur as the interaction progresses. Kendon's example of the structure of greeting interactions [Kendon 90] may be viewed in this way (although Kendon's own perspective on the way that interaction is structured seems closer to a locally structured view [Kendon 80]): in this example a greeting can be divided into three primary phases: distance salutation, approach and close salutation. Each of these phases has particular behaviors associated with it and once a greeting is begun then each phase will be carried out in turn, in a global view of interactional structure. In the present case, however, this view of interaction seems difficult to apply: as the above examples indicate, the interactions between child and robot are inherently freeform and we are left with the task of clearly defining the detailed structure of a very broad, loosely defined interaction ('play') or of preparing a number of simpler structures reflecting events that tend to occur, such as the chase game played by child A. While this is an option, it would be preferable to take a more flexible approach that does not require us to detail a set of actions that are supposed to occur from beginning to end. General limitations in using this 'global' approach are considered in [Ogden & Dautenhahn 00].

4.2 Locally Structured Interactions

In a local view of interactional structure, by contrast, actions are influenced by immediately preceding actions. Overall structure is not considered and we have little concern with 'expected future actions' beyond the immediate next action. Also, while not denying the importance of the history of an interaction to the current action we can note that in many cases the immediately preceding action is the one of primary importance (e.g. if a question is asked, most of the time we will expect it to be answered), allowing us to focus in most cases on immediately preceding and following actions. A view of interaction as locally structured is prevalent in the field of conversation analysis [Psathas 95]: in this view any given action tends to constrain the likely action taken by other interactants, as in the case of questions tending to prompt answers. This view seems promising in that actions can be considered as simple action-response pairs without having to consider questions of the type of interaction that is occurring: this both simplifies the structures that will need to be developed and broadens their applicability. This view also seems likely to result in more robust interactive behavior: if an interaction is expected to follow a given structure from beginning to end and it then deviates from that structure, it seems that this will be harder to recover from than an inappropriate response to an isolated action. In the locally structured case repair may or may not occur, but if it does not the interaction can continue without concern for what this means for some broader abstract structure. To give an example, we might have a rule that if the child approaches the robot, the robot will retreat at a similar speed to that at which it was approached (this rule is partially implicitly implemented in the existing programming of the robot, in that the robot will retreat if it is too close to an obstacle): a rule such as this allows chasing games to emerge without the need to specify an explicit 'structure of chasing' from the outset. To view the actions of the children in these terms, when child B points at the robot, this serves to focus the attention of the children on the robot. This is an example of immediate, local structure: the rest of the interaction is not relevant to the effect of this pointing action.

4.3 Meta-communicative Aspects of Interaction

Our second means of simplifying the interaction structures is to consider metacommunicative aspects of interaction only. Generally speaking, *metacommunicative* refers to the 'syntax' of communication as opposed to the 'semantics'. Before proceeding it is necessary to clarify in more detail what is meant by metacommunicative. We can divide the communication in interaction into two orders of communication: the first is the explicit transfer of information between interactants e.g. the *content* of what is being said or the *meaning* of a gesture (such as "OK", or "Goodbye"). The second is what we refer to as metacommunicative and is about how the semantic component of the interaction is *structured*: for example, it is concerned with where turns are taken or identifying who is involved in a given interaction: unlike communicative actions involved in the transmission of semantic content, metacommunicative actions are often engaged in without any conscious awareness on the part of the sender and / or the receiver. It may be that such features of interaction as a sense of involvement in interaction and the perceived quality of an interaction are due to or at least reflected in metacommunicative behaviours [Cappella 97]. We are therefore investigating interactional structures with a simple turn-taking nature: a good example of the kind of interaction that we are interested in is described in [Dautenhahn 99b] which studies temporal coordination of movements in robot-human interaction. There is no explicit semantic 'meaning' communicated in such interactions but nonetheless an interaction can be said to be taking place. In terms of generating robotic behavior we primarily focus on nonverbal communication ('body language'), which has a significant meta-communicative component. Examples of interaction with no verbally communicative component from the preceding section include the cases of the 'chasing' game and the 'following' game, both engaged in by the first pair of children in the preceding section. Here, we can assume that there is no attempt to communicate anything but an interaction is nonetheless taking place.

In addition to inspiring our approach to interactional structure, conversation analysis provides us with a means of assessing interactions between the children and the robot. Conversation analysis is a qualitative approach that produces detailed descriptions of interactive (not necessarily verbally communicative) behavior at a micro level. This allows detailed analysis of an interaction in the context in which it occurs and also allows structures present in the interaction to be discovered without any theoretical presuppositions being made: in this way what is actually present in the interaction can be discovered without concern that we are being guided to see structure that is not really present by theoretical assumptions. In the case of the above interactions this allows us to see structures and details that might be missed at a higher level of description. In addition to conversation analysis other, more quantitative, approaches are available: for instance, the Theme package [Magnusson 96] finds statistically significant temporal structure in behavior. We see these approaches as complementary and hope to explore the use of each to determine its suitability for the present project.

Detailed analyses of the structure of interactions such as those presented above will allow us to determine the kinds and complexities of interactions that are occurring. This, in turn, will allow us to develop the 'social interaction skills' of the robot so that children can get engaged in more advanced interactions with the robot. Even with macro-level descriptions such as those above we can get some idea of how to extend the robot: for example, if it is revealed that a child plays a game of 'chase' with the robot, as child A does in section 3.2.1 above, we can develop the robot to extend this interaction, both making it more rewarding for the child and increasing the complexity of the interaction, which will give the child experience in dealing with more advanced interactions. For example, we could extend the chase game so that the robot both runs from the child and, from time to time, turns round and pursues the child instead, or so that the speed with which the robot retreats varies according to the speed with which the child approaches. Of course it will be necessary to be careful in developing such extensions: if it seemed likely that the child would be disturbed by being chased by the robot then such an extension would obviously not be appropriate. Similarly, if we wish to encourage two children to cooperatively play with the robot we could set up the robot to preferentially respond to one child rather than the other at different points in the interaction – this may encourage different kinds of interaction than that described in Sect. 3.2.3 while encouraging the kind of interaction described in Sect. 3.2.1: equally, we could create try to create games that involve two cooperating children. While macro-analysis of the interactions allows us to propose rough advancements such as these, the micro-analyses should both inform the detailed design of such advancements and will likely expose further ways in which the interactive capabilities of the robot may be usefully extended.

5 Conclusions

This article reported on trials that investigated the role of a mobile robot as a social mediator in autism therapy. When pairs of two children were playing with the robot interesting interaction structures were observed, such as instruction, cooperation, possibly learning by imitation, and others. Results so far are very encouraging, although more trials will be needed before we can draw general conclusions and long term studies need to show whether the two-children-one-robot scenario has any positive impact on the children's social or other skills. The integration of knowledge on interactional structures within this scenario is hoped to help us in future work on the analysis and design of robot-human interactions.

In the general context of robots that can be integrated in human society such work points towards a new potential role that robots might adopt in human society, beyond tools and appliances, a fundamentally social role. The design of such systems in different application domains such as education and therapy will require a detailed analysis of robot-human relationships, how humans perceive and respond to robots [Bumby & Dautenhahn 99], and how a social robot, within the constraints of its particular 'purpose' for which it was designed, could meet the cognitive and social needs of human beings.

Acknowledgements. This project is supported by EPSRC GR/M62648 and conducted with the assistance of Radlett Lodge School of the National Autistic Society. The Labo-1 robot used in this research is donated by Applied AI Systems, Inc.

References

1. Aibo, URL: http://www.aibo.com Last referenced 23rd March 2001.
2. Arkin, R. C. (1998) *Behavior-based robotics*. MIT Press, Cambridge, Mass.
3. AURORA, URL: http://www.aurora-project.com/ Last referenced 23rd March 2001.
4. Billard, A. (2000) Play, dreams and imitation in Robota. Proc. *Socially Intelligent Agents - the Human in the Loop*, AAAI Fall Symposium, Technical Report FS-00-04, AAAI Press, pp. 9-12.
5. Blocher, K. H. (1999) Affective Social Quest (ASQ) *Teaching emotion recognition with interactive media and wireless expressive toys*, MSc Thesis, MIT, USA.
6. Bobick, A. F., Intille, S. S., Davis, J. W., Baird, F., Pinhanez, C. S., Campbell, L. W., Ivanov, Y. A., Schütte, A., & Wilson, A. (1999) The KidsRoom: A perceptually-based interactive and immersive story environment. *Presence: Teleoperators and Virtual Environments* 8(4), pp. 369-393.
7. Breazeal, C. & Scassellati, B. (1999) How to build robots that make friends and influence people. *Proc. IROS99*, Kyonjiu, Korea.

8. Bumby, K. & Dautenhahn, K. (1999) Investigating Children's Attitudes Towards Robots: A Case Study, *Proc. CT99, The Third International Cognitive Technology Conference*, August, San Francisco.
9. Button, G. (1990) Going up a Blind Alley: Conflating Conversation Analysis and Computational Modelling. In Luff, Gilbert and Frohlich (eds) *Computers and Conversation*, Academic Press Limited, London, UK.
10. Cappella, J.N. (1997) Behavioral and Judged Coordination in Adult Informal Social Interactions: Vocal and Kinesic Indicators. *Journal of Personality and Social Psychology* 72(1), pp. 119-131.
11. Charitos, D., Karadanos, G., Sereti, E., Triantafillou, S., Koukouvinou, S. & Martakos D. (2000) Employing virtual reality for aiding the organisation of autistic children behaviour in everyday tasks, *Proc. 3rd Intl. Conf. Disability, Virtual Reality & Assoc. Tech*, Alghero, Italy 2000, pp. 147-152.
12. Chen, A. H. A. & Bernard-Opitz, V. (1993) Comparison of personal and computer-assisted instruction for children with autism. *Mental Retardation* 31(6): 368-376.
13. Colby, K. M. & Smith, D. C. (1971) Computers in the Treatment of Nonspeaking Autistic Children. *Current Psychiatric Therapies*, 11, pp. 1-17.
14. Colby, K. M. (1973) The Rationale for Computer-Based Treatment of Language Difficulties in Nonspeaking Autistic Children. *Journal of Autism and Childhood Schizophrenia*, 3 (3), pp. 254 - 260.
15. Cooper, M., Keating, D., Harwin, W. & Dautenhahn, K (1999) Robots in the Classroom - Tools for Accessible Education. *Proc. AAATE Conference 1999, The 5th European Conference for the Advancement of Assistive Technology*, C. Bühler and H. Knops (Eds.), November, Düsseldorf/Germany, IOS Press, pp. 448-452.
16. Dautenhahn, K. (1997) I could be you - the phenomenological dimension of social understanding. *Cybernetics and Systems Journal*, 28(5), 417-453.
17. Dautenhahn, K. (1998) The Art of Designing Socially Intelligent Agents - Science, Fiction, and the Human in the Loop. Special Issue "Socially Intelligent Agents", Applied Artificial Intelligence Journal, Vol 12, 7-8, October- December, pp. 573-617, 1998.
18. Dautenhahn, K. (1999) Robots as Social Actors: AURORA and the Case of Autism. *Proc. CT99, The Third International Cognitive Technology Conference*, August, San Francisco.
19. Dautenhahn, K. (1999b) Embodiment and Interaction in Socially Intelligent Life-Like Agents. In: C. L. Nehaniv (Ed.): *Computation for Metaphors, Analogy and Agent*, Springer Lecture Notes in Artificial Intelligence, Volume 1562, Springer, pp. 102-142.
20. Dautenhahn, K. & Werry, I. (2000). Issues of Robot-Human Interaction Dynamics in the Rehabilitation of Children with Autism. *Proc: From Animals To Animats, The Sixth International Conference on the Simulation of Adaptive Behavior (SAB2000)*, 11 - 15 September 2000, Paris, France.
21. Dautenhahn, K. & Werry, I. (2001) The AURORA Project: Using Mobile Robots in Autism Therapy. *Learning Technology online newsletter*, publication of IEEE Computer Society Learning Technology Task Force (LTTF), Volume 3 Issue 1, January 2001, ISSN 1438-0625.
22. Druin A. & Hendler J., Eds, (2000) *Robots for Kids: Exploring New Technologies for Learning*. Morgan Kaufmann Publishers.
23. Jordan, R. (1999) *Autistic Spectrum Disorders - An Introductory Handbook for Practitioners*. David Fulton Publishers, London.
24. Kendon, A. (1980) Features of the Structural Analysis of Human Communicational Behaviour. In: von Raffler-Engel (ed) *Aspects of Nonverbal Communication*, Swets and Zeitlinger, Lisse, Netherlands, pp. 29-43.

25. Kendon, A. (1990) A Description of Some Human Greetings. In Kendon, Conducting Interaction: *Patterns of Behavior in Focused Encounters*, Cambridge University Press, Cambridge, UK.
26. Machado, I. & Paiva, A. (2000) The child behind the character. *Proc. Socially Intelligent Agents - The Human in the Loop*, AAAI Press, Technical report FS-00-04, pp. 102-106.
27. Magnusson, M.S. (1996) Hidden Real-Time Patterns in Intra- and Inter-Individual Behavior: Description and Detection. *European Journal of Psychological Assessment* 12(2), pp. 112-123.
28. Marsella, S. C., Johnson, W. L., & LaBore, C. (2000) Interactive Pedagogical Drama. *Proc. of the Fourth International Conference on Autonomous Agents*, June 3-7, Barcelona, Spain, ACM Press, pp. 301-308.
29. Michaud, F., Clavet, A., Lachiver, G., & Lucas, M. (2000) Designing toy robots to help autistic children - An open design project for Electrical and Computer Engineering education. *Proc. American Society for Engineering Education*, June 2000.
30. Moore, D. (1998) Computers and people with autism. *Communication*, Summer 1998, pp. 20-21.
31. Murray, D. (1997) Autism and information technology: therapy with computers. In: S. Powell and R. Jordan: *Autism and Learning: a guide to good practice*. London: David Fulton, pp. 100-117.
32. Nadel, J., Guerini, C., Peze, A. & Rivet, C. (1999) The evolving nature of imitation as a format of communication, In: J. Nadel and G. Butterworth (Eds.) *Imitation in Infancy*, Cambridge University Press, pp. 209-234.
33. Ogden, B & Dautenhahn, K. (2000) Robotic Etiquette: Structured Interaction in Humans and Robots. Proc. SIRS2000, Symposium on Intelligent Robotic Systems, Reading, UK, pp 353-361
34. Ogden, B. & Dautenhahn, K. (2001) Embedding Robotic Agents in the Social Environment. *Proc. TIMR 2001, Towards Intelligent Mobile Robots: The 3rd British Conference on Autonomous Mobile Robotics and Autonomous Systems*, Manchester, 5th April 2001.
35. Ogden, B. & Dautenhahn, K. (2001) Interactive Vision from the Top Down: Interactional Structure Applied to the Identification and Interpretation of Visual Interactive Behaviour. To be presented at *Gesture Workshop 2001, The 4th International Workshop on Gesture and Sign Language Based Human-Computer Interaction*, 18th-20th April 2001, City University, London, UK .
36. Panyan, M. V. (1984) Computer technology for autistic students. *Journal of Autism and Developmental Disorders* 14(4): 375-382.
37. Papert, S. (1980). *Mindstorms: Children, Computers, and Powerful Ideas*. Basic Books, New York.
38. Parsons, S., Beardon, L., Neale, H. R., Reynard, G., Eastgate, R., Wilson, J. R., Cobb, S. V. g., Benford, S. D., Mitchell, P. & Hopkins, E. (2000) Development of social skills amongst adults with Asperger's Syndrome using virtual environments: the 'AS Interactive' project. *Proc. 3rd Interl Conf. Disability, Virtual Reality & Assoc. Tech*, Alghero, Italy 2000, pp. 163-170.
39. Plaisant, C., Druin, A., Lathan, C., Dakhane, K., Edwards, K., Vice, J.M., Montemayor, J. (2000) A Storytelling Robot for Pediatric Rehabilitation. *Proc. ASSETS '00*, Washington, Nov. 2000, ACM, New York.
40. Powell, S. (1996) The use of computers in teaching people with autism. In: *Autism on the Agenda: Papers from a National Autistic Society Conference*. London: NAS, 1996.
41. Psathas, G. (1995) *Conversation Analysis: The Study of Talk-In-Interaction*. Sage Publications Inc, Thousand Oaks, California, USA.

42. Russo, D. C. & Koegel, R. L. & Lovaas, O. I. (1978) A Comparison of Human and Automated Instruction of Autistic Children. *Journal of Abnormal Child Psychology*, June, 6 (2), pp. 189-201.
43. Schank, R.C. & Abelson, R. (1977) *Scripts, Plans, Goals and Understanding*. Laurence Erlbaum Associates Inc, Hillsdale, NJ.
44. Schulte, J., Rosenberg, C. & Thrun, S. (1999) Spontaneous Short-term Interaction with Mobile Robots. *Proc. ICRA '99, 1999 IEEE Int. Conference on Robotics and Automation*, May 10-15, 1999, Marriott Hotel, Renaissance Center, Detroit, Michigan, USA.
45. Strickland, D. (1996) A virtual reality application with autistic children, *Presence: Teleoperators and Virtual Environments* 5(3), pp. 319-329.
46. Weir, S. & Emanuel, R. (1976) Using Logo to catalyse communication in an autistic child. DAI Research Report No. 15, University of Edinburgh.
47. Werry, I. & Dautenhahn, K. (1999) Applying robot technology to the rehabilitation of autistic children. In *Proc. SIRS99, 7th International Symposium on Intelligent Robotic Systems* '99, pp. 265-272.
48. Werry, I., Dautenhahn, K. & Harwin, W. (2001) Evaluating the response of children with autism to a robot. To appear in *Proc. RESNA 2001, Rehabilitation Engineering and Assistive Technology Society of North America*, Friday, June 22 - Tuesday, June 26, 2001, Reno, Nevada, USA.

The Cognitive Effects of Delayed Visual Feedback: Working Memory Disruption While Driving in Virtual Environments

Philip N. Day, Patrik O'Brian Holt, and George T. Russell

Image Systems Engineering Lab, Dept. of Computing & Electrical Eng., Heriot-Watt University, Edinburgh EH14 4AS, UK
{ceepnd,ph,gtr}@cee.hw.ac.uk
http://www.cee.hw.ac.uk/isel/

Abstract. Experiments that have been conducted using virtual reality based driving tasks in order to investigate the effects of delayed visual feedback are described. Results show that visual interference tasks and visual delays both have a similarly detrimental effect on performance. A potential explanation for the cognitive failures involved in delayed visual feedback is outlined; namely a disruption to working memory. In addition to this theoretical and experimental work a review of pertinent literature is also included along with a brief summary of our experience of using virtual reality technology in human factors experimentation.

1 Introduction

Technologies in the engineering sector such as the remote operation of vehicles, simulators and early fly-by-wire systems can introduce delays into feedback to the operator (which is commonly visual feedback). This problem has been noted in the past, in the context of differing subject areas. In fact, the observation that delayed visual feedback may have a significantly detrimental effect on human performance in operator-control systems has been made frequently during the last 50 years. There is available a substantial but somewhat dated literature on the effects of delayed feedback but there seems to be a lack of work on the cognitive nature of the errors induced by delayed feedback. Much of the work has been to find partial solutions to the problems that delays introduce without fully understanding these problems. This is perhaps an understandable approach as the technology has often improved to the point where delays are no longer evident in the system (such as in modern fly-by-wire systems). However, there are still some areas in which delays cannot be avoided, such as submersibles that use sonar for sensing, or communication with equipment in space exploration (for example, the Mars Rover). Even terrestrial communications across networks or telephone lines often introduce significant delays. Investigation into the reasons for delays causing difficulty is therefore still of relevance today, even with the high processing speed of modern computers.

M. Beynon, C.L. Nehaniv, and K. Dautenhahn (Eds.): CT 2001, LNAI 2117, pp. 75–82, 2001.

1.1 Delayed Sensory Feedback: A Brief Review

The earliest observation of the detrimental effects of delayed sensory feedback that could be found appears in the Foxboro study (1945) although delayed feedback was only mentioned in passing. Warrick (1949) performed the first detailed study of the effects of delayed feedback in the context of a tracking system. Results from this study showed an inverse linear relationship between the delay and log of time-on-target. Others also noted these effects in the context of perception of visual delay (Biel and Warrick 1949) and manual tracking systems (Garvey et al. 1958).

Following on from these studies, early studies in hearing and speech disorders made use of delayed auditory feedback with the following effects being noted: artificial stutter, decrease in reading speed, pausing or complete blocking of speech and the introduction of errors (Lee 1951). This work was replicated and extended by others, most notably by Smith (1962) who investigated delayed auditory and visual feedback.

From this work by Smith it was noted that visual feedback delays are known to be detrimental to the performance of operators with effects including overcompensation, lack of trust in the feedback and confusion and disorientation (Smith and Smith 1987).

These studies were concerned with specialised physiological areas such as speech disorders or displaced vision. However, in recent times there has been a significant growth in the use of human operator control systems. Although such 'operator in the loop systems' are now commonplace a number of potentially important issues remain unresolved. One such is the effect and control or compensation of delayed visual feedback. These effects of delayed visual feedback can be seen potentially in a number of different and disparate application areas, for example, remote control over long distances, remote control with sonar, remotely controlled aircraft and laparoscopic surgery.

2 Working Memory Disruption

2.1 Experimental System Model

In order to investigate the cognitive nature of the errors introduced into performance as a result of delayed visual feedback, the authors developed a model of the experimental system involved in studying delayed visual feedback including both the physical artifacts and the user cognition involved as shown in Fig. 1.

This model was originally derived from experiments conducted using physical vehicles, cameras and computers as delay systems. However, this model has been found to also hold for experiments conducted using non-physical artifacts, with participants controlling a vehicle in a virtual environment where both the vehicle and the terrain to negotiate are computer generated (Day et al. 1999b).

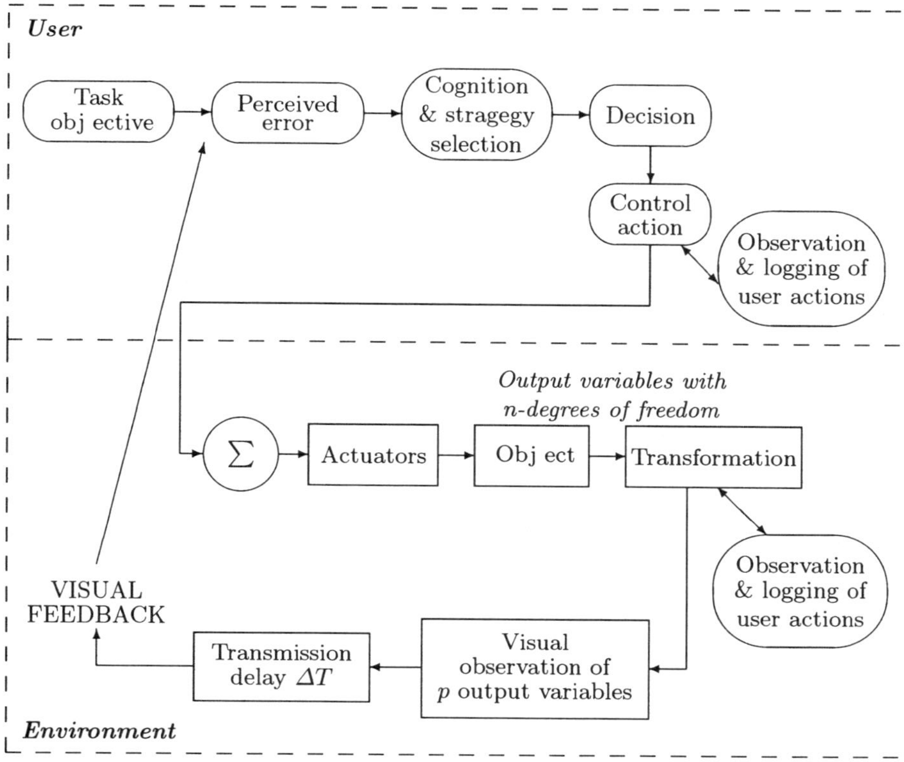

Fig. 1. Experimental system model

2.2 Model of Working Memory (WM)

From the understanding gained from using this model, it was decided that as the cognition involved appeared to be closely linked to the visual-motor nature of the tasks used, it would be sensible to further investigate the nature of cognition involved in the context of these demanding visuo-motor tasks (of driving). This is in contrast to the approach that is sometimes taken by experimental psychologists of separating out one performance criteria to be tested. If one performance criteria was separately tested, it might well mean that the complex interactions between all the tasks involved in driving would not be captured in the laboratory experiments.

The hypothesis was adopted that the errors due to delayed visual feedback can be explained at least in part as due to a disruption of working memory. This paper adopts the model of working memory proposed by Baddeley (1986) with a central executive and separate visuo-spatial and articulatory portions. One of the significant factors in using this model is that previous experiments have already ascertained that visual interference (due to low frame rates in the visual feedback) shares similarities with visual delay (Day 1999a, 1999b,

2000). Baddeley's early experiments (1975, 1980) were also concerned with visual interference, thus implying that this work may have relevance to the study of delayed visual feedback effects.

Baddeley's investigations into the effects of visuo-spatial interference were inspired by the difficulty that he experienced in driving a car while listening to a game of American football (Baddeley 1986, pp. 110–111). For this reason, tasks were designed that replicated this effect with the complex nature of driving being simulated by the pursuit rotor task, and the mental visualising element of listening to a game of sport being modeled by letter tracing tasks.

As a result, it was decided to extend this work of Baddeley's by using the same letter tracing, but substituting a driving task for the pursuit rotor tracking task (a reasonable step as the pursuit rotor task was selected in order to simulate the task of driving). These experiments were then conducted in order to test the hypothesis that the detrimental effects of visual delays and visual interference tasks can both be explained in terms of working memory disruption.

2.3 Using the WM Model

In order to investigate this hypothesis, experiments were designed and conducted using immersive virtual reality (VR) in which participants drove a virtual vehicle along a track towards a target under three levels of delay. In addition to this driving task in half the trials participants had to perform the verbal letter tracing tasks designed by Brooks (1968) and subsequently used by Baddeley (1975,1980). In these tasks, the subject is shown a block capital letter, with the bottom left hand corner marked with a star. The task is to traverse the letter from memory categorising each corner as a point on the extreme top or bottom or as a point in between. In each case the subject starts at the corner marked with an asterisk and continues in a clockwise direction around the letter responding "yes" if the corner in question was at the top or bottom and "no" otherwise. Hence for the letter F, the response would by "yes, yes, yes, no, no, no, no, no, no, yes". This is a visuo-spatial task. The letters used all were: F, G, M, N, W and Z. All have ten data points (corners) and are displayed as shown in Fig. 2.

The experiments therefore followed a within-users 2*3*5 factorial design (with two visual interference settings, three delay settings and five tracks). Delays of 0, 400 and 800 ms were introduced in the experiment, along with additional visuo-spatial interference tasks as described above. The experiments were performed using an HP 725/75 UNIX workstation with a single pipe PV10

Fig. 2. Visual Interference Letters

(ProVision 10 virtual reality accelerator) from Division, dVISOR head-mounted display, Division 3D mouse and Polhemus magnetic tracking. Software was coded using Division's dVISE/dVS software - version 3.1. Delays were introduced electronically using custom built hardware designed and manufactured by the Department of Computing and Electrical Engineering of Heriot-Watt University for delaying signals transmitted from the Division peripherals to the HP workstation. Control of the system was achieved by head movements and button clicks on the 3D mouse.

36 people participated in the experiment of which 28 were male and 8 female. Age and gender were not found to have a significant effect on the results. All participants had normal or corrected-to-normal vision. No participant reported any physical disabilities that could affect their ability to control a vehicle.

3 Results

As one might expect, as summarised in Table 1, performance decreased for both visual delays and the interpolated visual interference tasks of letter-tracing.

Results showed that visual interference tasks had significant effects on targeting errors ($F = 13.41, p < 0.005$), mean tracking errors ($F = 4.78, p < 0.05$), and total tracking errors ($F = 11.63, p < 0.005$). Completion times were not significantly affected by the visual interference tasks. This agrees with the results of Baddeley's work thus showing that the use of driving in a virtual environment is an appropriate extension to the original task of pursuit rotor tracking.

In a similar manner delays were found to have significant effects on completion times ($F = 12.26, p < 0.001$), mean tracking errors ($F = 6.14, p < 0.005$) and total tracking errors ($F = 17.61, p < 0.001$). Targeting errors were not significantly affected by delays.

These results demonstrate that overall driving performance (measured by tracking errors) was affected in a highly similar manner by visual interference tasks and delays. This therefore strengthens our hypothesis that visual delays cause disruption of visual working memory (in a similar manner to visual interference tasks). The results also suggest that people sacrificed speed for accuracy in the presence of visual delays but did not under visual interference. It should be noted that the visual interference tasks had such a high impact on participants that some did not notice the presence of visual delays at all.

Table 1. Performance summary (mean values)

	0 ms	*400 ms*	*800 ms*	*no letters*	*letters*
times (s)	22.23	23.13	25.51	22.82	24.43
targeting errors (m)	0.28	0.31	0.50	0.15	0.57
mean tracking errors (m)	1.78	1.82	1.86	1.78	1.86
total tracking errors (m)	420.60	460.50	534.10	424.10	519.40

In secondary analysis the track complexity was found to have a highly significant effect on completion times ($F = 26.70, p < 0.001$), mean tracking errors ($F = 906.98, p < 0.001$) and total tracking errors ($F = 168.05, p < 0.001$), but did not have a significant effect on targeting errors. The interaction between visual interference tasks and delays was not found to have a significant effect on times, targeting or tracking errors. The interaction between visual interference tasks and track complexity did not have a significant effect on times or targeting errors, but did have a significant effect on mean tracking errors ($F = 7.87, p < 0.005$) and total tracking errors ($F = 5.59, p < 0.005$). The interaction between delays and track complexity was found to have a significant effect on completion times ($F = 3.63, p < 0.005$) and total tracking errors ($F = 3.42, p < 0.05$) but not on targeting or mean tracking errors. The interaction between visual interference tasks, delays and track complexity was found to have a significant effect on mean tracking errors ($F = 3.42, p < 0.01$) but not on times, targeting or total tracking errors.

4 Discussion

As can clearly be seen from the given results, driving performance is affected in a similarly detrimental manner by both visual interference tasks and visual delays. For this reason the hypothesis, that both can be explained in terms of visuo-spatial working memory disruption, appears to hold. However, there do appear to be slight differences in the strategy employed by the participants to compensate for the both, although this may simply be explained by the fact that participants noticed the visual interference tasks more than the visual delays, and many reported afterwards that they didn't fully concentrate on the visual interference tasks when driving performance was obviously suffering. The visual delays could not be ignored in such a manner. This also ties in with the fact that completion times appeared to be unaffected by the visual interference tasks.

Further work is therefore needed in order to further investigate whether existing knowledge on visuo-spatial working memory (in particular, the disruption of working memory by such mechanisms as visual interference) can be applied to methods of compensation (such as intelligent software agents) for delayed visual feedback. Further analysis of the data already collected may also yield important clues as to exactly how performance is affected.

The use of VR for human factors experimentation has already been shown to have great potential, due in part to the extremely large amounts of data that can be collected automatically (Day et al. 2000). For example, as a result of running these experiments using computer based virtual environments, the authors now have a detailed set of data describing human performance under delayed visual feedback. An additional advantage is that, despite the heavy cognitive load placed on participants with experiments lasting between 45 and 75 minutes, they appeared to enjoy the experience and therefore such factors as boredom or fatigue were reduced. The attractiveness of VR had the added benefit that volunteers to participate were found readily and it was not necessary to cajole people into taking part, as is sometimes the case when there is no financial incentive for volunteering.

Some users found the high cognitive load to be quite tiring and stressful. Two participants also found the delayed feedback to induce symptoms similar to motion sickness such as slight nausea and increased sweating which agrees with the finding that time-lagged simulations can cause cybersickness (Bailey and Knotts 1987, Frank et al. 1987, Liu et al. 1993). Apart from these two people, the only other complaint that participants had was that the headset was too bulky and rather heavy which is easy to rectify by using more modern VR hardware.

References

Baddeley, A.D., Grant, S., Wight, E., Thomson, N.: Imagery and visual working memory. In Rabbitt, P.M.A., Dornic, S. (eds.): Attention and Performance V. Academic Press, London (1975) 205–217

Baddeley, A.D., Lieberman, K.: Spatial working memory. In Nickerson, R. (ed): Attention and Performance VIII. Erlbaum, Hillsdale NJ. (1980) 521–539

Baddeley, A.D.: Working Memory. Clarendon Press, Oxford, England (1986)

Bailey, R.E., Knotts, L.H.: Effect of time delay on manual flight control and flying qualities during in-flight and ground-based simulation. In Proceedings of the AIAA Flight Simulation Technologies Conference **87-2370** (1987) 30–38.

Biel, W.C., Warrick, M.J.: Studies in perception of time delay. The American Psychologist **4** (1949) 303

Brooks, L.R.: Spatial and verbal components in the act of recall. Canadian Journal of Psychology **22** (1968) 349–368

Day, P.N.: An investigation into the effects of delays in visual feedback on real-time system users. In Sasse, M.A. and Johnson, C. (eds.): Human-Computer Interaction – INTERACT '99. IOS Press, Oxford (1999a) 674–675

Day, P.N., Holt, P., Russell, G.T.: Modeling the Effects of Delayed Visual Feedback in Real-Time Operator Control Loops: A Cognitive Perspective. In Proceedings of the XVIII European Annual Conference on Human Decision Making and Manual Control. Group D Publications Ltd (1999b) 70–79

Day, P.N., Holt, P., Russell, G.T.: A Comparison of Real and Virtual Worlds for Human Factors Experimentation. In Hollands, R. (ed.): Proceedings of the Seventh UK VR-SIG Conference UK VR-SIG (2000) 159–165

Foxboro Co.: Studies in aided tracking. Nat. Def. Res. Comm. Rep. No. 25, Mem. to Div. 7 (1945)

Frank, L.H., Casali, J.G., Wierwille, W.W.: Effects of visual display and motion system delays on operator performance and uneasiness in a driving simulator. Human Factors **30**(2) (1987) 201–217

Garvey, W.D., Sweeney, J.S., Birmingham, H.P.: Differential effect of "display lags" and "control lags" on the performance of manual tracking systems. Journal of Experimental Psychology **56**(1) (1958) 8–10

Lee, B.S.: Artificial stutter. Journal of Speech and Hearing Disorders **16** (1951) 53–55

Liu, A., Tharp, G., French, L., Lai, S., Stark, L.: Some of what one needs to know about using head-mounted displays to improve teleoperator performance. IEEE Transactions on Robotics and Automation **9**(5) (1993) 638–648

Smith, K.U.: Delayed Sensory Feedback and Behaviour. W.B. Saunders Co., Philadelphia (1962)

Smith, T.J., Smith, K.U.: Feedback-Control Mechanisms of Human Behaviour. In Salvendy, G. (ed.): Handbook of Human Factors. John Wiley, New York (1987) 251–293

Warrick, M. J.: Effect of transmission-type control lags on tracking accuracy. USAF Air. Mat. Com. **5916** (1949)

Embodiment, Perception, and Virtual Reality

Melanie Chan

Leeds Metropolitan University
Calverley Street, Leeds, LS1 3HE
United Kingdom
m.chan@lmu.ac.uk

Abstract. Virtual reality flight simulators and architectural walk-through models appear to be based on the notion of reproducing reality or embodied presence. Theme parks and computer entertainment games however use virtual reality technologies to construct elaborate fantasy worlds. Human beings have continually tried to explore varying dimensions of consciousness through incantations, meditation, prayer, dream states, drugs and fantasies. Cultural forms such as film, literature, art and virtual reality could be considered as a means of producing virtual realities which allow the exploration of consciousness in particular ways. Each of these cultural forms operates slightly differently due to the potential and limitations of different media. The following discussion suggests that virtual reality is a specific cultural form which enables us to extend our human embodied condition by enhancing our capacities to think and act creatively.

1 Introduction

Instruments of Mind is a conference which brings together debates from different academic disciplines to gain a holistic approach to the relationships between consciousness and cognitive technology. The core question of the conference is the ways in which technology contributes to the making of meaning. Symbolic systems of signification, such as language, mathematics, music and art enable us to construct and disseminate meaning. These signification systems are culturally, historically and socially specific, both the symbols and the medium used in the communication process have a bearing on the production of meaning and understanding. Whilst current debates around cognition and technology may seem particularly pressing, technological devices and practices have been used to construct and disseminate meaning throughout history. Technologies such as moveable type, printing, the telegraph and telephone have shifted notions regarding time, space, distance, working patterns and personal relationships.

In contemporary, Western, urban cultures it is possible to encounter a plethora of different images during our everyday activities. We may encounter: advertisements, signs, packaging, videos, film, computer games and web-sites on a daily basis. Some of these images may employ sophisticated visual languages and techniques to produce an illusion of reality. Viewers may become accustomed to the visual signs and techniques used to reproduce the illusion of reality and this may have a bearing on the

M. Beynon, C.L. Nehaniv, and K. Dautenhahn (Eds.): CT 2001, LNAI 2117, pp. 83-94, 2001.

ways in which they respond to virtual reality technologies. To a certain degree our experiences are historically and culturally determined, by knowing what certain objects are and what they can do. We also use cultural signifying processes and tools to extend our abilities. There is a feedback system between cognition, signification and technology. In other words, we construct tools to shape our environment but in turn they also shape our cognitive abilities.

Virtual reality is a term which is used to describe computer generated three dimensional environments: an abstract data space. The development of virtual reality technology can be traced to flight simulators that were produced in the 1930's by the US military in order to provide training for fighter pilots. The US Advanced Research Projects agency (ARPA) continues to invest in virtual reality technologies in such areas as the development of tank simulations (SIMNET) and combat scenarios. Ivan Sutherland, a doctoral student at Massachusetts Institute of Technology is credited with the development of the first computer generated virtual environment, called *The Ultimate Display*, in 1965. Now, at the beginning of the twenty-first century there are many different types of virtual reality systems in use. The significant processing power of Silicon Graphics Visualization Systems tends to be used by research groups or major companies but there are also an increasing amount of virtual reality enthusiasts building their own systems from graphics cards, customized glasses and headsets. Viewers interact with the virtual environment using such devices as goggles, trackballs, gloves, wands, and head mounted displays.

Virtual reality is now regarded as a useful tool in architecture, design, education, business, health care and as an artistic medium. Research into the operational aspects of virtual reality technologies, in computing and engineering, has increased in recent years (Cope, Parker, Wilkinson, & Dullinger [1]; Harrod, Wilkinson, & Cope [2]). Attention has also been given to the social and cultural analysis of new technologies such as the internet and e-mail, one of the most influential works in this area is by Sherry Turkle's *Life on Screen: Identity in the Age of the Internet* [3]. Although Turkle mentions the visual relationships in chat rooms on the Internet, she mainly focuses on text based communication. Virtual reality technologies may attempt to reproduce social interaction through the use of avatars or other graphic symbols representing gestures and facial expressions. Kirstie Bellman and Christopher Landauer have also studied social interaction in virtual environments such as (multi user domains) MUDS [4]. Researchers such as Judith Donath, at the Massachussetts Institute of Technology (M.I.T.) are also involved in the development of graphic interfaces that can be used as means of providing social interaction in virtual environments [5].

The work that is conducted in virtual reality research and development is certainly impressive. It is not my intention to diminish the work that has been done. Nevertheless there are many ontological philosophical questions underlying virtual reality which have yet to be studied in great detail. For instance, how close is the relationship between the computer/viewer in virtual reality? Do users alter their behavior in order to interact with the virtual environment? Could virtual reality alter the concept of human embodied existence? There also seems to be specific issues arising from the concept of immersion and whether virtual reality prioritizes sight. Kristine Nowak and Frank Biocca have also explored the role of the virtual body in virtual environments in the paper they presented at the 3rd International Cognitive Technology Conference CT99 [6].

Artists use pictorial conventions that are historically and culturally specific. Painting, for example, provides a way of exploring our imagination by producing images based on the relationships between light, shade, texture and colour. Sculpture is a visual and tactile medium employing such materials as stone, marble, copper, and wood. Contemporary artists use a variety of different media including photography, video, film and computer graphics. Virtual reality differs from other media, such as film and video, because the user's body is crucial to the engagement with the image. With virtual reality the viewer can also be thought of as a user of the technology because the simulated images change in response to their movements. Talking about the use of virtual reality as an artistic media, Timothy Druckery states 'what we traditionally identify as a picture no longer represents a window but rather a door into an immaterial space of data' [7].

The fantasy elements surrounding virtual reality outstrip what is currently possible but it could also be argued that this actually drives research and development. Science fiction films and novels give the impression that the complete immersion of consciousness within virtual reality is possible. Films such as *The Lawnmower Man, Strange Days* and *The Matrix* have all generated interest in virtual reality, The science fiction writer William Gibson has also popularised cyber-culture and virtual reality through novels such as *Neuromancer* [8]. Science fiction films and novels could be interpreted metaphorically as a means of engaging with some of the complex issues surrounding virtual reality. They also seem to captivate a burgeoning desire to transcend the body. On the other hand, it could be contended that science fiction films and novels overestimate the potential of virtual reality technologies. Jaron Lanier, an artist, musician and computer programmer enthuses about virtual reality as a creative art form but is also cautious about the limitations of the technology. In 1995, when virtual reality was entering mainstream cultural discourse he pointed out the limitations of this technology. He was particularly critical that virtual reality offers users a transcendent or even para-normal experience, yet the idea of transcendence persists [9].

The discussion begins by providing a brief exploration of the discourses surrounding virtual reality. The first section makes a connection between the dualistic philosophy of mind versus body and virtual reality. The discussion will then move on to consider contemporary philosophical and scientific discourses such as evolutionary psychology and neuroscience. A contrasting argument to neuroscience will be offered through the examination of sensory motor and phenomenological discussions of perception and embodiment. Finally there will be a discussion of the ways in which virtual reality may be used as a creative medium. To limit the scope of the discussion I will refer to the work of one artist, Char Davies because her work particularly highlights the relationships between perception and embodiment. Of course there are many others who are working with virtual reality technologies in particular at the Centre for Advanced Inquiry into the Interactive Arts CaiiA at the University of Wales, The Banff Centre in Canada and at the Massachussets Institute of Technology (M.I.T.).

2 Mind vs. Body?

The impulse to be immersed in virtual reality is often connected to the philosophical concept of mind and body separation, which is deeply embedded in Western philosophical paradigms. The French philosopher Rene Descartes (1596-1650) founded his philosophical principles on the notion that 'I am thinking therefore I exist' [10]. According to Descartes it is possible to pretend that one no longer has a body but it is impossible to doubt one is thinking. The question is what does Descartes mean by thinking? If he means mental activity then could this include fantasies and dreaming? Is Descartes referring to conscious rather than unconscious thought? The ontological implications of Descartes philosophical position are that the self is constructed by identifying with consciousness. Our thoughts determine who we think we are. It is possible to imagine or pretend that we do not have a body but actual existence without the body is impossible. The idea of pure thought, which is separate from the body, seems to be connected to virtual reality, consciousness and computers. Indeed references to Descartes work can be found in many contemporary studies of virtual reality and information technologies (Tofts & McKeigh [11], Hillis [12], and Goldberg [13]).

To gain a deeper understanding of Descartes' philosophical stance, it is important to place his work in an historical and culturally specific framework. It is also useful to remember that our contemporary interpretations of Descartes' work are limited because they are based on translations and copies taken from his notebooks. The term philosophy as used by Descartes is not comparable to the academic discipline we have today. He uses the term to encompass the study of science, physics, chemistry and biology. Descartes work also differs from contemporary philosophy because he postulates a transcendental source, God, as the ultimate grounding term for his philosophical standpoint. In Descartes theoretical system human existence has two components, the body and mind. The term mind as used in his work can refer to consciousness and to the soul. He describes the body as an extended substance which can be measured, but asserts that the mind is immeasurable. The mind for Descartes is transcendental, when death occurs it is freed from the body but continues to exist.

According to Descartes there are four faculties that enable us to investigate the world these are: the intellect, memory, imagination and sense perception. The intellect according to Descartes operates according to rules of rationality and logic. He is however, particularly suspicious of the knowledge gained through the senses. The senses can be easily fooled, so the knowledge we gain from them is dubious. Whilst he is cautious about the knowledge gained through the senses he does value the ability to see. Put otherwise, sight is prioritized over the other senses.

Descartes claimed that mathematics offers the most accurate and reliable form of knowledge because it is based on logical principles. An understanding of mathematical principles is important to the study of virtual reality technologies because they underpin the algorithms of the computer code which produces the virtual environment. The Greek philosopher Plato (427 BC – 347 BC) contended that mathematics provides the key to reality. Indeed mathematics comes from the Greek word *mathesis* meaning learning. The mathematical realm, according to Plato, was beyond our everyday experiences of the world around us. For Descartes space is something that can be mapped and quantified in mathematical terms, Descartes used mathematical principles to develop a means of mapping and representing three-dimensional space.

Cartesian co-ordinates provide a system of mathematical representation that refers to height, width and depth.

Empty space does not exist for Descartes. He asserts that even when we think something is empty, it is still full of some sort of substance (such as air). According to Descartes if one thing moves then the space it leaves is immediately filled with something else. The production of virtual reality environments also seems to be underpinned by a similar view of space. The virtual environment is constructed using a mesh or grid and each point is filled with information. When an object moves from one position to another the set of mathematical co-ordinates representing those positions also changes. In virtual reality environments the viewer/user is also represented in mathematical terms and their movements are tracked according to co-ordinate changes.

Paradoxically virtual reality combines mathematical principles and sensory stimulation. The mathematical principles driving virtual reality would be in accordance with the ideas of Plato and Descartes. Virtual environments are abstractions generated by mathematical processes whereby space is constructed according to a set of numerical co-ordinates. It seems that one of the assumptions underlying virtual reality is that mathematics explains the physical world and that the knowledge it provides could be used to simulate reality by mathematical means. Nonetheless from a Cartesian view, the knowledge we gain from the virtual environment could be dubious because it is primarily received through the senses.

3 Evolutionary Psychology, Perception, and Embodiment

In *How the Mind Works*, Steven Pinker proposes that our assumptions about the world around us are, to a certain extent, based on evolutionary principles [14]. Hubert Dreyfus also describes the evolutionary and the cultural dimensions of our experiences. He says there are three main ways in which the body copes with the similarities and differences we encounter during the course of our everyday life [15]. First the structure of the brain and the way in which it operates limits our experiences. The order and frequency of our experiences are also important. How often do we confront similar situations or experiences? The third limitation concerns satisfaction. Does the action we take produce a satisfying outcome. Dreyfus stresses that satisfaction is not simply a case of pleasure versus pain, rather it is a kind of coping with the situations that are presented to us.

Pinker claims stereo-vision is a way of enhancing our survival skills because each eye sees a slightly different impression of the world. There are three-dimensional objects in the world but what we actually see are different surfaces and textures that allow us to ascertain how objects are positioned. Light also enters each pupil at a slightly different angle and this provides the basis for the measurement of proximity and distance. Pinker explains that in the past having two eyes was thought to be the result of the harmonization of a symmetrical body. Two vantage points could however be regarded as providing an evolutionary advantage.

Pinker gives a plausible explanation as to why perceptive faculties are developmental in human beings and other mammals such as monkeys and cats. He says at birth the face is not fully developed, it continues to grow and this pushes the eyes further apart. Neurons detect the difference between the eyes, as the information from them changes with their movement. When this particular phase of development ends (at about three months of age in human babies) the tracking of the changes taking place also ceases. According to Pinker as the faculty of sight develops each neuron involved in the process begins to respond to a particular eye. It results in a situation whereby the neuron will only respond to the eye it prefers.

The eyes are about two and a half inches apart, to look at the same part of the visual field they move in certain ways. The closer the object the more our eyes cross. Pinker says the focusing mechanism (the amount of light entering the lens) and crossing of our eyes work together producing a reflex action. When focusing on something close our eyes converge, looking into the distance they become parallel. If virtual reality head-mounted displays project the same image to each eye this can cause eyestrain. Looking straight ahead is what we do when we are looking at distant rather than close objects, so the resulting images would look blurred.

One way of overcoming the fusion of cross-eyes is to position two very small pictures, each representing the way in which the eyes see the world. The images are then covered with a glass lens which focuses them, which saves the eye from straining. Glasses with special panes made of liquid crystal displays which act as shutters can also be used. The shutter allows one image to be seen by one eye, then the other shutter opens so the second eye also sees an image. Virtual reality is dependent on the way in which the sense of sight operates because it requires a refresh rate of 20 frames per second. At a speed of 20 frames per second or more the human eye is fooled into believing that a succession of still images is moving. Virtual reality glasses and headsets can be set up so that the alternation between the opening and closing of the two shutters is so fast that we cannot detect it.

It is possible that that there may be different neurons involved in the analysis of slant, tilt, colour, surface, texture and boundary. Could the neural activity involved in the perceptive process be reproduced using computerized neural nets? Neuroscientists such as John Donoghue (Brown University, USA), Philip Kennedy and Roy Bakay (Emory University, USA) are working on various ways of recording and reproducing neural activity. Research into the use of implants and visual prosthesis could have implications for virtual reality technologies [16]. For instance viewers might become even more intimately connected to the virtual environment through the use of implants. On the other hand, the operation of brain appears to be more subtle and complex than a parallel computation, so it may not be possible to reproduce perception using neural nets.

4 Perception and the Sensory Motor Capacities of the Body

Our field of vision is limited in several ways, it is not possible to see objects in their entirety; rather we see only the parts which are facing us directly. To see other parts of our environment it is necessary to move our eyes, head or body. Another alternative is to move the objects, for example, we may grasp an object and rotate it in our hands. At a certain distance our vision becomes hazy and blurred. Furthermore, our field of vision is limited to what we are currently, consciously processing. Yet our experience of seeing seems to be constant. We do not appear to be looking at a series of discrete pictures of the world. Because our bodies move and gain more information about the world around us we have the impression that we do see everything.

Researchers such as Kevin O'Regan, at the Universite Rene Descartes and Alva Noe, University of California, postulate that we do not have a picture of the world inside our heads, the world is the only representation that exists [17]. Seeing is a way of exploring the representation that is the real world.

The sensory motor capacities of the visual apparatus include the eye, retina, cornea and visual cortex. Perception refers to the ways in which we try and make sense of what we see through a process of categorization and making generalizations. We tend to make hypotheses about the world around us, test, confirm, or refute them. Sometimes we can be wrong and our eyes are fooled. It could be argued that reality, even virtual reality, does not involve the decoding of sensory messages. Another explanation is that we actually construct our world through our active involvement within it. Sensations are interconnected and are part of our exploration of the world. Sensation is a noun meaning feeling or awareness which may be considered as an activity, an event, rather than a thing. The structure of the sensory motor features of the human body in conjunction with the brain produces a wide range of sensory experiences.

There are two elements to the sensory motor capacities involved in perception, first there is the visual apparatus itself; secondly there is the intrinsic properties of objects. The eye is a specific organ which operates differently from the other sensory modalities. The eyes rotate, they are spherical, light bends when entering the pupil and we blink. The world around us changes as our eyes move but these changes do not appear to be random otherwise the world would appear chaotic and confusing. It seems that the movements that occur are operating under the laws which are connected to the sensory motor capabilities of the human body. The movement of our eyes across the surface of an object stimulates certain photoreceptors. Noe and O'Regan note that if we look at a straight line our eyes move along it and the same photoreceptors are stimulated. The intrinsic qualities of the straight line and the way in which the eye moves across it determines our overall perception an understanding [18].

In the course of our lives we explore the world using our sensory motor capabilities and encounter many different people, places, and things. These experiences will probably motivate us to react in a certain way. Part of the process of mastering the sensory motor capacities of perception is to draw upon these prior experiences. We learn from experience that certain objects look a certain way, or do particular things. The world appears to act as an external memory stimulating our prior knowledge, perhaps by presenting us with similar scenarios. Yet we also seem to have the capacity to adapt to changes in our visual field by drawing on prior experiences of coping in the world and managing uncertainty.

5 Phenomenology

A phenomenological approach might offer an alternative way of thinking about the potential of virtual reality and sensory immersion. From a phenomenological perspective attempts to simulate reality using virtual reality could be reductive because the sensory richness of our everyday experiences in the world cannot be reproduced. Phenomenology is a philosophical theory which can be traced to the German philosopher Edmund Husserl (1859-1938). Later writers such as Martin Heidegger (1889-1976) and Maurice Merleau Ponty (1908-1961) have built upon and challenged Husserl's work. Though there are various facets within phenomenology the common ground between them is the emphasis given to a pre-existing, ontological world. The ontological world exceeds human reflection and provides the very ground for all existence. Heidegger and Merleau Ponty refer to the study of our ontological existence as 'being-in-the-world'. The experience of seeing appears to be both immediate and spontaneous, yet as Merleau Ponty points out 'nothing is more difficult than to know precisely what we see' because it seems so natural and self-evident [19].

According to the philosophical principles of phenomenology, the thinking being brings the concept of the self into existence. Consciousness is what produces the experience of an outer, pre-existing world. Phenomenology does not privilege consciousness or inner reflection since this would be a form of idealism but it is not empiricist either. The ontological world and our experiences are not the same. There is a gap between the act of perception and the analysis which takes place after the original event. There is no such thing as an image on the retina, nor is there an image in the mind. What we have is a perceptual experience, not an object. We perceive height, width and depth all at once, not as separate entities.

It could be maintained that virtual environments limit our habitual sense of engaging with the world because this technology prioritizes sight. The way in which we see the world around us is linked to our bodily shape, abilities by experience and previous interactions with the environment . For example we tend to measure how long it would take to walk down a long road, or how far we need to reach to grab something. Simulating the sense of touch, of reaching out and grabbing objects remains underdeveloped. The use of data gloves attempt to provide a tactile sense of feedback in virtual environments. They can however be used as a means of grabbing or controlling rather than a means of navigating and exploring the virtual environment. It is hard to distinguish between different stimuli, of knowing when one form starts and another ends. In our everyday experiences of the world our senses intertwine and we respond through our bodies to the situations which are presented to us. We could be presented with complex, changing perhaps even life threatening situations. In virtual environments it may appear that we respond to what is pre-programmed and is not life threatening. Yet if the virtual environment simulates a compelling illusion of a threatening situation it is possible that the user/viewer may become emotionally and physiologically distressed.

6 Signification, Representation, and Meaning

Virtual reality could be thought of as a signification system which allows the extension of our embodied capabilities rather than separating the mind and body. Symbolic systems of communication such as art, music and language enable us to extend our cognitive abilities, by providing a means of communication and social interaction. Crucially there is always a gap between ontological reality and representations of reality. Language represents reality through tropes (figures of speech) such as metaphor and metonym. Images are constructed by a visual language which includes: color, line, shade, and texture. Mathematics is also a language operating within certain parameters. The software used to generate virtual environments is underpinned by mathematical language. The software then produces the graphical tools necessary for the production of virtual images so the construction of virtual environments is both enabled and limited by the relationships between mathematics and images.

One of the driving forces of Western art from the Renaissance to the late nineteenth century was the impulse to represent the illusion of reality. During the Renaissance artists used a perspective grid or mesh to map a particular viewpoint. The problem with perspective is that the illusion only works when seen from the same position as the artist. The frame of a perspective painting or drawing also indicates that it is somehow screened off from reality. Film can also produce a compelling illusion of reality. Whilst film audiences are invited to engage with a sequential narrative with virtual reality the user/viewer is actively composing a narrative in collaboration with the environment. Representing the illusion of reality using computer graphics is time consuming and requires a substantial amount of processing power. Although computer graphics have improved in recent years, are such attempts at realism actually necessary? For example, sometimes images that provide the mere suggestion of familiar objects can provoke a response from the viewer. The renowned art historian E.H. Gombrich states that 'it is the power of expectation that moulds what we see in life no less than art' [20]. A few simple pencil lines, for example, can evoke familiar shapes and be meaningful to us.

The artist Char Davies is using virtual reality to explore the relationships between perception and embodiment. Davies trained as a painter before becoming the Director of Research at Softimage Inc. She is now undertaking doctoral research at the Centre for Advanced Inquiry into the Interactive Arts CaiiA at the University of Wales. She makes it clear that she does not 'feel comfortable with the idea of using technology to leave the body behind' [21]. The graphic language she uses produces semi-transparent, figurative and abstract images. Although working with light rather than pigment, works such as *Osmose* and *Ephemere* evoke a painterly aesthetic. Davies also strives to produce works which challenge ingrained habits of seeing and interacting with the world. Her worked in underpinned theoretically by philosophers such as Gaston Bachelard (1884-1962) and Martin Heidegger (1889-1976). She uses virtual reality to suggest the possibility of shifting our mental awareness by changing spaces. There is a tendency to take our being-in-the world for granted. To what degree do we notice our everyday surroundings? Our habits, beliefs and values all intersect the experience of seeing. Pausing and reflecting on these ingrained habits can enable us to see even our habitual surroundings in new ways.

Throughout history and in different cultures, people have associated certain spaces such as churches, temples, caves, mountains or particular landscapes with the sacred or spirituality. These spaces could provoke a sense of meditative reflection or peace for those who visit them. Char Davies uses virtual reality to evoke a sense of the sacred. She admits there may be tensions between her aims and cultural associations of technology as a means of power, control or even destruction. Her work also appears to be based on Eastern philosophy practices such as Taoism and meditation. Davies produced the art direction for a virtual environment called *Osmose* in collaboration with John Harrison, Georges Mauro, Rick Bidlack and Dorota Blaszczak. The work was produced using Softimage® 3D modeling, animation and development software and Silicon Graphics Onyx 2 Infinite Reality system. *Osmose* is a rich metaphorical work that highlights the relationships between technology, human beings and nature. The work was difficult to produce because each frame had to be carefully constructed with a refresh rate of $1/30^{th}$ of a second. *Osmose* was exhibited at galleries in Montreal, New York, London and Monterrey. Reviews of *Osmose* have also appeared in *Wired, Art in America*, *World Art* and *Metropolis.*

The use of the user's breath is a key consideration in *Osmose.* Davies is a keen scuba diver and her diving experiences provided the inspiration to use virtual reality as a means of constructing immersive, imaginary spaces. A harness is used to provide a connection between user's and the virtual environment. Sensors track the movement of the spine as well as the contraction and expansion of the chest. Inhaling enables upward movements, exhaling produces downward movements and tilting provides horizontal movement. The user/viewer also wears a head mounted display which provides a 360 degree image of the virtual environment. There is a philosophical dimension to the use of the breath in *Osmose.* In Buddhist practices the breath is a means of overcoming the usual dualism of subject/object. Breathing connects the self and the outside world, transcending the boundaries of our individual bodies.

Osmose includes several virtual environments or worlds: a Cartesian grid, a forest, leaf, clearing, pond, abyss, subterranean Earth and cloud. There is also a text-based environment displaying the works of Martin Heidegger and Henry David Thoreau (1817-1862). Entering *Osmose*, users/viewers first encounter the Cartesian grid, indicating the geometrical architecture of the virtual environment. They can then explore other worlds composed from both figurative and abstract forms. It seems that some sense of figuration is needed to help the user/viewer orientate themselves in the virtual world. The experience of exploring the virtual environment can be disorientating without some sense of horizontal and vertical relationships. *Osmose* does however convey some sense of fluidity between figure and ground. For instance, users/viewers are able to move through objects dissolving the barriers between the self and the world. The colours used in *Osmose* are luminescent, subtle rather than the bright tones usually associated with computer games. Sound is also used to compliment the images in *Osmose.* The sounds change in accordance with the position of the user/viewer and their movements in the virtual environment. Sound is not used as means of conveying realism rather it attempts to stimulate the imagination. Samples of both male and female voices are used to suggest rather than recreate human presence.

Ephemere continues the exploration of metaphorical relationships between nature and the human body. The environment provokes an awareness of the interconnectedness of the body, organs, blood vessels and bones with the Earth. *Ephemere* consists of three main environments the landscape, earth and body. These environments have a temporal dimension, as the user/viewer moves through the landscape the images change to evoke a sense of dawn, day, evening and sunset. Seasonal changes are also evoked as the figurative representations of the environment change from spring, summer, to autumn and winter. The temporal dimension of the environment is a reminder of the transient and precious qualities of all life. The environments in *Ephemere* also respond to the viewer/user in ways which highlight temporality and change. If viewers/ users stare for a long time at one particular image they may see seeds sprout or the emergence of flowers. Patient and contemplative viewing is rewarded. To summarise Davies' work shows an alternative to the frantic pace of most computer entertainment games, which are often based on a simple narrative of hunt, kill or die.

7 Conclusion

From a philosophical standpoint virtual reality technologies highlight the complex relations between embodiment and perception. For some virtual reality reflects the Western dualisms of mind versus body and subject versus object relations. The discussion has shown the ways in which virtual reality is founded upon mathematical principles, which are usually associated with logic and rationality. Nevertheless the technology can be used to explore other aspects of consciousness such as creativity and the imagination. I find Char Davies' work interesting because it shows that although virtual reality stems from a military and industrial social system it can be used as means of contemplation and reflection.

The transient aspects of *Ephemere* and the luminosity of the figure and ground relationships in *Osmose* seem to provoke a reconsideration of subject and object relations. Indeed recent developments in scientific theories such as quantum physics suggest that there is no clear distinction between the subject and object of investigation. Yet the changes taking place in scientific discourses were pre-empted by Eastern philosophies such as Buddhism and Tao which have been part of the cultural practices of different groups in Tibet, India and China for over two thousand years.

To conclude, it is necessary to state that whilst environments such as *Ephemere* and *Osmose* may provoke contemplation for some people, others may have quite different experiences. These environments are also transient to the extent that they are programmed to last for fifteen minutes. After that time the viewer/user must return to the everyday world around them. Davies has warned that she does not want her work to be considered as spiritual virtual reality because it could be used as a means of escapism. In the end, what I feel is important about her work is that it does act as catalyst for users/viewers which may allow them to pause and reflect upon their everyday experiences and relationships to perception and embodiment.

References

1. Cope, N., Parker, S., Wilkinson, S., & Dullinger, D. 'A Multilevel Simulation System for Teaching Manufacturing Systems Engineering' in *Proceedings of the CISS – First Joint Conference of International Simulation Societies*, August, Zurich, Switzerland pp. 501-505, 1994.
2. Harrod, S., Wilkinson, S., & Cope, N., Development of an Interactive Manufacturing Cell Using VRML (Virtual Reality Modelling Language) & JAVA' in 6th *European Concurrent Engineering Conference 1999 (ECEC'99): From Product Design to Product Marketing*, 21-23 April, Erlangen, Germany, SCS, pp. 199-203, 1999.
3. Turkle, S. *Life on the screen: identity in the age of the Internet*, London: Weidenfeld & Nicolson, 1996.
4. Bellman, K. and Landauer, C. 'Playing in the MUD: Virtual Worlds are Real Places', *Applied Artificial Intelligence*, 14 (1), pp. 93-123, 2000.
5. Donath, J. Sociable Information Spaces, presented at *the Second IEEE Intl workshop on Community Networking*, Princeton, New Jersey, June, pp. 20-22, 1995.
6. Nowak, K. and Biocca, F. "I think there is someone else here with me!" The role of the virtual body in the sensation of co-presence with other humans and artificial intelligences in advanced virtual environments. *Proc CT99 "Networking Minds" 3rd International Cognitive Technology Conference*, San Francisco 11-14, 1999, pp291-302.

7. Druckery, T. *Iterations: The New Image*, London and Cambridge Mass: M.I.T. Press, P62, 1993.
8. Gibson, W. *Neuromancer*, London, Harper Collins, 1995.
9. Frenkel, Karen A. 'A Conversation with Jaron Lanier' in *Interactions*, July 1995, pp47-72.
10. Descartes, R. *The Philosophical Writings of Descartes Vols. 1 and 11*, transl. John Cottingham, Robert Stoothoff and Dugald Murdoch, Cambridge, Cambridge University Press, 1995.
11. Tofts, D. & McKeigh, M. *Memory Trade: A Prehistory of Cyberculture*, Australia, Interface, 1998.
12. Hillis, K. Digital *Sensations – Space, Identity and Embodiment in Virtual Reality*, Minneapolis and London, University of Minnesota Press, 1999.
13. Goldberg, K. ed, *The Robot in The Garden – Telerobotics and Telepistemology in the Age of the Internet,* Cambridge Massachusetts & London: MIT Press, 2000.
14. Pinker, S. *How the Mind Works*, London: Penguin, 1997.
15. Dreyfus, H. 'The Current Relevance of Merleau-Ponty's Phenomenology of Embodiment' in *Perspectives on Embodiment*, Honi Haber and Gail Weiss (eds.) New York and London: Routledge, 1996.
16. Rizzo, J.F. and Wyatt, J.L., 'Prospects for a Visual Prosthesis', *Neuroscientis*t, Vol.3, pp. 251-262, July 1997.
17. O'Regan, K.J. & Noe, A. *A Sensorimotor Account of Vision and Visual Consciousness,* [Internet] http://nivea.psycho-univ-paris5.fr. Accessed Nov 2000.
18. Ibid.
19. Merleau-Ponty, M. *Phenomenology of Perception*, transl. by Colin Smith, first published 1962, New York and London: Routledge, p. viii, 1999.
20. Gombrich, E.H. *Art and Illusion – A Study in the Psychology of Pictorial* Representation, first published 1960, Oxford:Phaidon, p. 188, 1977.
21. Davies, C. 'A Breath of Fresh Air' in *The Guardian*, Thursday November 21st, p. 17, 1996.

Freeing Machines from Cartesian Chains

I. René J.A. te Boekhorst

Department of Information Technology
University of Zürich, Switzerland
teboekhorst@smile.ch

Abstract. The impact of technology on thinking about behaviour has shifted from mechanistic descriptions towards the computational stance of cognitive science and classical Artificial Intelligence. All these approaches share an output-oriented black-box rationalism, which is also the foundation of neo-Darwinistic accounts of behaviour. To gauge the limitations of this type of explanations and of ethological methods in particular, I analysed the behaviour of simple robots as if they were living creatures. This revealed interesting patterns but did not take the lid of the black box. The self-organized cooperative behaviour of the robots could only be understood if feedback from environmental changes was considered. Furthermore, the robots were not designed by "engineering from scratch" or a "problem-solving approach", but instead by an almost task-free attitude without preconceptions like "imperfect design" and "behavioural errors". This questions the use of a priori stated "costs" and "benefits", and thus is at odds with the starting points of normative and rationalistic theorizing.

1 Introduction

1.1 Behavioural Sciences: From Mechanistic Motivations to Computing Cognition

Advances in technology have often paved the way for a mechanistic interpretation of observed natural phenomena. Mechanistic explanations and technological interpretations of animal and human behaviour have been proposed since Descartes stated position that animal and human bodies are "nothing but a statue or machine made of earth".[1] Descartes' mechanical biology is echoed in the mechanistic jargon of the ethological theories of Lorenz [25] and Tinbergen [37] [14]. Classical ethology, however, is not truly Cartesian, but rather reflects a struggle between vitalistic and mechanistic explanations of behaviour [20][2]. Although the classical ethologists

[1] All citations from Descartes in this paper are quoted by Ablondi [1]

[2] Note, however, that Lorenz' distinction between "motivation" on the one hand and the reflex machinery on the other hand as "two absolutely heterogeneous factors" (Lorenz [26], p. 251) that are equally important, can be seen as an extension of Descartes' dualism from people to animals [24]. I argue that this extension became especially clear when ethologists started to use computer metaphors (cf. [15],[16]) to distance their vision from Lorenzian theory.

M. Beynon, C.L. Nehaniv, and K. Dautenhahn (Eds.): CT 2001, LNAI 2117, pp. 95-108, 2001.

certainly tried to free their language as much as possible from anthropomorphic expressions, they did not aim to picture animals as mere automatons. On the contrary, by introducing a spontaneous renewal of action specific "energy", that "drives" the animal to search actively for "stimuli" (by means of "appetitive behaviour") that "releases" the specific "fixed action pattern", their theory was an alternative to the view of behaviour as being just a collection of simple stimulus-response reflexes. With their quasi-technical vocabulary the ethologists aimed to arrive at an objective, physiology-like type of explanation. According to Crist [14], the inexorable if unwitting consequence of applying this technical language was the mechanomorphic portrayal of animals.

We also have to view the early development of ethology in the context of two important theoretical advances in engineering. The mechanistic language, the notion of negative feedback (between the executed behaviour and the action specific energy that drives it) and its typical black box approach almost predestined early classical ethology to be associated with cybernetics and information theory (for cybernetic models of behaviour: [34], [31]]; for applications of information theory in ethology, see [27]). Shannon's information theory [35], maybe partly via its contribution to the development of the modern computer, has influenced those who embraced information processing and algorithmic computation as the paradigm for understanding cognition (see for example [13], [19]). Later developments in classical ethology have also taken up a connection with this paradigm. To distance themselves from a number of flaws of Lorenz's theory (discussed in [20]) ethologists have advocated a "programming description" of behaviour. By doing so, they emphasised a distinction between "software" (motivational algorithms) and "hardware" (physiological) explanations ([2], [21]), reminiscent of the Cartesian disembodiment of the mind, and thus developed models of behaviour that share certain similarities with those of cognitive psychology and classical Artificial Intelligence. All these models have in common a functionalistic (in the sense of Putnam [33]) and rationalistic stance. In the form of concepts borrowed from classical micro-economic theory, this rationalism is found back in neo-Darwinian explanations of behaviour [6]. These developments have resulted in the prevalent ethological approach to explain as well the cognitive (proximate) causation as the evolutionary (ultimate) consequences of behaviour from a functionalistic-rationalistic perspective [6].

1.2 Thought Experiments in Favour of Technological Models

As befits the output-oriented, black box rationalism of mechanistic and computational models, the classical arguments in support of them are typically deduced from thought-experiments. The first one probably has to be attributed to Descartes. In the *Discourse on Method*, Descartes writes: " I made special efforts to show that if any [automatons] had the organs and outward shape of a monkey or some other animal that lacks reason, we should have no means of knowing that they did not possess entirely the same nature as these animals". The same type of reasoning was applied some 300 years later by Turing to address the controversy of human versus machine intelligence.

Another aspect of the interpretation of machine performance is illustrated by what Braitenberg [10] calls "experiments in synthetic psychology". Braitenberg conceptualised hypothetical, self-operating vehicles that exhibit intricate behaviour based on a surprisingly simple architecture. He describes how people are likely to interpret the behaviour of these machines as “cowardly” or “aggressive” when they watch them moving around. Our own experiments with “didabots” (to be described below) and earlier work with physical animal-like robots by Grey Walter [17] bear out this observation. When onlookers are asked to infer *the mechanism* steering the didabots, they almost always resort to programming descriptions or other instances of introspective rationalism (or, to what Kennedy’s ([24], pp. 80) characterises as “non-physiological subjective concepts that we employ in conscious analytical thought”). Such reactions point to a fundamental anthropomorphic subjectivism. Humphrey [23] might call them examples of “transactional thinking”, i.e. reflecting our tendency to socialize with certain wholly inanimate materials as a by-product of our evolutionary heritage as a social species. Given this bias, it is questionable whether claims about similarities between machine performance and animal behaviour from a human observer are to be taken at face value. A valid evaluation of Descartes’ test for machine behaviour (and Turing’s test for machine intelligence as well), therefore seems to require comparisons based on *objective* descriptions of animal (human) and machine performance. Some researchers advocate the use of ethological methods to obtain such descriptions, not so much to check the tests, but, more ambitiously, to arrive at explicit explanations for specific activity patterns. In section 2, I will report such an endeavour and (in section 3) discuss what might be learned from studying (Braitenberg) machines as if they were animals.

1.3 Objections

The shortcomings of a technologically oriented view (whether based on analogies with mechanical or computing engines) of behaviour are tied to what are typically regarded as desirable properties of a machine. Since a machine is engineered so that it will be easy to repair and behaves reliably and predictably, (or, in case of a typical digital computer, operates in an algorithmic fashion and hence delivers output independent of physical substrate or temporal coherence), the derived models for behaviour tend to share one or more of the following characteristics: decomposable, discrete, deterministic, disembodied and linear. However, the real world is convoluted, non-linear, physically constrained and messy. Attempts to fit technological metaphors and rationalistic theories to these properties of the real world may thus lead to overly complex and contrived explanations [6].

These shortcomings have been addressed mostly by criticising the computational approach of classical A.I. It has been pointed out that robots designed according to the principles of classical A.I. failed to perform meaningful behaviour in the real world [11, 32]. By concentrating on hardware rather than on algorithms, a mechanical approach regained popularity under the banner of "New A.I". But note that criticising computation alone does not necessarily free A.I. from its Cartesian heritage as long as the behaving subject is still viewed as a mere automaton. Proponents of "New A.I" have suggested various routes out of this dilemma. For one thing, they advocate that artificial machines should be "biologically inspired". This means that discoveries and theories from ethology (or broader biological disciplines) should be used as guiding principles in the design of robots ([3], [28], [32]). Furthermore, it has been proposed that machines should be evaluated as if they are living systems (cf. [36]). This stance is reflected in the repeated appeal to use "ethological methods” to study the behaviour of autonomous robots [20]. However, until now this request has not yet been taken up in artificial autonomous agents research. What an ethological analysis of robot behaviour might look like is the topic of the next section.

2 The Application of Ethological Methods to Study the Behaviour of Robots

2.1 Ethological Methods

To gauge the effects of studying artificial agents as if they were living animals, I applied ethological methods to analyse the data from two robot experiments. But first of all, it should be made clear what is meant by "ethological methods". The starting point of all ethological analysis is a secure basis of description. This basis finds its roots in Lorenz's theoretical framework, which – by its focus on "fixed action patterns" – accommodates the dissection of the continuous stream of an animal's behaviour into identifiable units. Typically, ethologists do this by monitoring intact animals under natural conditions and by recording detailed observations of specific patterns of activity on film and video. The data are then worked out using certain statistical techniques (i.e. PCA, factor analysis, cluster analysis, information theoretical statistics and Markov modelling) to reveal temporal patterns in the sequence of activities (for an overview of the ethological application of these techniques, see [22]). Such descriptions are often complemented by controlled experiments in which a given type of behaviour is studied in one situation and then compared with other activities, situations and species. Last but not least, ethologists stress that behaviour should be viewed in the context of the environment to which it has been adapted.

2.2 The Robots

The experiments were done with "didabots", simple robots specially constructed for didactical purposes at the AILab (Department of Information Technology, University of Zürich) by Marinus Maris (for detailed specifications, see [30] and [32]). Didabots are actually realized Braitenberg vehicles, with two sensors at each side and two at the front. The sensors detect infrared light that is reflected from objects after being radiated by IR transmitters (situated just below the sensors). The sensors are connected to the (two) motors in such a way that if the robot detects an object on one side, the motors force forward movements on the wheels at the corresponding side and backward movements on the wheels at the opposite wheel. As a consequence, the robot will turn away from the obstacle. We found that didabots most of the time avoid obstacles, but now and then do collide with objects when these are so small as to fall outside the range of sensory detection. If these objects were also light, the didabots would shift them over a short distance and in this way change locally the structure of their environment. To study the possible feedback on the behaviour of the robots, we exploited this "imperfection" by setting the weights of the frontal IR-sensors to zero. In addition, the robots are equipped with 6 ambient light detectors connected to the motors in such a way that the didabots are attracted to each others switched on light-bulb.

2.3 Experimental Setup

The behaviour of the didabots was studied in two sets of experiments. The first one was performed to obtain a detailed description of the various activities of the robots and will form the focus of this paper. The second set, described in detail by Maris & te Boekhorst [30], illustrates how controlled experiments can shed light on certain conspicuous aspects of the robot-environment interactions.

The first experiment was carried out with five didabots that moved around in an arena (230 x 260 cm) containing 25 randomly dispersed polystyrene white cubes of 8 cm on a side. The experiments were observed by a "bird's eye" colour Charge Coupled Device camera (hung straight above the arena) that sent its images to the computer. The experiment took 25.5 minutes and was videotaped for a thorough ethological analysis (to be described below) of the behaviour of the didabots. To characterize the structure of the environment, we counted the number of cubes along the wall and the frequency of clusters of cubes of all sizes every fifteen seconds. Cubes were considered to belong to a cluster whenever the distance to their nearest neighbours was smaller than the size of a cube. All contacts between robots and cubes were recorded continuously and summed for each interval of fifteen seconds and stored together with time stamp and the data about the spatial configuration of the cubes.

The second set of experiments was conducted to investigate the effects of two factors (environmental structure and the group size of robots) on the interactions between the robots and their environment. To this end we varied the number of cubes (12 or 25) and the number of robots (ranging from 1 to 5). Each combination of factors was replicated three times, giving 2×5×3 = 30 trials. Each trial lasted 30 minutes during which the above-mentioned type of data was collected. For these experiments, the ambient light sensors were inactivated.

2.4 Ethological Analysis of Didabot Behaviour

The Behavioural Repertoire. As a first step, an ethogram was established by classifying the activities of the robots into 35 easily distinguishable behaviour elements. Next, a continuous spoken protocol (recorded on a portable cassette recorder) of the videotaped behaviour of one particular robot (marked for that purpose) was transcribed and worked out. The activities were written down in the sequence of their occurrence and grouped in 102 intervals ("samples") of 15 second each. From this listing, a number of measures were calculated. To get an impression of the diversity of the behaviour during the experiment, I counted for each sample (j) (with j = 1, 2, ..., 102): a) the frequency of activity types broken down by context (i.e. classified as robot-robot activities and robot-object activities and b) the *relative dominance* of activity types expressed as the summed frequency of the most often occurring activity type in the sample divided by the summed frequency of all activities.

A conspicuous feature of the time series of these measures is the sudden drop in robot-object activities around the seventieth sample (Fig. 1). Also note that robot-object activities after the break appear on average less often than before. The frequency of robot-robot activities and the relative dominance index, however, are higher at the end of the experiment (i.e. after sample 60).

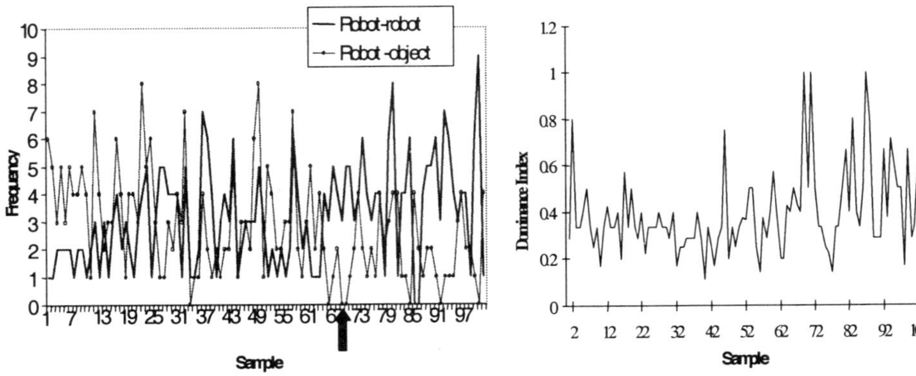

Fig. 1. Frequencies of robot-robot and robot-object Activities (left) and the Dominance Index (right)

Dimension Reduction. To assess the temporal structure of didabot behaviour, the frequencies of transitions from one activity to another activity were tabulated in a r × c matrix whose ij-th entry refers to transitions from activity i to activity j. The row corresponding to the i-th activity is called the sequential profile of that activity. Sorting the activities from highest to lowest marginal totals revealed that those occurring less than 10 times are mainly responsible for the majority of the empty cells in the transition matrix. Since a large number of empty cells seriously bias the statistical procedures to follow, I decided to lump these rare activities as a "rest" category. The 12 remaining behavioural elements were investigated for similarities in their sequential profiles. When two activities A and B have identical sequential profiles, they can be said to take up exchangeable positions in the behavioural stream. In that case elements, which more often follow A, also more frequently follow B and the sequential profiles of the two will be perfectly correlated. The ethological interpretation is that A and B are steered by the same underlying motivational system and it is for these reasons that ethologists often reduce the dimensions of transition matrices by means of factor analysis. Such an analysis applied to the didabot data showed that the 12 remaining elements could be combined into five major factors (Fig. 2).

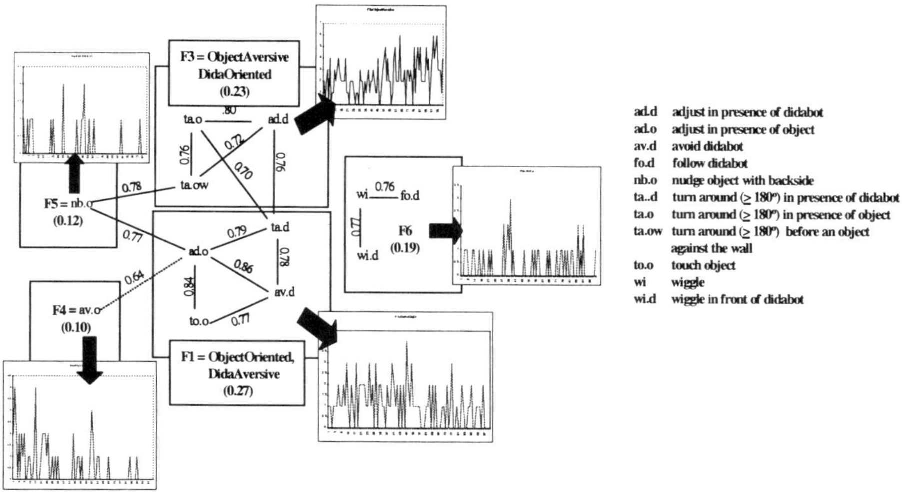

Fig. 2. Results of a factor analysis of the sequential profiles of the 12 most frequent activities. The proportion of variance explained by each factor is given in brackets. Numbers along the connecting lines are Pearson coefficients for correlations among the activities. The time series are the frequencies of activities associated with the factors for each 15 s interval ("samples")

In the protocols, I replaced the original activities by the factor on which they loaded with a coefficient of 0.70 or higher. From the resulting sequences the frequency of these substituted factor identities was calculated for each interval of 15 seconds. From Fig. 2 it can be seen that the frequency of activities related to factor F3 (ObjectAversive, DidaOriented) increases somewhat over time, whereas those loading high on factor F1 (ObjectOriented, DidaAversive) and factor F4 (AvoidObject) decrease especially at the end of the trial. Also note the high occurrence of factor F5 (NudgeObject) and the activities making up factor F2 (Following Dida and Wiggling) around the middle of the experiment.

Sequential Analysis. The five behavioural factors are the basis for further sequential analysis by modelling their transition rates (= transition frequency of $F_i \rightarrow F_j$ relative to the total frequency of F_i) as a first-order Markov chain. For this, I estimated the transition rates for two sub periods of 175 samples each, one well before the increase in the dominance index and the frequency of robot-robot activities (around sample 60) and one after the decrease in robot-object activities at sample 69 – 71. The first 11 samples and sample 69 – 71 were excluded from analysis. Deviations from first-order dependency in the sequence of acts were investigated by applying a χ^2 test for first- against second-order dependency (described in [18]) and the results of the analysis are depicted in Fig. 3. Looking at the transition rates (larger than 0.1) from sub period I to II, the most salient features are: 1) the decrease of the number of transitions; 2) the reduced number of activities with auto-transitions (i.e. that follow themselves in time); 3) the ever stronger central position of activities associated with F3 and 4) the increasing auto-transition rate of activities that make up F3. Furthermore, the χ^2 values are not significant, suggesting at most a first-order sequential dependency.

These results support the distinction of an end phase in which the focal didabot interacted less often with objects, became more "social" but also behaviourally less diverse.

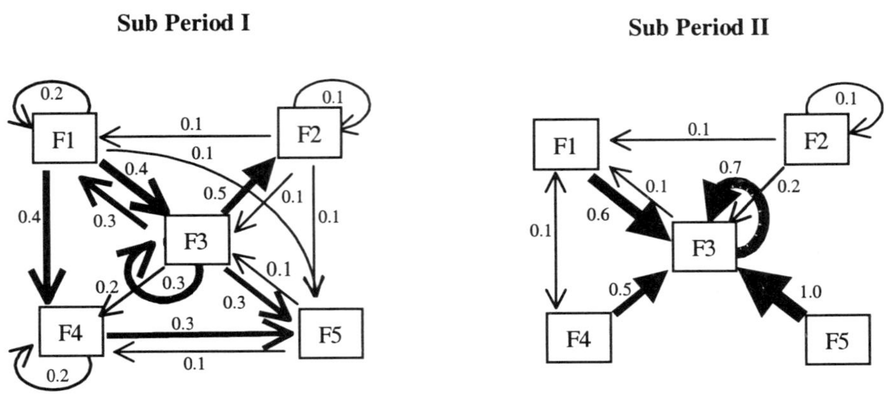

Fig. 3. Transition rates for activities associated with the 5 factors for two sub periods

The Effect of Environmental Structure. I measured environmental structure by counting for each cluster of objects (defined in section 2.3) the number of "fellow-members" of each object. Summing the outcomes over all clusters gives Lloyd's index of mean crowding, $m = \sum_i^c N_i(N_i - 1)/n$, where N_i is the size of the i-th cluster, n is the total number of objects and c the number of clusters. Dividing this measure by the average cluster size results in Lloyd's index of patchiness (p). These indices are widely used in ecological studies to the estimate the degree of clumping of organisms and have the advantage of being scale independent. I calculated m, p and the average cluster size for each sample of 15 seconds. The results are plotted in Fig. 4 and show that all three indices increased up to sample 60. After that, both mean crowding and mean cluster size remained the same, but the patchiness dropped abruptly to its minimum. Observing the course of the process readily provides the explanation for these patterns: the didabots shift the cubes together, thereby increasing the patchiness, until the majority of them are joined in one big central cluster and the remainder are pushed against the wall. This also clarifies much of the behaviour of the robots: after having tided up their arena, their movements were less interfered by the obstacles and hence they interacted more often with each other. However, two questions remain: how robust are the results and how does the heap building come about in the first place?

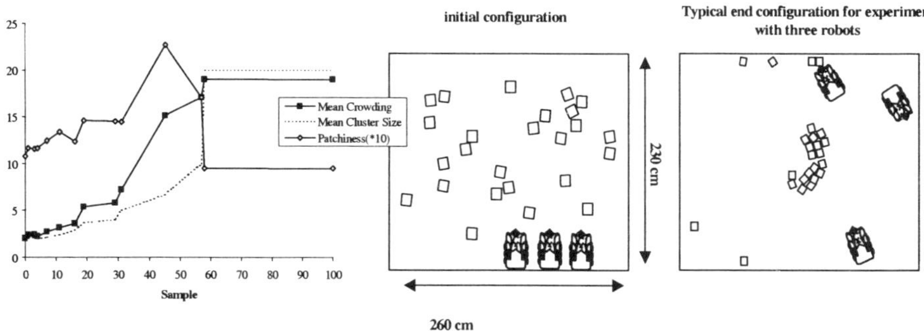

Fig. 4. Measures for clumping of the environment (left) and the typical begin- and end configuration of objects in a didabot experiment

Controlled and Replicated Experiments. The first of these questions is addressed by the controlled and replicated experiments in which we varied both the number of blocks (12 or 25) and the "group size" of the robots (from 1 to 5). The most important result is that the final number of clusters and mean cluster size depend significantly on both factors. When presented with only 12 cubes, on average 2.3 small clusters (mean size = 4 cubes per cluster) emerged independent of the number of robots. With 25 cubes, a mean of 3.4 clusters was formed which contained on average 7.92 cubes, but in this case the group size of the robots was critical (these results are statistically significant after using a 2-way ANOVA with repeated measurements and are described in detail in [30]). Using one robot never resulted in one central cluster and applying five robots only rarely so. The best results in this respect were obtained with groups of three or four robots (Fig. 5).

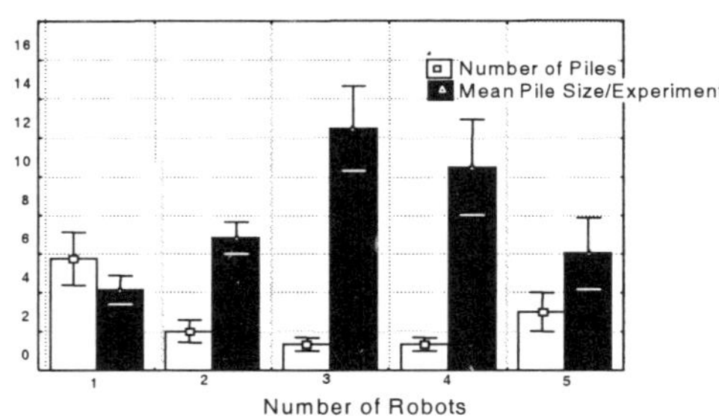

Fig. 5. Mean number of clusters and average cluster size in experiments with 25 objects and 1-5 robots

3 Discussion

In a letter to Mersennes, dated 30th july 1640, Descartes wrote: "(S)uppose that we were equally used to seeing automatons which perfectly imitated every one of our actions that it is possible for automatons to imitate; suppose, further, that in spite of this, we never took them for anything more than automatons; in this case we would be in no doubt that all the animals which lack reason were automatons too". How do the results of the ethological analysis stand up against Descartes' statements? A comparison between the *type* of outcomes presented here with those of analyses on the social behaviour of great apes ([4], [5], [7], [8], [9]) would reveal no striking differences. However, this does not necessarily imply that animal behaviour is just as simple as the performance of a machine, but rather that under certain circumstances the behaviour of machines is much more complex than is commonly appreciated. Furthermore, to arrive at the conclusion that animal behaviour in general can be understood in material terms is one thing, coming up with materialist explanations for specific patterns of action – such as heap building – is quite another. To gauge if ethological methods are of any help in this respect, let us see how an ethologist would deal with the outcomes of the didabot experiments.

If the didabots were living creatures, how might a classical ethologist interpret the results of our experiments? Clearly, without sufficient energy (almost empty batteries) only incomplete heap building would take place and because the same is true when there are only a small number of cubes at hand, the conclusion must be that both internal and external stimulation are crucial. The density of cubes, however, is not the only "sign stimulus" that releases heap building. Also social enhancement is involved because a solitary didabot will not do a complete job. Furthermore, the wandering about of the robots before they push objects is suggestive of "appetitive" behaviour, and the shifting itself has a "consummatory" effect in that it stops after the obstacles have been put together or shifted against the wall. This can be seen as a form of negative feedback, the essence of many cybernetic interpretations of "goal-directed behaviour".

If you are a cognitive ethologist, you might take up the cybernetic argument and focus on the "rules" underlying the goal-directed "decisions" of a didabot. These rules can be represented algorithmically as a chain of "IF ... THEN... ELSE" statements, telling the robot what to do under what conditions. Because the actions can be grouped in "motivational classes" (the factors F1 to F5), the rules might work in a hierarchical fashion. For instance, there could be a condition (such as a sufficient number of cubes and robots) that first brings the didabot in a certain motivational state (say, F1 = "ObjectOriented, DidaAversive"), after which a second rule activates a particular element of lower order belonging to that class (for example "touch object"). A closer inspection of the didabots reveals that they most often touch and push single objects, suggesting that the didabot, in addition to the activation rules, possesses an internal representation about the size of objects. This representation could be the norm value of a set point which - when larger than the sensed diameter of an obstacle - switches on the pushing command. The norm value may be set by natural selection. A neo-Darwinist colleague of yours would readily incorporate this suggestion in an evolutionary scenario. Her side of the story would concentrate on the costs and benefits associated with the decision to push or not to push. This decision is supposed to be linked to a genetically determined trait, let's call it the Object

Avoidance Ability (OAA). This trait is expressed differently in individual didabots so that two extreme "strategies" ("bumpers", characterized by a low OAA, versus "dodgers") can be distinguished. The observation that a given strategy or OAA is more common than others is seen as optimisation by natural selection and the explanation would take the form of a cost-benefit trade-off argumentation. The benefit of object-avoidance is obviously the prevention of damaging collisions. However, avoiding obstacles is costly because it requires extra travel time. The optimal OAA, which is perceptibly the set point of the cognitive explanation, can be found as the maximum difference between the estimated benefits and costs of object avoidance plotted as a function of the OAA. The evolutionary hypothesis also explains why didabots build heaps. By doing so, they minimize both the probability of future collisions and energy expenditure: after having cleaned up the area, the robots have to meander less to avoid the objects.

Support for the suggested "genetic" basis would come from applying a genetic algorithm to the design of the didabot. Starting with vehicles that have all their network weights intact, it would soon be discovered that heap-builders differ by just one random mutation from perfect "dodgers". And as long as it is not known that this single "knocked-out gene" represents the weight of the frontal sensor it could be argued that the gene for cooperation has been isolated.

The lack of the frontal sensor is indeed the essence of the true mechanism. Because of this, a didabot is unable to react to objects right in front of it. Consequently, the didabot collides with such an object and pushes it along until another obstacle (a wall, another object or another didabot) is detected. Due to the subsequent avoidance movement the shifted object is left behind. If the object is deposited close enough to another, they form a constellation that is large enough to be detected and avoided by the robots. In this way pairs are formed. The increased patchiness of the environment improves the ability of the robots to bypass collisions. When only one robot is employed, the environment is soon structured sufficiently for the didabot to maneuver almost without hitting cubes. One single robot therefore never builds a single heap. Because clusters happen to grow predominantly by adding *single* blocks to existing "seeds", their formation depends critically on the availability of single cubes. These are supplied when using more than one robot: due to mutual avoidance movements, we observed that the robots now and then "erroneously" break up small clusters. Once a heap of sufficient size is formed, the movements of the didabots appear to become more regular in that they tend to circle around the cluster, which in turn lowers the chance of destruction. However, when too many robots are around their mutual avoidance movements become so frequent that also larger heaps are destroyed. This explains the "optimal" group size for forming one single, large cluster. To summarize, due to *constrained* information the robots patterned their environment, which in turn affected their behaviour in such a way that it reinforced further patterning of the environment. This positive feedback between macro-patterns and the interactions on the micro-level responsible for that pattern, damped by (behavioural) "error", is the hallmark of self-organization.

My conclusion from comparing the imaginary accounts with the explanation based on our observations is that current theories from cognitive science and ethological methodology are unable to identify the mechanism responsible for the collective activities of the didabots. To a certain extent this may be due to the fact that the statistical tools typically used by ethologists assume linearity in the data, which clearly is the wrong starting point for understanding the behaviour of the didabots and of living creatures in general. Applying suitable non-linear statistical methods would, however, solve only part of the problem. In my opinion it is unlikely that inspirations from ethological theories will yield agents with behaviour of the intricacy New A.I hopes for as long as these theories themselves are either rooted in techno-mechanistic biology or reflect the Cartesian spirit of the functionalistic computational stance borrowed from classical A.I.

The way out, recognized by many advocates of "New A.I", is to treat agent-environment systems as non-linear, dynamic, self-organizing systems and it would be a good thing if more ethologists would do this too. Self-organization is indeed exactly what drives the heap building of the didabots and - as far as I know - has not been so convincingly demonstrated in robots before. Why is it that so little is achieved in this respect? Maybe this has to do with the fact that many of us are computer scientists and engineers. Therefore we may not be enough acquainted with biology to recognise that biological theories are often more inspired by engineering and computer science than these theories can inspire us. Even more important, we might not have tried hard enough. Creating conditions for self-organisation requires a *very* radical change in the attitude of (computer) scientists and engineers, because it implies distancing from task-orientedness, giving up an overly rationalistic stance, and disposing of the holy cow named "Information Technology". For example, the didabots in our experiment were not designed to clean up the area. Their object collecting ability appears to be the result of "imperfect" design (and hence limited information) and behavioural "errors". Had we endowed them with a frontal sensor, they would have been well behaving, but incredibly dull, object avoiding vehicles that would never accomplish anything social.

References

1. Ablondi, F.: Automata, Living and Non-Living: Descartes' Mechanical Biology and His Criteria for Life. Biology and Philosophy 13 (1998) 179-186
2. Baerends, G. P.: The Functional Organisation Of Behaviour. Animal Behaviour 24 (1976) 726-738
3. Beer, R.D., Chiel, H.J., Sterling, L.S.: A Biological Perspective on Autonomous Agent Design. Robotics and Autonomous Systems 6 (1990) 169-186
4. te Boekhorst, I, J. A.: Social Structure of Three Great Ape Species: An Approach Based on Field Data and Individual Oriented Models. Ph.D Thesis. University of Utrecht, the Netherlands (1991)
5. te Boekhorst, I, J. A, van Oorschot, I., de Jongh, T.: Social Structure of a Group of Captive Lowland Gorillas (Gorilla gorilla gorilla). Acta Zoologica et Pathologica Antverpiensia 81 (1990)
6. te Boekhorst, I.J.A., Hemelrijk, C.K.: Nonlinear and Synthetic Models for Primate Societies. In: Kohler, T.A., Gummerman, G.J. (eds.): Dynamics in Human and Primate Societis. Agent-Based Modeling of Social and Spatial Processes. Oxford University Press, Oxford (1999) 19-44
7. te Boekhorst, I.J.A., Hogeweg, P.: Self-Structuring In Artificial "Chimps" Offers New Hypotheses For Male Grouping In Chimpanzees. Behaviour 130 (1994) 229-252.
8. te Boekhorst, I.J.A., Hogeweg, P.: Effects Of Tree Size On Travelband Formation In Orang-Utans: Data-Analysis Suggested By A Model Study. In: Brooks, R.,Maes, P. (eds.): Artificial Life IV. Bradford Books, MIT Press Cambridge (MA) (1994) 119-129

9. te Boekhorst, I. J. A., Schürmann, C. L., Sugardjito, J.:Residential Status And Seasonal Movements Of Wild Orang-Utans In The Gunung Leuser Reserve (Sumatera, Indonesia). Animal Behaviour 9 (1991) 1098-1109
10. Braitenberg, V.: Vehicles. Experiments in Synthetic Psychology. Bradford Books, MIT Press Cambridge (MA) (1984)
11. Brooks, R.: Intelligence Without Reason. In: Proceedings of IJCA-91. Morgan Kaufman, San Mateo (CA) 25-81.
12. Brooks, R.: The Relationship between Matter and Life. Nature 409 (2000) 409-411
13. Chomsky, N.: Syntactic Structures. Mouton, The Hague (1957)
14. Crist, E.: The Ethological Constitution of Animals as Natural Objects: The Technical Writings of Konrad Lorenz and Nikolaas Tinbergen. Biology and Philosophy 13 (1998) 61-102
15. Dawkins, R.: Hierarchical Organization: A Candidate Principle For Ethology. In: Bateson, P. P. G., Hinde, R. A. (eds.): Growing Points in Ethology. Cambridge University Press, Cambridge (U.K.) (1976) 7-54
16. Dawkins, R.: The Blind Watchmaker. Penguin Books, London (1986)
17. Grey Walter, W.: An Imitation of Life. Scientific American 182 (1959) 42-45
18. Haccou, P., Meelis, E. Statistical Analysis of Behavioural Data. Oxford University Press, Oxford (1992)
19. Harnad, S.: The Symbol Grounding Problem. Physica D 42 (1990) 335-346
20. Hendriks-Jansen, H.: Catching Ourselves in the Act. Situated Activity, Interactive Emergence, Evolution and Human Thought. Bradford Books Press, MIT Cambridge (MA) (1996)
21. Hinde, R.A.:Ethology: Its Nature and Relations with Other Sciences. Oxford University Press, Oxford (1982)
22. van Hooff, J. A. R. A. M.: Categories And Sequences Of Behavior: Methods Of Description And Analysis. In: K. R. Scherer, K.R., Ekman, P. (eds): Handbook of Methods in Nonverbal Behaviour. Cambridge University Press, Cambridge (UK) (1982) 362-439
23. Humphrey, N.:The Social Function Of Intellect. In: Bateson, P.G., Hinde, R. A. (eds.): Growing Points in Ethology. Cambridge University Press, Cambridge (U.K) (1976) 303-317.
24. Kennedy, J. S.: The New Anthropomorphism. Cambridge University Press, Cambridge (1992)
25. Lorenz, K.Z.: Ueber den Begriff der Instinkthandlung. Folia Biotheoretica 2 (1937) 17-50
26. Lorenz, K.Z.: The Comparative Method in Studying Innate Behaviour Patterns. Symposia of the Society for Experimental Biology 4 (1950) 221-268
27. Losey, G.S.: Information Theory and Communication. In: Colgan, P. (ed.): Quantitative Ecology. Wiley & Sons, New York (1978) 44-78
28. Maes, P.: Modeling Adaptive Autonomous Agents. Journal of Artificial Life (1994) 135-162
29. Mataric, M.J.: Integration of Representation Into Goal-Driven Behavior-Based Robots. IEEE Journal of Robotics and Automation 8 (3) (1992) 304-312
30. Maris, M., te Boekhorst, I. J. A.: Exploiting Physical Constraints: Heap Formation Through Behavioral Error In A Group Of Robots. In: Intelligent Robots and Systems, Proceedings of the 1996 IEEE/RSJ International Conference on Intelligent Robots and Systems (IROS96) part III. Osaka (1996) 1655-1661
31. McFarland, D.: Feedback Mechanisms in Animal Behaviour. Academic Press, London (1971)
32. Pfeifer, R., Scheier, C.: Understanding Intelligence. MIT Press, Cambridge (MA) (1999)
33. Putnam, H.: Philosophy and Our Mental Life. In: Philosophical Papers 2: Mind, Language and Reality, Cambridge University Press, New York (1975) 291-303
34. Rosenbluth, A., Wiener, N., Bigelow,J.: Behaviour, Purpose and Teleology. Philosophy of Science 10 (1943) 18- 24

35. Shannon, C.E., Weaver, W.: The Mathematical Theory of Communication. University of Illinois Press, Urbana (1949)
36. Steels, L.: The Artificial Life Roots of Artificial Intelligence. Journal of Artificial Life 1 (1994) 89-125
37. Tinbergen, N.: The Study of Instinct. Clarendon Press, Oxford (1951)

The Relationship between the Arrangement of Participants and the Comfortableness of Conversation in HyperMirror

Osamu Morikawa[1] and Takanori Maesako[2]

[1]Research Institute for Human Science and Biomedical Engineering, AIST
1-1-1 Higashi, Tsukuba, Ibaraki ,305-8566, Japan
[2]Faculty of Human Sciences, Osaka University
1-2 Yamadaoka, Suita, Osaka 565-0871, Japan
morikawa.osamu@aist.go.jp, maesako@mcgyver.hus.osaka-u.ac.jp

Abstract. HyperMirror is a new type of video conversation system which does not simulate face-to-face conversation in real space. In real space, people may feel that a relative positional relationship to the other person is comfortable and sometimes that it is not. They seem to feel a similar relationship also in HyperMirror. In this paper, we observe the relationship between arrangement of participants on the HyperMirror screen and comfortableness of conversation by changing position of the camera and the participants' standing positions. We find two facts; in the HyperMirror screen, they feel at ease to speak when they are near or look toward their partner, and it is more important that they look toward their partner than that they are looked toward.

1 Introduction

The HyperMirror displaying one's self-image is a new type of video communication system, it does not imitate face-to-face conversation [8, 9]. In a HyperMirror conversation (Fig. 1) all participants see the same image, displaying themselves and their partners in the same room on the screen. So that, the positional arrangement of the participants and the items displayed on the screen can be used in the communication. For example, a pointing at objects on the screen is possible. Moreover, there are no walls between the participants, they may move freely within the conversation space on the HyperMirror screen.

Through the experience with the HyperMirror for four years, it appears that there are arrangements in which the participants feel at ease to speak, while in some arrangements they find it uncomfortable to speak. In the participants' report, they said, "As you don't look at me, I feel you don't talk to me although you seem to talk to me." "I don't feel you speak heartily." It seems to be a similar sense to a disagreement of gaze. In this paper, we study the relationship between the arrangement of participants on the HyperMirror screen and the comfortableness of conversation by changing position of camera and partner's standing position.

M. Beynon, C.L. Nehaniv, and K. Dautenhahn (Eds.): CT 2001, LNAI 2117, pp. 109-116, 2001.

2 The Relationship between the Standing Point and the Facing Direction on the HyperMirror Screen

Usually in video conversation, people look at their partner on a video monitor when they speak. Therefore, when the listener on the screen is near the speaker's camera, the full face of the speaker will be shown. But if the listener is displayed, the farther from the speaker's camera, the more of the speaker's profile will be shown on the screen.

In HyperMirror, the self's mirror reflection is presented with his/her partner on his/her own screen. Let's assume that there is a camera on the left side of a screen, on which there are two listeners' (A and B's) reflections (Fig. 2). When a speaker looks at A's reflection, the speaker's reflection is profiled to the right a little. On the HyperMirror screen, when the speaker's mirror reflection is on the left of A's reflection (inside of zone-1), it looks to see listener A or B. The difference between looking toward A and B shows the difference of the speaker's face direction. When the speaker looks to A's reflection, the speaker's reflection is almost full face. When the speaker looks to B's reflection, the speaker's reflection is profiled a little.

When the speaker's reflection is in the zone-2 (between A's reflection and B's), the speaker's reflection always looks to see B even though the speaker looks at A or B. Even though the speaker looks at A's reflection, the speaker's reflection looks away from A.

When the speaker's reflection is in the zone-3 (on the right side of A's and B's reflections), the speaker's reflection looks away from A and B.

Fig. 1. HyperMirror

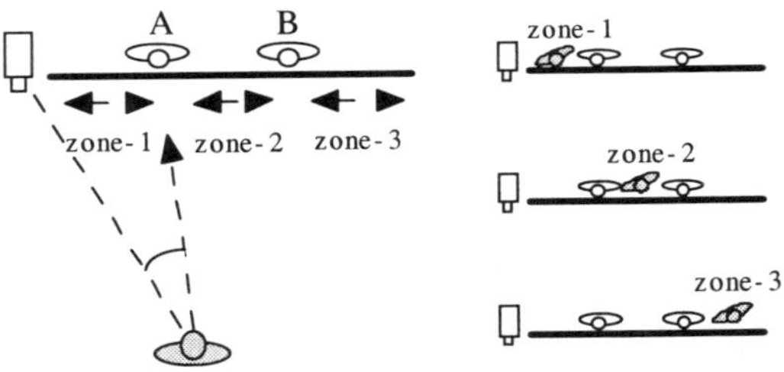

Fig. 2. Presentation zone and its meaning

2.1 Experiment 1

In HyperMirror, since one's own image is displayed on a screen, the facing direction has different meanings depending on where the self-image appears on the screen. We carried out experiments examining how the comfortableness of talking varies depending on the relative positions of the listener and the position of the displayed self-image. The subjects were 36 (18 female) people, their ages ranging between 10 and 60 years. Each experiment was conducted with two acquainted subjects.

The equipment was a “2 site chromakey” version of the HyperMirror (Fig. 3). A projector, a screen and a camera were installed in both rooms, the background of the second room was a simple blue curtain for chromakey synthesis. The screens were 90cm high, 120cm wide, standing 90cm above the floor. The cameras were placed 150cm high in the two rooms, on the left side of the screen in Room-1, and on the right side of the screen in Room-2. The distance between the camera's location and the center of the screen was 70cm. The standing positions of the subjects were 300cm in Room-1 and 360cm in Room-2 from the center of the screen. Three positions were examined in each room, one facing the center of the screen, one 80cm to the left and one 80cm to the right from the center. In this paper we denote the left, center, and right positions by l, c, r (in Room-1), and L, C, R (in Room-2), respectively.

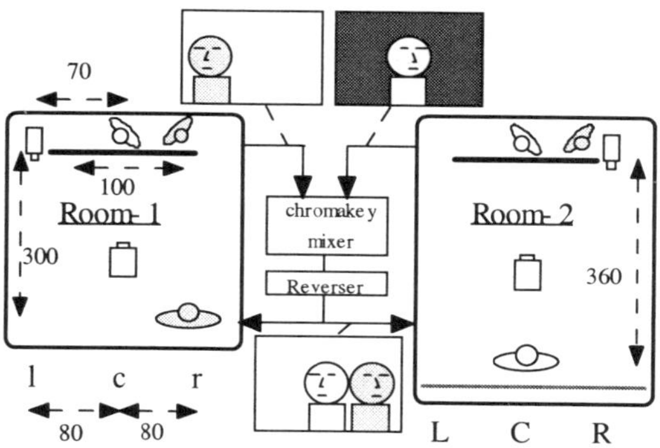

Fig. 3.. HyperMirror system used experiment.

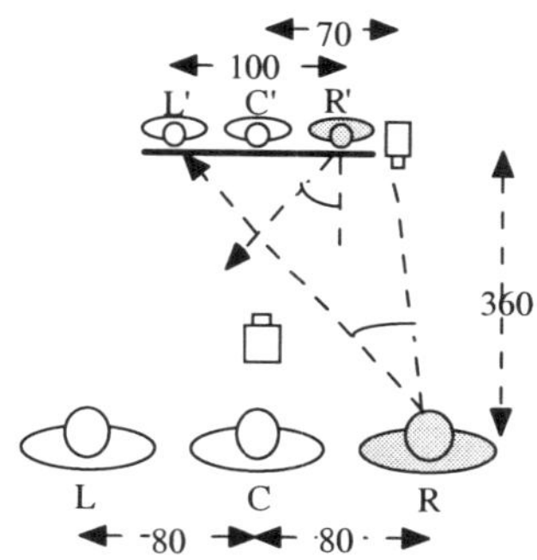

Fig. 4. Angle of the faces of the reflection

Although the distances between the camera and the subjects were different between the two rooms, we adjusted the camera to project images of same size and at the same position on the two screens. The cameras were adjusted to show images of 210 cm width with the 3 standing positions at 50 cm intervals on both screens. The precision of both cameras and screens were of NTSC-TV quality. The HyperMirror video signal was a chromakey synthesis of the video signals from the cameras in each room. It was sent to each room and the reverse image of it was projected on each screen.

As the distances to camera are different between in Room-1 and Room-2 the angles to full face are different between them. Let us assume that the subject at R (in Room-2) sees the reflection at L' on the screen who is standing at l (in Room-1) (Fig. 4). The camera in Room-2 is 360cm forward and 10cm = (80-70) left. On this occasion when subject at R looks at the camera, the reflection is a full face, so that the face of reflection turns left a little. By simple calculation, the angle is 18 degrees (full face is 0 degrees). So, the mirror reflection of the subject at R looks to left side of the screen in 18 degrees. Similarly, the angles of the faces of the reflection at R, C or L when s/he sees L', C' or R' can be calculated. By calculation, they are 3, 13 and 22 degrees in Room-1 and 3, 11, 18 degrees in Room-2.

To make the subjects familiar with the environment, after a brief explanation of the HyperMirror system, they experienced the handshaking over the screen, finger pointing at objects in the shared space of the HyperMirror environment, and were allowed to freely communicate with their partners for 5 minutes. Next, the subjects were instructed to move into the directed positions, shake hands, greet each other over the screen and rate the comfortableness of conversation in the given arrangement on a

7-degree-scale (very natural / comfortable, natural / comfortable, a little natural / comfortable, neither natural / comfortable nor unnatural / uncomfortable, a little unnatural / uncomfortable, unnatural / uncomfortable, very unnatural / uncomfortable). The experiments were carried out in six different arrangements of standing positions, in random order. After the evaluation of six positions, the subjects expressed their impressions freely.

2.2 Results

After translating the subjective evaluation to a scale running from -3 to +3 (-3 = unnatural or uncomfortable, +3 = natural or comfortable), statistical analyses were performed. Analysis showed that subjects gave a positive evaluation when the image of a subject in Room-1 was on the left side of the partner's image on the screen {lR,cR,lC}. In these cases, they were facing their partner on the screen. Negative evaluations were given when a image of a subject in Room-1 was on the right side of the partner's image on the screen {rL,rC,cL}. In these cases, they did not face their partner on the screen. The worst evaluations were given for the cases where they were standing far away from their partner. The evaluations can be divided into 3 categories (see Fig. 5): uncomfortable for speaking {rL}, mixed feeling {rC,cL}, and comfortable for speaking {lR,cR,lC}. The differences among the three groups were statistically significant (Student's t-test, p less than or equal to 1%). There was no statistical evidence for differences within each group.

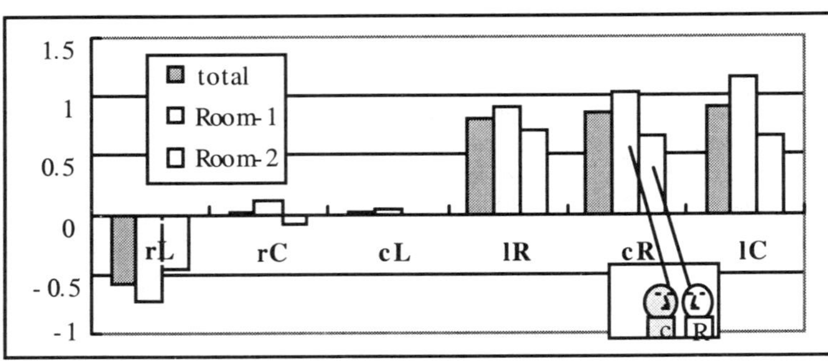

Fig. 5. Subjective evaluations of experiment-1

Comparing Room-1 and Room-2, subjective evaluations tended to be more extreme for Room-1 than for Room-2. In Room-1, the distance from the subject to the camera was shorter than in Room-2, causing a greater change in facial angle between the three standing positions than for Room-2. This was reflected in the subjective evaluations. Room-1 tended to get a lower evaluation on the uncomfortable condition {rL}, where subjects are separated in left to right on the screen and see outside each other , than Room-2, and higher evaluations on the comfortable conditions {lR,cR,lC} than Room-2.

3 Experiment 2

According to previous experiment, there were higher naturalness or comfort scores when subjects look to their partner than when they look to away from their partner. Then we experiment on the comfortableness when the directions of both people are same.

This experiment was carried out in the same rooms as the previous one, only the camera in Room-2 was placed on the left side of the screen this time. This way, all participants were displayed facing to the right regardless of their standing point. Therefore, it was impossible for two participants to be displayed facing each other. In all cases, the participant on the left side was shown facing the participant on the right, who again was displayed turning slightly to the right. The procedure of the experiment was the same as in the previous one.

3.1 Results

Translating the subjective evaluation to a 7-degree-scale for statistical analysis was done in the same way as for the previous experiment (Fig. 6).

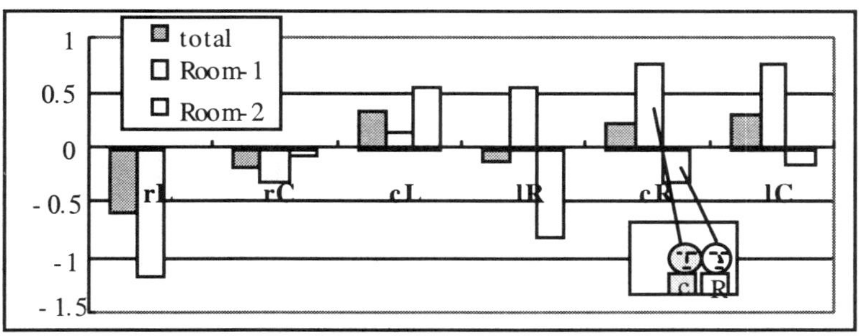

Fig. 6. Subjective evaluations of experiment-2

The tendency of the subjective evaluations was different from the previous experiment, as can be seen in Fig. 6. The evaluations did not seem to depend on whether the subjects were on the left or right side of the screen, but rather, on the distance of the subjects from each other. When the separation distance on the screen was large {rL,lR}, the subjective evaluations were low. When subjects were standing close together, the evaluations were lower when they were further from the camera {rC,cR} than when they were nearer to the camera {cL, lC}. That is, a higher score was given to a nearly full face view than to a profile. Although some of the standing position arrangements received high subjective evaluations in Experiment 2, they were lower than the natural/comfortable arrangements in Experiment 1 {lR,cR,lC}. There were no statistically significant differences between Experiment 1 and Experiment 2 because of the large variance in Experiment 2.

4 Discussion

In the first experiment, the conversation was felt natural when both participants were mutually facing towards each other, no matter what the separating distance was. The conversation was felt equally natural in cases of standing side by side (80cm distance, if converted to real space) and cases of standing distantly (160cm distance, if converted to real space). However, when the participants' images were shown facing away from each other, although it was not very unnatural when the participants stood side by side, it was felt very unnatural when they stood distantly.

In the second experiment, the participant standing on the left was always displayed on the screen facing the other participant, who was displayed facing out to the right (both of them were displayed facing to the right). In each arrangement, the subject standing on the left felt less unnatural than the other participant standing on the right. Moreover, the results for "same role"arrangements (relative left-right positions of the participants) revealed similar tendencies for both participants.

In HyperMirror conversation a participant plays the roles of a speaker and an audience at a same time because his/her reflection is displayed on the screen and s/he sees it. Subjects evaluate the HyperMirror picture as audiences. In short, the evaluation of the HyperMirror picture depended greatly on how one's own image was displayed. It appears to be more important for the comfortableness of the conversation that one's own image faces the partner than the partner faces one's direction.

A participant tolerates that partner's reflection does not face her/his own reflection. S/he wants that her/his own reflection faces the partner's reflection [2, 3, 7]. But it is not satisfying in some conditions. In the condition, where the participant's own reflection does not face to partner when s/he looks at the partner's reflection on the screen and s/he cannot see the screen when her/his reflection faces to partner's reflection on the screen. That is, s/he cannot play the role s/he wants. This gap between the role a participant can play and the role s/he wants to play causes uncomfortableness [1]. A gaze direction expressed by the body posture which is observed in many HyperMirror conversations seems to be based on this gap.

In both experiments, the differences of evaluation scores were distinguished in Room-1, probably because the shorter distance between the participants and the screen leads to drastic changes in the angle of one's facing direction and the direction of the camera's focus. This also proved that the way one's own image is displayed has more effect on the comfortableness of conversation than the way the partner is displayed.

These results reflect human behavior in sociopetal/sociofugal settings in real space. In real space, when people want to speak, they face toward the partner and choose a position to see her/him in central vision. Consequently, it is a sociopetal setting. On the other hand, when people happen to be standing together by mere chance and they don't have a desire to talk, they behave to discourage interaction and choose a position not to see the other in central vision. Consequently, it is a sociofugal setting.

People choose an arrangement which best suits their desire to talk. Conversely, sometimes the arrangement makes an influence on the conversation of people. In sociofugal settings, people see their partner peripherally, whether s/he is out of their sight or not. Either way, it is easy to ignore the presence of other people, and it is natural not to have a conversation.

In HyperMirror communication, even if people are in a sociofugal arrangement on the screen, they can see their partner in their central vision. The visual information is equal to that of the sociopetal arrangement. That is, the following real world causal relationship that we can get through the arrangement does not hold in HyperMirror space:

Sociopetal arrangement --> people see each other in their central vision --> people want to talk.

Sociofugal arrangement --> people do not easily see each other in their central vision --> people avoid talking.

The difficulty that people feel in talking in the sociofugal arrangement on the HyperMirror screen is not caused by physical elements of the visual information. It is caused by people's knowledge of sociofugal arrangements in the real world. The recognition of the sociofugal arrangement is a cognitive schema. People feel that they cannot talk easily in that arrangement even though they can see their partners in their central vision.

Although a speaker does not always see a listener during conversation, s/he sees the listener at some important points to know the listener's statement. Similarly, a listener does not always see the speaker during conversation, s/he looks at the speaker or s/he nods his/her head to show her/his listening to or being interested in the speech. These actions are useful not only for a speaker and a listener but also for onlookers. These actions identify who is a listener and a speaker [4, 5]. This role of the actions is important when more than two people have a conversation. It may be difficult for them to identify a speaker and a listener at the moment they see the picture. But observation for a while shows who joins in the conversation. When more than two people have a conversation, a participant becomes a speaker, a listener, and an onlooker or an audience. Moreover, in HyperMirror a speaker has an audience viewpoint because s/he sees her/his own reflection on the screen during conversation, and that the speaker has such a viewpoint is shown to play an important role in smooth conversation in HyperMirror.

Another example is that people try to keep an appropriate space with the partner on the HyperMirror screen [9]. This action is similar to taking personal space in the real world that causes discomfort if invaded [6].

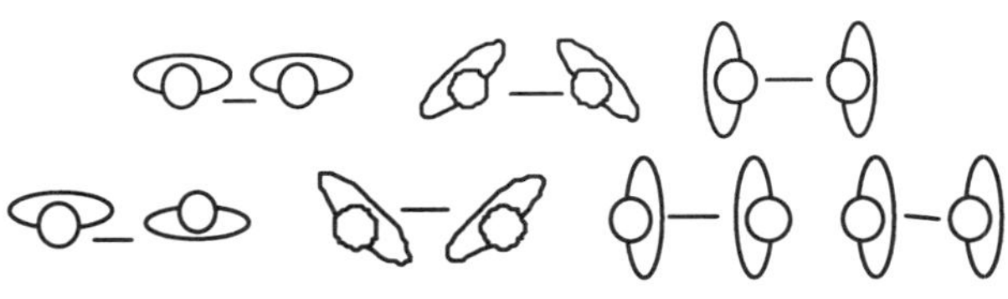

Fig. 7. Sociopetal (upper) and sociofugal (lower) arrangements

5 Conclusions

In this paper, the relation between arrangement of participants and comfortableness of conversation is described.

HyperMirror is a new type of video mediated communication system different from face to face conversation in the real world. In the real world, the arrangement of participants influences the participants' sight. Then in sociopetal arrangement the conversation may take a lively turn, while in sociofugal arrangement the conversation is controlled automatically with the decrease of common space of their sight. But when the arrangement of participants doesn't influence the participants' sight in HyperMirror space, an arrangement has a similar meaning to participants as it has in the real world.

In Experiment 1, people felt comfortable talking when they faced their partner, whether they were close together or far away from each other. When they were not looking at their partner, they must be near to each other on the screen to feel comfortable. This feeling is similar to the sociopetal arrangement in the real world. In contrast, when they did not face their partner, people felt uncomfortable because of a real world sociofugal arrangement. In addition, the result of Experimental 2 showed that although it is important for participants to have their partner looking at them on the HyperMirror screen, it is more important that the self reflection looks one's partner.

It turns out that the negative evaluation of the sociofugal arrangements on the HyperMirror display were not caused by the physical elements of the conveyed visual information. It is rather the recognition of the sociofugal arrangement as a cognitive scheme, which lead to uneasiness in the conversations.

The HyperMirror system was designed on the principle that optimum utilization is based on an understanding of its differences from the real world. On the other hand, people tend to use their knowledge of the real world in HyperMirror space. Thus, the improving HyperMirror requires not only research of the system but also of human cognition.

References

[1] Argyle, M. & Cook, M. *Gaze and Mutual Gaze*, Cambridge University Press (1976).

[2] Exline ,R.V. "Explorations in the process of person perception: Visual interaction in relation to competition, sex, and need for affiliation", *Journal of Personality*, 3 1,1-20 (1963).

[3] Exline, R.V., Gray, D., & Schuette, D. "Visual behavior in a dyad as affected by interview content and sex of respondent", *Journal of Personality & Social Psychology*, 1, 201-209 (1965).

[4] Gibson. J.J. & Pick A.D. "Perception of another person's looking behavior", *American J. Psychology*, 76,386-394 (1963)

[5] Goodwin, C. *Conversational Organization: Interaction between Speakers and Hearers*, Academic Press, New York (1981)

[6] Hall, E., *The Hidden Dimension*, (1970 In Hidaka, T., & Sato, N., trans. in Japanese), Misuzu Shobo (1966).

[7] Kendon, A. "Some functions of gaze direction in social interaction", *Acta Psychologica*, 26, 22-63 (1967).

[8] Morikawa, O. & Maesako, T.; HyperMirror: a Video-Mediated communication system, CHI'97 extended abstracts, 317-318 (1997)

[9] Morikawa, O. & Maesako, T.; HyperMirror: Toward Pleasant-to-use Video Mediated Communication System, CSCW'98, 149-158 (1998).

Extended Abstract

Mapping the Semantic Asymmetries of Virtual and Augmented Reality Space

Frank Biocca[1], David Lamas[2], Ping Gai[1], and Robert Brady[1]

[1]Media Interface and Network Design (M.I.N.D.) Labs
Dept. of Telecommunication
Michigan State University
Biocca@msu.edu
[2]Universidade Fernando Pessoa
Multimedia Resource Center
Porto, Portugal

1 Introduction

This article reports on a key experiment in a series of experiments that attempts to map some of the perceptual, cognitive, kinetic, and semantic properties of virtual space. This study is part of the Mobile Infosphere Test Bed which is part of an effort to design and study the use of body-surrounding interfaces for wearable computers with augmented reality displays.

2 The Mobile Infosphere Test Bed

Augmented reality displays allow for the superimposition of static 2D overlays or 3D computer graphic objects onto the physical environment surrounding the user (Azuma, 1997; Barfield & Caudell, 2000; Behringer, Klinker, & Mizell, 1999). Wearable computers support the widespread diffusion of continuous, personal computing across work and home space and across time in the personal and work day (Barfield & Caudell, 2000; Chris Baber, 1998; Starner, 1999). Together, wearable computers and augmented reality displays embedded in a heterogeneous, mobile computing network infrastructure potentially turn these computing resources into rich data intensive tools for highly-contextualized, immersive, collaborative computing.

Information appliances[1] abound today but still provide litle or no integration among themselves. These are just a few examples (Want & Borriello, 2000):

---

1 An information appliance is an appliance that can provide specialized access to information (Want & Borriello, 2000).

M. Beynon, C.L. Nehaniv, and K. Dautenhahn (Eds.): CT 2001, LNAI 2117, pp. 117-122, 2001.

- Electronic books;
- Portable global positioning devices;
- Internet-enabled cellular phones;
- WebTV and home entertainment;
- PDAs enabled with wireless connectivity;
- Embedded Web servers;
- Smart rooms; and
- Wearable computers.

The Mobile Infosphere interface seeks to merge these into a common interface.

The Mobile Infosphere test bed at Michigan State University is designed to explore mobile interface issues, implementations, and design options. The project designs, tests, and evaluates wireless networks, software, and interfaces. With a focus on human factors and the mobile user, the mobile Infosphere test bed assesses the human-computer interaction dynamics between users and mobile data, quality of service and methods of information delivery, and interface design for immersive mobile collaborative users of wearable computers connected to augmented reality, heads-up displays.

3 Theoretical and Design Issues of the Single User Augmented Reality Interface

In our previous research we examined ways in which networked virtual environments can support human communication (Biocca, 1995). Most recently we have examined how multisensory and multimodal information presented in embodied computing environments (Biocca, 1997; Biocca, Kim, & Choi, in press) can amplify human intelligence and task performance (Biocca, 1996) while minimizing unwanted perceptual or cognitive effects of an augmented reality system (Biocca & Rolland, 1998). Currently, we have completed some patents on a new projective head-mounted display designed ultimately for use in immersive, networked rooms and mobile augmented reality systems (Biocca & Rolland, 2000).

A goal of the Mobile Infosphere project is to pursue this research with rigorous human-computer interaction studies using mobile augmented reality, wearable prototype designs. The primary objective of the mobile infosphere project is to provide the HCI theory, guidelines, and prototype for a high bandwidth mobile computer interface. The design and HCI studies have examined ways to present information using body-stabilized augmented reality displays to determine basic design principles for presenting fields of information to users in augmented reality environments.

Because a great deal of human cognitive capacity is allocated to perceiving, using, and remembering the spatial relation of objects in the environment relative to the individual (Bryant, 1992; Tversky, 1997; Tversky, 1998), our current approach is to find ways to make optimal use of human spatial cognition for information manipulation and storage in a mobile augmented reality environment.

Below is an outline of the interface prototype design.

4 Theoretical Issues: Leveraging the Power of Human Spatial Cognition for Mobile Computing

- Leverage spatial cognition: Derive interface techniques that leverage powerful spatial cognition and spatial cueing when navigating augmented reality and mobile computing environments.
- Determine Tool and Object Layout: Determine interaction patterns and parameters for optimal ergonomic use of 3D layout of tools, applications, and content organized within a spatialized display around the body of the user (egocentric coordinate system).
- Map human egocentric space: Experimentally map asymmetries in the semantic, haptic, and perceptual properties of human egocentric space so that these can guide how to map procedural, semantic, and episodic knowledge to egocentric and exocentric spatial coordinate systems.

5 Design Objectives: Create a "Body-Surround" Spatial Interface for Fully Mobile Collaborative Users

What are the basic requirements for mobile, immersive, information rich augmented reality interface? If we assume that the interface is designed to support a fully mobile, physically-active, continuous-information user:

- Non-interference with Physical Mobility and Engagement: Mobility and user multitasking indicates that the interface needs to provide fields of virtual information to a user under conditions of high physical mobility without overly taxing the user's perception of the environment and attention to their task.
- Immersive multi-format data: If information is to be arrayed around the user's moving body, this suggests the need for multi-positioned video windows and 3D icons.
- Memory and Task Support: The system will need to assist the user's memory for procedures (procedural memory) and context specific reference information (semantic memory). It must make that data accessible, while the user is in full motion.
- Context sensitivity: The interpretation of sensor data such as GPS and user behavior is used to help establish geography, task, and user state context.
- Continuously Record and Interpret Behavior: Because the interface is with the users at all times, the system can automatically augment memory and record key information about what the user is doing, where they are, and who they are interacting with.
- Current Design and Implementation of User Interface for the Mobile Infosphere: In the mobile infosphere, data are organized around body-centered information fields and environment-centered information fields. These refer to the local coordinate system for the array of 3D tools and data objects in the user's augmented reality environment.

5.1 Body Centered Information Spaces

The body-centered space is organized into three fields. These are described in greater depth in the full article:

- Heads Up Displays: Because this is an area of focal attention, the data displayed here are either system alerts or continuously monitored information such as navigation data.
- Body stabilized data hemisphere: This field is stabilized around the central axis of the body and displays iconic representations of key data objects.
- Body surface tool belts: This area is anchored close to the body surface to make use of proprioceptive cues. It contains data handling and processing tools.

5.2 Environment Anchored Information Spaces

Using GPS coordinates, fiducial markers, or indoor tracking systems some information fields are centered around the physical environment. These consist primarily of two classes.

- Object centered information fields: This augmented reality information is anchored around the axes or surfaces of objects.
- Large unit fields such as those for rooms or buildings: Information is centered around world coordinates of large units.
- Global coordinate fields: This field carries broad information such as information making use of GPS and positional information.

6 Mapping Asymmetries in the Semantics and Memory for Space

We report on experiments that seek to explore the relation of egocentric spatial organization with meaning. Namely, these experiments looked into:

- The variation of meaning in relation to location;
- The performance of memory recall in relation to location.

In an augmented reality system, such as the Mobile Infosphere, information and tools can be mapped to and around the body. Nevertheless not all locations are equally valuable based on the properties of human factors such as visual attention, semantic properties, ability to remember the location, and accessibility to the eye and hand.

Our main goal with these experiments is to gain valuable insight on design guidelines that will help us build such a 3D interface.

As different locations in the space around the body appear to have different perceptual, cognitive, semantic, and motor properties, our main hypothesis is that there is an optimal arrangement of tools and information around the body for a user of an augmented reality system such as the Mobile Infosphere.

6.1 Variation of Meaning

Virtual environment spatial organization has strong relations with the real world. As such we should expect to find, in virtual environments, the same meaning and connotations associated with specific locations that we find in the real world. For instance, a face looking down on us may make us feel inferior in some way. On the other hand, if we look down on somebody we tend to feel more powerful. It is also possible that these variations of meaning differ between objects and sentient or anthropomorphic entities.

Our hypothesis in the experiment about to be presented is that different locations around our body have different meanings.

6.2 Method

6.2.1 Participants

Participants were undergraduate student volunteers receiving class credits as a reward for participating. All subjects were right handed.

6.2.2 Apparatus

Virtual reality system: Participants used an immersive VR system powered by an SGI Onyx Reality Engine and equipped with a V8 head-mounted display, Fakespace pinch gloves, and Polhemus tracker.

Body-centered Virtual Environment: In order to measure meaning around our body, a body centered virtual environment was designed. A field of ten positions was fixed to the central axis of the body. These ten positions surround the left, right, and front of the body as well as near and far. These positions were used to rotate the location of objects. Two objects were used, one was a 3D face and head, the other a neutral blue sphere. The environment was used to allow subjects to experience both an abstract object with the shape of a sphere and a sentient looking object with the shape of a face in ten positions around the body. (Figure to be inserted).

Semantic Differential Questions: Four semantic differential questions were used to probe meaning over the different combinations of the 10 positions in space and the two stimuli.

Spatial Ability: Subjects were measured as to spatial ability using the cube comparison scales of the French Test.

Measure of presence: We used the standardized ITC measure of physical presence.

6.2.3 Procedure

After being greeted and completing human subjects forms, participants completed the spatial ability measure. Subjects were then led into the experimental room and assisted in placing the immersive virtual headset,and pinch gloves. As soon as the subject was ready, the test environment was initiated and for each of the 20 manipulations, the subject's answers to the four questions were recorded on a 4 by 20 grid.

Right after the subject answered the last question on the last manipulation, the test environment was stopped and the subject moved to another room. Subjects were administered the presence questionnaire.

6.3 Results

Results to be reported in the final paper.

Acknowledgements. This work was supported by the National Science Foundation Grants IIS 00-82743ITR,and CISE #9911123, the MSU Foundation, and the Ameritech Endowment

References

Azuma, R. T. (1997). A survey of augmented reality. Presence: Teleoperators and virtual environments, 6(5), 355-385.

Barfield, W., & Caudell, T. (2000). Fundamentals of wearable computers and augmented reality. Hillsdale, NJ: Lawrence Erlbaum.

Behringer, R., Klinker, G., & Mizell, D. (1999). Augmented Reality: placing artificial objects in real scenes. A K Peters Ltd.

Biocca, F. (1996). Intelligence Augmentation: The vision inside virtual reality, Cognitive Technology:In Search of a Humane Interface (pp. 59-73).

Biocca, F. (1997). The Cyborg's Dilemma: Progressive Embodiment in Virtual Environments. Journal of Computer-Mediated Communication, 3(2).

Biocca, F., Kim, J., & Choi, Y. (in press). Visual touch in virtual environments: An exploratory study of presence, multimodal interfaces, and cross-modal sensory illusions. Presence.

Biocca, F., & Rolland, J. (1998). Virtual eyes can rearrange your body: Adaptation to visual displacement in see-through, head-mounted displays. Presence: Teleoperators and Virtual Environments, 7(3), 262-277.

Biocca, F., & Rolland, J. (2000). Teleportal face-to-face system, Patent pending (Patent Application (6550-00048; MSU 99-029).

Biocca, F., & Levy, M. (1995). Communication in the age of virtual reality. Hillsdale, NJ: Lawrence Erlbaum Press.

Bryant, D. J. (1992). A spatial representation system in humans. Psycholoquy (http://psycoloquy.92.3.16.space.1.bryant).

Chris Baber, D. H., Lee Cooper, James Knight, Brain Mellor. (1998). Preliminary Investigations into the Human Factors of Wearable Computers. Paper presented at the Proceedings of HCI '98: People and Computers XIII.

Starner, T. (1999). Wearable computers and contextual awareness. Unpublished Dissertation, MIT, Cambridge, MA.

Tversky, B. (1997). Remembering Spaces. In E. Tulving & F. M. Craik (Eds.), Hanbdook of Memory (pp. 1-21). New York: Oxford University Press.

Tversky, B. (1998). Three dimensions of spatial cognition. In M. Conway & S. Gathercole & C. Coroldi (Eds.), Theories of memory (Vol. II, pp. 259-275). London: Taylor & Francis.

Want, R., & Borriello, G. (2000). Survey on information appliances. IEEE Computer Graphics and Applications(May/June), 2-9.

Presence and the Role of Activity Theory in Understanding: How Students Learn in Virtual Learning Environments

Anne Jelfs and Denise Whitelock

University College Northampton, Boughton Green Road, Northampton, NN2 7AL, U.K.
I.E.T, Open University, Walton Hall, Milton Keynes, MK7 6AA, U.K.
anne.jelfs@northampton.ac.uk, d.m.whitelock@open.ac.uk

Abstract. To know where they are in the environment, humans rely on their senses for information. If the environment is artificially generated then it raises the question as to what information is needed to allow humans to know their location in the environment. This paper looks at the role of desktop Virtual Environments as conceptual learning tools in science and the notion of 'Presence' within these types of environments, plus how Activity Theory can help in understanding how students learn in Virtual Learning Environments. Our research looked at how students' understanding of science in Virtual Learning Environments could be enhanced and could potentially increase the pedagogic value of the learning experience. Our findings indicate that Activity Theory and the role of artefacts impact on human interaction, which in turn leads to cognitive change.

1 Introduction

Virtual environments were first developed from 3D cinema in the 1950's and 60's (see Rheingold 1991 for a detailed description of the history). It is only with the recent growth in technology and the emphasis that the aerospace industry has placed on its value, that has seen the growth in this technology for educational purposes. In fact it the success of virtual environments in areas such as pilot training which has led to the development of virtual environments for other educational settings (Kruegar 1982). These types of environments offer the student the opportunity for 'hands on' learning and the opportunity to meet situations where it is either too expensive or dangerous to allow students to try out the roles they want to learn. Virtual environments also have the advantage that they give the student time to reflect on their course of action (Schank 1995). Virtual Learning Environments (VLEs) allow students to return to a previous position and see their current result, which in turn allows the student the opportunity to approach the situation differently a second time. It can also induce a feeling of 'Presence or Telepresence' (see the work of Kalawsky 1993 and Sheriden 1992). To know where they are in the environment, humans rely on their senses for information. If the environment is artificially generated then it raises the question as to what information is needed to allow humans to know their location in the environment and have a more tacit feeling of presence within it. This paper looks at the role of desktop Virtual Environments as conceptual learning tools in Science and the no

M. Beynon, C.L. Nehaniv, and K. Dautenhahn (Eds.): CT 2001, LNAI 2117, pp. 123-129, 2001.

tion of 'Presence' within these types of environments, plus how Activity Theory can help in understanding how students learn in VLEs.

Presence does not refer to one's surroundings as they exist in the physical world, but to the perception of those surroundings (Steuer 1992). Steuer refers to telepresence as the extent to which one feels present in the mediated environment, rather than the immediate physical environment. This means that the dependent measures of virtual reality must all be measures of individual experience providing an obvious means of applying knowledge about perceptual processes and individual differences in determining the nature of virtual reality (Steuer 1992). The more an individual is aware of the interface then the harder it will be to achieve a high level of telepresence. To lessen the awareness of the interface, there needs to be increased level of presence. This sense of Presence can be interpreted in the notion of being physically present in the 'virtual' world when it is presented to the user. It is the feeling of presence which needs further study to investigate the contribution it makes on educational performance and, to attempt to do this, we conducted a study with post-sixteen students at a local comprehensive school. One of the problems posed for students is that conceptual knowledge and understanding of science raises difficulties. Our research looked at how students' understanding of science in Virtual Learning Environments could be enhanced. The motivating effect of such environments has also been discussed by Whitelock and Scanlon (1996) in a collaborative physics simulation. Whitelock et al discuss the role of Presence, representational fidelity and control in designing effective learning environments and our empirical studies (Whitelock & Jelfs 1999a, Whitelock & Jelfs 1999b, and Jelfs & Whitelock 2000) have shown that students learning within virtual environments were affected by their notions of engagement, presence and previous game playing experience.

Science is a basic subject in the National Curriculum in the United Kingdom, and we are interested in how to improve the quality of the teaching and learning experience in this discipline, whether at undergraduate or school level. At the same time there is increased access to computing equipment in schools and universities, and the reduced cost of Information and Communication Technologies (ICTs) to consumers. In fact ICTs are also part of the curriculum in British Education, and one of the targets is to equip students with the basic skills in the use of computers. There has been a growth in interest in young learners and how they interact with Virtual Environments and in fact Grove and Williams (1998) suggest that:

'when young learners use a virtual environment for the first time, their understanding of it will be based on their past experience and knowledge of IT and VR.' p.176

Students past experience is, more likely than not, based on games and Grove and Williams question whether it is beneficial to encourage children to see the experience of virtual learning environments as a game. There is however, the alternative point of view put forward by Schank (1995), that when a student is doing something that is fun, s/he can be learning a great deal without having to notice it.

'The problem is to change the skills to be learned from hand-eye co-ordination tasks to content-based tasks, where one needs to know real information to accomplish one's goal on the computer.' p. 97

According to Subrahmanyam & Greenfield (1996), although software developers have tried to attract girls to video games, they have remained largely a male province. One reason offered is that girls do not like the violent, aggressive games that are commercially available. One factor suggested by Subrahmanyam and Greenfield is that boys and men are willing to use a more 'trial and error' approach to game playing. This is not so apparent in girls and women who appear to prefer more predictable games with rules and patterns. Interest surrounding this notion formed part of our current research, which focuses on Activity Theory.

Activity Theory (Leont'ev 1978) proposes a notion of mediation where all human experience is shaped by the tools and sign systems we use. Founded in the work of Vygotsky (1962) it is based on activity as a psychological explanation where equipment mediates activity connecting the individual to the world of things. This means that for students their experience is also shaped by the tools used, which in educational technology is the computer. Nardi (1997) suggests that activity cannot be understood without understanding the role of artefacts in everyday existence, especially the artefacts integrated into social practice. This notion affects the role of computers in a learning situation, and the need to understand the concepts behind computer interaction. Humans usually use computers as a way to achieve their goals (Kaptelinin 1997), whether this is game playing or educational experiences, not because they wish to interact with a computer.

Together with our interest in the use of audio and Activity Theory we developed a research interest into how girls related to Virtual Learning Environments within the Science discipline and also the discussion on whether female students do in fact have different self-concepts and have a perception of science as a 'male' subject (Lee, 1998).

At the UK Open University (UKOU), one of the world's largest Distance Education providers, the potential of desktop Virtual Environments became a viable teaching method. Although fully immersive systems are not a practical means of tuition for the university, it does not preclude the use of desktop environments as learning environments. In fact Watts (1998) points out that it is only in the past couple of years that there has been a large enough number of powerful PCs on the market to allow the technology to be available to a significant number of users. Previously the computer market was not in a position which would have enabled the Open University to make the use of virtual environments a viable proposition, but now with the growth in computer ownership and the increased specification of many home computers the situation has changed.

The potential for cognitive learning through virtual learning environments (VLEs) was identified by the UK Open University's Science Faculty, where VLEs were used in its Level One (foundation level) course. The VLEs were provided on one of the CD ROMs which students could use in their studies. One of the environments which students used was called the Atlantic Ridge, which required the students to explore, via a submersible, the bottom of the ocean. Students could explore the terrain for geological structures and biological life in seven major locations along the Ridge. Movement through the environment was generated by the computer 'mouse' manipulating the submersible. We used this particular program for the post-sixteen female students as it was easy to use and had been thoroughly tested for its pedagogical value.

2 Methodology

The empirical study involved Year 12 female students, aged between 16 and 17 years. Parental permission was obtained as well as the co-operation of the school, so that we could video record the interactions. At the outset students were asked to complete a pre-test on their cognitive understanding of the geology of the ocean floor, then they were given a series of tasks to complete and a post-test was administered. Students were requested to move around the environment to become acquainted with the location and the movements of the submersible. After a brief session of 'getting to know' the environment, students were presented with a printed list of locations and biological life which they had to look for and comment on. Their conversations were recorded for later transcription. It was felt that this was an important issue for our understanding of 'presence', because as Lombard & Ditton (1997) report, 'presence' is a psychological state that typically is best measured via subject self-report (although observation of involved media users might also be a useful indicator). To research the potential for school students to use VLEs we combined Activity Theory and a qualitative observational method.

To get to know about people's images necessitates attempts to collect verbal accounts of people's phenomenal experience (Richardson 1999). Due to these difficulties we decided to conduct a qualitative study and record both on video and audio the participants' interactions. This would allow for the discovery of different conceptions of reality, and how particular phenomena are perceived (Marton 1988). Therefore a qualitative methodology was used to probe students' understanding of the learning experience in a phenomenological tradition. This included video and audio recording, as well as pre- and post-test questionnaires and a previous computer experience questionnaire.

3 Procedure

Sixteen female participants took part in the study. They were all aged between 16-17 years and in full time education at an all girls comprehensive school. Prior to our visit to the school we asked students to complete a brief questionnaire concerning their previous computing and game playing experience. The students were then allocated to two groups, so that prior experience was controlled as a variable and students were then paired in 'friendship' groups, so that they felt comfortable in the collaborative situation within the unknown domain. The pairs of students were randomly assigned under two different conditions, one group had enhanced audio feedback and the other had minimal audio feedback. They were then required to locate certain aspects of the environment and to discuss with each other the important salient features. All the tapes were transcribed for analysis.

4 Findings

All of the students had previously used a computer, either a PC or a Macintosh. The students were asked to rate their ability to use a computer and their game playing skills. As can be seen from the table below, the students rated themselves as having fairly low game playing abilities. This substantiates to some extent the suggestions made by Subrahmanyam and Greenfield, that game playing does not attract females in the same ways as males. However, of the sixteen students 6 were studying Advanced level computing and were therefore very experienced computer users.

	Computer Ability	Game playing Ability
Expert	0	0
Advanced	5	1
Competent	7	6
Novice	4	8
Never used	0	1

From analysis of the transcripts we have found that students collaborated and supported each other well in these environments, enabling the co-construction of knowledge. However, when one student was seen as the more able person to manipulate the mouse, then that person did take control. This did not mean that the second student did not contribute, we found that they were willing to chat about the environment without necessarily taking control of the mouse. Sometimes when one student could not skilfully move the submersible, then the other person took over. A high degree of 'Presence' can afford maximum interaction with a program, but does not always lead to critical reflection. Some students when deeply immersed relied on their partners to respond to the needs of the activity.

Students completed a pre- and post-test on their understanding of the North Atlantic Ridge. From the data collected we looked at the cognitive learning that had taken place and found that students did learn substantial levels of details about the ocean floor. Comparisons between the pre- and post-tests found that 50% of the students had an increased understanding of the marine life which exists at the Atlantic Ridge, plus a further 50% had increased their general knowledge about the environment. 25% had also increased their understanding of the geology of the Ridge. Further research needs to be conducted on the comparison between learning in Virtual Leaning Environments and learning through traditional methods such as face-to-face tuition and text. More importantly is the finding that students report that audio feedback provides the 'feeling of presence' more that any other parameter. It is the audio feedback that provides aid to navigation, tells the user they are in a dynamic environment and also provides an emotional response.

5 Conclusions

Our findings so far indicate that prior experience of game playing and previous interactions with computers affect the notion of presence and engagement. Students with higher levels of computer and game playing ability used the computer to explore the terrain, whereas the students who rated themselves as lower in ability concentrated more on the content of the locations. We have depicted our conclusions in the following way.

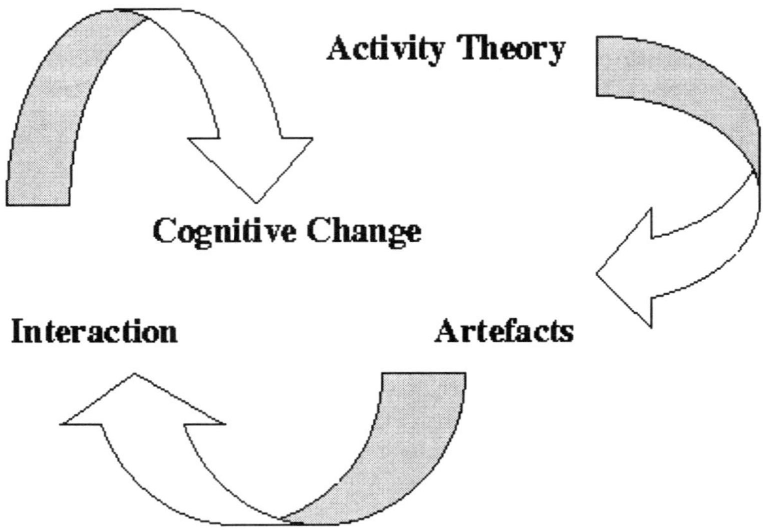

The spiral illustrates our conclusions in the use of Activity Theory to understand the role of artefacts and how artefacts impact on human interaction, which in turn leads to cognitive change. This confirms some of the ideas of the Human-Computer Interaction proponents of Activity Theory who suggest that understanding the histories of both artefacts and users can throw light on the meaningful use of cognitive tools. We are continuing our research in this area and are currently reviewing the impact that audio has on student notions of Presence within Virtual Learning Environments.

References

1. Grove, J. & Williams, N. (1998) *IT for Learning Enhancement.* M. Monteith, Swets & Zeitlinger, Lisse: 176-183.
2. Jelfs, A. & Whitelock, D. (2000) The notion of presence in virtual learning environments: what makes the environment 'real'. *British Journal of Educational Technology* Vol 31 (2) 145-152

3. Kalawsky, R. (1993) *The Science of Virtual Reality and Virtual Environments.* Addison-Wesley.
4. Kaptelinin, V. (1997) Computer-Mediated Activity: Functional Organs in Social and Developmental Contexts. pp 45-68 in *Context & Consciousness. Activity Theory and Human-Computer Interaction*, ed. Nardi, (2nd Ed) MIT Press.
5. Kruegar, M. (1982) *Artificial Reality.* New York, Addison Wesley.
6. Lombard, M. & Ditton, T. (1997) *At the Heart of It All: Tech Concept of Presence Journal of Computer Mediated Communication* 3 (2) September.
7. Lee, J.D. (1998) 'Which Kids can "Become" Scientists? Effects of Gender, Self-Concepts, and Perceptions of Scientists', *Social Psychology Quarterly* Vol. 61 No. 3, pp. 199-219.
8. Leont'ev, A.N. (1978) *Activity, Consciousness & Personality.* Translated by M.T. Hall. Prentice-Hall.
9. Marton, F. (1988) Phenomenography: A research approach to investigating different understandings of reality. *Qualitative Research in Education. Focus and Methods.* R. Sherman and R. Webb. London, The Falmer Press: 141-161.
10. Nardi, B. ed (1997) *Context & Consciousness. Activity Theory and Human-Computer Interaction.* (2nd Edition) MIT Press.
11. Rheingold, H. (1991) *Virtual Reality.* London, Simon & Schuster.
12. Richardson, J. T. E. (1999) *Imagery.* Hove, East Sussex, UK, Psychology Press.
13. Schank, R. and Cleary, C. (1995) *Engines for Education.* Hove, UK, Lawrence Erlbaum Associates.
14. Sheridan, T.B. (1992) 'Musings on Telepresence and Virtual *Presence' Presence. Teleoperators and Virtual Environments* Vol. 1 No. 2 MIT Press.
15. Steuer, J. (1992). Defining Virtual Reality: Dimensions Determining Telepresence. *Journal of Communication* 42 (4): 73-93.
16. Subrahmanyam, K. & Greenfield, P.M. (1996) Gender and Computer Games, in *From Barbie to Mortal Kombat,* J. Cassell & H. Jenkins, eds. Cambridge, Mass. MIT Press pp. 46-71.
17. Vygotsky, L. (1962) *Thought and Language.* Translated from the Russian and edited by E. Hamsman and G. Vankan, Cambridge, Mass. MIT Press.
18. Whitelock D. & Scanlon E. (1996) Motivation, Media and Motion: Reviewing a Computer Supported Collaborative Learning Experience. *Artificial Intelligence in Education Conference*, Lisbon, October, 1996, pp.276-283.
19. Whitelock, D. & Jelfs, A. (1999a) Understanding the Role of Presence in Virtual Learning Environments. *2nd International Workshop on Presence*, Exeter, England, 6-7th April .
20. Whitelock, D. & Jelfs, A. (1999b) Examining the Role of Presence in Virtual Learning Environments. *European Conference for Research on Learning and Instruction 99,* Goteborg, Sweden, 25th - 27th August.

Experiment as an Instrument of Innovation: Experience and Embodied Thought

David C. Gooding

Science Studies Centre, Department of Psychology,
University of Bath, Bath BA2 7AY, UK

Abstract. Traditional dualist assumptions about how humans acquire and represent knowledge of the world support theories that deal in second- or third-order representations of their subject matter, such as images, diagrams, equations and theories. These accounts ignore the processes whereby such representations are achieved. I provide examples from the work of Michael Faraday which show that abstraction, or discerning patterns and structure in phenomenological chaos depends on the development observational techniques, which I characterize as cognitive technologies.. The examples also illustrate the importance of human agency to the process of making representations of the world which enable scientists to understand and think about it.

1 Introduction

It is widely supposed that thought precedes and anticipates action, that conjectures precede experiments designed to test them and that science provides a rational understanding of nature, which technologies merely embody. Thus, in science as in other human activities, tools and technologies are consequences of creative, intentional, purposive thought but have no role in enabling or shaping thought. The pre-eminence of mind is also expressed by the priority assigned to representations and reasoning processes in cognitive psychology and, similarly, to ideas and arguments in the history of science, with its corollary that technologies emerge from the application of scientifically discerned principles. Until recently the design of intelligent systems also presupposed that functional competence could be achieved by providing a system with abstract representations of objects and processes in an environment and with the means of reasoning with and about these.

This view implies that systematic, rational thought is or can be separate from the world that it seeks to understand, manipulate or control. It is not easy to show how profoundly mistaken this view is. One source of difficulty is that the priority of mental representations over physical manipulations presupposes a mind-body dualism that has an established philosophical pedigree. This dualism is associated with Descartes, whose optical ray-diagrams of eyeballs gave us some of the earliest explanations of how images of the material world are re-created in the theatre of the mind. There they can be explored and manipulated in thought, by methods perfected and popularised

M. Beynon, C.L. Nehaniv, and K. Dautenhahn (Eds.): CT 2001, LNAI 2117, pp. 130-140, 2001.

by Galileo, who embedded them in thought-experimental narratives. In the work of Galileo, Newton and Einstein it appeared that thought could transcend the limitations of embodied, sensory experience. Experiments and mere technological extensions to our senses could only follow the lead of ratiocination. The demonstrative force of thought experiments is taken by some to show that Plato was right: the human mind can directly intuit the nature of reality [1], [2]. If this view of knowledge and how we acquire it is correct then, as Dreyfus [3], Edelman [4] and others have argued, the anticipatory, instructional approach to artificial intelligence should have been far more successful than it has been.

This constraint is illustrated by an important feature of the history of science. The sciences frequently run up against the limitations of a way of representing aspects of the world -- from material objects such as fundamental particles to abstract entities such as numbers or space and time. One of the most profound changes in our ability to describe aspects of experience has involved developing new conceptions of what it is possible to represent. Examples include the invention of Euclidean geometry as a way of reasoning consistently about spatially extended objects; Cartesian co-ordinates which linked geometrical and algebraic reasoning; the calculus, which provided a way of reasoning about continuously changing variables; the theory of probability as a way of calculating rather than estimating the likelihood of events, and diagrammatic techniques in quantum physics that represent properties of mathematical descriptions which themselves already transcend the phenomenological or visual approach of classical physics. Computer-based simulation methods may turn out to be a similar representational turning point for the sciences [5], [6]. An important point about these developments is that they are not merely ways of describing. Unlike sense-extending devices such as microscopes, telescopes or cosmic ray detectors, each enabled a new way of thinking about a particular domain.

The need for a new way of thinking lay outside the domain and could not be met from existing intellectual resources and practices. Developments such as these show that anticipation, the application of prior conceptions is limited. Yet, having dismissed a rationalist view of thought we should not embrace its empiricist counterpart: it is just as naive to suppose that fundamentally new conceptions are given in experience.

This raises an important question:

> How are new representations invented and justified, and how do they become part of an effective system of communication and action?

Dualist and representationalist philosophies fail to recognise the role of human agency as the use of cognitive technologies in the construction of meaning, so they cannot address this question. This provides another reason why traditional views of the priority and self-sufficiency of thought must be wrong [7].

In this chapter I am concerned with this second difficulty with dualist assumptions about how humans acquire and represent knowledge of the world. Such theories deal in second- or third-order representations of their subject matter such as images, diagrams, equations and theories, without understanding how such representations are achieved. My explanation of this difficulty turns on several points: first, once they have been introduced and made a part of our communicative practices, the representational capability of an image, symbol or other token of meaning appears self-evident, in rather the same way that, say, a geometrical demonstration or the conclusion of a thought-experiment do. Second: this self-evidence is not given by intuition or some direct correspondence of tokens to objects in the world, it is accomplished by human action. This action usually involves devices and techniques as well as words, images or other tokens of meaning. Finally -- and paradoxically -- a representation succeeds as an instrument for thinking and communicating only insofar as its constructed-ness is concealed or made transparent.

We are still some way from achieving such transparency for information and computing technologies [8]. This is partly due to a lack of understanding of ways in which cognition and communication depend on embodied interactions with objects. It may help, therefore, to approach the problem with a more prosaic example of the construction of new representational techniques to communicate new experience. The example also illustrates how important agency is to the process of making representations of the world which enable us to understand and think about it. I shall focus on the constitutive role of human agency. However it is important to note that the techniques and instruments of creative thinking can acquire an agency of their own. Where created as instruments of communication or dissemination of new experience, it is essential that they become autonomous.

2 Making New Meanings

It is fast becoming a commonplace that thought is socially and culturally situated [9], [10] and that its efficacy must be explained by reference to the fact that it is embodied [11], [12], [13]. Historians and others now portray scientific work as a large array of loosely connected cultures, each with its sources of expertise, linguistic and technological practices and technologies [14], [15], [16]. They emphasise the local and self-vindicating character of scientific results [17], the obstacles to communicating the techniques and knowledge needed to replicate new results [18], and the considerable effort required to translate these into context-independent, general or universal facts of nature [19].

To answer the question raised in the previous section we need to consider how scientists accomplish the move from the particular, local and personal context of the laboratory or research group to the public domain of shared representations, methods and standards of evaluation. Accounts of this process vary according to the scientific field studied and the disciplinary emphasis of the researcher. Psychologists, historians, and sociologists emphasise different aspects but I believe that all would agree on the following features: the need for mastery or expertise in a particular domain; sharing or combining expertise from different domains; the interaction or integration of distinct and sometimes incommensurable methods of representation; the importance of contingencies or local factors and of personal factors which vary from one research site to another; the close interaction of investigative and communicative techniques with the content of what is experienced or disclosed; that such techniques and local knowledge always remain open to scrutiny and criticism (especially when replication fails, as it often does); the construction of reliable ways of reproducing findings (such as well-defined mathematical techniques or tests, 'black-boxed' or off-the-shelf technologies and so on); the institutionalisation of procedures for the evaluation and dissemination of expertise (and since the early 20th century, of the organisation and management of these processes on an industrial and global scale) and, finally, the importance of agency to the construction and communication of facts, concepts and other findings. As I remarked earlier, this agency is both human and technological. In short, meaning or significance is closely connected to context, and cannot finally be divorced from it. In the terms of a familiar philosophical distinction, justification cannot be made independent of discovery.

Many of these points can be illustrated by examples from the work of one of the world's most productive scientists, Michael Faraday. We tend to rank Faraday highly because of the extraordinary number of his discoveries and the many scientific fields they contributed to [20], yet one of Faraday's most important achievements was to develop, early on, a style of investigation that lent itself to accessible communication and demonstration of his findings. The novel phenomena he produced became part of a substantial challenge to the prevailing views of electricity and magnetism, based on Newtonian-Laplacian mathematical methods and concepts of forces as quasi-material substances. This challenge depended in part on the close integration of his methods of investigation and methods of representation, both of which drew on images, techniques and devices that were widely used at the time [21]. In focusing on cognitive aspects of his work, it is important to remember that Faraday was just as closely acquainted with the social world of ideas, opinions, methods and material resources of his time as he was with the details of the physical world that he investigated.

2.1 Integrating the Modes of Perception

The constructive role of graphic representation is a particularly important feature of the cognitive dimension of scientific discovery. Understanding how scientists use objects to invent and enhance the representational capability of words and visual images opens up the complexity of their investigative practices. I shall consider a typical example of Faraday's use of images, with a view to characterising the changing use of images in the construction of new scientific objects.

Elsewhere I have labelled interpretative images and their associated linguistic framework as 'construals' [22]. This term denotes proto-interpretative representations which combine images and words as provisional or tentative interpretations of novel experience. The important point about this experience is that it is being created it through the interaction of visual, tactile, sensorimotor and auditory modes of perception together with existing interpretative concepts including mental images. The power of these construals or word-image hybrids is that that they integrate the different types of knowledge and experience. The ability to integrate information from various sources is crucial to scientific inference [23]. Faraday's detailed records of his laboratory work show how visualization works in conjunction with sensorimotor awareness (proprioception or kinaesthetic awareness) to produce representations whose cognitive (generative) and social (communicative) functions are intextricably linked. Mental models having such integrative power could not have been developed purely in the mind's eye, that is by anticipation, ratiocination or from passive, visual perception.

Simon's distinction between internal and external representations is largely taken for granted in artificial intelligence research [24]. To avoid isolating visual representations either from other modes of perception or from other perceivers it wise to suspend the 'internal-external' distinction because it presupposes a dualist (mind-world) view which a better understanding of the functions of visual imagery will render obsolete. Simon emphased the importance for effective external representations for successful reasoning and problem solving because he recognised the importance of the environment of a system. But the environment is, increasingly populated by artefacts which function as records and as guides for reasoning procedures that are too complex to conduct solely with internal or mental representations. In this way we are continually enhancing the capacity of our environment for creative thought, by adding new cognitive technologies.

Representations must be 'externalised' if they are to communicate well enough to enable discussion and criticism. Just as a picture or a data-plot may be worth a thousand words (or data-points), so a few words or symbols may eventually come to express many thousand visual, auditory or tactile experiences. 'Externalized' representations can take many forms: verbal accounts, drawings, apparatus, photographs, experimental narratives, databases -- all feature in the whole range of knowledge-making processes. However, if we consider representations only as *external* or as *end products* there is no hint of their constructive, integrative and cognitive role as *construals*.

2.2 Faraday's Cognitive Technologies

To illustrate these points I turn to Faraday's bench-top explorations of electromagnetism. Like his mentor Humphry Davy, Faraday construed many of his experiments as showing a temporal slice -- a 'snapshot' -- of the effect of some more complex but hidden, physical process. In response to Oersted's discovery that a current-carrying wire has magnetic properties Faraday and Davy had by September of 1821 developed experimental methods of integrating discrete experimental events (or rather, integrating the images depicting them). Electrical and magnetic effects are mixed in a way that the eye simply cannot see. So, Davy and Faraday combined discrete images obtained over time into a single geometrical structure; they also created a physical structure of sensors with which to record the effects of a single event at different points of space [25]. A typical procedure involved carefully positioning one or more needles in the region of a wire, connecting the circuit to a battery and observing the effect on the needle(s). Similarly, continuous exploration of the space around the wire would produce many discrete observations of needle positions. Davy and Faraday combined these results into a single model, a three-dimensional representation of the magnetic effects of the current. A structure of needles arranged in a spiral around the wire and examined after discharging a current through it, gave a three-dimensional magnetic 'snapshot' of the magnetizing effect of the current. Another setup, a horizontal disc with needles arranged around its perimeter, emerged from a set of temporally distinct observations, which this setup integrates into a single spatial array.

These objects are complexes of material things, active manipulations, effects and proto-interpretations or construals of the outcomes. Davy and Faraday devise and perform them in order to understand the observable, two-dimensional patterns of magnetised needles and iron filings as spatial or temporal sections of processes whose complexity or speed placed them beyond the reach of unaided observation. From these map-like images they hoped to develop structures that could explain effects at every place of action. They would then apply the structural model to the interpretation of other phenomena. Their construals pass quickly into interpretations that seem clear, even self-evident. For example, Davy concluded from an experiment in which a battery was discharged through a vertical wire passing through a cardboard disk on which either steel needles or iron filings had been arranged: "It was perfectly evident ... that as many polar arrangements may be formed as chords can be drawn in circles surrounding the wire; and so far these phenomena agree with [Wollaston's] idea of revolving magnetism ... " [26]. Here Davy articulates verbally a visual inference made from a two-dimensional arrangement to a four-dimensional process (a 'revolving' structure of 'magnetism').

Davy and Faraday's structures and patterns explained nothing in themselves. Once they had been identified as features of a process, however, they could suggest and guide further exploration of structures hidden from view. These should manifest themselves as other (new) patterns. An important example is Davy's explanation of phenomena observed in an experiment carried out in May of 1821. Assisted by Faraday, he passed a current through a vacuum to produce a luminous glow discharge. Davy reported that when "a powerful magnet [was] presented to this [luminous] arc or column, having its pole at a very acute angle to it, the arc, or column, was attracted or repelled with a rotatory motion, or made to revolve by placing the poles in different positions, according to the same law ... as described in my last paper." [27]. Davy and Faraday construed this process in terms of hidden, real-time (4-D) processes involving 3-D structures.

Faraday later developed this approach with other devices to 'extend' his ability to analyse high frequency processes. Where the discrimination Faraday sought exceeded the capacity of practised manipulation before unaided senses, he made devices and procedures to extend his sensory and discriminatory powers, for example, to 'slow down' the high frequency processes that might produce an appearance of structure or of motion. This is an example of inferring a structure and process from the features of a pattern and from its behaviour (under manipulation), where the process is rendered sensible by a sense-extending device. The spinning, toothed discs used in his work on optical perception are an example of the extension of visual perception by apparatus.

A related method was to reproduce patterned appearances by means of mechanical simulations. Where he could simulate some aspect of a natural phenomenon by a high speed mechanical process, Faraday took this to be a fair indication as to the nature of that process. Typical simulations were the toothed wooden wheels whose rotation could reproduce apparent rotation of the apparent discs of aquatic animalcules (observed by Leeuwenhoek in 1702, but shown by Faraday in 1831 to be progressive undulations in their *cilia*) and his simulation of the appearance of the surface of a vibrating fluid using a perforated silver plate [28].

3 Visualization as a Process

It is clear that effective visualization requires an intimate knowledge of a particular domain. Faraday's records of his work suggest a more precise account of the way in which he developed the representational power of a construal into an interpretation or model.

3.1 Dimensional Enhancement and Reduction

There is first a *reduction* of complex phenomena to an abstract image (usually a pattern or set of patterns, such as a magnetically induced distribution of iron filings). The image is then *enhanced* by 'adding' dimensions, first to create a three dimensional structure which can be drawn and then -- where a causal explanation is sought -- further enhancement by constructing a real-time, four dimensional process model. I call the progression from two- to four dimensions, *dimensional enhancement*. The process is more complex than this summary suggests, since the 2-D images with which the process begins are construals or partial abstractions, dimensionally-reduced representations of complex real-time experience. *Dimensional reduction* is always necessary when recording real world processes as, say sketches in a notebook. Dimensional enhancement therefore depends on a prior abstraction or *reduction*.

A second feature is that in all cases the initial enhancement is followed by a consolidating move in which the originating 2-D image(s) and new ones are derived from the 4-D process model. *Consolidation* involves reducing the complex images from four-dimensions to two. Dimensional reduction is therefore used in both the construction and the consolidation stages. In the latter reduction enables dissemination (say, of predictions or observed results) in the form of printed diagrams. A search for new effects predicted by the model might typically involve 2-D patterns rather than full-blown 4-D processes because new observational techniques would have to be devised in order to observe these. The consolidation stage is analogous to prediction and retrodiction as inferences on sentential representations. Thus it resembles a deduction, albeit one accomplished through manipulating objects that are neither propositions nor symbolic representations.

These features of the process highlight three different roles for images, each corresponding to a different stage of the process of constructing a new representation and integrating it into an argument:

(i) they may be instrumental in generating new representations or in extending the use of existing ones.

(ii) they symbolize an integrated model of a process that involves many more variables than the eye or the mind could otherwise readily comprehend. In these two cases visualization is essential to the construction of interpretative and analytical concepts.

(iii) they enable empirical support for the theory embodied by the model, usually through the dissemination of images in 2-dimensional form. Here the visualization of observations or data assists a verbal argument that may have been developed by non-visual means.

3.2 From Cognitive Technology to Material Artefact

The movement between patterns, structures and processes is characteristic of Faraday's experimental reasoning. It is clearly at work in his record of the day's work that led to the 'rotation apparatus' which was the first electric motor. By September of 1821 when he returned to the examination of wire-needle interactions, Faraday had become skilled in making these apparatus-based spatio-temporal transformations. The important point is that to start with there is no clear distinction between the 'internal' or 'mental' representations and the 'external' or 'material' devices used to generate, refine and record these in his laboratory notes. The distinction between ideas and mental images on the one hand and material artefacts that embody them, on the other emerges only after he has produced a working prototype for the rotation motor. By examining the micro-structure of exploratory work we can show that representation involves dimensional reduction (whereby selected features are represented visually, as patterns), followed by enhancements leading to new, 3-dimensional configurations, reductions that generate predictions about new phenomena, and consolidation which establishes the derived structures as plausible explanations or realisations of the observed patterns. In this case visualization produced a new material artefact (an electric motor).

I have given a detailed account of this work elsewhere [7] so I shall include only the main points here. Faraday began his independent investigation of electromagnetic experiments by Oersted, Davy, Biot and others during the summer of 1821. In these experiments he developed the image of a circle, created by accumulating small motions of a magnetized needle near a current. His notes for experiments of 3rd and 4th September 1821 begin with a re-examination of the original magnet-wire interactions. This work used the circular image as an heuristic for subsequent exploration with more complex experimental setups.

Faraday first repeated the observation of the attractive and repulsive effects of the current, paying particular attention to the effect of position. Like Davy and W. H. Wollaston, he believed that the magnetism in the region of the wire was somehow structured. This appears to have been the main focus of this investigation. At this stage all he had to go on were the magnetization patterns he and Davy had produced earlier. The first set of sketches records a more detailed examination of the space around the current. In these sketches, arrows relate apparent needle motions to needle positions relative to the wire. His next drawing superimposes several sets of such observations into a single pair of diagrams. This actually *reduces* observed real-world process to a two-dimensional 'map'. It is important to note that Faraday's next move is to manipulate objects in this map. His next figure shows the same set of accumulated observations, but rotated through 90°. In practice it would have been very difficult to observe even one instance of this.

Thus far we have a complex set of observations reduced to a 2-D map, which is then manipulated (by mental rotation) to create a 3-D mental model of a whole set of needle-wire interactions. From this, Faraday used the 3-D representation to infer the possibility of motion in circles, and from there he moved on to design a working electric motor. This is achieved via another mental transformation. Instead of imagining a set of needle positions, Faraday uses a stationary needle as a kind of reference frame and imagines how the needle would tend to move if the wire were moved so as to occupy each of the positions shown in a diagram where the wire is vertical and indicated in section, i.e. at right angles to the page. These positions therefore fall on the circumference of a drawn circle. He infers that the needle would tend to move in a circle.

Faraday constructed an image of circular motion by the process I have labelled *dimensional enhancement*. He then used the circle as an heuristic for further constructive work which realized continuous motion of a wire as a phenomenon in the world. In the next stage of investigation the problem was to realize the right set of physical constraints. This was far from straightforward.

Faraday made several attempts before he succeeded in constructing a device that could realize circular motion as a phenomenon in the world. His record shows that one of his first set ups consisted of a fixed magnet and a wire suspended by floatation so as to be capable of motion. But this produced lateral or 'side to side' motion of the wire, reproduced a subset of the original, bewildering chaos of needle-wire motions. Faraday varied this set up, bending the wire into a crank and attempting to 'push' it by repositioning the magnet. Here again, his technique was to accumulate many discrete *actions* into a single process. The results are summarized by two images of circles surrounded by letters indicating the positions of north and south poles. (He noted that the magnet was held at right angles to the wire).

Motion of this wire (or 'crank') involved human intervention, but it showed him what had been wrong with his earlier setup. The magnet should be parallel to the current, not perpendicular to it. The wire had to be constrained yet free to move around, but Faraday now realized that wire and magnet could be aligned along a common axis along which the current could pass. This showed him where to position the magnet. Motion of the wire around the stationary pole would then follow.

This is the third major change to the configuration of his apparatus as recorded in his notes. He recorded this inference in words and sketched its outcome in the margin of his notebook. It is depicted in his notes by two circles, each with a single pole indicated at its centre. Faraday then made a schematic drawing of this configuration. This realized the hypothetical motions derived earlier, while the motions it produced could now be re-construed as tendencies to continuous motion, constrained by the physical setup.

4 Conclusion

Abstraction, or discerning pattern or structure in phenomenological chaos is bound up with observational techniques. In the examples given here, these techniques are defined in terms of an implied frame of operation. Together these make up a cognitive technology capable of generating new construals, interpretations, models and artefacts. We have been able to show how this frame places geometrical description of experimental manipulations into the same representational space as mental manipulations. The manipulation of 2-D visualizations generated general descriptions of temporal instants of processes (frozen or 'fixed' as patterns or structures). It may appear that Faraday's descriptions are 'direct' observations of phenomena, but from a cognitive standpoint there is no difference: both depend on the invention of manipulative and representational techniques which can share a common, representational space.

Faraday's records of his investigations show that during this work there is no clear distinction between the 'internal' or 'mental' representations and the 'external' or 'material' devices used to generate, refine and record these in his laboratory notes. Such distinctions between percepts and objects, images and their referents or between the conceptual and material domains emerge only as Faraday clarifies through his material and mental manipulations where the boundaries are to be drawn between phenomena, representations of them, and material artefacts which realize and reproduce those phenomena.

To focus on the visual mode of perception is not to suggest that this can work in isolation either from other modes of perception or from other persons as sources of ideas and of other experiences. The need to highlight the visual is a symptom of previous neglect, engendered by a preference for textual and numerical modes of representation and argumentation in our accounts of science.

It is worth noting that the contexts in which visual representations are used have also changed. The history of science shows that sensory modes of observing nature have been systematically displaced by (or delegated to) technologies that extend a particular sense (as microscopes and telescopes do) or that render a selected feature into a form accessible to one or another of our senses (as the thermometer, calorimeter, bubble chamber, X-ray imaging devices and the Geiger counter do). Increased technological and social complexity has drawn observation and experiment into 'labour processes' that also shape perception [6], [29]. The fact that such technologies successfully reduce complex manipulative processes to simple visual images suggests, misleadingly, their independence from the many techniques and types of knowledge needed to produce them in the first place. Nevertheless, observation continues to be an active process whose primary aim is to create inter-personal experience on the based on shared cognitive technologies.

References

1. Brown, J. R., *The Laboratory of the Mind*, Routledge, London (1991).
2. Penrose, R. *The Emperor's New Mind,* Vintage Science, London (1989).
3. Dreyfus, H. What Computers Still Can't Do, MIT Press, Cambridge, MA. (1997).
4. Edelman, G. *Bright Air, Brilliant Fire*, Penguin Science, London (1992).
5. Casti, J. *Would-be Worlds*, J. Wiley and Sons, New York (1997).
6. Galison, P. *Image and Logic*, Chicago University Press, Chicago (1997).
7. Gooding, D. *Experiment and the Making of Meaning*, Kluwer Academic, Dordrecht and Boston (1990).
8. Norman, D. *The Invisible Computer*, MIT Press, Cambridge MA. (1998).
9. Suchman, L. *Plans and Situated Actions*, Cambridge University Press, Cambridge and New York (1987).
10. Classen, C. *Worlds of Sense*. Routledge, London (1993).
11. Clark, A. *Being There*. MIT Press, Cambridge, MA. (1997).
12. Johnson, M. *The Body in the Mind*, Chicago University Press, Chicago (1987).
13. Lakoff, G. and Johnson, M. *Philosophy in the Flesh*. Basic Books, New York (1999).
14. Gooding, D., Pinch, T. and Schaffer, S., *The Uses of Experiment*, Cambridge University Press, Cambridge and New York (1989)..
15. Pickering, A., ed., *Science as Practice and Culture*, Chicago University Press, Chicago (1992).
16. Galison, P. and Stump, eds., *The Disunity of Science*, Stanford University Press, Stanford (1996).
17. Hacking, I. *The Self-Vindication of the Laboratory Sciences*. In: Pickering, A., ed., Science as Practice and Culture. Chicago University Press, Chicago. (1992), 29-64.
18. Collins, H. *Changing Order*. Sage Publications, London and Beverly Hills (1995).
19. Latour, B. *Science in Action*. Harvard University Press, Cambridge, MA. (1987).
20. Cantor, G., Gooding, D. and James, F. *Faraday*, Humanities Press, Atlantic Highlands, N. J. (1996).
21. Gooding, D. 'Magnetic Curves' and the Magnetic Field: Experiment and Representation in the History of a Theory. In: Gooding, D. *et. al.*, eds., *The Uses of Experiment*, Cambridge University Press, Cambridge and New York (1989), 183-223.
22. Reference [7], chapters 1 and 2.
23. Gooding, D. Creative Rationality. Towards an Abductive Model of Scientific Change. *Philosophica*, 58 (1996) 73-101.
24. Simon, H. A. *The Sciences of the Artificial*, MIT Press, Cambridge, MA. (1981), p. 153.
25. See: reference [8], chapter 2.
26. See reference [25].
27. See reference [25].
28. Tweney, R., Stopping Time, *Physis*, 20 (1992) 149-164.
29. Rasmussen, N. *Picture Control*, Stanford University Press, Stanford (1997).

Can We Afford It? Issues in Designing Transparent Technologies

John Halloran

Interact Lab, School of Cognitive and Computing Sciences,
University of Sussex, Falmer, Brighton, East Sussex BN1 9QH, UK
johnhall@cogs.susx.ac.uk

Abstract. Situated Action claims that action is guided by an environment featuring objects, including technologies, whose use is transparent to us. Work on affordances - perceived properties of objects which, in the context of a given course of action, tell us directly what those objects are for - helps explain this. It is argued that an affordance helps make a technology transparent because it is a function which has become enculturated; that is, it has overlearned, shared significance for action. However, sometimes technologies feature functions beyond what we can easily call affordances, which are less culturally familiar to us. This paper gives examples of such functions; considers how technology functions, including affordances, become enculturated; and looks at implications for designing transparent technologies.

1 Introduction

One idea the phrase 'cognitive technology' suggests is that in human action, a technology can do the thinking for us. For example, in crossing a busy road, we act on a green light without needing to think it through: the light tells us what to do. A traffic light, then, is an example of a 'transparent technology': one whose use in action is clear to us. How might we understand how technologies become transparent, in such a way that we can derive principles for designing transparent technologies?

Norman (1988) introduces a distinction between 'knowledge in the head' and 'knowledge in the world'. This distinction reflects that certain of the knowledge relating to action is internal – in the head. For example, we may have explicit knowledge about how to cross a road. However, other knowledge is reified, projected onto objects out there in the world: traffic lights. When knowledge is in the world, we need to do less internal cognitive work, because cognition is 'offloaded' onto objects external to us (Scaife and Rogers, 1996). Thus, technologies can come to act as repositories of praxis whose significance for action is transparent to us.

How do people come to able to interpret the meaning of technologies, such as traffic lights, in such a way that the need for conscious cognition in action is obviated? Norman's discussion suggests there might be an interaction between internal cognition and technologies in the world which could be important to analyse. However, Situated Action (SA) (Suchman, 1987) claims that the relevant resources for action tend to be in the world rather than the head, carrying information which we can perceive and act on immediately, in such a way that internal cognition is not a critical

M. Beynon, C.L. Nehaniv, and K. Dautenhahn (Eds.): CT 2001, LNAI 2117, pp. 141-148, 2001.

level of analysis. To design transparent technologies, the implication is that rather than asking questions about internal cognition, we should concentrate on the properties of technologies which allow them to communicate their use to us directly.

How might we characterise the properties of technologies that show us how they can be used? One approach might be to invoke the concept of *affordances*. For Gibson (1977; 1979), an affordance is a biological link between environmental information and human action involving no cognition other than that involved in 'direct perception'. Gibson gives examples including walls, whose solidity means we do not walk into them; logs, which because of their shape and height allow us to sit; and steps, which facilitate ascent. If we could 'afford' technologies – build affordances into them – we might be able to make them transparent in a similar way. Norman (*op. cit.*) addresses this issue, and suggests design guidelines for technology affordances: for example, that keys should fit locks in one way only ('physical constraining'); that switches should be placed on an array to reflect the position of the device - for example lights or hotplates - they are controlling ('logical constraining'); and so on.

However, some technologies are *not* transparent to us, and this may be for reasons other than the absence of appropriate affordances for action. Suchman's work (*op. cit.*) shows how two people struggle to use a photocopier. This example shows that technologies can exhibit dynamic, sequenced behaviour; and also communicate with us by issuing instructions. The reason for the users' difficulties appears to be that they are not familiar with these functions. This suggests that in designing transparent technologies, we have to consider a range of action-facilitating functions including, but going beyond, what we can call affordances; and how they can become familiar.

This raises questions about how far we might we be able to 'afford' technologies – that is, make them transparent by building in appropriate affordances for the actions they are designed to support. A Gibsonian affordance is biological, and so should be fixed in human behaviour, transcending culture. But Norman argues that the sorts of affordances he is discussing, although they appear as immediate as Gibson's, are realized against a background of shared cultural knowledge. This suggests that Norman's concept of affordance really names an *enculturated function*. This phrase is intended to capture the idea that rather than there being a biological link between information and action, the meaning of the properties of a technology for action has been learned through experience of activity in a culture. In this paper, ideas from activity theory are used to consider how a range of enculturated functions, including affordances, emerge out of interaction between culture, internal cognition ('knowledge in the head') and technologies ('knowledge in the world').

2 Simple Technologies: Affording Staplers

What makes a technology transparent? Possibly, that it has the right kinds of affordance for the action we want to carry out. For example, Fig. 1 shows a standard stapler. With this stapler, there is only one place the paper can be put. Therefore, this stapler is transparent on how the paper is introduced to it; its affordance in terms of placing the paper is unambiguous. Perceived properties of this technology, then, enable us to use it.

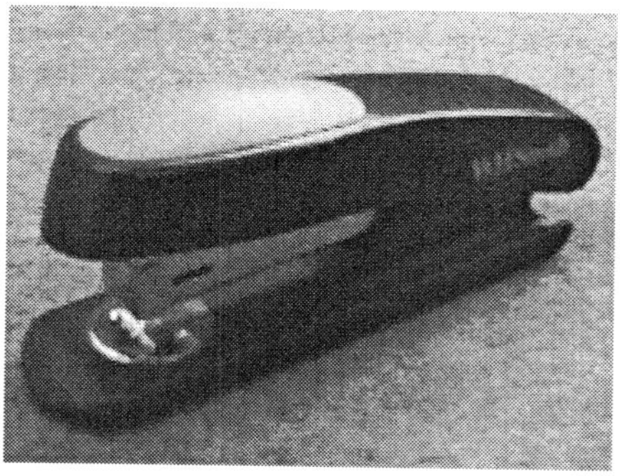

Fig. 1. Standard stapler

Fig. 2. Crocodile stapler

Figure 2 shows a 'crocodile' stapler. Unlike the standard stapler, this has two openings. The staple tray hangs to form the crocodile's 'tongue', and we can introduce paper above or below it. Where should we put the paper? This is an example of ambiguous design. It can be 'cured' through physical constraint so that it features, like the standard stapler, just one opening (the 'tongue' should stick to the roof of the crocodile's mouth).

What happens when affordances are ambiguous? The crocodile stapler may give us pause for some thought about where we should put the paper, especially if we have put it into the wrong slot and failed to get a stapled document. Some conscious cognition may be involved – realizing this is the wrong slot, so we should try the other, for example. In other words, we have to invoke knowledge in the head consciously to work out how this technology works. However, on this example, conscious cognition is *only* involved because the artefact is ambiguously designed in terms of its affordances. Otherwise we would not need to think about how to do the stapling. Thus, the solution in terms of design is to remove the ambiguity; that is, build in an unambiguous affordance. This would result in a transparent technology: one which communicates with us directly and non-consciously about how it is to be used.

The implication of this is that if technologies are appropriately 'afforded' (i.e., have unambiguous affordances), knowledge in the head, or internal cognition, is not required. Gibson's view is that an affordance is a biological link between information and action which is evolutionarily conferred, making object function clear to us through 'direct perception'. Norman extends Gibson's work by discussing how we can afford technologies. However, as we have seen, Norman's affordances differ from Gibsonian affordances in one crucial respect: they are enculturated, depending on a history of culturally relative learning.

In fact, in considering the two staplers, the affordances we are talking about are not Gibsonian. We would have no idea where to put the paper in either machine if we did not already know what paper is, and how and why it can be joined. This suggests that technology transparency is not necessarily given by an environment of resources which are transparent in Gibson's sense, but through experience of activity. However, for many simple, culturally familiar activities like stapling, the affordances involved may seem, phenomenally, very like Gibson's, since the activity is so familiar to us. In other words, identifying these culturally overlearned properties and building them into technologies might mean knowledge in the head never seems necessary to achieve something like stapling. In terms of design, it seems this technology can be afforded.

3 Complex Technologies: Problems with a Photocopier

The concept of affordances – whether biological (Gibson) or enculturated (Norman) - provides a way into thinking how and why technology features can appear transparent to us as users, without the need, apparently, for conscious cognition. However, there may be reasons apart from absence or ambiguity of affordances why some technologies are difficult to use. A stapler does not only afford entering paper; it is also squeezable. Which should we do first? The sequence of actions necessary to staple paper is not something that can be immediately derived from the technology – that is, it is not directly afforded. Thus, what a technology has to provide to enable action is not only appropriate affordances, but ways of helping us perform a sequence of actions. In addressing this issue, Norman moves beyond discussion of affordances into 'action cycles', which depend on our having a goal and a plan of action. Thus, internal cognition re-enters the debate on situated action. As well as affordances, we need the technology to support our goals and plans.

What happens when a technology has to be designed in respect of an action cycle? Suchman (*op. cit.*: 118-178) offers an extended example of technology-supported cooperative work in which two people are attempting to make five double-sided photocopies of a bound document. Instructions for completing this sequence of actions are grouped as displays whose appearance on the photocopier interface is occasioned by hardware sensors being appropriately detected by the machine (implying these are activated by users). Completion of the last action prescribed by a display initiates a new display which will be appropriate given that the right hardware sensors have been activated - that is, that the users are following the procedure as required by the machine.

To make their copies, the users have to make unbound master copies from the bound document, and then copy these. The users successfully make their set of masters. They remove the masters, but to signal to the machine that this stage is complete, they have to open the copier. They fail to do this because it is perceived as unnecessary. As far as the users are concerned, they both know the unbound master copies have been made, and this knowledge ought to be shared by the machine, especially since the users have removed them. However, it is not. The photocopier prompts the users to continue making their masters. And, despite having already made them, the users continue to do this. The machine's set of instructions remains inside an inappropriate iteration – inappropriate given the users' intentions. In the end, the users fail to make their double-sided copies, and give up.

The problems arise, according to Suchman, because the machine presents sets of displays which humans interpret in terms of sequential implicature (Grice, 1975). When an instruction is received by a person, it is assumed to have a rational relationship to the previous one, and the sequence of instructions is perceived as leading to the person's goal. In fact, the machine is not producing such a set of instructions, but just cycling through an iteration which is repeated according to which of its sensors are activated (or not). This sort of iteration construct, argues Suchman, is not familiar in human communication and so cannot be understood by the photocopier users. The principle of sequential implicature has such power that the users continue to interpret the next instruction as necessary to reach their goal, and thus continue making masters although this has already been done, believing the machine will eventually lead them where they want to go.

A major motivation for Suchman's work on situated action is a rejection of cognitivist planning models which attempt to support action by representing it in terms of abstract plans featuring computational constructs like iteration which appear to be assumed to be adequate to support action regardless of whether users are novices. When technologies are built on such models, rather than facilitating action cycles, Suchman argues, they can defeat them. Suchman points out that because the photocopier produces sequenced behaviour and communicates with users, it is seen like a human interlocutor. However, it violates the kinds of communicative protocols (including sequential implicature) and action sequencing humans beings are used to. Thus the photocopier fails to support an action cycle.

4 Can We Afford Complex Technologies? Affordances and Enculturated Functions

The idea of affordances is, generally, that a technology communicates its use to us directly by virtue of our perception of its physical properties. However, the discussion above suggests that there are at least two further sorts of function technologies need to exhibit: support for action cycles; and human-appropriate communicative protocols. Neither of these latter two sorts of function are easy to characterize as affordances, since affordances appear to relate to discrete events in an activity rather then the coordination of these events (for example, designing a telephone keypad so that buttons match functions does not tell us how to use a phone). Because the action cycles associated with photocopying are more complex than for stapling, how far we could improve the usability of the photocopier in terms of disambiguating its affordances is not nearly as obvious as for the crocodile stapler: we also need to consider, if we drop planning models, how action cycles can be supported. An immediate implication of Suchman's discussion is that the photocopier interface needs to be rebuilt to respect principles of sequential implicature. However, an important issue remains: the users Suchman focuses on appear never to have carried out their task before. If, notwithstanding criticism of the planning model, situated action does depend on internal cognition like goals and plans, a further problem arises, which is how the technology itself can help develop this internal cognition. This implies that in designing transparent technologies, the properties we need to consider go wider than just affordances, and should include those related to support for action cycles both in the sense of facilitating previously existing goals and plans as well as promoting appropriate new ones.

How might we address these issues? We have seen that Norman's affordances are enculturated – that is, realized against a background of shared cultural knowledge. Not only this, the discussion of staplers shows that knowledge about how to coordinate affordances in action – that is, knowledge of action cycles - is also enculturated: we know what a stapler is for because are familiar with the activity it supports and are part of a community where this practice is familiar. Thus, what helps the activity become familiar is not just technology affordances or implementation of abstract plans but a range of enculturated functions including affordances. In designing transparent technologies, we need to consider how enculturated functions can emerge and how much this depends on the technology itself. Activity Theory (AT) can help us think about this.

AT (Vygotsky, 1978; Leont'ev, 1982) takes as its basic unit of analysis, the 'activity'. An activity consists of a set of coordinated 'actions' and 'operations'. AT holds that activities, in part or in whole, can become non-conscious through a process called 'operationalization'. When an activity is 'operationalized', actions and action sequences, including our use of technologies, are largely automated. Operationalization occurs through repeating sets of coordinated actions which are initially conscious and attentional, for example keying in the number on a mobile phone, until using the array of buttons becomes automatic and we do not have to think about where we are putting our fingers. When this happens, a technology is transparent.

This implies that activities involving technologies have to be learned by individuals. This can be conscious and effortful, as we saw in the photocopier example. This means that internal cognition remains an important level of analysis. However, activities need not be operationalized from scratch. It seems unlikely that in using a mobile phone for the first time, we would need to spend time puzzling over how to use the button array, because this part of the activity is already operationalized: the same button array appears in many activities. The button array, in fact, represents individual affordances which are coordinated together through the activity of keying-in which motivates the realization of those affordances: the button array transcends affordances and is an enculturated function.

What this suggests is that operationalization involves, as well as internal cognition, a background of culturally familiar activities which overlap and make use of the same technology functions. How far a technology is transparent to us in a particular activity, then, relates to how far we are also familiar with other activities which make use of similar functions. We can think of a button array, and its use, as a 'generic function'; one whose operationalization is achieved through repetition across parallel activities. What this means is that operationalization is not just an individual effort, but is promoted through membership of a culture, and by engagement in the routine activities that membership entails. This implies that in designing transparent technologies, it is not just design parameters, for example physical or logical constraining, that are important, but also considering how culturally generic technology functions which may feature already are.

The implication of this discussion is that a major way an enculturated function can emerge is through its reappearance across different activities. Thus, a technology can offload internal cognition through affordances but also through recapitulation of enculturated functions which do not require new learning. How might we use these ideas to generate recommendations for redesigning Suchman's photocopier to make it more transparent? We will consider this question by contrasting the photocopier with an ATM (cash) machine.

An ATM machine shares the features of dynamically sequenced behaviour depending on our input, as well as communications, with the photocopier. Unlike the photocopier, it is hard to imagine that most users would have a problem using an ATM. One difference between an ATM and the photocopier is the way each machine formats its communications, and how these are coordinated with its behaviour. It is possible, with the photocopier, to send its interface into an infinite loop, whereby the interface toggles between two different instructions. This appears not to be possible with an ATM. One reason the ATM does not do this may be because it is designed to preserve principles of sequential implicature, whereby machine communications reflect the ways we interpret the communications of other people. We can regard sequential implicature as a form of enculturated function; one which is familiar from a myriad of human interactions in everyday life. One reason the ATM works, then, appears to be that it embodies this enculturated function in ways the photocopier does not.

At the same time, the ATM machine recapitulates other enculturated functions like button arrays and menu screens. The menu screen in particular offers the choices we might expect if withdrawing money in person: we know how to respond to these. Thus, enculturated functions can emerge not just through recapitulation but also through activity overlap. The photocopier menu is far more complex and there is no analogous activity overlap. This reflects that being part of a community of practice where similar activities are frequently carried out, perhaps in different modalities, is an important promoter of enculturated functions: indeed, we can imagine that certain users, for example print shop staff, *would* be able to successfully carry out the activity of making double-sided copies, notwithstanding the idiosyncratic behaviour of its interface, by virtue of their membership of a culture which repeats such activities frequently.

The implication of this discussion is that enculturated functions emerge through reappearance across a wide range of activities with which we are culturally familiar. Enculturated functions can range from affordances to action cycles. How far a technology is transparent depends on how far it recapitulates functions known from other activities. Where it does not, as with the photocopier, there is additional learning load, and greater pressure to preserve, for example, physical constraining and instructional clarity in order to promote the internal cognition which may not yet exist.

5 Conclusion

The question that motivates this paper is whether we can afford technologies: that is, make technologies transparent to us by building in affordances. The answer tends to be negative, for two reasons. First, the work on affordances for technology reflects that their significance for action is culturally relative. We cannot build functions into technologies and expect those technologies to be transparent if the users of the technology have not already realized the meanings of these functions through their own activity. Second, affordances are only one type of technology function. Others include support for action cycles through sequenced behaviour and communications. These may be unfamiliar to users. To be able to make technologies transparent, then, we need a better understanding of how a whole class of enculturated functions emerges, and what are the principles by which they can be preserved or promoted by technologies. An implication for cognitive technologies in the sense of technologies which can extend human cognition is that technology transparency is very much related to previous cultural experience, which places constraints on how far technologies really can promote radically new forms of cognition.

Acknowledgements. My thanks go to Yvonne Rogers and Mike Scaife for helpful advice and comments.

References

1. Gibson, J. J. (1977) The theory of affordances. In R. Shaw and J. Bransford, Eds. Perceiving, Acting, and Knowing: Toward an Ecological Psychology. Hillsdale: Lawrence Erlbaum Associates.
2. Gibson, J. J. (1979) The Ecological Approach to Visual Perception. Boston: Houghton-Mifflin.
3. Grice, H. P. (1975) Logic and conversation. In P. Cole and J. Morgan, Eds. Syntax and Semantics Volume 3: Speech Acts. New York: Academic Press.
4. Leont'ev, A. N. (1982) Problems of the Development of the Mind. Moscow: Progress.
5. Norman, D. A. (1988) The Psychology of Everyday Things. New York: Basic Books.
6. Scaife, M., and Rogers, Y. (1996) External cognition: how do graphical representations work? International Journal of Human-Computer Studies, 45, 185-213.
7. Suchman, L. (1987) Plans and Situated Actions: The Problem of Human-Machine Communication. Cambridge: Cambridge University Press.
8. Vygotsky, L. S. (1978) Mind in Society: The Development of Higher Psychological Processes. Cambridge: Harvard University Press.

"The End of the (Dreyfus) Affair": (Post)Heideggerian Meditations on Man, Machine, and Meaning

Syed Mustafa Ali

Computing Department
The Open University
England, UK
s.m.ali@open.ac.uk

Abstract. In this paper, the possibility of developing a Heideggerian solution to the Schizophrenia Problem associated with Cognitive Technology is investigated. This problem arises as a result of the computer bracketing emotion from cognition during human-computer interaction and results in human psychic self-amputation. It is argued that in order to solve the Schizophrenia Problem, it is necessary to first solve the 'hard problem' of consciousness since emotion is at least partially experiential. Heidegger's thought, particularly as interpreted by Hubert Dreyfus, appears relevant in this regard since it ostensibly provides the basis for solving the 'hard problem' via the construction of artificial systems capable of the emergent generation of conscious experience. However, it will be shown that Heidegger's commitment to a non-experiential conception of nature renders this whole approach problematic, thereby necessitating consideration of alternative, post-Heideggerian approaches to solving the Schizophrenia Problem.

1 Introduction

According to Janney [1, p.1], "an underlying assumption of Cognitive Technology [is] that computers can be regarded as tools for prosthetically extending the capacities of the human mind." On this view, Cognitive Technology is not concerned with the replication or replacement of human cognition - arguably the central goal of 'strong' artificial intelligence - but with the construction of cyborgs, that is, cybernetic organisms or man-machine hybrids, in which possibilities for human cognition are enhanced [2]. However, it may be necessary to reconsider this position in order to address what might be referred to as the 'Schizophrenia Problem' associated with human-computer interaction. Janney describes the essence of this problem as follows: "As a partner, the computer tends to resemble a schizophrenic suffering from severe 'intrapsychic ataxia' – the psychiatric term for a radical separation of cognition from emotion. Its frame of reference, like that of the schizophrenic, is detached, rigid, and self-reflexive. Interacting in accordance with the requirements of its programs, the computer, like the schizophrenic, forces us to empathize one-sidedly with it and communicate with it on its own terms. And the suspicion arises that the better we can do this, the more like it we become." [1, p.1] Crucially, on his view, intrapsychic ataxia, is "a built-in feature of computers." [1, p.4] Notwithstanding the intrinsic na-

M. Beynon, C.L. Nehaniv, and K. Dautenhahn (Eds.): CT 2001, LNAI 2117, pp. 149–156, 2001.

ture of the Schizophrenia Problem, Janney remains optimistic about the possibility of its (at least partial) 'solution' within Cognitive Technology as is evidenced by his intent 'to encourage discussion about what *can* be done in Cognitive Technology to *address* the problems pointed out [emphasis added]." [1, p.1] As he goes on to state, "an important future goal of Cognitive Technology will have to be to encourage the development of computer technology that reduces our need for psychic self-amputation." [1, p.5]

While concurring with Janney that "a one-sided extension of the cognitive capacities of the human mind – at the expense of the user's emotional and motivational capacities – is technological madness" [1, p.5], it is maintained that if the Schizophrenia Problem is to be 'solved' - by which is meant elimination and not mere reduction of the need for psychic self-amputation – it will be necessary for Cognitive Technology to reconsider its position on the issue of replication of human cognition and emotion. Although efforts are underway in this direction, it is suggested herein that they are unlikely to prove ultimately successful. This is because the Schizophrenia problem can be shown to be intrinsically, if only partially, related to the 'hard problem' of consciousness [3], that is, the problem of explaining how ontological subjectivity (or first-person experience) can arise from an ontologically objective (or non-experiential) substrate. For example, Picard has argued that the problem of synthesizing emotion can largely be bracketed from the problem of explaining (and possibly synthesizing) consciousness. However, as she is careful to point out, consciousness and emotion, while not identical, "are closely intertwined". While current scientific (specifically, neurological) evidence lends support to the view that consciousness is not necessary for the occurrence of all emotions, Picard concedes that emotional *experience* "appears to rely upon consciousness for its existence." [4, p.73] If consciousness is necessary for emotional experience, then in order to solve the Schizophrenia Problem, Cognitive Technology must first solve the 'hard problem'. This would seem to suggest that, contrary to one of the underlying assumptions of Cognitive Technology, replication of mind (cognition and emotion) – arguably the central goal of AI (or Artificial Intelligence) – constitutes a necessary condition for IA (or Intelligence Augmentation).

In this connection, it might be argued that the thought of the German phenomenologist Martin Heidegger (1889-1976) - more specifically, that aspect of his early thinking concerned with the being (or ontology) of human beings as interpreted by Hubert Dreyfus [5] - is highly relevant to Cognitive Technology in that it appears to suggest how the Schizophrenia Problem can be solved. According to Dreyfus, Heidegger holds subjective experience to be grounded in, and thereby emergent from, a more primitive existential experience - *Dasein* or being-in-the-world - that is ontologically prior to subjectivity and objectivity. If Dreyfus' Heidegger is correct, then the Schizophrenia Problem is solvable because the 'hard problem' can be solved by constructing artificial *Dasein*s capable of generating consciousness as an emergent phenomenon.

In this paper, it will be argued that appealing to Heideggerian thought in the context of attempting to solve the Schizophrenia Problem associated with Cognitive Technology is problematic on (at least) three counts: First, Dreyfus' interpretation of Heidegger, or rather, technologists' selective appropriation of Dreyfus' interpretation of Heidegger, while (possibly) illuminating from a technological standpoint, can be shown to be distorting when viewed from the perspective of Heidegger scholarship. Crucially, this fact may be of more than merely academic significance; second, Heidegger's commitment to an empirical-realist conception of nature as intrinsically non-experiential can be shown to undermine the possibility of a Heideggerian 'emergentist' solution to the 'hard problem'; third, it is suggested that because the technical construction of artificial systems - in this instance, synthetic *Dasein*s - occurs under an implicit subject-object (or artificer-artifact) orientation, the primitive components of such systems will necessarily stand in extrinsic (or external) relation to each other. This is of critical significance since Heidegger holds that beings are relationally-constituted, thereby entailing a commitment to an ontology grounded in intrinsic (or internal) relationality.

In closing, it will be argued that since Heidegger cannot solve the 'hard problem', it is necessary to look elsewhere for a solution to the Schizophrenia Problem. In this connection, Whiteheadian panexperientialism seems promising in that it appears to solve the 'hard problem'. However, this is at the price of a commitment to an ontology grounded in intrinsic (or internal) relationality which undermines the possibility for constructing artificial *Dasein*s capable of consciousness, thereby rendering the Schizophrenia Problem unsolvable.

2 'The Dreyfus Affair'

Determining the implications of Heidegger's thought for Cognitive Technology is arguably as difficult a task as determining his standing in Western academic philosophy: On the one hand, Heidegger is (generally) regarded as an intellectual charlatan of consummate proportion (and extremely dubious moral standing) by members of the Anglo-American philosophical establishment; on the other, he is (largely) revered as a genuinely original thinker who has contributed both profusely and profoundly to the enrichment of Continental philosophy. Similarly, on the one hand, Heidegger's later thought, in particular, his assertion that "the essence of technology is by no means anything technological" [6, p.4], has been regarded by anti-technologists as establishing grounds upon which to mount a universal critique of technology; on the other hand, certain Heideggerian insights have been embraced by technologists in an attempt at resolving intractable problems of long standing. Although the claim that Heidegger has contributed significantly to the debate on the meaning and scope of technology is not, in itself, in question, determining the precise nature of his contribution(s) – in the present context, the implications of his thought for the development and critical evaluation of Cognitive Technology - is problematic because there are many ways to interpret and appropriate his meditations on this issue by appealing to different 'aspects' and 'phases' of his phenomenological inquiry into being.

In this connection, Dreyfus' [7] seminal critique of 'GOFAI' (Good-Old-Fashioned-Artificial-Intelligence), which makes extensive use of the 'existential analytic of *Dasein*' (that is, the situated analysis of the onto-phenomenological structures of human being) presented in Heidegger's *Being and Time* [8] in order to contest the sufficiency of disembodied, a-contextual, symbolic computation as a means by which to instantiate real yet synthetic intelligence, has played an important, perhaps even decisive, role in motivating practitioners to consider engaged, embedded, and non-representational approaches to computing grounded (at least partly) in Heideggerian thought. It is crucial to appreciate at the outset that Dreyfus' approach to AI critique was philosophical and not technological, being driven by a desire to draw attention to the perceived failings of an extant technology. Dreyfus' primary concern was not - and, arguably, could not be, given his lack of technical expertise - to develop technological solutions to the problems of AI; this task was left to the technologists among his later followers. Connectionist approaches to consciousness [9] and cognition [10], robotic approaches to artificial life [11] [12], and the (re)conceptualisation of the information systems paradigm in terms of communication rather than computation [13] [14] have all benefited from Dreyfus' engagement with Heidegger.

There are (at least) two points to note in connection with the above: First, 'The Dreyfus Affair' - that is, Dreyfus' engagement with Heidegger on the one hand, and with the AI community on the other - provides a relatively recent example of the social determination of technology, the specifically philosophical character of the determination calling into question more conventional theses on technological determinism; second, perhaps what is most significant and yet often overlooked, is the fact that Dreyfus' critique of AI was only finally acknowledged and subsequently integrated into technology theory and practice because it *could* be so incorporated. In short, it is maintained that Dreyfus - and thereby Heidegger - was eventually taken seriously by technologists because his interpretation of Heidegger allowed the technological project to continue.

While this appears to reverse the order of determination described previously, it is not this fact that is especially interesting since the reflexive nature of the relations of determination between society and technology has long been appreciated by sociologists and philosophers of technology. Rather, what *is* interesting, is the fact that Dreyfus' critique was ultimately regarded as both valid and relevant because it was (taken to be) grounded in an instrumentalist-pragmatist interpretation of Heidegger's thought, an interpretation coming out of the 'sensible' early period that preceded 'The Turn' in his philosophy when concern (allegedly) shifted from existential analysis of the meaning of being (as intelligibility) to historical inquiry into the truth of being (as 'concealing unconcealment'). However, as Blattner [15], Fell [16], and, significantly, Dreyfus [17] himself, have all shown, instrumentalist, pragmatist and/or behaviourist interpretations of Heidegger's thought are both limited and 'dangerous' because partial and hence, distorting. It is, therefore, somewhat ironic that Dreyfus, who has been charged with misappropriating Heidegger's thought, for example, with respect to the question of whether or not Heidegger is a naturalistic anti-representationalist [18], himself ends up being misappropriated by practitioners of technoscience (AI, A-Life etc).

Philosophically-speaking, 'The Dreyfus Affair' appears to be over.

3 Heidegger and Cognitive Technology

The implications of the end of 'The Dreyfus Affair' for Cognitive Technology are somewhat unclear since it is possible that Dreyfus' interpretation of Heidegger remains practically relevant despite its philosophical shortcomings. For example, the application of Heideggerian thought to Cognitive Technology with the latter interpreted as artificial (or synthetic) means by which *meaning* might be extended in the interaction between humans and machines appears warranted given (1) the identification of being with intelligibility or meaning [5], viz. *Sein* as *Sinn* (sense), (2) the reciprocal relational-grounding of being and *Dasein* (or being-in-the-world), (3) the ontological priority of *Dasein* over the conscious subject, and (4) the onto-phenomenological claim that being-with (*Mitsein*) other *Dasein*s is a constitutive existential structure of *Dasein*. This is because (1)-(4) ostensibly provide the foundations of a framework for solving the Schizophrenia Problem by allowing for an emergentist solution to the 'hard problem' that can be implemented by natural and artificial (or synthetic) *Dasein*s alike. On this basis, it might be argued that it is necessary to shift the goal of Cognitive Technology from constructing 'instruments *of* mind' - what Heidegger would call *Zuhandenheit* which Dreyfus [5] translates as 'availability' (or 'ready-to-hand') in reference to *Dasein*-centric, pragmatically-functional, transparent 'equipment' (*Zeug*) - to emergent construction of minded-instruments, that is, 'instruments *with* mind'.

4 Heidegger and the 'Hard Problem' of Consciousness

As Schatzki [19] has shown, Heidegger is an *empirical* realist: On his view, what something is 'in itself' is what it is independently of its *actually* being encountered by a *Dasein*. (Kant, by contrast, is a *transcendental* realist: On his view, what something is 'in itself' is what it is independently of any *possible* knowledge of it.) It is crucial to appreciate that empirical realism entails that the being of all beings, both human and non-human, is, in principle, publicly accessible to *Dasein* because this fact assumes critical significance when the 'other-minds' problem, that is, the problem of determining whether or not other beings are capable of consciousness (first-personhood, ontological subjectivity, private experience), is considered. The (later) Heideggerian solution to this problem involves recognizing the following as existential facts: (1) being-with other *Dasein*s is a fundamental (or constitutive) structure of *Dasein*; (2) *Dasein* (as being-in-the-world) has primacy over consciousness; (3) both *Dasein* and consciousness are linguistically-constructed. On this basis, the 'other-minds' problem is discharged by observing that because (1) *Dasein*s share language and (2) there are a plurality of *Dasein*s, therefore, a plurality of consciousnesses (or minds) is possible. However, it is important to draw out the full implications of this approach to solving the 'other-minds' problem: Heidegger is forced to conceive subjectivity in objective (or public) terms because, on his empirically-realist view, the subjectivity of a subject is disclosable, in principle, to and by other subjects. Since it is only *Dasein*s that share language, only *Dasein*s can become consciousnesses (or first-person, private subjects). Crucially, on his view, nature as it is 'in-itself' (that is, independent of *Dasein*) discloses itself 'in a barren merciless' as ontologically objective and hence, 'absurd' or meaningless. Heidegger [8] insists that this view of nature is not grounded in a value judgement but reflects an ontological determination that follows from the fact that it is *Dasein* alone who gives being (intelligibility or meaning) to beings [16]. However, this position is contestable on (at least) four grounds:

First, it is not at all clear that consciousness is a (purely) linguistic phenomenon, more specifically, an emergent linguistic artifact.

Second, more importantly, it does not follow from the fact that since *Dasein*s are the only beings that share language, therefore only *Dasein*s are capable of conscious (or at least some degree of private, subjective) experience. According to Krell [20], life may constitute a sufficient existential condition for being a 'clearing' or 'opening', that is, a space of possible ways for things (including human beings) to be. While it might be conceded that the being (sense or meaning) of beings disclosed by *Dasein* is of a significantly higher order than that disclosed by (other) beings themselves, it simply does not follow from the shareability of language peculiar to *Dasein* that disclosure of being by other beings is impossible; human-centred meaning is not necessarily coextensive with meaning as such. In short, Heidegger's position appears untenably anthropocentric.

Third, the view that nature is fundamentally 'vacuous' or non-experiential is an assumption which is undermined by the empirical fact that while experiential beings are definitively known to exist, it is unclear whether any non-experiential beings have, in fact, ever been encountered [21].

Finally, Heidegger's dualism of meaningful subjects and meaningless objects gives rise to the 'hard problem' [3], that is, the problem of explaining how ontological subjectivity can arise from an ontologically objective substrate. Heidegger cannot avoid this problem because his empirical realism commits him to the view that science can, in principle, causally explain how things came to be the way they are [5]; clearly, this includes explaining how the brain - which Globus [9] identifies as a necessary condition for *Dasein* - can give rise to consciousness. Emergentist solutions to the 'hard problem', which view consciousness as an irreducible systemic property arising from the interaction of components, none of which possess this property - or properties categorially-continuous with this property - in isolation or in other systemic complexes, are problematic because they disregard the principle of ontological continuity, arguably a cornerstone of scientific naturalism [21].

5 Post-Heideggerian Ontology and Cognitive Technology

It appears then that Heidegger's engagement with Cognitive Technology, at least with respect to the relevance of his thought to the Schizophrenia Problem, is, like 'The Dreyfus Affair', at an end. Principally, this is because Heidegger cannot solve the 'hard problem' due to what is, somewhat ironically, a phenomenologically-unsound (mis)conception of nature as intrinsically non-experiential. Thus, if the Schizophrenia Problem is to be addressed, it is necessary to consider 'post-Heideggerian' conceptions of the being of nature.

On Whiteheadian panexperientialism [21], for example, nature is held to be relationally-constituted and experiential at its most primitive ontological level. However, this does not imply that all beings are experiential in the same way (that is, ontological monism does not entail ontical monism); rather, certain complex beings enjoy a higher-level of experience relative to simpler beings. In addition, all complex beings belong to one of two kinds, experiential 'compound' (or 'societal') individuals or non-experiential aggregates, depending on the nature of their internal (or constitutive) relational organisation.

Crucially, if Whiteheadian panexperientialism is the way that nature is in-itself then the possibility of constructing an artificial *Dasein* is radically undermined because artificing (construction, making) involves an orientation in which 'subjects' stand in ontological opposition to 'objects' [6], thereby 'rupturing' [22] the nexus of internal (subjective, constitutive) relations constituting natural beings so as to establish - more precisely, impose - external (objective, non-constitutive) relations between 'primitives' (components) in the synthetic systemic complex. To the extent that *Dasein* is, ontically-speaking, a natural phenomenon[1], its being must be internally-constituted; however, artificial systems are externally-constituted which implies that they cannot provide the necessary ontical (causal) substrate for *Dasein*. In short, genuine *Mitsein*, arguably a necessary condition for an emergentist solution to the 'hard problem' and, thereby, to the Schizophrenia Problem associated with Cognitive Technology, cannot be generated technically.

In conclusion, it appears that Cognitive Technology is faced with a choice: Either consider alternative substrates to the computer, which is the paradigmatic exemplar of the synthetic [23], or accept that the Schizophrenia Problem is intrinsically unsolvable and concentrate on "finding out where the prosthesis 'pinches', so to speak [since] progress will depend on discovering and describing the sources of sensory and psychic irritation at the human-computer interface." [1, p.5].

References

1. Janney, R.W. (1997) The Prosthesis as Partner: Pragmatics and the Human-Computer Interface. J.P. Marsh, C.L. Nehaniv & B. Gorayska (Eds.), *Proc. Second International Cognitive Technology Conference CT'97: Humanizing the Information Age*, IEEE Computer Society Press, 1-6.
2. Haraway, D.J. (1985) Manifesto for Cyborgs: Science, Technology, and Socialist Feminism in the 1980's. Socialist Review 80: 65-108.
3. Chalmers, D.J. (1996) *The Conscious Mind: In Search of a Fundamental Theory.* Oxford, Oxford University Press.
4. Picard, R.W. (1997) *Affective Computing.* Cambridge, MIT Press.
5. Dreyfus, H.L. (1991) Being-in-the-world: A Commentary on Division I of Heidegger's Being and Time. Cambridge, MIT Press.
6. Heidegger, M. (1977) *The Question Concerning Technology and Other Essays.* Translated by W.Lovitt. New York, Harper & Row.
7. Dreyfus, H.L (1972) What Computers Can't Do: A Critique of Artificial Reason. New York, Harper & Row.
8. Heidegger, M. (1927) *Being and Time.* Translated by J.Macquarrie and E.Robinson. Oxford, Blackwell.
9. Globus, G.G. (1995) *The Postmodern Brain.* Philadephia, John Benjamins.
10. Clark, A. (1997) Being There: Putting Brain, Body, and World Together Again. Cambridge, MIT Press.
11. Wheeler, M. (1996) From Robots to Rothko: The Bringing Forth of Worlds. In The Philosophy of Artificial Life. Edited by M.A.Boden. Oxford University Press, 209-236.
12. Prem, E. (1997) Epistemic Autonomy in Models of Living Systems. *In Fourth European Conference on Artificial Life.* Edited by P.Husbands and I.Harvey. Cambridge, MIT Press, 2-9.
13. Winograd, T., Flores, F. (1986) Understanding Computers and Cognition: A New Foundation for Design. Reading, Addison-Wesley.
14. Coyne, R.D. (1995) *Designing Information Technology in the Postmodern Age: From Method to Metaphor.* Cambridge, MIT Press.

---

[1] On empirical-realism, specific natural (biological) conditions are necessary (yet insufficient) for *Dasein* [5] [9].

15. Blattner, W.D. (1992) Existential Temporality in Being and Time: (Why Heidegger is not a Pragmatist). In *Heidegger: A Critical Reader.* Edited by H.Dreyfus and H.Hall. Oxford, Blackwell, 99-129.
16. Fell, J.P. (1992) The Familiar and The Strange: On the Limits of Praxis in the Early Heidegger. In *Heidegger: A Critical Reader.* Edited by H.Dreyfus and H.Hall. Oxford, Blackwell, 65-80.
17. Dreyfus, H.L (1992) Heidegger's History of the Being of Equipment. *In Heidegger: A Critical Reader.* Edited by H.Dreyfus and H.Hall. Oxford, Blackwell, 173-185.
18. Christensen, C.B (1998) Getting Heidegger Off the West Coast. *Inquiry* 41, 65-87.
19. Schatzki, T. (1982) Early Heidegger on Being, The Clearing, and Realism. *In Heidegger: A Critical Reader.* Edited by H.Dreyfus and H.Hall. Oxford, Blackwell, 81-124.
20. Krell, D.F. (1992) *Daimon Life: Heidegger and Life-Philosophy.* Bloomington and Indianopolis, Indiana University Press.
21. Griffin, D.R. (1998*) Unsnarling the World-Knot: Consciousness, Freedom and The Mind-Body Problem.* Berkeley, University of California Press.
22. Ladrière, J. (1998) The Technical Universe in an Ontological Perspective. *Philosophy and Technology* 4 (1), 66-91.
23. Ali, S.M. (1999) *The Concept of Poiesis and Its Application in a Heideggerian Critique of Computationally Emergent Artificiality.* Ph.D Thesis. Brunel University.

New Visions of Old Models

Igor Chimir[1] and Mark Horney[2]

[1]Institute of Information Technologies
Odessa, Ukraine
chemyr@te.net.ua
[2]University of Oregon
Eugene, OR, USA
mhorney@oregon.uoregon.edu

Abstract. This article addresses the issue of building models of cognitive processes suitable for computer simulation. The approach is the application of an object-oriented modeling technique to models in Cognitive Psychology using the Unified Modeling Language (UML) as a basic formalism. The idea is displayed by an elaboration of a UML-description of the spatial structure of the model of focused auditory attention hypothesized by Broadbent (1958).

1 Introduction

We share the viewpoint that a barometer of how deeply cognitive psychologists understand a human mind is their proficiency at simulating it in computer systems. If a cognitive model provides an accurate, coherent, and complete account of an ability, then cognitive psychologists should be able to use this model to build a simulator that performs this ability as humans perform it. However, programmers interested in cognitive models easily observe the considerable distance between the presentation of models in Cognitive Psychology and the efforts needed to transfer these models into computer programs. One barrier to be overcome in this task is the ill-formalized means used by cognitive psychologists in presenting their models. As a rule they use natural language and graphical charts in the "input – process – output" paradigm. Our experience in giving courses on Cognitive Psychology to students from Software Engineering departments suggests that computer oriented simulation is absolutely essential, not only from the methodological point of view, but also to satisfy the natural desire of these particular students for computer prototyping of the systems they are studying. This article demonstrates the capabilities of the graphical Unified Modeling Language [1], first as a means for formal description of cognitive models, and secondly as a tool for elaboration of the specifications needed for further computer simulation.

We find that the process of transforming traditional descriptions of cognitive models into UML notation has an independent methodological value. Natural language descriptions are fraught with uncertainty and fuzziness, which leads to multiple of interpretations of the same description by different individuals. The constraints of a unified description forces the UML modelers to reduce uncertainty, clarify their thinking, and in fact invent their own version of the model. Thus, the *process of invention permanently follows the process of study*. We have also found that the process

M. Beynon, C.L. Nehaniv, and K. Dautenhahn (Eds.): CT 2001, LNAI 2117, pp. 157-163, 2001.

of transforming a natural language description into UML notation serves as a "methodological magnifying glass", allowing us to "see" what is hidden beneath the uncertainty of a natural language description. In the process of the UML modeling one can bring this magnifying glass closer to the object or move it away by means of a hierarchy of models from those structurally simple depicting only key concepts, to those more sophisticated.

One of the known models of focused attention has been chosen to demonstrate our proposed approach. Despite of the fact that some authors consider the phenomenon of attention as a part of control of information processing [2], studying models of attention is a necessary component of many modern textbooks on Cognitive Psychology [3]. We have chosen Broadbent's model of attention [4] to serve as an example of the UML modeling process. This does not imply that we consider this model the most adequate to the phenomenon of attention. We selected it because it is well known and simpler than the integrated models that appeared later, e.g. [5,6]. We also note that UML is not the only unified notation for describing complex systems. However, other well-known notations, such as semantic networks [7], Petri nets [8], production systems [9], etc. have one common shortcoming. They force modelers to think about and subsequently represent a system in terms of only one abstraction. However, for deep and coherent understanding, multiple abstractions are needed. We consider UML an attempt to *model systems by means of the composition of related abstractions,* in which each abstraction gives a detailed description of only one aspect of the modeled system.

2 Broadbent's Model of Attention

Data from experiments studying the ability of people to focus attention during the process of sound perception (dichotic listening experiments) were first generalized by Broadbent [4] in his hypothesis, later called Broadbent's model of attention. According to Broadbent, the information-processing system of a human has limited capacity, and so a filter is required in order to prevent the system from becoming overloaded. Information about the different stimuli in the environment is initially stored briefly in a sensory buffer. Information about one of the stimuli is then selected for further processing by being allowed to pass through the filter. Information about the other stimuli resides in the sensory buffer for a short period of time, and then decays unless selected by the filter. Broadbent formulated his hypothesis as a set of principles, which are general enough to be considered as general principles of information processing by the nervous system of a human. These principles are often used in creating an information-flow diagram in the "input-process-output" paradigm, representing a full cycle of information processing by the human nervous system. Figure 1 depicts our version of this diagram.

Broadbent's model presupposes that units depicted in Fig. 1 not only store and transport the information but also *transform* it. It is convenient to consider two classes of information transformation realizing within Broadbent's model: (1) *sensory transformation* (before filtration), and (2) *postsensory transformation* (after filtration). Sensory transformation, transfers the flow of external sensory events into the code needed to store them in the sensory buffer. In subsequent publications Broadbent [10] defines the essence of postsensory transformation as *sensory event recognition*. The limited capacity channel's functional task is to recognize sensory events and transform them into a code appropriate for the long term storage. Therefore we can interpret the limitation of the channel as its ability to recognize only one sensory event "at once."

The filter works according to the principle "all or nothing." Within approximately 1 – 2 seconds the filter accomplishes full-scale switching from the informational segment, which corresponds to the right ear to the segment, to that which corresponds to the left ear. If information was not "filtered" from the sensory buffer during the time, which is less than time of natural attenuation, then it disappears irretrievably.

Information in the long term storage is kept in a form, which allows to account for correlation of events following one after another. Broadbent characterized this correlation by means of conditional probabilities. An organism uses information from the long term storage to exert influence on the environment by effectors. It is supposed that this influence is differential and depends on such characteristics of the sensory event as those indicating danger to the organism.

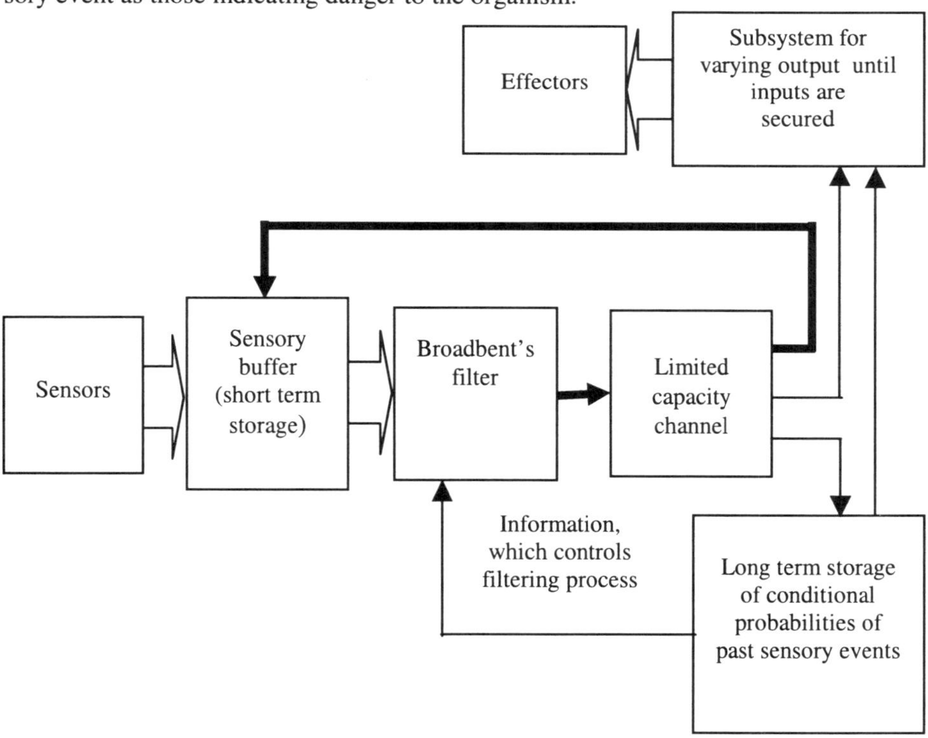

Fig. 1. Information-flow diagram, which represents the full cycle of information processing by the nervous system of a human in accordance with Broadbent's (1958) model of attention

3 Object-Oriented Presentation of the Structure of Broadbent's Model of Attention

The model offered by Broadbent is characterized by a high degree of uncertainty, which prevents an immediate synthesis of its structure into strict UML notation.

Therefore, a certain preliminary translation is needed. This translation will depict the system as a set of classes, which, taken as a whole, define the functional abilities of the model.

A key concept of Broadbent's hypothesis is the sensory event. A sensory event is a fragment of the auditory signal, which can be unambiguously categorized in the process of postsensory transformation. Consider a sensory event as an auditory signal, which hereafter will be categorized as a word. Define two classes of sensory events: (1) *routine* sensory events (`RoutineEvents`), and (2) *suspicious* sensory events (`SuspiciousEvents`.) An organism automatically focuses attention on suspicious events, interrupting processing of the routine events, because suspicious events are potentially dangerous. Using the concept of "priority" we can say that suspicious events always have higher priority than routine events. A distinguishing feature of suspicious events is their physical characteristics: high intensity or frequency of the sound. An example is the 400 hz signal in Cherry's experiment on dichotic listening tasks [11]. In this experiment a subject stably detected a 400 hz signal, transmitted into the left ear while his attention was focused on the perception of the text, transmitted into the right ear. The border between routine and suspicious events is fuzzy and depends on the context. The same event in different contexts subjects detect as routine or suspicious. For instance, in above mentioned Cherry's experiment some subjects do detect female voice messages transmitted into the left ear and some subjects do not.

Sensory transformation converts a sensory event into a sensory segment class (`SensorySegment`). Sensory segment succession is memorized in the sensory buffer in the form of a sensory array class (`SensoryArray`). If we limit Broadbent's model by the frames of focused auditory attention, we have to consider two classes: `R_SensoryArray` and `L_SensoryArray` corresponding to the right and left ears sensory buffers. It is clear that our classification of sensory events can be extended also on sensory segments. So, we can define classes: `RE_SensorySegment` and `SE_SensorySegment`. The UML Class Diagram in Fig. 2 depicts relationships between classes of segments and segment arrays. In the diagram the property of natural attenuation of the information in sensory buffer is modeled by `+ lifecycle` attribute of `SensoryArray` class and inherits by all other classes of the model.

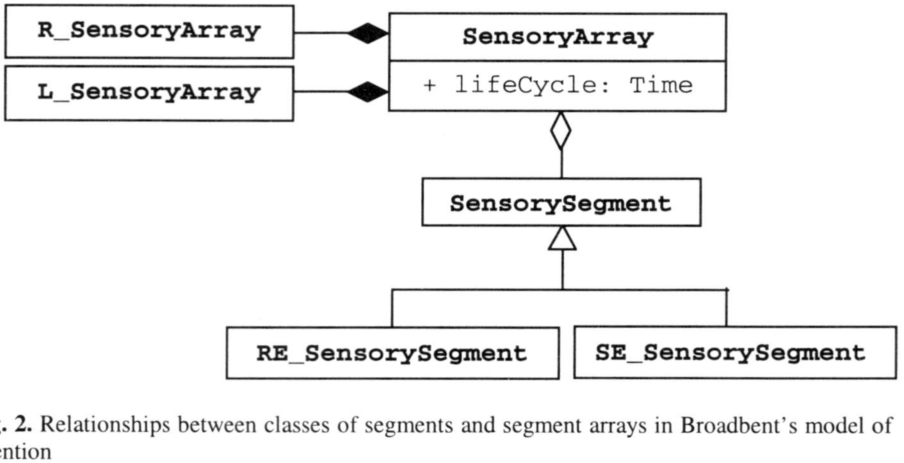

Fig. 2. Relationships between classes of segments and segment arrays in Broadbent's model of attention

Postsensory transformation of routine events is not random but directed towards achieving a certain goal. An example of a goal-driven routine postsensory transformation is the process of filtration from a "drive state," described by Broadbent [4] in his principles 5 and 7. In this example, the goal is to eliminate the feeling of starvation. Thus we can say that *the process of selection and recognition of the flow of sensory events is a part of a more general process of step-by-step "motion towards the current goal.*" The goal driven motion is realized by a strategy of controlling the filtering process, which is in good correspondence with Neisser's theory of the perceptual cycle [12]. At every step a sensory event, which has been just recognized, is compared with *schemas "anticipated" at this step*, and the selection of the next sensory event is determined by the "proximity" of the recognized event and one of the anticipated schemata. Control of the filtering process based on knowledge in the long term storage. We model the long term storage, following Neisser's terminology, by the `CognitiveMap` class with an `AnticipatorySchemata` subclass.

When a suspicious event is detected, the filter switches and suspicious event selection is realized automatically. Consider that every sensory organ has its own Detector of suspicious events, and model all detectors of all sensory organs by an `SE_Detector` class. So, in the general case, Broadbent's *filter is under control during the process of sensory transformation and also during the process of postsensory transformation.* `CognitiveMap` is a source of control messages while filtering the flow of routine sensory events and SE_Detector class is a source of control messages while filtering suspicious events.

This explication is enough to create a UML diagram of the spatial structure of Broadbent's model of attention.

The structure in the Fig. 3 follows from Broadbent's principles and reflects its ideas, but from our point of view is more constructive than the source text and can be develop towards a more concrete version.

To simplify Fig. 3, some classes were omitted, For instance: `RE_SensorySegment`, `SE_SensorySegment`, and `AnticipatorySchemata`, and all classes other than the Filter are presented in abbreviated form. The Filter is described in full form by sets of attributes and operations, which can be divided into groups "serving" relationships between the Filter and other classes.

Relationships, which realize the control function of the Filter from the side of Detector and Cognitive Map (`SE_Control` and `RE_Control`) are served by the following sets of attributes and operations: `arrayIndex`; `segmentIndex`; `setArray()`; and `setSegment()`. Relationships, which realize the filtering function are served by the following sets of attributes and operations: `switchingCycle`; `selectionCycle`; `switchArray()`; and `selectSegment()`.

We can expand this representation detailing classes and its descriptors. However, for this demonstration of the fundamental abilities of UML modeling of cognitive process, we considered that the diagram in Fig. 3 is sufficient.

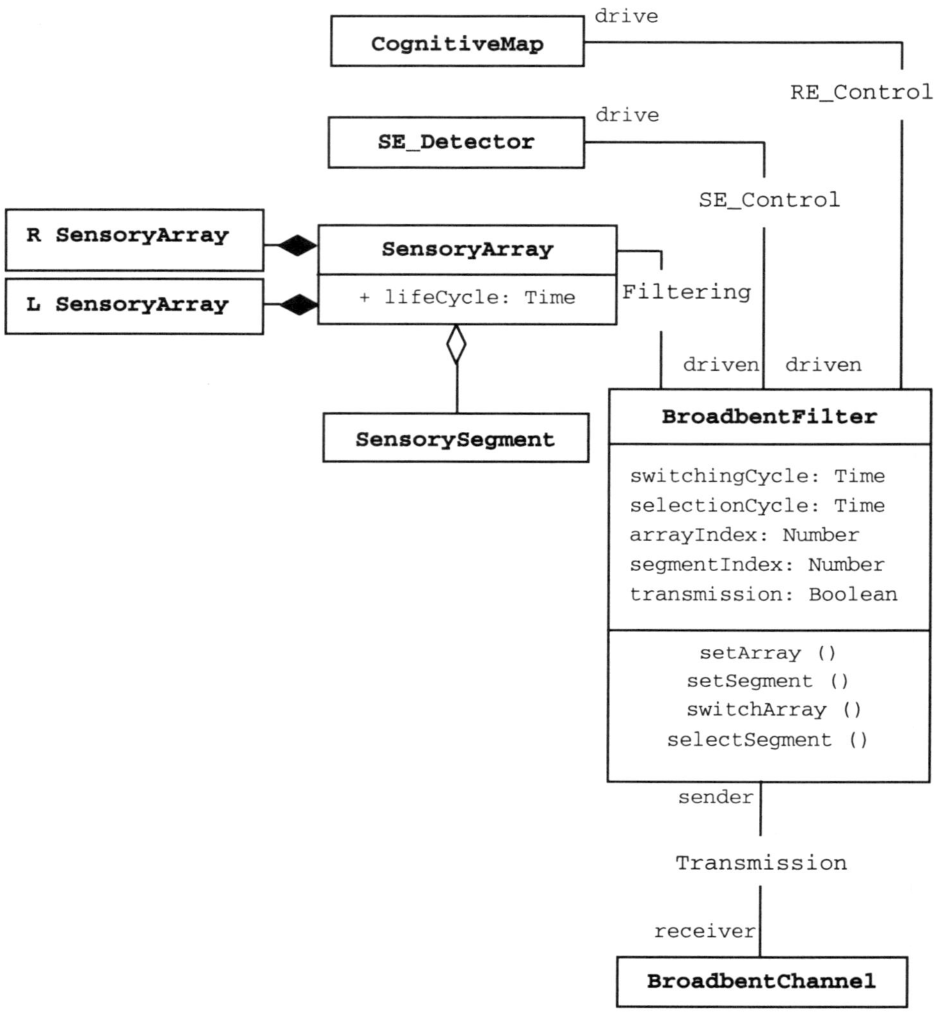

Fig. 3. Class diagram, which depicts the structure of Broadbent's model of focused attention

4 Discussion

In developing this UML model, an interesting side effect emerged. We began with the idea of using UML notation to model an original verbal description. In doing this, we have uncovered new elements. The ideas of routine and suspicious sensory events, of priority evaluation for modeling of "conditional probabilities," and the idea of "anticipated schemata" accounting for the internal states of the organism are examples. The anticipatory schemata class presupposes that at every step of solving a problem, involving perception of the environment, an organism selects objects of attention according to certain expectations regarding those objects. Anticipatory schemata do not represent the contents of the whole long-term memory, but rather only the current step. It seems that an effective procedure allowing the "calculation" of anticipatory schemata for the next step does not exist. Instead of calculation we can consider a certain class of relationships, which provide direct access to the anticipatory schemata of the next step. This class can model both the Neisser's perceptual cycle [12] and the problem solving activity.

The UML diagrams presented here are not yet sufficient for programmers. They do not, for instance, contain information about behavioral and controlling processes. How does a class, for example, "know" when, and what operator should be activated? They do not describe the process of interaction between classes. However, UML offers some types of diagrams for modeling classes' behavior and interaction. Perhaps the most useful are the `Activity Diagram` and `Sequence Diagram`.

In conclusion let us reiterate that UML modeling has the obvious advantages already mentioned for researchers. We have also suggested that an advantage of this technique is that the *process of creating* the diagrams is itself a useful exercise, providing a methodological magnifying glass, that clarifies thinking and leads to more accurate, adequate and complete accounts of a human abilities. We would also like to suggest this technique as an instructional method for both students of cognitive psychology and of computer science. It provides a concrete learning environment allowing for the step wise development of concepts and skills where students can constantly check their understanding through development of their own simulations. These, if working properly, should re-generate the data found in the research underpinning their courses. In future work we hope to develop a textbook to this topic.

References

1. Booch, G., Rumbaugh, J., & Jacobson, I., (1999). *The Unified Modeling Language User Guide*. Addison Wesley Longman.
2. Barsalou, L. W., (1992). *Cognitive Psychology. An Overview for Cognitive Scientists*. Lawrence Erlbaum Associates.
3. Eysenck, M. W., & Keane, M. T., (1999). *Cognitive Psychology. A Student's Handbook*. Psychology Press.
4. Broadbent, D. E. (1958). *Perception and Communication*. Oxford: Pergamon.
5. Kanheman, D (1973). *Attention and Effort*. Englewood Cliffs, NJ: Prentice – Hall.
6. Cowan, N. (1988). Evolving Conceptions of Memory Storage, Selective Attention, and Their Mutual Constraints within the Human Information-Processing System. *Psychological Bulletin*, 104, 163 – 191.
7. Collins, A. M., & Loftus, E. F. (1975). A spreading-activation theory of semantic processing. *Psychological Review*, 82, 407 – 428.
8. Rene, D. (1992). *Petri net and Grafcet: tools for modelling discrete event systems*, Prentice Hall.
9. Holland, J., Holyoak, K., Nisbett, R., & Thagard, P. (1986). *Induction: Processes in Inference, Learning and Discovery*. Cambridge, MA: MIT Press.
10. Broadbent, D. E. (1971). Decision and stress. London: Academic Press.
11. Cherry, E. C. (1953). Some experiments on the recognition of speech with one and two ears. *Journal of the Acoustical Society of America*, 25, 975 – 979.
12. Neisser, U. (1976), *Cognition and Reality*. W. H. Freeman and Company, San Francisco.

Abstract

Victorian Data Processing – When Software Was People

Martin Campbell-Kelly

Dept of Computer Science
University of Warwick, U.K.
mck@dcs.warwick.ac.uk

To most people the phrase *Victorian Office* conjures up an image such as in Dickens' *A Christmas Carol* – with Bob Cratchit, the solitary clerk, seated on a high stool, quill-pen in hand. Indeed, many Victorian Offices were like this, but by the 1850s a quite different type of office was emerging – the industrialized office employing several hundred clerks. These offices were the ancestors of the modern computerized bureaucracy. In these huge organizations, clerks performed tasks that would later be done by office machines, and are today performed by computers.

In this paper the historical and economic context of the development of the industrialized office will be described. The paper will include a "tour" of a number of Victorian data processing organizations: the General Register Office, the Railway Clearing House, the Central Telegraph Office, the Prudential Assurance Company and the Post Office Savings Bank. The data processing techniques and labour processes will be explained, and the changing gender and structure of the workforce described.

The Victorian Office can be viewed from today's perspective as a collection of human agents, obeying procedures stored in the organizational memory. Parallels are drawn between today's software and Victorian clerical processes in terms of dependability, evolutionary change, and the boundaries of determinacy and autonomy. Some general conclusions will be made about the nature of "information revolutions."

M. Beynon, C.L. Nehaniv, and K. Dautenhahn (Eds.): CT 2001, LNAI 2117, p. 164, 2001.

On the Meaning of Computer Programs

Josh Tenenberg

Computing and Software Systems, University of Washington, Tacoma, 1900 Commerce St, Tacoma WA 98402-3100, U.S.A.
jtenenbg@u.washington.edu

Abstract. This paper explores how computer programmers extract meaning from the computer program texts that they read. This issue is examined from the perspective that program reading is governed by a number of *economic* choices, since resources, particularly cognitive resources, are severely constrained. These economic choices are informed by the reader's existing belief set, which includes beliefs pertaining to the overlapping and enclosing social groups to which the program reader, the original programmer, and the program's users belong. Membership within these social groups, which may be as specific as the set of programmers working within a particular organization or as general as the members of a particular nation or cultural group, implies a set of shared knowledge that characterizes membership in the social group. This shared knowledge includes both linguistic and non-linguistic components and is what ultimately provides the interpretative context in which meaning is constructed. This account is distinguished from previous theories of computer program comprehension by its emphasis on the *social* and *economic* perspective, and by its recognition of the similarities between computer program understanding and natural language understanding.

1 Program Readers

Computer programs are sequences of instructions that direct the operation of a computer. Programs are written in a programming language and are interpreted by one or more language translators into machine language and converted into electrical energy so as to query and change the energy state of the underlying computer hardware. Programs in execution can be viewed as carrying out functionality from the perspective of their role within a human and social context, such as word processing, graphical manipulation, and accounting.

If computers were programmed by the gods, perfectly and without cost, then there would be no need for people to read programs. But programs are written by people, and in order to fix errors introduced during software development or to add new functionality desired by software purchasers, programmers must read and understand the program texts written by other programmers.

The original programmer thus writes for two very different audiences – people and computers. This "original programmer" may in fact be a large group of people but for consistency will be referred to in the singular for the balance

M. Beynon, C.L. Nehaniv, and K. Dautenhahn (Eds.): CT 2001, LNAI 2117, pp. 165–174, 2001.

of the paper. The original programmer adds particular syntactic expressions because of her understanding that the human reader, but not the computer reader, has beliefs, intentions, goals, desires and preferences. Further, the computer will always (short of hardware failure) read and interpret the instructions in its program while the human reader might decide that reading certain program segments is not worth the cognitive effort. In this paper *program readers* will refer to people and not computers unless otherwise indicated.

The reader undertakes their reading task with a set of existing beliefs and with a set of programming artifacts that includes the program itself and perhaps other documents such as requirements and design documents, internal memos, and technical documentation. The reader performs her actions within a sociocultural embedding in an organization, community and society. Meaning construction involves a change in the belief state of the program reader. The reader can choose from a set of actions in order to alter her belief state. Such actions include purely internal cognitive events such as recall and inference, as well as events that have external components, such as speaking with other programmers or reading program and documentation texts.

In ascribing meaning to the expressions in a program, a reader must determine both how the expressions will affect the underlying computer state, that is, its actual behavior, and how the sequence of computer state changes are related to issues of human concern. Each of these tasks will be examined in turn.

2 Syntax and Semantics: The Traditional View

The *syntax* of a language specifies the set of all legal sentences in the language. Modern programming languages fall into the language class called *deterministic context free languages* (DCFL's). The meta-language for describing DCFL's, the *context-free grammars*, describe the atomic units of the language as well as how these atomic units may be combined via a set of rewrite rules. The context-free grammars are *generative*, in that from a finite set of atomic units and a finite set of recursively specified rewrite rules, an unbounded number of legal programs can be described. The use of context-free grammars to describe programming languages, rather than more expressive meta-languages (such as *phrase-structure* grammars) is an engineered choice, since the described DCFL's balance the need for expressivity against the need for fast, automated translation to the machine language of the underlying computer hardware.

The *semantics* of a programming language, as the term is used in computer science, refers to the way in which the underlying computer state changes as a result of expression execution. This semantics is compositional in that the rewrite rules can have corresponding semantic rules; the semantics of a compound expression is determined by the semantics of its components and the semantics of the composition operations. Because the underlying semantics relates to computer state changes over time, the meta-languages generally used for describing semantics have been a combination of formal state-based dynamic logic and informal natural language. As a result, communicating the semantics of programming languages has led to a higher level of ambiguity and misunderstanding within groups of program language users than for communicating the syntax of programming languages.

Programming languages also provide mechanisms for the introduction of new linguistic entities. For example, the following defines a new linguistic term `sum` in the Java programming language.

```
double sum( double[] A ) {
    float total = 0;
    for ( int i = 0; i < A.length; i++ )
        total += A[i];
}
```

The central idea behind these language extensions is to enable the definition of new *abstractions*. As Guy Steele writes [19, pp.xv-xvi] "The most important concept in all of computer science is abstraction. ... Abstraction consists in treating something complex as if it were simpler, throwing away detail. In the extreme case, one treats the complex quantity as atomic, unanalyzed, primitive." Programming languages are extended in order that atomic expressions can stand for larger syntactic complexes. In the Java example above, the expression `sum( S )` stands for the sum of the elements in the sequence `S`, i.e., $\sum_{i=0}^{n-1} S[i]$. "Naming is perhaps the most powerful abstraction notion we have, in any language, for it allows any complex to be reduced for linguistic purposes to a primitive atom" [19, pp.xv-xvi]. Named abstractions, such as *subroutines* (as in the Java `sum` example) or *objects*, are supported by all modern programming languages. In this way, programming languages can be extended to arbitrary levels, where complexes at one level become the atomic units at the next higher level through abstraction and naming. Any particular program is thus expressed at a number of different linguistic levels provided by the base language and each of its defined extensions.

3 Programming Languages as Social Constructs

Any particular programmer will belong to a number of overlapping and enclosing programming communities. Acculturation into a community will involve learning the set of linguistic abstractions shared by members of this community along with the associated knowledge that the abstractions stand for. Researchers examining programmer cognition have referred to such shared abstractions as *plans* [18]. Additionally, software practitioners have codified many of these abstractions in framework libraries such as C++'s *Standard Template Library* (STL) [14] and in repositories of micro-architectures called *patterns* [9]. These range from the most general abstractions that transcend programming language differences and are common to most trained programmers, (e.g. the *binary-search* routine), to abstractions common to object oriented programmers (e.g. the *iterator* pattern), to abstractions used by programmer subcultures, e.g. users of Java's Collection classes, users of a program library built by a specific company, or users of a library built for a single project. The individual programmer may even have a number of abstractions that only they themselves use. As Harold Abelson points out [8, Forward], "Perhaps the whole distinction between program and programming language is a misleading idea, and that future programmers will see themselves not as writing programs in particular, but as creating new languages for each new application."

The names used to describe abstractions are important to human readers but of no consequence to the computer since people are able to transfer semantic knowledge associated with particular names acquired through acculturation in non-programming social settings. For example, the standard meaning of *search* in English is to look for something, and naming a computational abstraction "search" provides the reader with a strong indications of its functionality. Using names that stand for real-world concepts can thus help program readers understand the meaning of programs. But other names without real-world referents can, through social habit and convention, come to stand for particular computational abstractions, such as Lisp's *cdr* or SQL's *clob*. Similarly, terms that do have real-world referents in natural language can have such meanings over-ridden by their use within programmer communities. For example, *push* refers to a computation that places an object on top of a stack, as opposed to "push the box out of the doorway" in everyday usage. The meaning that a reader accords to such expressions will have much more to do with such things as the level of standardization of the the named abstractions, the extent to which the reader has been acculturated into the language community, and the reader's beliefs about the original programmer's acculturation into this language community, rather than to any similarity of meaning between the computational abstraction and real world operations.

This acculturation occurs explicitly through instruction as well as individual study using professional journals, textbooks, and programs written by others. But a significant amount of the acculturation happens through communication, feedback, practice, and observation within the programming setting itself. Programmers code together, look critically at one another's code, engage in online discussion groups, and attend professional meetings, workshops, and conferences. Perhaps much of the success of *pair programming* [21] (one of the central components of Extreme Programming [3]) is due to the rapid acculturation and implicit knowledge transfer that occurs when programmers work in close contact with one another.

4 Real World Models and Shared Knowledge

Syntactic constructs in computer programs refer not only to the programming objects common to programmer communities, such as numbers, lists, and functions, but also to entities in the everyday world, from the employees and payrolls of an accounting system to the paintings and painters of an art museum's inventory system.

To facilitate a reader's understanding, the writer chooses some of the linguistic expressions so as to make explicit the program's function within the real-world context. For example, a programmer modeling biological phenomena might name some of the computational objects "locus", "genome", and "crossover" in order to establish the real-word context and mapping for these terms.

A reader's interpretation of the linguistic expressions in the program text crucially depends upon the shared knowledge between program writer and reader about the real-world. With respect to natural language understanding, James Allen writes [1, 548]

> shared knowledge ... is the knowledge that both agents know and know that the other knows. Shared knowledge arises from the common background and situation that the agents find themselves in and includes general knowledge about the world (such as how common actions are done, the standard type of hierarchy classifications of objects, general facts about the society we live in, and so on). Agents that know each other or share a common profession will also have considerable shared knowledge as a result of their previous interactions and their education in that field. ... While individual beliefs may play a central role in the content of a conversation, most of the knowledge brought to bear to interpret the other's actions will be shared knowledge.

Programmers are members of various overlapping and enclosing social groups within the larger society – for example, professional organizations, civic clubs, local communities, and national and ethnic cultural groups. These social groups have shared knowledge that is learned as part of the acculturation process within the group. Language that is specific to members of the group stands for the shared background knowledge of its members and provides an efficient means for discussing such knowledge. That is, acculturated users know not only the jargon, colloquialisms and idioms of the social group, but understand the concepts and knowledge underlying the terms, using them appropriately and understanding their appropriate use by others. As De Mauro comments on Wittgenstein's ideas concerning socially shared language [7, 53-4]:

> But in the measure in which you belong to my own community, you have been subjected to a linguistic and cultural training similar to my own and I have valid grounds for supposing that your propositions have a similar meaning for both of us. And the 'hypothesis' which I make when I hear you speak, and which you make speaking to me, is confirmed for both of us by both your and my total behavior.

Each individual may belong to many such "communities", each with its own linguistic and cultural training.

One of the perennial difficulties of software development is that programmers may not be members of the same social groups as the software *users*, the people who will interact directly with the program after it is developed. It is no surprise, then, that determining the software *requirements*, i.e. what the software is intended to do, accounts for a significant proportion of the software development budget and that a large percentage of software errors can be traced to errors in the requirements (up to almost 50% by some estimates [2]). The process of determining requirements involves a transfer of knowledge from users and clients (those who pay for the software) to programmers. This process is time consuming, error prone, and costly because not only is the sheer quantity of knowledge that programmers must acquire significant, much of this knowledge is implicit and taken-for-granted by the users, acquired by them via the informal, context-embedded processes described above for programmers. There might thus be a vast language and culture gap to bridge between the users – ranging from doctors and accountants to dancers and photographers – and the software developers.

Practices of placing expert users into software development organizations for the duration of a development project – one of the operating principles of Extreme Programming [3] – should enhance knowledge transfer by lowering communication costs and increasing communication bandwidth. What we may see in the future is an increasing movement of personnel in the opposite direction, where software developers join the embedding user organization for the duration of the software lifecycle, exploiting the fact that much of the knowledge about a program's meaning is encoded only in the neurons of the users and programmers.

5 The Cognitive Economics of Meaning Construction

The knowledge content of a message can far exceed the information-theoretic limit imposed by the number of bits used to encode the message, due to the immense amount of extant shared knowledge that a message can activate in the reader's mind. In his writings on cognitive sociology [6], Cicourel uses the term *indexicality* to refer to the aspects of language "that require the attribution of meaning beyond the surface form" since linguistic expressions serve as indexes for mental encodings of previous experiences.

The central economic choice of a writer of natural language text, then, concerns what to explicitly include in the text and what to leave out of the text, i.e., its degree of indexicality. In other words, what knowledge and abstractions can the writer assume that the reader already possesses? The writer trades text size against the risks and costs associated with ambiguity and misunderstanding; short texts are preferred to large texts, all other things being equal, since people are under various selection pressures to manage their own resources efficiently.

As with natural languages, programming languages are also indexical, since meaning construction requires that readers possess both a model of the executing hardware and a model of the embedding social context in which the program executes. Writers of computer programs make similar, though not identical kinds of choices as writers of natural language text. The difference concerns the fact that programs are read by computer as well as human readers. Programmers are constrained to use only those linguistic abstractions for which there exist explicit translations (via the above described extension mechanisms, and/or through the existing interpreters and language translators) to the underlying machine language of the executing hardware. Because programs are written at a number of different levels of description the program writer has considerable latitude in choosing the set of linguistic abstractions and the associated names within their programs.

There are three important reasons why a program reader makes economic choices when reading and interpreting a program. First, the understanding task itself is known to be computationally intractable [22]. That is, no efficient algorithm exists guaranteeing that meaning can always be correctly discerned. Second, the human brain has particular limitations, for example with respect to memory and processing speed that constrain the manner and rate with which inferences can be made. And third, human tasks are performed within a social and economic environment that limits the expenditure of resources on any particular task.

Not only must the individual program reader efficiently manage their internal cognitive resources, but they must also take account of the external environment in order to estimate costs associated with different knowledge acquisition actions and values associated with possible outcomes of these actions. Examples of external action-cost constraints include the software and hardware systems available to the reader for executing and maintaining the program, the presence of other personnel with expertise related to the task, communication technologies and policies that enable the sharing of data and knowledge, project development practices that provide a documented historical trajectory of the program's evolution, and opportunities for further education and training related to the task at hand. Subjective outcome values, though particular to the reader, will certainly be influenced by such things as her individual value system, beliefs about the institutional and social tolerance for errors, and beliefs about the economic climate in which the organization operates.

As a consequence of these economic constraints, we can conjecture that readers employ understanding processes that allow them to tradeoff the amount of resource that they devote to the understanding task – most importantly *time* – against the level or quality of meaning that they construct. We can take this process to be approximately monotonic, i.e., more resource will in general produce more and better understanding. Without such a process, the reader would have no basis for expending further resources in pursuit of greater understanding.

How much and what kinds of resource a reader devotes to any particular reading episode will depend upon their tradeoff of the perceived costs and benefits as mentioned above associated with their different action choices, and the level of understanding that they believe they possess at different times during the problem solving episode. Empirical studies supporting this conjecture indicate that programmers read only a portion of the program text related to their task rather than the entire program – the so-called *as-needed* reading strategies [13]. Further, we can expect program readers to exploit their shared knowledge. This prediction is consistent with the study described in [12], where programmers used the abstraction structure and names to determine which particular parts of the program to read and which to ignore: "Subjects spent a major part of their time searching for code segments relevant to the modification task and no time understanding parts of the program that were perceived to be of little or no relevance. ... Subjects hypothesized relevance based on their knowledge about the task domain and programming in general. Subjects used procedure and variable names to infer functionality. ... While looking for code subjects guessed correctly the names of procedures they had not seen." The program reader's economic choices therefore concern determining the level of description at which the program text should be read and the level of understanding that must be achieved in order to carry out the task at hand.

Several studies have attempted to determine if readers traverse program text and construct mental representations of the text by starting at lower levels of program description and moving to more abstract levels, or by movement in the opposite direction, from abstractions to concrete descriptions [15,16,4]. The above discussion, however, implies that there is no such fixed strategy; rather, a reader might move in either direction depending upon estimates of the costs and benefits of their action choices at any given time. This is not to say that readers employ a strict decision-theoretic policy, such as that described in [11]. Such a strategy violates the computational constraints mentioned above, since this meta-cognitive activity, i.e., enumerating preferences and evaluating expectations, is itself too costly an activity to perform optimally. Nonetheless, we can expect some type of *minimal*, or *bounded* rationality, as proposed by Cherniak [5] and Simon [17], that enables agents to pursue preferred world states to the extent that they are known, but in an approximate and heuristic fashion. This is consistent with reports by von Mayrhauser and Vans [20] in observing program understanding behavior among experts in large-scale comprehension tasks, which they describe as moving in either the upward or downward direction opportunistically.

6 Conversational Maxims and Cooperative Social Norms

An additional factor that helps a reader gain computational efficiencies is the extent to which the reader believes that the writer intends her text to be understood. Grice [10] argues that hearers in natural language conversation assume that speakers follow cooperative social norms. These norms can be viewed as a set of implicit rules, which Grice called *conversational maxims*. These maxims, as summarized in [1, p566] are:

Maxim of Quality – Do not say things for which you lack evidence.
Maxim of Quantity – Make your contribution as informative as required, but not overly informative.
Maxim of Manner – Avoid obscurity of expression and ambiguity.
Maxim of Relation – What you say should be relevant to the current topic.

With respect to programs, these maxims are generally satisfied in the organizational and social settings in which programs are produced. The Maxim of

Quality is met since programs must be translatable and executable in order to provide functionality. The Maxim of Quantity is met when programs are written at different levels of abstraction, so that readers can balance information content with resource expenditure and the amount of knowledge that the reader brings to the reading task. The Maxim of Manner is met when the original programmer uses public, shared language, such as that codified by standards committees and user groups, instead of private, ad-hoc language that will take the reader longer to decode. And the Maxim of Relation is met when the program writer structures their code into cohesive units, e.g. subroutine libraries, objects, frameworks, plans, and patterns. That is, the goal structure provided by programming plans provides cohesiveness and "topicality" to program text.

Studies by Soloway [18] confirm that readers have strong expectations that program writers follow such relevance constraints, which he termed *discourse rules*, and comprehension was negatively impacted when programs violated these discourse rules. Although program writers are not obliged to follow these maxims, it would nonetheless be surprising for writers to violate the very principles that provide such efficiencies in their natural language communications.

7 Summary

Programs are written so as to be both executable by computers in order to carry out useful work, and to be read by other people who must maintain the programs in order to fix errors and to extend the program's functionality. In order to construct meaning from a program the program reader makes economic choices about her actions where the costs and benefits are influenced not only by cognitive constraints but also by the organizational and social context in which the program-related activities occur. This context affects the costs that the reader assigns to the different actions available to her, as well as to the values associated with the different expected outcomes of performing these actions. Of fundamental importance is the extent to which the reader believes that she shares common knowledge with the program writer, both in programming and application domains. This common knowledge is associated with the different social and language-using groups to which the reader and writer belong. Group-specific language is used to economically index the large quantity of group-specific knowledge that provides the interpretative context for meaning construction. Following cooperative conversational maxims, program writers exploit shared knowledge and language by using the abstraction and naming mechanisms of programming languages to express programs at a variety of different levels. Program readers likewise exploit this shared knowledge and language as well as the cooperative communicative intent of the writer to balance the level of meaning that they construct against the resource constraints under which they operate.

References

[1] James Allen. *Natural language understanding.* Benjamin Cummings, 2nd edition, 1995.

[2] Victor Basili and Barry Perricone. Software errors and complexity: An empirical investigation. *Communications of the Association of Computing Machinery,* 27(1):42–52, 1984.

[3] Kent Beck. *Extreme programming explained.* Addison Wesley, 2000.

[4] Ruven Brooks. Towards a theory of comprehension of computer programs. *International Journal of Man-Machine Studies,* 18:543–554, 1983.

[5] Christopher Cherniak. *Minimal rationality.* MIT Press, 1986.

[6] Aaron Cicourel. *Cognitive sociology: Language and meaning in social interaction.* Penguin Education, 1973.

[7] T. De Mauro. *Ludwig Wittgenstein: His place in the development of semantics.* Reidel, D., 1967.

[8] D. Friedman, M. Wand, and C. Haynes. *Essentials of programming languages.* McGraw Hill, 1992.

[9] E. Gamma, R. Helm, R. Johnson, and J. Vlissides. *Design patterns: Elements of reusable object-oriented software.* Addison-Wesley, 1994.

[10] H. P. Grice. Logic and conversation. In P. Cole and J. Morgan, editors, *Syntax and Semantics,* volume 3. Speech Acts, pages 41–58. Academic Press, 1975.

[11] R. Jeffrey. *The logic of decision.* McGraw-Hill, 1965.

[12] J. Koenemann and S. Robertson. Expert problem solving strategies for program comprehension. In *ACM Human Factors in Computing Systems CHI'91,* pages 125–130, 1991.

[13] D. Littman, J. Pinto, S. Letovsky, and E. Soloway. Mental models and software maintenance. In Soloway and Iyengar, editors, *Empirical studies of programmers.* Ablex publishing corporation, 1986.

[14] D. Musser and A. Saini. *STL tutorial and reference guide.* Addison-Wesley, 1996.

[15] N. Pennington. Comprehension strategies in programming. In Olson, Sheppard, and Soloway, editors, *Empirical studies of programmers, second workshop.* Ablex publishing corporation, 1987.

[16] Teresa Shaft and Iris Vessey. The relevance of application domain knowledge: The case of computer program comprehension. *Information systems research,* 6(3):286–299, September 1995.

[17] H. Simon. *Models of bounded rationality.* MIT Press, 1958.

[18] E. Soloway, B. Adelson, and K. Ehrlich. Knowledge and processes in the comprehension of computer programs. In Chi, Glaser, and Farr, editors, *The nature of expertise.* Erlbaum, 1988.

[19] G. Springer and D. Friedman. *Scheme and the art of programming.* McGraw Hill, 1989.

[20] A. von Mayrhauser and A. Vans. Program comprehension during software maintenance and evolution. *Computer,* pages 44–55, August 1995.

[21] Laurie Williams and Robert Kessler. The effects of "pair-pressure" and "pair-learning" on software engineering education. In *Proceedings of the 13th Conference on Software Engineering Education and Training,* pages 59–65, 2000.

[22] Steve Woods and Qiang Yang. The program understanding problem: Analysis and a heuristic approach. In *Proceedings of the 18th International Conference on Software Engineering (ICSE-96),* pages 6–15, Berlin, Germany, 1996.

Sense from a Sea of Resources: Tools to Help People Piece Information Together

Aran Lunzer and Yuzuru Tanaka

Meme Media Laboratory, Hokkaido University, Sapporo 060-8628, Japan
{aran,tanaka}@meme.hokudai.ac.jp

Abstract. Spurred on by the eager adoption of XML, the world appears to be on the verge of a revolution in the ease with which information resources from diverse, remote providers can be brought together into new assemblies, expressing new concepts. To benefit from this revolution we will need frameworks that help providers to organise their information and enable access to it, and tools that will help would-be users of these resources to find and combine the pieces that they want. We report our ongoing research on the Topica framework and the Context Workbench, that together address these new challenges in a spirit of helping information users to make their own sense out of the sea of possibilities.

1 Introduction

A person's representation in a payroll system is not a convincing model of a person. Nor does it have to be; to automate a payroll process, all that is needed is certain administrative information relevant to each employee's relationship with the company. Elsewhere there may be medical, academic, financial, even hobby-club membership records for the same person – but traditionally these would be unconnected realms of information, each suited to the particular enquiries, decisions, and calculations needed within its domain.

However, this picture of isolation is changing, driven by the explosive growth of networks in general, the Internet in particular, and especially by the commercial interest in bridging between the islands of information belonging to potential trading partners. Recently effort has focussed on the development of semi-structured information representations such as XML, setting standards for the names of business-related entities and for the values they can take. Such standardisation allows information designers working independently to create resources that can be meaningfully connected to or compared with each other.

Resource integration is likely to be of interest outside the commercial world, too, where the growth of the World-Wide Web (WWW) is changing people's access to everyday information. Instead of being able only to consult carefully edited mass-media sources, people now have access to an unruly sea of information on unlimited topics and of unlimited variety of quality and bias. And although there may be no business cases to drive the standardisation of these resources, even without rigorous standards there are likely to be domains in

M. Beynon, C.L. Nehaniv, and K. Dautenhahn (Eds.): CT 2001, LNAI 2117, pp. 175–189, 2001.

which simple forms of coordination are possible. An obvious example is the correlation of dates, allowing daily records of one form to be matched up with those of another – such as, perhaps, cinema guides with weather forecasts.

We believe that effective coordination of resources depends on the availability of interactive tools that support users in bringing together the items that are of interest, and working with them as connected structures.

In the following section we explain our ongoing work on establishing an information architecture suited to ad-hoc assembly of resources. Section 3 then introduces a novel framework for organising and providing access to these resources, and Sect. 4 describes the kinds of interactive operation that will enable people to make effective use of that framework. Finally, in Sect. 5 we discuss related work and our development challenges.

2 The 'Meme Media' Approach to Expressing Resources

The primary issue in providing support for people to perform ad-hoc assembly of information is deciding on an appropriate form for that information to take. For many years, the work at this laboratory has been motivated by an interest in developing a form of computer-held information suited to the implications of Richard Dawkins's *meme* concept [3]. Each meme – a cultural gene – would have to be expressed as a structure that could be replicated, recombined with other memes, and pitched into competition with other memes for survival as a member of the *meme pool* (cf. gene pool) of their host society.

The approach that we adopted is a building-block architecture, in which computer-held resources are wrapped in a way that allows them to be combined, transported, and re-edited. We often refer to the overall approach as 'meme media' – a combination of expressions and facilities appropriate to the handling of memes. Given the possibilities for commercial activities based on the trade of memes, the meme pool can also play the role of a *meme market*.

Research into meme media and meme market software has been pursued here since 1987, leading first to development of the 'IntelligentPad' architecture [21, 23], based on a two-dimensional graphical representation of meme components, and later the 'IntelligentBox' architecture [16] in which the components have three-dimensional appearance. In each case, users can combine components using direct-manipulation operations on screen – in IntelligentPad this appears as the 'pasting' of one card-like 'pad' onto another to become its child – and can choose to establish information-passing connections between parent and child through named 'slots' that represent the external interfaces for each component.

The international distribution infrastructure for a meme pool has emerged conveniently in the growth of the World-Wide Web (WWW), although to help users exchange pads in their native form currently requires some specialised server and browser mechanisms, that we refer to collectively as the Piazza architecture. Each site – also called a *piazza* – supports drag-and-drop manipulations not just for copying pads from the piazza into a user's local environment for personal use, but also to move pads in the other direction, allowing instantaneous publishing of the user's own pads in the publicly accessible repository; Piazza is thus one of the new generation of server technologies that allow people browsing the Web to add their own information to remote sites. In particular, given the availability of gateway pads that act as links from one piazza site to another, we foresee users establishing their own piazzas to act as galleries or outlets for pads, then installing gateways to these sites in other, centralised piazzas where potential customers are likely to be browsing.

2.1 Applications of Meme Media

Our group and a number of collaborators from academia and industry have developed various experimental applications that illustrate the fundamental activities involved in using meme media – namely, encountering existing assemblies of meme components, taking one's own copies, modifying the resources or the way they are composed, then making the new compositions publicly available in turn.

One domain for such activity is in the distribution and exchange of scientific results. A group within the Faculty of Physics at Hokkaido University has investigated the use of IntelligentPad and Piazza for the international publication, distribution and exchange of nuclear reaction experiment data. Because pads can have active behaviours, Piazza can be used to distribute not just data but also pads that wrap simple analysis tools, perhaps developed by the experimenters themselves. The snapshot in Fig. 1, for example, shows an exchange centred around a scatter-plot tool that supports the overlaying of results from multiple data sets. Thus a group of scientists might receive some data from a remote collaborating team, copy a supplied tool and apply it to their own results, then re-publish the data, along with some added comments and results from their own similar experiments, for further review and update by other collaborators and peers. Such accelerated exchange and feedback is highly desirable for the encouragement of interdisciplinary, international distribution and evolution of intellectual assets in general.

Another enthusiast for meme-media ideas is Seigo Matsuoka, a prolific writer and commentator on society-related topics including the creation, editing and use of information. Matsuoka is also the leader of the Editorial Engineering Laboratory (EEL), a private research organisation that is, among other things, embarked on a long-running project to encourage the traditionally reticent Japanese public to play a more active role in the flow of information within their country, experimenting for themselves with the way ideas from diverse sources can be re-edited to make new sense.

The EEL's first use of meme media was in the design and construction of *The Miyako*, a digital archive system for Kyoto cultural heritage. In addition to providing the front-end for an elegant browsing interface to a large database of photographs, writings and recordings relating to Kyoto, the use of IntelligentPad technology paved the way for future versions in which users would be able to perform personal re-categorisations of the resources, or even to add their own.

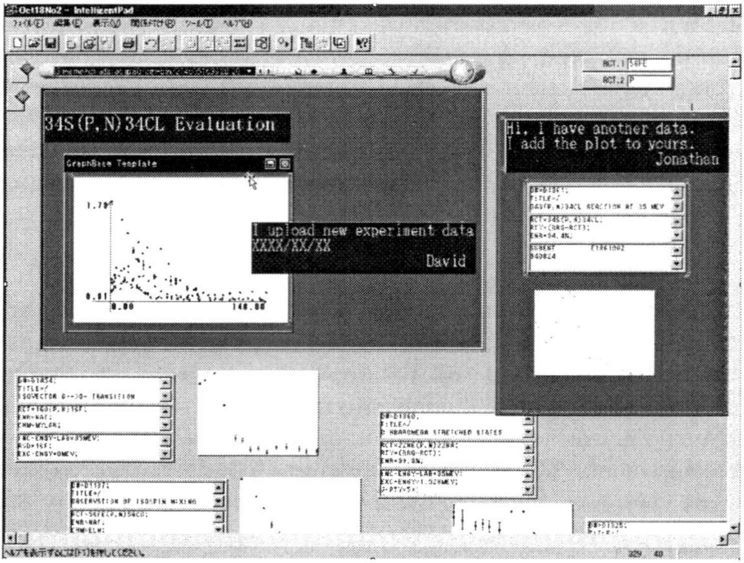

Fig. 1. Piazza applied to the distribution and reuse of nuclear-reaction experiment data and tools. Here the upper-left quadrant of the workspace is a portal onto a publicly shared piazza; resources can be dragged freely between here and the surrounding local workspace.

A more ambitious experiment is an Internet-based environment called Meme Country, which began active operation in Japan at the start of last year. This is a joint project in which the EEL is backed up by a number of individuals and companies, providing ideas and content as well as technical support. Like other long-lived online community sites, Meme Country establishes a virtual country in which Internet users are free to register and participate from time to time from the comfort of their own computers. Meme Country's focus is on contribution and exchange of knowledge: ideas, designs, even jokes. Its activities, often cleverly designed to mask their underlying computer technology, make extensive use of 'meme cards', based on IntelligentPad pads, to which words, pictures and other resources can be added. These cards can be transferred between users, stored in centralised or personal sites, and supplemented with comments and links that are themselves publicly viewable. The year-long first phase of operation saw more than 1,500 Meme Country 'citizens' participate in activities ranging from simple daily word games to remote-learning courses in fields including rhetoric and industrial design. A few participants worked so hard that the EEL decided to offer them jobs!

3 The 'Topica' Approach to Resource Organisation

As became clear early in the rising tide of the WWW, a tremendously successful medium for exchanging ideas leads to a tremendous challenge in organising and providing access to the large numbers of heterogeneous, independently created pieces of information available through that medium. Standard organisation mechanisms such as lists and directories, stretched and adapted in various ways, have certainly met with some success in cases such as Yahoo. But clearly there is the scope for richer organisational approaches, offering something beyond drill-down navigation among hierarchical sub-groups.

We decided to draw a parallel with the world of consumer products, and its well-established traditions of offering information that is not focussed on a single type of product, but on a group of related products – for example, products that have the same context of use. Modern supermarket layout depends fundamentally on such techniques, as does the layout of department stores, malls, and even of towns as a whole; likewise the design of documents such as brochures, catalogues, and magazines loaded with advertising material. As consumers we have thus become accustomed to helpful (albeit commercially motivated) providers of resources offering their information in places where it is convenient for us to find it. Given that online environments offer rich possibilities for creating novel, dynamic forms of virtual space, how might such spaces be used in offering products to consumers who roam the virtual world?

Below we lay out the principles that we are applying in our novel framework for spatial organisation or and access to information resources. We call this framework *Topica* [22], the title of one of Aristotle's treatises on logic.

3.1 Resources That Appear in Many Contexts

A fundamental principle of the Topica framework is that any given resource can belong simultaneously to multiple contexts. This is enabled by always referring to resources using pointers, or URIs; responsibility for the storage of their content – be it text, video, a relational table, or whatever – lies outside Topica.

This approach has various advantages over the hierarchical resource management provided by most of the standard computing platforms. For example, say you have a number of resources (typically, files) relating to the presentations you have given at conferences. A strict hierarchical approach to file management requires that you choose some fixed way to organise them. Perhaps you decide on using folders to group the conferences by year, then provide a second level of folders containing all the materials relating to a given conference. Or perhaps you prefer a subdivision first by name of conference (so that all Cognitive Technology conferences are together), then by type of material (call for participation, draft submission, reviewers' notes, and so on). Unfortunately, though, even an arrangement that you find is useful much of the time can be frustrating on other occasions – such as when you quickly want to see all the drafts you submitted in a given year. And as Dourish et al. [4] point out, attempting to work around this by using the shortcut, link, or alias mechanisms allowed by various platforms brings further challenges.

It is preferable to let users develop as many alternative arrangements for the information as they feel are useful. On any given occasion of interaction a user can choose the arrangement that suits the goals of the moment, safe in the knowledge that the resources that will be arrived at are independent of the navigational approach taken to reach them. Assistance can be provided for maintaining the consistency of the alternative structures when additions or removals are made, for example with the help of active rules.

3.2 Topoi: Storage Locations with Meaning

Items are stored in a topica document[1] by being registered at distinct storage locations defined within it. Each location is referred to as a *topos* (plural *topoi*). Intuitively, a number of different objects all being held at the same topos should signify that those objects have something in common, although beyond this there are no strict conventions regarding how topoi should be defined or used. Many objects can be held at a given topos, and a given object might be held simultaneously at multiple topoi.

If a topica document is considered as analogous to a shop, then the topoi serve as that shop's shelves. Products of the same type, or serving a similar purpose, may be laid out together at a given topos. The relationship between the topoi may also be significant, as is the relationship between shelves in a shop – shelves may be grouped into categories, dividing the shop broadly into a number of sections. Consistent use of arrangement both within and between the topoi can be used to help guide consumers to the products that interest them – just as in a shop the shelves near the floor may be dedicated to large, everyday items whereas those around eye level are reserved for luxury or speciality goods, while within some section of shelving the items may be arranged according to a common convention such as alphabetical ordering.

So one topica document providing access to the conference-related materials, for example, might be arranged in a tabular layout divided across the display according to year, and down the display to provide topoi containing the various types of material. Documents representing other organisations of the same information would differ in their choice and layout of topoi.

3.3 Inter-topos Coupling

One feature of Topica that takes advantage of the flexibility of a virtual domain, as compared with physical space, is the ability to make invisible connections

[1] We refer to resource units as *documents* in a broad sense; a document's content need not be textual, but might be an image, a video, an active interface, or a collection of references to other documents. Thus documents are accessed through other documents, just as WWW link collections are themselves pages that can be linked to from other pages.

between an item at one topos and other, related items at other topoi. As a dramatic example, imagine a food store in which someone (perhaps the owner) had, as a service for customers, established connections between the different items on offer. A shopper might be able to choose some hors d'oeuvres items at the delicatessen section, then ask for all the shelves magically to clear themselves of all but the foods that would make appropriate continuations for a meal. Different hors d'oeuvres pave the way to different meals.

For a more down-to-earth illustration, consider a topica document that behaves like a data-entry form. Figure 2 shows a template for the management of invitation letters, based on the assumption that the essential layout and content are fixed while just the names and circumstances change. Topoi are defined for the variable items such as the date of the letter, the recipient, and the subject to be discussed. However, clearly the values at these topoi are no longer independent: they represent explicit instances of letters, each with a particular date, recipient, topic and so on. This form of dependency can be regarded as a coupling between specific values at the various topoi; for this simple one-to-one form of mapping, the coupling could be modelled as relational tuples whose positions correspond to the topoi.

Note that the existence of coupling is a structural property of a resource, and does not in itself dictate what forms of interaction with that resource may be offered; that is the responsibility of the browsing tools that are used to view it. The hors d'oeuvres scenario is an instance of what we refer to as the *focus* operation, by which a user chooses from among the available values at one topos, and the resource's presentation is changed so that the other topoi now offer only

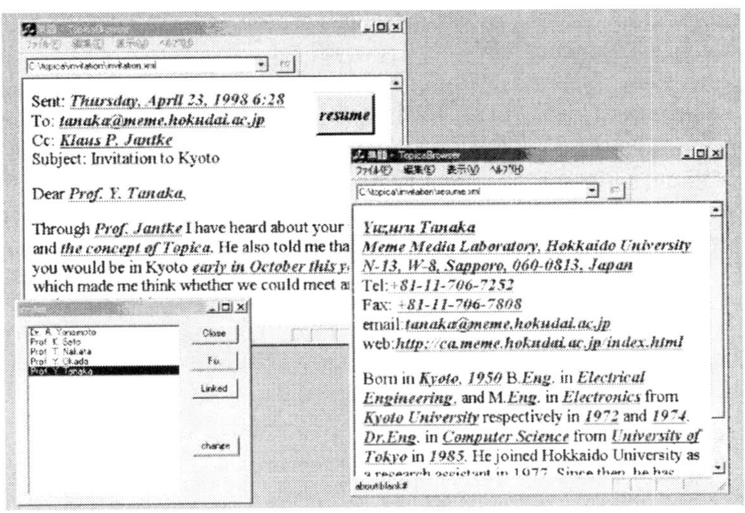

Fig. 2. A topica document (upper left) for handling invitation letters, including attachment of the invitee's CV – also expressed as a topica document (lower right).

those values that are consistent with the chosen value(s). For the invitation-letter repository, such an operation would support straightforward form-based query: choose a date, and see only the people and discussion topics mentioned in letters sent on that date. The interaction mechanisms are also responsible for helping users to create new couplings: to add a new instance of invitation letter, for example, a user would fill in the various topoi with appropriate values then invoke an operation to register a new coupling between them.

3.4 Navigation and Schemata

The invitation-letter topica document shown above also has a topos that is used to hold the CV of the person being invited. The values at the CV topos might themselves usefully be topica documents, expressed as templates with topoi as the fields for recording personal details and career information. When a user asks to look at such a value, a new topica-document display will be opened; we consider this an action of *navigating* to the resulting topica document, analogous to following a link to a Web page. Having done so, the user is again confronted by a topica document that can perform as a template, and this offers the opportunity to find other ways of filling in that template – in this case, other CVs. But the user has arrived at this document through a particular navigation path; should the values that are available be affected by the path that was followed?

We propose, firstly, that the ability to use a topica document as a template for accessing other document instances should depend on all those documents being tagged as conforming to the same *schema*. The schema defines details such as the names of some set of topoi that must be included in the document. Then a user viewing some topica document serving as a CV could ask to have this document function as a template, through whose topoi all the values held in documents using the same CV schema could be seen.

Secondly, we support the automatic recording of a path of navigation between topica documents. Through this facility, a user who arrives at a CV as a result of navigating through an invitation letter could ask to change it into a template view containing only the CV-schema documents that are themselves also held within invitation letters – specifically, within documents that conform to the same invitation-letter schema.

4 Topica Interaction through the 'Context Workbench'

As mentioned earlier, Topica's guidelines for creating a repository out of resources connected in various ways do not in themselves dictate the forms of interaction that may be applied to those structures. In this section we introduce a number of interactive facilities that form the core mechanisms of the Context Workbench – just one approach to providing user support for browsing and manipulating Topica-structured resources.

4.1 Pivoting between Multiple Contexts

The Context Workbench must support what we refer to as the *pivot* operation, for switching among the various contexts in which a given resource object participates. For example, consider the camera-ready version of a paper for an upcoming conference. This may be referred to in a number of topica documents, such as: a collection of all materials related to that conference; a list of the author's publications for this year; reading materials for a departmental lecture course; a cache of recently modified resources to be copied during the next system backup. Having arrived at the paper through navigation to any one of these contexts, it may be useful for a user to see the various other contexts in which it appears, and to be able to move rapidly among them to find the other resources that are related in each case.

As far as possible, the mechanism for switching between contexts should help the user to grasp the distinction between the unchanging central resource (or possibly group of resources) and the changing surroundings that reflect the new context. The suggested name 'pivot' reflects the idea of the central resource remaining stationary while other information moves around it – or perhaps, in practice, as the old view fades away to be replaced by the new one. Among the challenges in presenting this operation is that of how to handle cases that might involve tens or hundreds of available alternative contexts.

4.2 Correlated Manipulation and Rearrangement

There are many kinds of user operation that can make use of inter-topos coupling of resources, such that an operation applied to a value at one topos is applied in a correlated way to its corresponding values at other topoi. In the context of the invitation-letter example we described the *focus* operation – being able to choose a subset of values at one topos and thus filter the content of all other topoi to just those values that are consistent with that subset. But instead of simply retaining or removing values from view, coupling may be used to support forms of *brushing*, as commonly offered in visualisation systems – i.e., being able to tag one item in some way (such as by changing its colour), and have all its coupled resources automatically tagged in the same way.

Rather than just the properties of individual elements, we can also propagate group properties, such as the relative placement of a number of items. If a user asks to sort the dates at the invitation-letter 'date' topos into ascending order, for example, the values at the other topoi could also arrange themselves into their corresponding date-dependent order. We can even provide correlation of arbitrary spatial rearrangement: consider someone who maintains a shopping list that couples shops to be visited (at one topos) with the items that are currently needed from them (at another). To plan shopping outings, it may be useful to cluster the shops informally according to which ones it would be convenient to visit in one trip – perhaps to fit in with regular weekday or weekend errands. If this clustering of the shops can be automatically propagated to create a clustering of the items that are to be bought, this could make it easier to grasp which of them could be obtained in each of various possible outings.

4.3 Derivation or Lookup of Related Resources

By connecting resources to others that deal with the same entities, users can make the system display additional relevant information. For example, consider a resource listing second-hand cars that are on sale in some neighbourhood. Typically such a list would provide only basic information about each car, such as the make, model, year of manufacture, and current mileage. Further details of interest to a prospective buyer – such as each car's typical performance and fuel consumption, and perhaps qualitative information such as reliability and owner satisfaction – may be found in separate resources provided by manufacturers and consumer organisations. We would like to support people in obtaining the relevant additional information for each vehicle on offer, either by direct lookup or by a simple calculation (for example, to generate an estimate of the annual fuel costs based on the buyer's typical driving needs).

Essentially what we need to do is extend the inter-topos coupling mechanism to work at the inter-document level. The ease with which this can be done depends on whether the objects that the resources refer to in common (for example, the various car manufacturers) have been specified using directly comparable value domains. Where that is the case, a user can request the equivalent of a relational join between the topoi in their respective documents. Where such a join cannot be defined directly, the resources can be connected instead with the help of another document that uses inter-topos grouping to connect the values used in one resource with their equivalent descriptions in the other.

One feature enabled by correlated manipulation is a form of aliasing – the opportunity to manipulate some set of unfamiliar objects using a set of more familiar objects as handles. For example, in Japanese administration it is still standard practice to record dates using the emperor-based method of naming the year (2001 is Heisei year 13, or H.13) – so how might one partition a set of such dated items according to, say, Western-calendar decades? As a first step one would need a topos in which the various emperor-year values were connected to the sets of events falling in those years. Then one could create (or perhaps find, since it is a standard form of conversion) a mapping resource that couples each Western-calendar year with its equivalent(s) in the Japanese system. Through this mapping, the original items can be manipulated using the new encoding.

4.4 Explicit Support for Resource Inconsistencies

Building information models that represent real-world entities always involves taking subjective decisions on where and how to simplify the rich detail of those entities. Therefore only in a model that has been designed throughout by a single individual or organisation can one hope to find a consistent set of compromises. Combining resources produced by diverse organisations will almost inevitably involve the appearance of some inconsistency or outright disagreement in the information models. In the car-resource example, fuel consumption figures published in different places for the same model of car may differ according to the particular tests that have been run, or the grade of fuel being used. In other

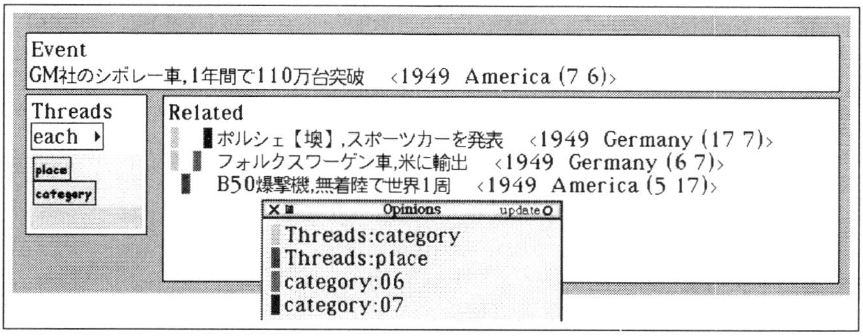

Fig. 3. Parallel derivations in HIBench. The cell called Related is deriving its contents by applying the active matchers in the combinator cell Threads to the reference event currently held in Event. Because Threads specifies both 'place' and 'category' matching, and because the reference event belongs to two categories (7 and 6), matching events are found for each category as well as for the place (America). The three results are marked with coloured tags whose meanings are listed in the 'Opinions' view. (Category 7 covers corporate business, while 6 is international trade. The reference event here comments on Chevrolet production. The first related item (a match on category 7) is an announcement by the Porsche company; the next tells of the export of Volkswagens; the last match is a round-the-world flight by an American B50 bomber.)

cases there may be facts that are openly disputed – such as different political views regarding which country a particular city is a part of. We would like to help users to handle these cases explicitly, seeing and understanding the different values, how they arise, and the different effects they have on any derived results.

Providing support for dealing with such inconsistency has been the motivation for our work on *subjunctive interfaces* [7,8]. In essence, a subjunctive interface provides the framework within which a user can specify multiple provisional values for any given setting, can pursue simultaneous, parallel derivations on the basis of those different values, and can view the various outcomes in juxtaposition to enable their comparison.

Figure 3 shows a display detail from HIBench, a simple tool that we built to try out the subjunctive-interface mechanisms that will be needed for context workbenches. HIBench was inspired by, and is based around, a repository of information items representing historical events, which forms the main content of [10][2]. Each event comprises a short news-headline-like description, a date, one or more locations (typically countries) affected by the event, and one or more categories (denoting fields such as politics, education, various branches of science and technology, literature and other arts, sport etc.) assigned by the repository's

[2] In total there are approximately fifty thousand items; we used just the latter half of the repository, covering events in the 20th century. All information in the repository is in Japanese; the English place-name translations seen in the screen shot were built in to assist non-Japanese explanation.

editors. HIBench allows construction of a processing structure made up of cells whose contents are derived using scripts that refer to other cells. One form of derivation is the use of queries to reveal the context of an event in terms of other nearby events that match it in some way – such as by having occurred at the same location, or by belonging to one of the same categories. In the figure we see a case in which a user has requested two separate types of matching (based on location and on category), and where the reference event in fact belongs to two separate categories. Thus three separate queries are automatically generated and processed. To help the user understand how the different results arose, each is tagged according to the derivation path that generated it.

As well as HIBench-like support for specifying alternative queries, a general Context Workbench must allow the specification of alternative resources to be combined, and provide juxtaposed display of the alternative results. Our goal is to make this interface consistent with the correlated-arrangement facilities described in Sect. 4.2, so that the user can encode the layout of diverse results by applying spatial arrangement of the resources used in deriving them.

5 Discussion and Related Work

The long-term direction of our meme-media research is clearly in tune with the call by Shneiderman [20] for computer scientists to build 'environments that would empower personal and collaborative creativity by enabling users to: collect information from an existing domain of knowledge, create innovations using advanced tools, consult with peers or mentors in the field, and then disseminate the results widely.' In particular, work towards frameworks for combining, editing then re-publishing resources has been placed firmly on the computing agenda by the rise to prominence of XML as a standard that can be revealed at the user level, rather than hidden within anonymously designed services; many groups are now working on user-oriented tools for assisting with such resource integration (e.g., [2,13]).

The assignment of unique identifiers to resources, to help in integrating a range of distinct locales that all refer to a single given resource, is also a key property in the construction of information repositories in the Placeless Documents project [4], and in Ted Nelson's recent work on ZigZag [14]. Each of these takes a different approach to the representation of resource collections and the mechanisms by which they may be constructed by users. In Presto [5], an experimental environment built on top of the Placeless Documents system, documents are gathered in *fluid document collections*. These are dynamically updated groups, defined by users in terms of a *query term* based on document attributes, and *inclusion* and *exclusion* lists to specify documents that must be included or excluded regardless of whether they match the query. By contrast, the only data structure supported by ZigZag is a form of object graph in which each resource can be chained – with a single forward link and a single back link – into any number of lists referred to as *dimensions*. Through the early interfaces for ZigZag one could only manipulate this structure a single link at a time, although there are no theoretical reasons against having an interactive layer that supports maintenance operations of a higher level, such as dynamically maintained collections. We are also following with interest the ongoing elucidation of information structure and interactive operations suitable for *context bases* [24] in a group led by Nicolas Spyratos, one of our external collaborators on the Topica project.

Many researchers have taken an interest in the potential benefits of tapping into users' perceptual abilities by supporting the arrangement of resources in a spatial manner. Barreau and Nardi [1] studied how people use spatial layout on graphical 'desktop' interfaces simply to help them remember where they put things. The Data Mountain [17] is one of the most recent tools to exploit this memory-assistance property; its features include the use of a pseudo-3D display of a sloped surface that appears to recede away from the user – reportedly allowing it 'to display more information without incurring additional cognitive load, because of pre-attentive processing of perspective views.' While these projects have mainly considered the placement of documents as isolated items, others have investigated specifically how spatial arrangement can be a way for users to build informal groupings of their resources, possibly as an intermediate step before deciding on and adopting a more formal categorisation. Mander et al. [9], for example, describe a tool supporting the creation and manipulation of 'piles' of electronic documents, intended to be like the paper piles that people often create in their offices. The VIKI [19] and VITE [6] systems take this idea further, incorporating an explicit model for the transition from informal to formal categorisation and offering spatial-pattern recognition mechanisms to assist the user in making such conversions.

Allowing users to create new compositions of information requires a flexible form of browser framework. We aim to provide a structure that is spreadsheet-like, in the sense that users should be able to define new cells for revealing information extracted or derived from some part of the underlying repository. However, these cells must be able to support the full richness of the layout and interactivity defined for the resources they are showing. So, like the embedding of plug-ins within page regions in a Web browser, the job of our framework is to allocate display space to resource-specific browsing components and to coordinate user actions within them such as object selection or rearrangement. There are various existing frameworks that offer guidance on how we might address this need: Snap-Together Visualization [15] handles distribution and coordination of a single table of relational data among arbitrary types of view that conform to a standard API comprising simple display and selection operations; the Visage project (e.g., [18]) provides more sophisticated coordination of views, the data elements they are displaying, and user-specified operations such as brushing and aggregation.

Finally – switching to a high-level view of what the Context Workbench facilities are intended to provide for users – we take some suggestions from the writings of Seigo Matsuoka [11,12] regarding worthwhile tool support for manipulating information. Firstly, the aliasing that will be enabled by our facilities for correlated manipulation using derived resources (Sect. 4.3) fits his suggestion of the value of letting individuals work with information using terms with which they are comfortable and familiar, rather than being forced to use some standardised, possibly rather abstract terminology – a need that echoes Saussure's distinction between *langue* and *parole*. Secondly, being able to map an arrangement that makes sense for some set of entities onto another set to which it is not directly applicable (our example was shops, and the items to be bought there) touches on what Matsuoka suggests as perhaps the principal ingredient of the communication of ideas: not the transfer of isolated concepts, but of patterns for relating concepts to each other. Finally, support for working with provisional or ambiguous values can be seen as relevant to what Matsuoka refers to as the *world model* used in expressing information, meaning a stated or implicit context that determines how a particular piece of information should be interpreted. While the parallels mentioned here are undoubtedly simplistic, by keeping these suggestions in mind we hope to maintain a longer-range view of what our tools are aiming to provide.

And we look forward to discovering other ways in which giving people the facilities to combine, filter and derive information will help them to make their own sense from the rising tide of resources.

References

1. Barreau, D. and Nardi, B.A.: Finding and Reminding: File Organization from the Desktop. ACM SIGCHI Bulletin **27**(3) (1995) 39–43
2. Ceri, S., Comai, S., Damiani, E., Fraternali, P., and Paraboschi, S.: XML-GL: a graphical language for querying and restructuring XML documents. In Proceedings of the 8th International World Wide Web Conference, Toronto, Canada. (1999)
3. Dawkins, R.: The Selfish Gene. Oxford University Press, Oxford (1976)
4. Dourish, P., Edwards, W.K., LaMarca, A., Lamping, J., Petersen, K., Salisbury, M., Terry, D.B. and Thornton, J.: Extending Document Management Systems with User-Specific Active Properties. ACM Transactions on Information Systems **18**(2) (2000) 140–170
5. Dourish, P., Edwards, W.K., LaMarca, A. and Salisbury, M.: Presto: An Experimental Architecture for Fluid Interactive Document Spaces. ACM Transactions on Computer-Human Interaction **6**(2) (1999) 133-161
6. Hsieh, H.-W. and Shipman, F.M., III: VITE: a visual interface supporting the direct manipulation of structured data using two-way mappings. In Proceedings of the ACM Conference on Intelligent User Interfaces (IUI '00), New Orleans, LA, USA (2000) 141–148
7. Lunzer, A.: Towards the Subjunctive Interface: General Support for Parameter Exploration by Overlaying Alternative Application States. In Late Breaking Hot Topics Proceedings of IEEE Visualization '98, Research Triangle Park, NC, USA (1998) 45–48
8. Lunzer, A.: Choice And Comparison Where The User Wants Them: Subjunctive Interfaces For Computer-Supported Exploration. In Proceedings of 7th IFIP Conference on Human-Computer Interaction (INTERACT '99), Edinburgh, Scotland (1999) 474–482

9. Mander, R., Salomon, G., and Wong, Y. Y.: A 'Pile' Metaphor for Supporting Casual Organization of Information. In Proceedings of the ACM Conference on Human Factors in Computer Systems (CHI '92), Monterey, CA, USA (1992) 627–634
10. Matsuoka, S.: The Longest Chronicle: History Informs. (in Japanese), NTT, Tokyo (1996)
11. Matsuoka, S.: Editorial Engineering of Knowledge. (in Japanese) Asahi Shimbun, Tokyo (1996)
12. Matsuoka, S.: Knowledge Editing Techniques. (in Japanese) Kodansha, Tokyo (2000)
13. Munroe, D. and Papakonstantinou, Y.: BBQ: A Visual Interface for Browsing and Querying XML. In Proceedings of Visual Database Systems (VDB5), Fukuoka, Japan (2000)
14. Nelson, T.H.: What's On My Mind. Invited talk at the first Wearable Computer Conference, Fairfax, VA, USA. `http://www.sfc.keio.ac.jp/~ted/zigzag/xybrap.html` (1998)
15. North, C. and Shneiderman, B.: Snap-Together Visualization: A User Interface for Coordinating Visualizations via Relational Schemata. In Proceedings of the 5th International Working Conference on Advanced Visual Interfaces (AVI 2000), Palermo, Italy (2000) 128–135
16. Okada, Y. and Tanaka, Y.: IntelligentBox: A Constructive Visual Software Development System for Interactive 3D Graphics Applications. In Proceedings of IEEE Computer Animation '95, Geneva, Switzerland (1995) 114–125
17. Robertson, G., Czerwinski, M., Larson, K., Robbins, D.C., Thiel, D. and van Dantzich, M.: Data Mountain: Using Spatial Memory for Document Management. In Proceedings of the 11th Annual ACM Symposium on User Interface Software and Technology (UIST '98), San Francisco, CA, USA (1998) 153–162
18. Roth, S. F., Chuah, M. C., Kerpedjiev, S., Kolojejchick, J. and Lucas, P.: Towards an Information Visualization Workspace: Combining Multiple Means of Expression. Human-Computer Interaction Journal **12**(1- 2) (1997) 131–185
19. Shipman, F.M., Marshall, C.C., and Moran, T.P.: Finding and Using Implicit Structure in Human-Organized Spatial Layouts of Information. In Proceedings of the ACM Conference on Human Factors in Computer Systems (CHI '95), Denver, CO, USA (1995) 346–353
20. Shneiderman, B.: Codex, memex, genex: The pursuit of transformational technologies. International Journal of Human-Computer Interaction **10**(2) (1998) 87–106
21. Tanaka, Y.: Meme Media and a World-Wide Meme Pool. In Proceedings of ACM Multimedia '96, Boston, MA, USA (1996) 175–186
22. Tanaka, Y. and Fujima, J.: Meme Media and Topica Architectures for Editing, Distributing, and Managing Intellectual Resources. In Proceedings of 2000 Kyoto International Conference on Digital Libraries: Research and Practice, Kyoto, Japan. (Post-conference proceedings will be published by the IEEE.) (2000)
23. Tanaka, Y. and Imataki, T.: IntelligentPad: A Hypermedia System Allowing Functional Composition of Active Media Objects through Direct Manipulation. In Proceedings of the IFIP 11th World Computer Congress, San Francisco, CA, USA (1989) 541–546
24. Theodorakis, M., Analyti, A., Constantopoulos, P. and Spyratos, N.: Querying Contextualized Information Bases. In Proceedings of the 24th International Conference on Information and Communication Technologies and Programming (ICTP'99), Plovdiv, Bulgaria (1999)

Beyond the Algorithmic Mind

Steve Talbott

The Nature Institute
101 Route 21C
Ghent, New York
U.S.A.
stevet@oreilly.com
www.natureinstitute.org

Abstract. The past few hundred years have seen an increasing commitment to abstraction as the primary instrument of cognition. One result of this commitment is the conviction that the world consists, ultimately, of nothing but structure — a conviction exemplified in the feeling that the machine is essentially the algorithm governing its operation. But the reduction of understanding to the grasp of manipulable abstractions is at work in our culture far beyond the notion of algorithmic machines. The reduction is evident even in practical activities such as farming, chemistry, manufacturing, and business. All of which poses a problem, since abstraction, by itself, cannot give us a world. What operations of mind give us a world from which to abstract? What mental activity is necessary to counterbalance the one-sided drive toward abstraction, rendering it healthy and constructive? Here I do no more than sketch the background against which the question can be acutely felt. Then I point in the briefest possible way to the direction from which I believe an answer can be found.

1 Introduction

You may have heard of the Game of Life. No typical computer recreation, it would bore most gamers to tears. Yet it has inspired researchers in a number of fields and has led to the creation of an entire new discipline known as Artificial Life.

The game divides your computer screen into a fine-meshed rectangular grid, wherein each tiny cell can be either bright or dark, on or off, "alive" or "dead." The idea is to start with an initial configuration of bright or live cells and then, with each tick of the clock, see how the configuration changes as these simple rules are applied:

- If exactly two of a cell's eight immediate neighbors are alive at the clock tick ending one interval, the cell will remain in its current state (alive or dead) during the next interval.
- If exactly three of a cell's immediate neighbors are alive, the cell will be alive during the next interval regardless of its current state.
- And in all other cases — that is, if less than two or more than three of the neighbors are alive — the cell will be dead during the next interval.

M. Beynon, C.L. Nehaniv, and K. Dautenhahn (Eds.): CT 2001, LNAI 2117, pp. 190-202, 2001.

You can, then, think of a cell as dying from loneliness if too few of its neighbors are alive, and dying from over-crowding if too many of them are alive.

2 The City through a Window

Now, what interested the early students of this game in the 1960s (preeminent among them, John Horton Conway) was the fact that, given well-selected initial configurations, remarkable patterns are produced. A "glider" might sail serenely across the screen. A "glider gun" might produce an endless series of gliders. Another entity might swallow up any glider that made contact with it, while itself remaining intact. There were static patterns, blinking patterns, rotating patterns, and forms that could evolve and even reproduce themselves in endlessly fascinating ways.

What is still more remarkable is the conclusion some researchers eventually drew from all this. Full of excitement as they watched their enchanted screens, they began to suspect that they were being initiated into the deepest secrets of biological evolution, of reproduction, and of life itself.

Christopher Langton, widely regarded as the founder of the discipline of Artificial Life, has related how the Game of Life played a seminal role in the development of his own thinking. Having dropped out of college, he was working as a programmer at Massachusetts General Hospital in Boston during the winter of 1971-72. He sat down to debug some code one day — actually, it was the middle of the night, his favorite time to work — and, as a diversion, he set the Game of Life running on the terminal in front of him. He would occasionally glance up from his program printouts to see all these beguiling shapes endlessly transforming themselves on the screen. Then one time — as he relates it — immediately after looking at the screen and then turning his attention back to debugging, "the hairs on the back of my neck stood up. I sensed the presence of someone else in the room". But when he looked around, no one else was there. He then looked back at the computer screen.

> I realized that it must have been the Game of Life. There was something *alive* on that screen. And at that moment ... I lost any distinction between the hardware and the process. I realized that at some deep level, there's really not that much difference between what could happen in the computer and what could happen in my own personal hardware — that it was really the same process that was going on up on the screen.

So this was his moment of epiphany, and his story continues:

> I remember looking out the window in the middle of the night, with all this machinery humming away [around me]. It was one of those clear, frosty nights when the stars were sort of sparkling. Across the Charles River in Cambridge you could see the Science Museum and all the cars driving around. I thought about the patterns of activity, all the things going on out there. The city was sitting there, just *living*. And it seemed to be the same sort of thing as the Game of Life. It was certainly much more complex. But it was not necessarily different in kind. (Quoted in Waldrop 1992, pp. 202-3)

You can almost see Langton's gaze shifting between the flat, two-dimensional screen in front of him and the nighttime view of Cambridge through the cold window. And you can feel the attractive power of the mental reduction, whereby all the blinking and streaming lights, all that urban activity, became for him little more than a shifting pattern of light on the window pane, much like the Game of Life.

It is indeed a remarkable reduction. It rather strikingly reminds one of the similar reduction that occurred in the development of perspective art during the fourteenth and fifteenth centuries. A critical point came when the artist of that day suddenly saw his canvas with entirely new eyes. He reconceived it as a window — a cross-section of the "pyramid of rays" extending from his eye to the surface of his subject. By this means he could reduce the scene, in all its variety and significance, to the geometrical purity of a set of points, lines, and planes "in perfect perspective". This reduction helped set the stage for the scientific revolution and the eventual turning away from qualities to quantities as a foundation for understanding.

Today, however, it's not just the visual aspect of the scene that is being reduced in this way, but its active coherence, all the intentions and behaviors it exhibits.

Peter Cochrane, head of research at British Telecom, puts the vision behind Artificial Life this way: "It may turn out that it is sufficient to regard all of life as no more than patterns of order that can replicate and reproduce" (Cochrane undated). And Langton himself has surmised that "Life isn't just *like* a computation, in the sense of being a property of the organization rather than the molecules. Life literally *is* a computation" (quoted in Waldrop 1992, p. 280).

3 Algorithms and Real Machines

All this illustrates, in however exaggerated a way, the decisive tendency of what I would call "the algorithmic mind", which is the mind predominantly at work in conventional science today and, indeed, throughout modern culture. We see a striving toward abstract pattern and structure, toward mathematics, logic, and syntax, toward empty formalism, at the expense of ... well, at the expense of what? That is the question.

One thing is clear at the outset. What "lives" so strikingly in those lawfully "evolving" patterns on the screen is neither more nor less than the internal coherence and necessity, the logic, of a set of articulated abstractions. One needs to distinguish this logical coherence and necessity from any actual thing or process in the world, any *embodied* dynamic. The distinction may seem trivial, and nearly everyone would readily acknowledge it when pressed, yet a case can be made that the routine forgetting of it provides a clue to the fundamental aberrations of contemporary thought.

Let me try to make the distinction clearer. The patterns of order and the computation that Cochrane and Langton find to be the essence of life are widely understood as *algorithmic*. In non-technical language, an algorithm (typified by a computer program) is "an infallible, step-by-step recipe for obtaining a pre-specified result" (Haugeland 1985, p. 65). There is a wonderful ambiguity, if not in the elegant mathematical theories supporting this definition, then at least in the common ways of thinking about algorithms. For we can easily regard the machinery carrying out the recipe as itself an expression of the algorithm, or else we can consider the algorithm solely as an abstraction. In the latter case, the algorithm "obtains a result" merely in the sense that 42 is a result obtained from 7 and 6: quite apart from any mind or mechanism *doing* a multiplication, there is a purely logical (mathematical) *relation* of multiplication by which 42 "results" from 7 and 6. But a conceptual relation among numbers, as such, is not so much an achieved result as, well, a conceptual relation.

If, on the other hand, we embody the algorithm in the stuff of the world — if we try to get something to *do* the multiplication, so that, in a dynamic and embodied sense, a result is obtained — then this result is no longer infallible. No one, after all, would dream of claiming infallibility for any conceivable mechanism. In the case of a computer multiplying 7 and 6, the electrical power might fail, I might spill my coffee into the circuitry, a bug in the supporting software might supervene, a giant meteor might strike the earth, or the hardware might — and over time certainly will — wear out.

Presumably, what we really mean by an "algorithmic machine" (such as a computer) is that, when it is functioning according to its design, we can abstract from its highly contingent corporeality the logical structure of the algorithm designed into it. Of the algorithm itself, then, we might say: it is an infallible recipe in the sense that, so long as something about the structure of the executing machinery maps properly to the structure of the algorithm, the abstract outcome of the machine's operation is specified by the algorithm.

In other words: so long as a given set of logical relations can be sustained in the stuff of the world, those very relations can be sustained in the stuff of the world. Since they can in fact be sustained for useful periods of time, this may be a truth deep enough to drive the marvelous technological successes of our society. But as a key to *understanding* the material world and its contents ("Life just *is* a computation"), it seems incomplete. What is it that gives substance and reality to an algorithm? And if we pay no attention to this ignored something, are we really prepared to reduce the world to the peculiar sort of mentality that abstractions are? To continue abstracting while losing sight of the world from which we are abstracting is to find ourselves, eventually, ensnared within a much-too-perfect conceptual web of our own weaving, bearing little relation to what is actually there.

4 There Is No Physics without a World

It's easy to slide unnoticeably and unjustifiably between the embodied and disembodied ways of thinking about algorithms, which encourages us to begin imagining that algorithms can explain the world. Referring to the Game of Life and the three-part law governing its performance, Daniel Dennett has remarked that "the entire physics of the Life world is captured in that single, unexceptionable law" (Dennett 1995, p. 167). Moreover, "our powers of prediction [regarding the Life world] are perfect: there is no noise, no uncertainty, no probability less than one" (Dennett 1991, p. 38).

But, as we have seen, the "unexceptionable law" is hardly a law of physics (what is the "physics" of a chain of logical propositions or a set of rules?), and it is a little odd to talk about our "powers of prediction" where only logical relations are in view. If, on the other hand, we *are* talking about a physical machine whose behavior we might venture to predict, then there is noise, no certainty, and no probability equal to one.

It is not that brilliant thinkers such as Dennett would fail to recognize this obvious truth. It's just that the truth doesn't seem to count for much in their thinking. The "something else" that enables us to talk about the phenomenal world instead of the pure thought relations of an assemblage of abstractions draws no particular attention from them.

The abstract tendencies of our thinking are evident in the way we increasingly take the logical structure of the machine's functioning to be the essence of the machine, or even the full reality of the machine. The hardware that happens to "instantiate" or "execute" the software algorithm becomes incidental to the aspects of the machine we are really interested in.

It is a commonplace of computing theory that an algorithm can be instantiated in hardware of wildly different sorts. That is, you can, in principle, rig up a computer from rolls of toilet paper (with each individual sheet treated as a memory location or part of an input or output tape), or from electromechanical relays, or from transistors in silicon. And it's true that when such differing devices are executing the same algorithm, there is, at a highly abstract level, an important sense in which they are all the same machine. But what about their differences? How much of the physical presence and reality of the machine is explained by the abstract, functional equivalence we happen to concern ourselves with?

If the algorithm is the full explanatory essence of the device — if, more radically (as Langton suggests) the algorithm just *is* the device — then we are left to conclude that the toilet paper-based machine just *is* the electromechanical relay-based machine just *is* the transistor-based machine. What seems to have flown out the window in this equation is any reality principle adequate to the full presence and substance of the world.

Given the preoccupation with abstract, formal structures, and given the prevailing ambiguity between these structures and physical reality, we can fairly ask: what is missing when we say, as many are tempted to say today, "Reality is ultimately nothing but structure"? What gives us real machines, real minds, a real world rather than an empty mesh of formal relations that are not the relations *of* anything existent in the material world?

5 Getting Precise about Nothing

To prevent misunderstanding: I am not about to suggest that, since real machines are never infallible algorithms, mechanistic models and robotic devices could never become, say, living, intelligent organisms. This would be to invite the immediate retort: "But living organisms — for example, human beings — are also fallible devices. They malfunction all the time, and eventually die."

No, the current issue is much more fundamental: How do you get from formal structures to a real world? It does no good to talk about wonderfully "effective" algorithms if these algorithms remain in the materially empty realm of pure abstraction. You can't get from abstraction to reality without recovering whatever it was you originally abstracted from — and this "whatever it was" receives almost no rigorous attention in the many disciplines enamored of formalism.

Putting it in slightly different terms: the quintessential formalisms are mathematics and logic, and you cannot get from the propositions of pure mathematics and logic to the real world. These propositions are as utterly empty as we can make them. This is their whole (extremely valuable) point: to be devoid of content so that they can put on display something like the pure possibilities of form. As Ludwig Wittgenstein has said,

> All propositions of logic mean the same thing, namely nothing.

Or, in Bertrand Russell's words, mathematical logic

> is the subject in which we never know what we are talking about. (Russell 1981, pp. 59-60)

Russell was one of the most accomplished logicians of all time, and he was not being facetious. Rather, he was rubbing our noses in the fact that when we finally achieve perfect formal precision, we have also achieved perfect abstraction. That is, we have abstracted ourselves clean out of this world, so that our perfect precision is precisely about nothing.

Einstein may have been tracking the same truth when he said,

> Insofar as the propositions of mathematics give an account of reality, they are not certain; and insofar as they are certain they do not describe reality. (Quoted in Kline 1980, p. 97)

Certainly the formal truths of logic and mathematics can be *applied* to the world — why should they not, since they were first abstracted from there? But they cannot by themselves *produce* a world. They do not and cannot by themselves mean anything concrete and particular for our actual existence. You cannot get from the empty structures, the logical forms, of Russell and Whitehead's *Principia Mathematica* to the birth of a child or the crumbling of a rock or the Treaty of Versailles. You have to add something else — something that holds within itself the entire content and substance, the givenness, of the world.

6 The Flight from Earth

The tendency to retreat from the world's substance and our own earthiness into an engagement with abstraction and algorithm alone shows itself in much of modern life.

It is, of course, not hard to find symptoms of world-abandonment in more exotic places — for example, in the Gibsonian disenfleshment of cyberspace, in the computational psychologist's reconceptualization of the human essence as a disembodied formalism, in the ever more acute yearnings for some kind of digital immortality — not to mention the Mars Society members who (in the words of founder Robert Zubrin) believe

> We need a central overriding purpose for our lives. At this point in history that focus can only be the human exploration and settlement of Mars .... You'd have to be made out of wood to not want to go to Mars.

But exotic dreams are one thing; what about the mundane existence of the chemist, the farmer, and the factory worker? Well, it may have escaped our notice, but the same fleeing of our own earthiness and materiality so evident in more levitated disciplines is equally apparent in every sphere of human activity, if only we begin to look for it.

A few years ago the chemist could synthesize maybe fifty to one hundred new compounds a year. Today, using silicon chip technology, she can produce fifty thousand new compounds every year. And it requires only a few months for a drug company to screen its entire library of, say, one million compounds against a target protein.

The chemist in such operations has no sensory experience whatever of the microscopically apportioned physical substances themselves; it's all a matter of instrumentation, database entries, and sophisticated computation. She is not familiarizing herself with earth; she is replacing it with abstractions. Marvelously refined and productive abstractions to be sure — I am not at all suggesting that the abandonment of earth deprives us of effective power. It would be truer to say that we are driven to the abandonment by our lust for power. It is easier to manipulate and control what we have not first tangibly experienced, understood, and loved.

As for the farmer, you will not be far wrong to imagine him high off the ground in the sealed, air-conditioned cab of an EarthMaster tractor, tooling across an endless Nebraska cornfield as the onboard software, communicating with a global positioning satellite, administers manufactured chemicals to the soil based on previous, acre-by-acre soil tests conducted in a far-off laboratory — all while he listens to country western music through his earphones or surfs the web. Not exactly "living close to the land". And you can be almost certain that his family tends no garden; they buy their food at the nearest Walmart or Safeway.

Perhaps the neighboring farmer raises chickens — each one de-beaked, doused in pesticides, restricted for life to a square foot or two of space, denied any fulfillment of its natural urge to explore, scratch, and peck — an unspeakable cruelty made possible by the fact that the farmer no longer lives with the animals he raises. His is a world of abstract factory inputs and outputs. It has to be so; no human being could comfortably face these abused animals in their own right.

Or take the manufacturing plant. If there's any place where we wrestle with the stuff of earth, surely this is it. *The Economist* describes typical automobile factories as

> ... chaotic places .... fork-lift trucks wend their way up and down the line .... noisy mess .... manic robots in a flurry of sparks and a deafening whirl of metal and machinery .... the clamor of vast presses stamping out body panels with 300-ton blows, the power of a jumbo jet taking off.

But what is interesting is how all this has been changing. The article goes on:

> Twenty years ago you could not see across the welding hall of the plant in Aurora, Illinois, because of the smoke. Today the welding plant is completely clear, the giant slabs of thick sheet steel are quietly cut into shape by high-voltage plasma guns, which produce a much more precise cut and no smoke.

So even our brutest working with material is becoming less brutely material today. The abstract patterns in the computer program activate the plasma gun, which in turn reproduces the patterns in the metal itself — all without anyone, or even any machine, having to bang away in an unseemly manner. We manipulate a few abstractions on a screen, and then hidden, precisely guided forces automatically reconfigure the stuff of the world — the metal is shaped, the DNA strand is cut, the chickens in their little boxes are fed, the bomb is dropped hundreds of miles away.

There's wonderfully effective manipulation in all this — and almost no experience of what it is we're manipulating (or killing). Our lives are navigations within a web of abstractions. It reminds me of Max Frisch's remark: Technology is the knack of so arranging the world that we don't have to experience it.

You will find the same story in every aspect of contemporary life. I have not even touched on the more obvious vocations. Software engineering (whose whole point, as the discipline is usually practiced, is to abstract computable patterns from the thick texture of life and then, to one degree or another, substitute those patterns for the living contexts they were drawn from). Global finance and investment (where trillion-dollar capital flows stream through the world seeking nothing more than their own abstract mathematical increase, with the actual impact on real people in real communities hidden from the investor). The computer-like functioning of major corporations, bent on nothing more than maximizing the bottom line (airline executives needn't understand much about airplanes, according to economist Alfred Kahn; after all, airplanes are just "marginal costs with wings"). The disappearance of the living organism from the sight of the genetic engineer and molecular biologist (leading to what ecologists David S. Wilcove and Thomas Eisner (2000) have referred to as "the demise of natural history").

7 Mental Engines of Abstraction

Our flight from materiality into abstraction soars most ambitiously in the evolutionary trajectory of the machine. Not only are our machines ceasing to mediate experiences of earth, as suggested above; they are themselves dematerializing before our eyes.

If there is an "archetypal" machine of our era, surely it is the computer. The first computers back in the 1940s were monstrous things, almost all hardware, and impressive for their ponderous, warehouse-filling bulk. The software was rather an afterthought, and was not sharply distinguished from the hardware; configuring it meant physically connecting a tangled nest of wires.

The situation is different today — and not just owing to the miracles of miniaturization. The computer is disappearing from sensuous experience in the same way the earth itself is disappearing — into abstractions such as information. As pointed out above, the machine, we've now come to think, just *is* its software. The hardware that "instantiates" the software may be faster or slower, but it doesn't affect the essential functioning. In this sense (which is the sense that increasingly holds our attention) the hardware is incidental. More and more the physical machine is seen as a mere substrate for the decisive abstractions we impress upon it. And we equate the essence of the machine with the abstractions.

Earlier, during the industrial era, the machine was often viewed as a grinding, perhaps enslaving material weight, dragging us down. This, however, is to misrepresent the matter. The machine may have dragged us down, but less because of its gross materiality than because of its disregard of *our* materiality, our embodiment. What oppressed us was the peculiar way the machine functioned — the way it disrupted and narrowed the material context of our work, cutting us off from any genuine conversation with the sensuous presence of the world. Our mechanized actions were stripped of their expressive, gestural qualities and finally reduced to the quantitative abstractions yielded by time and motion studies. The artisan gave way to the assembly line worker.

Far from chaining us to the tangible, inhabitable earth, the machine has been a prime instrument of our world-abandonment. But only now are we seeing the extreme realization of this truth.

Anyone who takes the mind seriously in its own terms is often scorned for believing there is a "ghost in the machine". But it is the machine itself that has dematerialized. Those who pursue computer models of the mind are the true believers in ghosts; the mind they look for is the mere ghost of a machine.

8 Atoms and Bits versus the World

Another common misconception also needs clarifying. Our flight from the world is not a flight from atoms to bits. Those who contrast atoms and bits miss the crucial truth: the world of atoms *is* the world of bits. The machines that have given us our streaming "bits" of information are also the machines that have given us our mechanistic models of the world, and the result is an atom that long ago began disappearing into the same formal void that has been swallowing up material machines and spitting out patterns of information.

When it is the glory of physics that it has become almost *nothing but* mathematical rigor, we can be sure, as Einstein's remark about mathematics and reality would suggest, that it is detaching itself from the world. Defending this detachment, the physicist Robert March has written,

> We should never have expected words born in the familiar world readily accessible to our senses, such as *particle* and *wave*, to perfectly describe the microcosm. The electron is what it is, and if the words we use to describe it seem full of paradox, *so much the worse for those words*. The equations have it pinned down neatly. (March 1977, p. 235)

But, as we have seen, equations alone pin down nothing. If, as March urges, we give up on the words and meanings derived from the world we have so far familiarized ourselves with — if we give up the effort to deepen and transform these meanings —

then we are left with a physics (or rather metaphysics) of "particles" that are fully as remote from material existence as the "bit" of the engineer. We embrace a complex of relationships while ignoring or denying whatever it is the relationships cover — whatever is doing the relating. We liken the world to a whirling dervish from which the dervish has disappeared, leaving only the whirl.

So it's not that we can substitute bits for the atoms dealt with by the physicist; it's that the bit of the engineer and the atom of the physicist have become almost one: abstract, insubstantial, statistically defined, and disconnected from the world of experience. Both atom and bit represent the end stage of the same long process whereby we have progressively lost the world.

Do not ask, then, whether we should exchange a world of atoms for a world of bits. Say, rather: "On the one side, a disincarnate system of atoms and bits; on the other side ... what? How do we gain a world we can actually experience?"

9 Reclaiming the World

Precisely because we've been losing awareness of the world as experienced content (as opposed to our awareness of its formal structure) it is nearly impossible to describe what is missing. What is missing is the very substance of the describable. What is missing is exactly what we have been losing and therefore also what we have been losing any developed means to express.

Here I can offer only a few aphoristic remarks — bare assertions, really — pointing to the gesture of mind that might counterbalance our well-developed but one-sided movement toward abstraction. Some of these assertions may seem crazy in light of the entrenched assumptions governing our current cognitive imbalance. But since their whole point is to question these assumptions, that is how they *should* seem. What remains is for the reader to decide whether the attempt to stand outside prevailing assumptions, if only for purposes of discussion, is worth the considerable effort it requires.

••• First, then, it needs acknowledging that our advanced abstracting skills are extremely valuable and should not be discarded. They give us our ability to stand apart from the world with a certain critical detachment, rather than simply being swept along by it. The point is only that abstraction cannot stand alone. We cannot abstract our formalisms from the world except insofar as we are first given a world to abstract from. And if we do not explicitly reckon with this given world, we cannot know what essential things are left out of our abstractions.

••• The imagination — our capacity for dealing with the world as image — is the counter to abstraction. Imaginal thinking gives us unities and wholes where abstraction analyzes and splits apart. This unifying activity is possible because of the way the elements of an image (as in a worthy work of art) interpenetrate each other, so that the whole can express itself through each part. Owen Barfield, the English philologist and preeminent student of the imagination, mentions three aspects of the imagination about which those who have studied it find themselves in considerable agreement:

> Imagination gives us a relation between whole and parts different from mere aggregation. Unknown in classical physics, this relation is "not altogether unknown in the organic realm. It has been said that imagination directly apprehends the whole as 'contained' in the part, or as in some mode identical with it." The hologram gives us an approach to this thought from the side of physics.
>
> Imagination "apprehends spatial form, and relations in space, as 'expressive' of nonspatial form and nonspatial relations." For example, in the human countenance we can read various interior relations of thought, feeling, and intention.
>
> Imagination operates prior to the kind of perception and thought that has become normal today. It functions "at a level where observed and observer, mind and object, are no longer — or are not yet — spatially divided from one another; so that the mind, as it were, becomes the object or the object becomes the mind." (Barfield 1965b, p. 127)

••• The imagination, in other words, works with qualities. Qualities have the ability to interpenetrate and influence each other. Red and yellow, considered as qualities rather than as wavelengths, interpenetrate to become orange, and even when the colors are kept side by side, the qualities of the one color influence the other. It is only by virtue of qualities that we can have integral unities, where the whole shines through the part. Every attempt at holism that is not thoroughly qualitative (you see such attempts in the excruciatingly abstract work of many complexity theorists) yields only a bogus holism where the whole is *not* manifest in the part.

••• The imagination, with its attention to the qualitative image, stands at the opposite pole from the mind's abstracting, logical activity:

> It is characteristic of images that they interpenetrate one another .... That is just what the terms of logic, and the notions we employ in logical or would-be logical thinking, must *not* do. *There*, interpenetration becomes the slovenly confusion of one determinate meaning with another determinate meaning, and there, its proper name is not interpenetration, but equivocation.... (Barfield 1977, p. 100)

••• Western science committed itself early on to ignore "secondary qualities" as opposed to primary ones. But it quickly turned out that all qualities were secondary so that, oddly enough, the only remaining primary quality was quantity. The banning of qualities from the objective world led naturally to Cartesianism.

••• Descartes decisively articulated the emerging subject-object, mind-matter dualism, whereby qualities were relegated to the subject/mind side of the cleavage. Today *everyone* claims to have transcended "Cartesian dualism". We do not transcend it, however, by accepting the false either-or it presents to us, and then allowing one or the other term to disappear or be swallowed up by its opposite. The root Cartesianism we really need to transcend lies in the initial formulation of a subjectivity without objectivity and an objectivity without subjectivity. Once we see our way past these incommensurables, we will realize that qualities exist not only in us, but also in the world.

••• That is, the radical truth we would accept if we were able to escape our own Cartesianism is that there is no subject except by virtue of the object, and there is no object except by virtue of the subject. The two are correlative and interdependent, rather like the poles of a magnet. The world — the objective world we hold in common — occurs within our experience, not outside it. This was a universally felt (if not philosophically articulated) truth until just several centuries ago.

Our inability today to enter into this truth — or even to *think* it coherently — reflects the cognitive one-sidedness, or mental limp, I have attempted to characterize in this paper. It reflects, that is, our resolve to ignore precisely those qualitative aspects of the world that most forcefully resist being categorized as wholly subjective or wholly objective. Unfortunately, this long ignoring of qualities as if they were cognitively insignificant (whatever their "poetic" value) does not leave us in a good position to recognize what they tell us about the world.

••• Perfectly mindless matter, free of all interior qualities, is an impossible thought, and no one succeeds in thinking it. Democritus' atoms had their inner capacity to "swerve", but

> Today the immaterial agent of change is more likely to be impounded in some such term as "tendency" or "pattern" or "mutation" (another way of saying "change") or "norm" or (in more up-to-date biology) "code," "message" or "information" — the whole change from e.g. a single cell to a complex living organism requiring no more than amino-acids and genes — *plus*, of course, an ability to code and decode, which last need not be unduly stressed. The trouble is, that particles *as such* ... cannot even arrange and rearrange themselves without more. Yet, if one credits them with immaterial "swerves" or "tendencies" and so forth, he has forgotten that those are the very things he was purporting to explain *by* them. (Barfield, 1971, p. 205 fn. 2)

••• The mutual implication of subject in world and world in subject is widely accepted regarding the macroscopic, or phenomenal ("appearing") world. We accept it by denying objectivity to this world, instead assimilating it to the subjective side of the Cartesian ledger. Then we picture an "objectively real" world of particles, waves, or whatever, lying *behind* the veil of qualitative phenomena. In practice, this projected objectivity either "goes metaphysical" or else becomes a mere replica, a strange double, of the appearing world. Is the "real" tree lying behind the appearances perceptible by us or not? If not, it looks suspiciously metaphysical; but if it *is* perceptible, then why do we think of it as more objective or real than all the other appearances? (Alternatively, why do we think of all the other appearances as *less* real than the ones we count as canonical?)

Qualities are what give us a world we can abstract from, and I have tried to hint at the fact that our inattention to them — our failure to bring qualities within the rigorous orbit of scientific method — leads to tremendous confusion about appearance and reality, about our own relation to the world, and about what it means to understand the world.

The person who, more than anyone else I know, has explored these issues extensively and in depth, is Owen Barfield. (See especially Barfield 1973 and 1965a.) For a succinct and accessible examination of some of the philosophical issues raised by the idea of a qualitative science (and also by the effort to transcend Cartesian dualism), see Brady 1998.

References

1. Barfield, Owen (1965a). *Saving the Appearances*. New York: Harcourt, Brace and World.
2. Barfield, Owen (1973). *Poetic Diction: A Study in Meaning*. Middletown, CT: Wesleyan University Press.
3. Barfield, Owen (1971). *What Coleridge Thought*. Middletown, CT: Wesleyan University Press.
4. Barfield, Owen (1977). "Lewis, Truth, and Imagination", in *Owen Barfield on C. S. Lewis*, edited by G. B. Tennyson. Middletown, CT: Wesleyan University Press.
5. Barfield, Owen (1965b). *Unancestral Voice*. Middletown, CT: Wesleyan University Press.
6. Brady, Ronald H (1998). "The Idea in Nature: Rereading Goethe's Organics", in *Goethe's Way of Science: A Phenomenology of Nature*, edited by David Seamon and Arthur Zajonc. Albany, NY: State University of New York Press.
7. Cochrane, Peter (undated). http://www.cochrane.org.uk/inside/quotes.htm.
8. Dennett, Daniel C. (1991). "Real Patterns", *Journal of Philosophy*, vol. 87, pp. 27-51.
9. Dennett, Daniel C. (1995). *Darwin's Dangerous Idea: Evolution and the Meanings of Life*. New York: Simon and Schuster.
10. Haugeland, John (1985). *Artificial Intelligence: The Very Idea*. Cambridge, MA: MIT Press.
11. Kline, Morris (1980). *Mathematics: The Loss of Certainty*. Oxford: Oxford University Press.
12. March, Robert (1977). *Physics for Poets*. Chicago, Contemporary Books.
13. Russell, Bertrand (1981). *Mysticism and Logic*. Totowa, NJ: Barnes and Noble.
14. Waldrop, M. Mitchell (1992). *Complexity: The Emerging Science at the Edge of Order and Chaos*. New York: Simon and Schuster.
15. Wilcove, David S. and Eisner, Thomas (2000). "The Impending Extinction of Natural History", *The Chronicle of Higher Education* (September 15), p. B24.

How Group Working Was Used to Provide a Constructive Computer-Based Learning Environment

Trevor Barker[1] and Janet Barker[2]

[1] Department of Computer Science, University of Hertfordshire, Hatfield, AL10 9AB, UK
T.1.Barker@herts.ac.uk
[2] Home Office Training, Queen Anne's Gate, London, SW1H 9AT, UK
Janet.Barker@Homeoffice.gsi.gov.uk

Abstract. In this paper, the development and evaluation of a computer based multimedia learning environment capable of supporting group working is reported. The application developed in this study was used in a Further Education (FE) college by tutors and students following a catering course. Learners engaged in the application, performing a range of tasks, based on group working and role playing both on and off-computer. Although no significant difference was found between performance in individual and group work it was found that learners and teachers valued the approach used and performed at least as well on the course than in previous years.

1 Introduction

Education in the UK is undergoing great change at the moment in the ways in which we teach and the ways we manage teaching and learning. Perhaps the most important change relates to the ways in which we use information and learning technology (ILT) in the education process. It is evident that currently there are many pressures in the FE and HE sector to move towards increased provision in online teaching and learning [1]. Brown [2] has identified several pressures that currently are driving HE towards the computer-based and online solutions. These include:

- Rapid growth in student numbers
- Variability in student intake
- Declining resources
- Conderns about quality of graduates
- Pressure for public accountability
- Concerns about relevance to industry
- Increased competition from other countries, universities and commerce
- Increased availability of networked resources
- Decreased costs of networked resources

Musselbrook et al. [3] note that in addition to increased student numbers, pressures on academic staff contact hours and decreases in the levels of public funding are also factors leading to the demand for increased online and computer-based provision.

M. Beynon, C.L. Nehaniv, and K. Dautenhahn (Eds.): CT 2001, LNAI 2117, pp. 203–213, 2001.

Several authors have noted that the increase in online provision has not always resulted in increased quality of materials provided, or in the provision of good quality educational systems. The impact of Information Technology in general on educational practice is less than has been predicted, even after ten years of investment in research, support and training according to Crook [4]

Teachers have in many cases been slow to accept and adopt computer based interactive systems in their teaching. Possible reasons for this are many. Some applications simply reproduce existing paper-based material in multimedia format and ignore the need for intrinsic motivation [5]. Interaction, it is claimed is an aid to learning, but is often about navigation and not directed specifically to learning [6], for example, distinguishes between real interaction and simple button pressing. Carswell et al. [7] consider that all too often, pedagogy is overshadowed by system design considerations in the development of online and multimedia learning applications.

2 Constructivist Approach to Learning

The constructivist approach emphasises the construction of knowledge and learner autonomy. A basic feature of constructivism centres on the learner's cognitive processing of activities such as organising, adapting, reordering and inventing. Therefore from a constructivist standpoint learning requires the learner to make personal sense of information provided, linking it to past experiences. The learners' previous experiences, their motivation and prejudices, social and emotional development as well as intellectual ability (internal factors) will all impact upon the individuals ability to learn. Features of the learning environment such as the opportunities, constraints and circumstances influencing their learning including other people are important (external factors). The links between the internal and external forces have been summarised by Ewing et al. [8]. The main features of constructivist learning include:

- Personal autonomy and control over learning
- Active involvement
- Collaboration with others
- Context
- Personal growth
- Perspective and understanding

3 Working in Groups

Group work is used in many training programmes and has become a common method in both educational and workplace training. The interest in groups within an educational setting has grown since the 1960s. Research originated in the 1940s where researchers such as Lewin [9] examined groups through the observation of group members behaviour and studying how delegates were able to transfer learning to their normal environment. English & Yazdani [10] suggest learning in groups is not merely a vehicle for learning, but also part of the learning process. They suggest that such skills do not just happen, they have to be fostered. New graduates are expected to have group working skills. Co-operative learning may be made more effective when explicit awareness of personal transferable skills is demanded and intentionally practised. They also found that:

- The facilitator is important as motivator mentor and mediator
- Groups need growth of mutual trust, understanding, honesty and respect for others
- Computer supported learning is better when it provides a solution to an identified problem in an existing environment.

Gibbs [11] identified debate, discussion, group work, resource sharing and vicarious learning as important features of online learning. Gibbs found that students learned from seeing each other's work. Gibbs also found that the environment was taken seriously by learners, that students showed awareness of audience of peers. Gibbs also concluded that the protected environment inherent in his study fostered experimentation. Group composition was also important. An understanding of who could see their work also encouraged experimentation by students.

Collaborative learning helps to promote communication and sharing of ideas which will in turn lead to more effective "sense making" by the individual. It also encourages the individual to be more flexible and to see alternative or parallel views of the world [12]. Group work has also been found to assist students who are less able through discussion work in mixed ability groups as they can facilitate and enhance the learning process [13]. Learning is seen as a constructive process that is developed through conversations and built in transitional communities where people construct knowledge as they talk together. Teaching and learning must, in part at least encourage new groups to form and to facilitate the process [14]. Collaborative working enables a subtle shift in the locus of classroom authority from the teacher to the student groups [15] to provide the learner with the ability to influence the classroom activities.

Collaborative learning encourages disagreement and questioning of the teacher and the task asked of them, as through this they begin to gain a greater understanding of the task and will aid them in completing the task well. It also assumes that there is seldom an absolute answer, and that "correctness" is not necessarily what is sought. It is a view of the world from the perspective of the group that is the desired outcome and not a perfect answer.

Collaborative working has been put forward by several authors as important in the learning process. [16], [17], [18]. Important elements in co-operative work are positive independence, individual accountability, face to face interaction, collaborative skills and group processing. Group working exercises in the application then, were designed to encourage those collaborative features listed above and hence the quality of learning.

As the power of computers has increased, so has the sophistication of computer delivered learning. De Diana and White [19], describe how Computer-Supported Collaborative (or Co-operative) Work (CSCW) and Computer-Based Learning (CBL) may be integrated to produce highly motivational interactive online learning environments. Approaches such as these are important in understanding how new technology may be effectively into learning.

4 Task-Based Learning

The multimedia course developed for this work had several significant features which were important for the study. Firstly it was centred around the use of tasks and questions, rather than the more traditional multimedia courses that delivered large amounts of information to the learner. Information was provided in the application in a form mostly to support learner activities. Secondly, group working was supported in the form of a range of off-computer tasks integrated into and supported by the multimedia materials. Thirdly, learners were provided with different views of the course, depending on their specific skills and abilities, their role in the group and the level at which they were able to engage with the materials.

Tasks were used to provide specific contexts for learning. Constructivist theories of learning have been described as the subjective construction of meaning from experience in specific contexts [20] and participation in active learning tasks may be seen as important in such a constructivist view of learning [21], [22]. It has also been suggested that tasks provide a means of engaging student's attention. Users of computer-based instruction packages should engage with the learning material. Frequent decision points are important as are games and simulations in which the results of decisions can be immediately seen [23]. The use of tasks and questions in a constructivist approach to learning has been shown to support ideas such as these [24].

It has been suggested that the task of building computer models may provide direct support for the construction of mental models. Wild [25], and Khan and Yip [26] see tasks involving free exploration and self-directed learning as important for the testing of such models. Khan and Yip suggest that for maximum effectiveness, task-centred instruction should be situated in tasks where knowledge is normally applied. The use of tasks to develop higher level cognitive skills in learning has been considered in the classroom by Felder and Brent [16]. Passive learning and an algorithmic approach toproblem solving were cited as among the reasons for high drop-out rates in science courses. In-class exercises investigated as an alternative approach included recall, stage setting and problem solving, and provide inspiration for the development of tasks for incorporation in multimedia learning packages. The use of questions also is important in developing good approaches [16].

In the applications developed for use in this study, questions ranged from simple multiple choice selections testing recall of simple information, to tests of the organization and structuring of complex ideas. Felder [17] describes how questions can be used in a range of ways to motivate learners by providing interesting challenges. The importance of learner control, collaborative working and differential paths in interactive learning materials has been emphasized by Stoney and Oliver [27].

Multimedia learning materials for six National Vocational Qualification (NVQ) units at level one in catering studies were developed for this study using methods described by Barker et al. [28]. Materials were differentiated according to the level of task available and the type of questions presented. Tasks at different levels were defined in relation to Bloom's taxonomy of learning levels [29] in which the first three levels are:

- Level 1: Knowledge, fact recall with no real understanding
- Level 2: Comprehension, the ability to grasp the meaning of material
- Level 3: Application, the ability to use learned material in new situations.

Group working was differentiated in much the same way. Learners were organised into groups of three or four by tutors. Individuals who might benefit from additional challenge were given greater responsibility in organising and leading group activities. Those learners who required additional support had less demanding roles within the group. For example, one task was to plan and prepare a meal based upon specified commodities. Another task was to perform a stock-taking exercise in their kitchen store room. Group members were responsible for specified activities within each task, for example, counting and recording actual stock levels, calculating theoretical stock levels and calculating the value of food used. After working individually through the related multimedia learning materials, the group met, firstly with the tutor, and later by themselves to discuss the task. The work of the group was then arranged by the students themselves, with support from the tutor and from the multimedia learning materials. Each activity in the task was supported at an appropriate level for the learner undertaking the task, by the tutor, by the learning materials and also by the group itself. Learners working at Bloom's level 3 would be expected to apply their new learning to new situations in the role of group decision makers. Group members working at lower levels would be expected to follow instructions and to make reports to the group about what they had been doing. At the end of the task, the group would report back to the tutor who appraised their performance and encouraged learners to reflect on the process. Computer-based questions were then answered, based upon the individual's role in the group.

5 Procedure

The application was installed on computer networks for use in open access learning centres in a college of Further Education (FE). Subject groups and tutors were given standard brief introductory talks prior to first use of the system. Tutors were given some additional training on the use of the materials prior to the first sessions with their groups. In all, nineteen learners in three groups took part in the exploratory study as shown in Table 1 below, though questionnaire data was only available for two groups (twelve learners in all).

Table 1. Group composition of participants in the study

Group	N	Mean Age	Gender M:F
Group 1	6	16.5	3:3
Group 2	6	17.1	4:2
Group 3	7	17.2	5:2

Learners followed the materials under the direction of their tutors. Studies lasted approximately twenty-six weeks, with students using the application between one and two hours each week. Data was collected at the end of the study. At stages throughout the course and at the end of the courses, tutors and students were asked to complete questionnaires and to participate in group discussions with designers and sometimes tutors. Written reports and summaries were made of these meetings. Tutors also completed a short report on the use of the software with particular reference to the underlying pedagogy used in the development of the materials.

5.1 Tutors' Attitudes

It has been suggested that one major impediment to the implementation of modern technology in education is the attitude of tutors [30]. Reasons for this fact range from feelings of threat, lack of involvement in the process, lack of technical support, lack of training and inadequate resources. Tutors' attitudes to the materials and their use were recorded at meetings throughout the course and in the form of a freely structured report at the end of the course. It was considered important to record tutors' attitudes to the materials and the methods throughout the study. It was hoped that by involving tutors directly in the study, not only would it be possible to gain valuable insight into the process, but also to have a positive influence on their attitude to the materials and the way they were used. One tutor involved in the study also worked with the software development team in developing the domain content of the learning materials, setting questions and tasks and in differentiating presentation for individual group members. It was hoped that by employing such an approach it might be possible to avoid many of the impediments to tutors' acceptance of modern technology identified by Fitzgerald et al. [30]

6 Results

Twelve learners completed the simple questionnaire at the end of the study, based on standard guidelines [31]. The results of the questionnaire for these learners are shown in Table 2 below. In general Table 2 shows that the learning materials were rated highly by learners. Most students indicated that they enjoyed working in groups (3.9) and that tasks completed away from the computer were enjoyable (3.4).

Table 3, below, provides an indication of performance on the course for the nineteen learners involved in the study, as measured by test scores in different sections of the course. Although activities were performed in groups, all testing was performed individually by students, both on and off the computer.

An Analysis of Variance performed on the data in Table 2 failed to show any significant difference between the conditions (p=0.52). It is interesting to note however, that despite the absence of significant differences between the conditions, scores on group tasks had a smaller standard deviation and range of scores than the other conditions. It is possible therefore, that working in groups on tasks benefits poor students, at the expense of better ones.

Pass rates for the objectives covered were higher for all areas than for similar students following the same courses by traditional methods in the previous year. Mean pass rates obtained in the previous year for similar objectives are shown in Table 4 below.

Table 2. Mean scores obtained in questionnaire on learner attitudes to the multimedia-delivered course (N=12)

Question	Average score (Range 1-5)
	1 not very , 5 very
How Interesting	3.8
How easy	4.1
How enjoyable	4.0
How much learned	3.6
How Useful were the following	
Working on your own	3.8
Working in pairs	3.6
Working in groups	3.9
Working with the tutor	3.2
Tests taken on the computer	2.9
Tests taken off the computer	3.1
Tasks done on the computer	3.4
Tasks done off the computer	3.5
Final Test or Examination	3.2
How Worried were you by:	
Using a computer	1.5
Using a mouse	2.4
Using headphones	1.5
Working in Learning Centre	2.9
Would you like to take another multimedia course in the future?	% YES 83%

Table 3. Performance on the multimedia course in the study as measured by scores on group and individual questions and tasks. (Standard Deviation given in brackets)

Number of Students	Number of students Passing Objectives	Average Scores on Group Questions	Average Scores on Group Tasks	Average Score on individual sections
19	(100%)	61%	62%	58%
SD		(12.82)	(10.10)	(13.32)
Range		34-84	48-80	40-83

Table 4. Performance on non-multimedia courses in the year prior to the study

Subject	Number of Students	Number of Students Passing
Catering Level 1	36	28 (78%)

The evaluation of performance on a learning application is a complex issue. In its simplest form, It may be measured by test scores, though it involves more than just this according to Reeves [32]. Although performance on the two courses are not strictly comparable it was nevertheless important to show that learners were at least as successful on the computer delivered course as on the traditionally delivered version. In general, students were successful in using the materials. Most students achieved as good as, or better than average marks and benefited from using the multimedia courses. By configuring tasks and questions at appropriate levels, improved test and task results were obtained it is argued.

7 Discussion

Support for group working has been shown to be important in learning with computers [19]. In this study learners working in groups were supported in task performance in real context by the use of a multimedia application. The learning environment provided in this way was based on task performance, group work and was constructive in its outlook. Although we were unable to show that performance was improved by group working on tasks (as measured by test scores), we were able to show that there were still many benefits of the approach. Student performance on the course was at least as good as in previous years and probably better. The small number of learners in the study and the reliance on individual assessment and testing methods may go some way to explaining the lack of a significant difference in mean test scores.

Tutors involved in the study reported that they were generally supportive of the off-computer activities, though some did mention the additional load on their time that this imposed. The use of the materials in a simulated vocational context was considered to be important by students and tutors alike. The ability to use materials and apply them in vocational context was especially welcomed by tutors. Constant links between the learning materials and activities taking place in the kitchen and store rooms were cited as important by catering lecturers. The materials were described as task-based and assignment-driven by one tutor who felt this was an improvement over most other materials available. An understanding of the need to base vocational materials on real assessment with real rewards in real contexts was an important outcome for the study. Fitzgerald and colleagues [30] found that best use of multimedia materials occurred where materials were fully integrated into learning.

Tasks were also used to develop problem solving skills in learners, to help in the development meta-level skills. The requirement to undertake off-computer tasks that involved real activities such as food preparation, stock control and explaining their work to others was intended to help develop these meta-level skills. A task-based approach was important in learner motivation, which was supported by requiring the active seeking of problem solving skills rather than a passive receiving of information presented by the multimedia application. Cognitive conflict was described by Piaget [33] as a student's drive to learn when his/her cognitive structures are insufficient to handle a situation.An important instructional principle adopted in the application was that cognitive conflicts should be introduced deliberately to motivate learning.

Motivation is improved according to Stoney and Wild [5], when there is optimal mismatch between the learners' current skills and knowledge and what needs to be learned. Tasks that are insufficiently challenging for an individual learner will de-motivate them [5], and tasks that are over demanding will de-motivate learners. When tasks and the learning environment are configured to suit the individual, then motivation is high. The use of group working was intended to increase motivation by providing support within the group so that learners may explore and accept challenge with confidence and security. The application of group work in real context also provided additional challenge for learners. There was evidence that learners on the application were well motivated. Learners pass their current level of achievement and enter their 'zone of proximal development' so that next time they do not require this additional support to achieve this level of understanding [34]. Students appreciated the possibility of trying tasks and questions and then being able to obtain more help if they needed it, from tutors as well as other group members and then to try again. In this way off-computer tasks were usually completed effectively and well.

In learning with computers there is a danger that learners might miss some of the motivation effects of working with and for others. They might also lose some of the other positive benefits of group working, for example the development of team-based skills. The approach adopted here, it is argued, was able to employ the best features of learning with technology, and working with others. Computer systems based upon instructivist learning may themselves be de-motivating and may not contribute to the development of mental models and higher-order thinking. The use of computers as an integral part of a learning system, based on group working and involving interaction with other learners was found to be effective at delivering high-quality learning.

Group working is seen therefore as a method of integrating computers into existing educational systems. This was of benefit to tutors and was considered to be an important feature of the application. Tutors reported that they not only valued the constructivist approach employed in the study, but there was evidence that they were motivated by it and that this had an influence on the quality of learning provided to students. In discussions, tutors who expressed positive attitudes to the course, stated that this was due in part to its constructivist features and how it engaged learners.

Group working has many benefits, but also there are difficulties and problems with the approach. There was evidence, for example, that the learning environment provided for group activities was not optimal, and may have had a negative influence on group working. Other problems centred around assessment and testing. Tutors expressed concern over assessing group work. It is likely that new methods will be needed.

Group working is a real life skill and needs to be taught, supported and fostered in education. Working with computers is an equally important life skill and should be supported in the same way. It is important that educators identify and overcome some of the problems that these approaches introduce into an educational system designed to assess and reward the individuals.

References

1. Dearing, R (1997), Higher Education in the learning society: *Report of the National Committee of Inquiry into Higher Education,* London: NCIHE Publications (HMSO), July
2. Brown, S. (1998), Reinventing the University , *Association for Learning Technology Journal,* 6(3), 30-37

3. Musselbrook, K., McAteer, E., Crook, C., Macleod, H. and Tolmie, A (2000), Learning networks and communication skills, *Association for Learning Technology Journal*, 8(1):71-79
4. Crook, C. K. (1997), Making hypertext lecture notes more interactive: undergraduate reactions. *Journal of Computer Assisted Learning*, 13, 236-244.
5. Stoney, S. and Wild, M. (1998), Motivation and interface design: maximising learning opportunities. *Journal of Computer Assisted Learning,* 14, 1, 40-50.
6. Hall, W. (1994), Ending the Tyranny of the Button.*IEEE Multimedia* 1(1):60-68.
7. Carswell, L., Petre, M., Woodroffe, M. and Stone, D. (1997). What's possible versus what's desirable in instructional systems: Who's driving and is the destination worth the journey? *Virtual campus real learning – ALTC-97.* University of Wolverhampton, UK, Sept. 15-17.
8. Ewing, J.M., Dowling, J.D., and Coutts, N. (1999), Learning Using the World Wide Web: A Collaborative Learning Event, *Journal of Educational Multimedia and Hypermedia*, 8(1):3-22.
9. Lewin, K. (1948). Frontiers of group dynamics. *Human Relations* 1:5-42.
10. English, S and Yazdani, M. (1999), Computer-supported cooperative learning in a Virtual University, *Journal of Computer Assisted Learning* 15(1):2-13.
11. Gibbs, G.R. (1999), Learning how to learn using a virtual Learning environment for philosophy, *Journal of Computer Assisted Learning*, 15:221-231.
12. Presselsen, B.Z. (1992). A Perspective on the evolution of cooperative thinking. In N.Davidson and F. Worsham (eds). *Enhancing thinking through cooperative learning.* New York: Teachers College.
13. Johnson, D. W. a. J., R. (1989*). Cooperation and competition: Theory and research*, Edna MN: Interaction Book Co.
14. Sennett, R. and Cobb, J. (1973). *The injuries of class.* New York: Knopf.
15. Bruffee, K. A. (1999). *Collaborative Learning.* 2nd Edition., The John Hopkins University Press, Baltimore and London.
16. Felder, R. M. and Brent, R. (1994). *Cooperative Learning in Technical Courses: Procedures, Pitfalls and Payoffs.* NSFDUE Grant DUE-9354379, October, 1994.
17. Brooks, J. G. and Brooks, M. G. (1993). *In search of understanding: The case for Constructivist Classrooms.* Alexandria, VA.
18. Felder, R. M. (1993). Reaching the Second Tier - Learning and Teaching in College Science Education. *JCST* March-April. 286-290.
19. De Diana, I. P. F. and White, T. N. (1994). Towards educational superinterface. *Journal of Computer Assisted Learning*, 10(2):93-103.
20. Somekh, B. (1996). Designing software to maximise learning. *Association for Learning Technology Journal*, 4(3):4-16.
21. Grabinger, S., Dunlap, J. C., Duffield, J. A. (1997). Rich environments for active learning in action: problem-based learning. *Association for Learning Technology Journal*, 5(2):5-17.
22. Park, I. and Hannafin, M. (1993). Empirically-based guidelines for the design of interactive multimedia. *Education Technology Research and Development*, 41:63-85.
23. Atkins, M. J. (1993). Theories of learning and multimedia applications: An overview. *Research Papers in Education*, Vol 8, No. 2.
24. Barker, T., Jones, S., Britton, C. and Messer, D. J. (1997b). The development of tasked based differentiated learning materials for students with learning difficulties and / or disabilities. *Proceedings of CAL-97 conference*, University of Exeter, March 1997.
25. Wild, M. (1996). Mental models and computer modelling. *Journal of Computer Assisted Learning,* 12(1):10-21.
26. Khan, T. and Yip, Y.J. (1996). Pedagogical principles of case based CAL. *Journal of Computer Assisted Learning*, 12(3):172-192.
27. Stoney, S. and Oliver, R. (1998). Interactive multimedia for adult learners: Can learning be fun? *Journal of Interactive Learning Research,* 9(1):55-82.
28. Barker, T., Jones, S., Britton, C. and Messer, D. J. (1997b). Creating Multimedia Learning Applications in a Further Education Environment. *Technical Report No. 271*, University of Hertfordshire, Division of Computer Science, January 1997.

29. Bloom, B. S. (1956). *Taxonomy of Educational Objectives.* Book 1, Cognitive Domain, David McKay Company, Inc. New York.
30. Fitzgerald, G. E., Wilson, B. and Semrau, L. P. (1997). An interactive multimedia program to enhance teacher problem solving skillss based on cognitive flexibility theory: Design and outcomes. *Journal of Educational Multimedia and Hypermedia*, 6(1):47-76.
31. Oppenheim, A. N. (1992). *Questionnaire design, interviewing and attitude measurement.* Pinter, London and Washington.
32. Reeves, T. C. (1992). Evaluating interactive multimedia. *Educational Technology* (May) 47-52.
33. Piaget, J. (1985). *The equilibrium of cognitive structures, The central problem in cognitive development.* University of Chicago Press, Chicago IL.
34. Vygotsky, L. (1986). *Thought and Language.* MIT Press, Cambridge, MA.

Neuro-Psycho-Computational Technology in Human Cognition under Bilingualism

Lydia Derkach

Dnepropetrovsk National University
Technology and Psychology Training Chair, 49010, Gagarin Ave., 68/2,
Dnepropetrovsk, Ukraine
lydia@lnd.dp.ua

Abstract. It is evident that in the new millennium many of the challenges we will face involve information processing complex dynamic systems. In this respect, the role of linguistic processes examined in the intact left and right hemispheres so far is of crucial importance for understanding the ways of predicting and controlling the cognitive processes underlying learning with the help of computational learning environments. The given paper reports on the findings of a long-term project aimed at investigating the role of both hemispheres in listening comprehension of aural texts by bilingual learners. It is also suggested that the right hemisphere contribution in the language use under bilingualism is due to the modular representation of language functions in the left and right brain. In summary it is concluded that profound understanding of the relationship between cerebral functioning and human psychic processes is of great importance for creating psychologically and neuropsychologically substantiated cognitive technology instruments of mind.

1 Introduction

It would appear self evident that computational technologies are a desirable characteristic of human cognition and these words are readily used by educators, psychologists, academics and teachers when the business of teaching is discussed. Indeed, Carlos Ovando suggests that, "...technology seems to be defining the future of our whole world in many ways" [1, p.79]. In view of this, it is surprising that so little is known about the effective methods of teaching which meet the requirements of modern neuro-psycho-computational attainments. This urgent necessity is reflected, in our view, in a strong desire of researchers to develop interdisciplinary links between various aspects of brain – study in order to answer the question: which innovative technology and how could be best adapted to human cognition?

M. Beynon, C.L. Nehaniv, and K. Dautenhahn (Eds.): CT 2001, LNAI 2117, pp. 214-225, 2001.

1.1 Cognitive Revolution and the Brain

"Cognitive revolution" and dominance of the "cognitive approach" in psychology have had detrimental effects on the discipline of psychology and interdisciplinary efforts that include behavioral and brain behavior conceptualization [2]. As most scholars believe the hemispheric brain asymmetry underlies all the personal traits of learners. According to Franco Fabbro, "knowledge of the brain can be useful not only for physicians, psychologists, and speech and language therapists, but also to teachers in general, irrespective of the level they teach. After all, every day they have to deal with one of the most typical features of the human brain, namely the ability to learn" [2 p.14]. Furthermore, a fundamentally important question is what can an ordinary classroom teacher do to individualize instruction [3].

1.2 Information Technologies and Bilingual Education

It is evident that in the 21-st century many of the challenges we will face involve information processing complex dynamic systems. In this respect, the role of linguistic processes examined in the intact left and right hemispheres so far is of crucial importance for understanding the ways of predicting and controlling the cognitive processes underlying learning with the help of computational learning environments.

Brown comments: "Bilingual teachers are interested in using technology but want to do so in a way that is consistent with their goals of encouraging students to actively and critically examine and question the world around them... But how to justify the greater use of machines that have traditionally placed the student in a passive role?" [5 p.178-179].

Similarly, C. Ovando stresses the importance of the computational technology in bilingual education and concludes that when teachers have the access to computers they often use it for "...individualized drill and practice activities with low-level cognitive demand, or as rewards for completing assignments, rather than an integral part of meaningful, complex thematic instruction" [1, p.79].

Developing this view, he believes that "using computers in instruction can expand students' language and academic skills through use of word processing software, spreadsheets, database software, communications programme, graphics packages, hypermedia, and access to telecommunications such as electronic mail and Internet" [6, p.119].

Clearly, the idea of microcomputer software for bilingual classes is intuitively appealing, it has face validity and is the term commonly used to denote instructional software that has been developed for bilingual learning. However, many bilingual teachers consider [7], [8], [9] it is also characterized by a very low-level in cognitive demand and tends to encourage passive learning environments, teachers need to develop a very critical, analytical perspective when reviewing software for use" [1, p.82].

Similarly, Hunt pointed out that "most of the software focused on typical vocabulary or grammar drill and practice. The very nature of drill and practice software runs counter to the natural acquisition approach for L2 instruction because it tends to present isolated, noncontextualized exercises that focus on accuracy rather than fluency" [10, p.8-9].

1.3 Cognitive Development and Interactive Discovery Learning

In view of this, an increasing understanding of cognitive development of a person under bilingualism through interactive discovery learning includes every aspect of the development of complex thinking skills and learning strategies across the four dimensions of language acquisition: linguistic, sociocultural, cognitive and academic. Learning strategies being used as techniques help students to understand, retain information and solve problems.

Following O' Malley and Chamot, learning strategies are defined as "the special thoughts or behaviors that individuals use to help them comprehend, learn, or retain new information" [11, p.1]. What follows is that learning becomes self-directed, faster, easier, more enjoyable and effective. In this respect the evident fact must be admitted that profound understanding of the relationship between cerebral functioning and human psychic processes is of great importance for creating psychologically and neuropsychologically substantiated cognitive technology instruments of mind.

What way neurological data influence cognitively complex bilingual education of learners and how one should take into consideration individual neurological differences of students while organizing information technology use is one of the principal aims of our longitudinal project.

2 Aims

It is our purpose with the given paper, through the suggestion of the innovative approach to language learning grounded on brain characteristics of the individual [12], [13], [14], [15], [16], [19], [20] to spark more interest among neuropsychologists and educationalists in the questions of providing the evidence and impetus for changes in the education system for the new century.

The given paper reports on the findings [14], [15] of a long-term project aimed at investigating the role of both hemispheres in bilingual language development which goes hand in hand with both cognitive and academic development, and is accompanied by a continuing development of thinking skills. In the current paper we will also suggest that the right hemisphere contribution in listening comprehension of aural texts under bilingualism is due to the modular representation of language functions in the left and right brain. Finally, we would like to suggest our speculations on the mind/machine interaction questions raised in our investigation.

2.1 Hypothesis

We hypothesized that multimedia is the most effective for students with right-brain dominance as the right hemisphere perceives and remembers visual, tactile and auditory images; it is more effective in processing holistic, integrative and emotional information [16]. Additionally, the focus of our research was on the study of the relationship between the level of prior knowledge (low, intermediate, high) and the effect of multimedia learning.

2.2 Computer Technology's Impact and Human Interaction in Education

Research and theory on continuing learning, in particular, [22], [23], [24] testifies that it is growing exponentially and the relative proportion of web based instruction is continuously increasing. Both educational and informational technologies – and especially multimedia over the past decade has challenged much interest among educationalists and researchers to the point where extensive review articles are now available outlining fundamental aspects of multimedia education, principles of instructional design for fostering multimedia learning.

Clearly, that continuing education requires profound skills of self-education. Moreover, the basic principles of continuing education which are realized both in teaching and learning presuppose:

-educational appropriateness of learning and training;
-openness and rigidity of learning;
-interactivity;
-planned character of material assimilation;
- individualization.

As the literature review shows, psychologists continue to investigate computer technology's impact on human learning and human interaction in education [25], [26], [27], [28], [29] and others. It is not surprising, however, that the advanced development of information technology and especially, multimedia, leads to the very fast emergence of a great number of educational on-line and off-line products and services [30]. Indeed, more and more individuals turn, for example, to the Internet for the solutions of their social, psychological and career problems [31]. Electronic interactions which take place via communication techniques, such as e-mail, bulletin boards, chat rooms, forums, telephony, videoconferencing, etc. encourage, as Kaas states, new ways of giving and receiving counseling [31, p.150]. However, the overall result of this extensive research activity has been a more realistic, precise image of a complex expertise and practical knowledge involved in continuing education generally.

3 Method

In view of this, it is surprising that so little is known about the nature of the relationship between the individual brain characteristics of a learner and the optimal cognitive strategies typical of the personality in the course of continuing bilingual education. Therefore, to maximize the effects of strategy instruction, the students should be provided with information on the use of the strategy as well as feedback on monitoring progress while using the strategy.

3.1 Interdisciplinarity, Multimedia Learning Strategies, and Brain Functioning

It would appear self-evident that interdisciplinarity is a desirable characteristic of the active, inquiry-based interactive learning which provides, in John Dewey's opinion [32], discovery learning for all students.

In terms of defining the problem, it could be said that one of the most controversial areas of inquiry in multimedia instruction and training of late, has been the study of right and left brain functioning in the process of utilizing multimedia learning strategies.

Although very little is known so far how the right and left brain integrate their processing capacities, investigators have begun recently to focus on more general neuropsychological and psychological mechanisms accounting for effective hemisphere interaction [3], [33]. In this respect, the role of technology impact on training and continuing education effect, examined in the right and left brain, is of crucial importance for understanding the ways of predicting and controlling cognitive processes underlying the continuing education.

3.2 Innovative Approach and Multimedia Learning Success

As we have already mentioned, the current research is aimed at presenting an innovative approach to brain functioning and hemispheric interaction in providing multimedia instruction and training in bilingual education.

The purpose of the given paper is, through the suggestion of neuropsychological (levels of information processing, that is, neurological profiles), psychological (learner models in continuing education and training) and methodological (computer-based multimedia development) research, grounded on a learner individual brain characteristics, to spark more interest among continuing education bilingual specialists in the question of fostering multimedia learning success via neuro-psycho-methodological approach. In other words, how to improve a continuing bilingual education effectiveness knowing the peculiar and dominant traits of the right and left brain of a student, as well as their combination – is also one of the trends of our longitudinal research.

3.3 Learners' Models and Bilingual Knowledge Acquisition

Thus, we were interested in designing learner models under continuing education and training which makes it possible to adapt instruction to individual progress in bilingual knowledge acquisition. In contrast to existing approaches to student modelling regarding intelligent tutoring systems [34], we propose our own model of a student which includes the neuro-psycho-methodological profile of a particular student who is pretty aware of the profile peculiarities alongside with the preferable, optimal, "comfort" for the given learner multimedia information technologies. The nature of such a profile incorporates the knowledge of cognitive learning styles and associated learning style preferences of a student which are due to the individual brain characteristics of a learner [35].

It is common knowledge that students who are successful learners spontaneously use cognitive learning strategies while learning a new information. The fact which is of primary importance for both the educators and learners, in our view, is that successful learners use cognitive learning strategies selectively to retain information and develop intellectual skills. In this framework, consequently, the model of a successful right-dominant or left-dominant learner, able of the selective use of cognitive learning strategies especially as these relate to cognitive styles and associated learning styles, is the future perspective of the given research.

3.4 Two Models of Students and Their Brain Preferences in Information Processing

According to our preliminary experimental data we suggest two types of models of the students at a continuing education under bilingualism which differentiate between their neuro-, psycho- and methodological peculiarities regarding their brain preferences in information processing. Consequently, we predicted that factors affecting students' preferences in selecting either traditional distance or Internet – assisted learning would be accounted for by the individual hemisphere differences. In light of this, we were eager to get the answers to the following questions:

-What have neurological issues got to do with continuing education of learners in general? In particular, how should one take into consideration individual neurological differences of learners while organizing distance learning in a bilingual environment?
-What is the most effective system of their diagnostics?
-How to improve the effect of distance learning knowing the peculiar dominant traits of the right and left brain as well as their combination?
-What are the optimal neurological implications to successful theory and practice of distance education?

In accordance with our previous research [12], [13], [14], [15], [16] we have concluded that differences in individual lateralization profile (that is, the dominant pattern of brain organization) reflect differences in cognitive processing of learners. Besides, we suggested that the individual lateralization profile might be defined as a complex integrative index of motor and sensory asymmetries typical of a student at a definite period of his ontogenetic development.

For singling out the typology of profiles we used several experimental techniques for measuring a student's profile (in scores), namely: Questionnaire [17]; Functional Probes [18] ; Test "Metagramma" [19]; Modification of the Associative Test [19], Self-Assessment Test "Brain Works" [21].

4 Results

With this background material in mind, we can turn particular attention to the analysis of the two lateralization profiles of information processing which we would like to describe. The first type of model of a student is represented by the COGNITIVE-LINGUISTIC LEARNERS who constantly demonstrate left individual lateralization neurological profile, that is, left – brain dominant students. While the second type - COMMUNICATIVE LEARNERS steadily demonstrate right individual lateralization neurological profile, that is, right-brain dominant learners.

It is relevant to ask why we focused only on the these two types of profiles but there are advantages to this. As with the discussion of the methodological problems of neurological types in continuing education and training, careful analysis of more than 22 individual lateralization profiles, discovered by different scholars, would be inappropriate for our purposes here.

In any system of categorization, according to P. Makin et al. [36], the reduction in the number of categories used increases the generality of those categories. This makes them easier to work with, but reduces specificity [36, p.61].

The implicit intention of our argument is to suggest that the notion of innovative approaches and technologies in continuing education and training instruction underlies much recent thinking about current psychology and neuropsychology.

While we have presented this line of reasoning couched in the given paper on the hemispheric interaction the method of individual lateralization profile applied by us was compared to the method suggested by M.Bryden "Handedness Inventory" [37] and Self-Asssessment Test "Brain Works" [21]. The experimental data proved the validity of the results obtained in both procedures.

Consequently, the two types of learners demonstrate the two types of information processing and represent, in our view, the two models of a student organizational behavior in a continuing learning. The most significant differences in information processing by the left-and right – brain-dominant students regarding READING AND WRITING are presented below.

4.1 Model of a Successful Right-Brain Student (Communicative Learner) in Continuing Learning

- Processing information is realized in patterns and subjectively
- Is sensitive to cooperative environments
- Is fond of problem-solving
- Learning is more optimal when referential processing (audio + text) is taking place
- Auditory explanation accompanied by the animation or picture design create better conditions for the remembering and retrieval of the information processed
- Pictures provide a context for understanding ambiguous text
- Multimedia is the most effective while presenting assembly instructions (text + pictures)
- Motion pictures are more preferable than still pictures in procedural information presenting (f.e.instructions for operating a device)
- Graphical presentations of information improve performance better than textual presentations
- Recognition accuracy of the material presented was 96% for pictures; 82% for sentences, 83% for words
- Recognition accuracy rates for pictures and text are better than for text alone
- Spatial information is optimally presented pictorially
- Sound appears to be an effective way to communicate small amounts of verbal information
- For recalling story details video with sound track appears to be effective

4.2 Model of a Successful Left-Brain Student (Cognitive-Linguistic) in Continuing Learning

- Processing information is realized sequentially, mainly objectively
- As a rule, a student is independent in cooperative interaction
- Is fond of problem-solving through intuition
- Learning is more optimal when referential processing (text) is taking place
- Verbal explanation accompanied by animation or picture design create better conditions for remembering and retrieval of the information processed
- Key words or ideas provide understanding of ambiguous text
- Short phrases in picture-phrase combinations or verbal captions are recalled better
- Still pictures are more preferable than motion pictures in procedural information presenting
- Textual presentations of information presenting improve performance better than graphical ones
- Recognition accuracy of the material presented was 94% for sentences, 91% for words and 97% for text
- Recognition accuracy rates for texts and key words are better than the text alone
- Spatial information is optimally presented using maps
- Visual mode appears to be an effective way to communicate small amounts of verbal information
- For recalling story details static pictures help a student to learn auditory, oral prose.

4.3 Psychological and Methodological Characteristics of Two Models of Learners

Regarding the specific conditions of continuing education and training we have obtained the following psychological and methodological characteristics of the two models of students in multimedia environments while performing problem-solving tasks, decision-making and using the preferable strategies of coping with them.

COGNITIVE-LINGUISTIC TYPE (LEFT-BRAIN STUDENTS).

The given type of learners (they usually prefer to operate with their right hand in many activities) make extensive use of left-brain strategies according to the functional asymmetry of the brain. Consequently, they prefer to deal with the problems which are solved in a logical way, sequentially, presented in a verbal but not picture mode. They are also very enthusiastic in searching precise facts and enjoy constructive tasks though dislike analysing diagrams, tables, charts. Students of the given type are quicker at summing up than at creating new, innovative ideas.

COMMUNICATIVE TYPE (RIGHT-BRAIN STUDENTS)

This group of students called left-handed (most of them are left-handed), who process the information with the help of right-brain strategies, are good at solving problems intuitively and, as a rule, possess a very strong imaginative thinking. They do enjoy inventions, searching for a principal idea, are extremely fond of the feeling of insight through the problem-solving situation.

Contrary to the right-handed students, left-handed learners prefer to work in groups which put forward idealistic goals in case-studies, chatting, videoconferencing, debates which makes it possible for them to demonstrate personal initiative.

5 Conclusions

Thus, the above mentioned two individual typical differences characterize two types of learners which are manifested in their general abilities, cognitive and learning styles in continuing education and which represent their comfortable zone of specialization and initiatives realization. A shorthand explanation of the principal difference between the two styles, according to P. Makin et al., [38] is in the fact that cognitive-linguistic learners, in terms of our classification, like doing things better while the communicative type learner prefers to do things differently [38, p.74].

And lastly, we were interested in the analysis of the relationship between the level of prior knowledge of a student, (low, intermediate, high), and the effect of learning in multimedia environments, grounded in the two models of the students that we have singled out.

Among 79 students in the experiment there were 15 students with low level of achievements, 46 with intermediate and 18 with a high level. It turned out during the 3-week period of a daily use of multimedia information that the impact on 15 low-prior-level students proved to be the most effective. The difference was about 92%. We support the opinion that multimedia presents the information in a more vivid way, makes it more obvious, helps low domain knowledge learners to integrate the prior and the new knowledge especially in communicative learners (89% in comparison with cognitive-linguistic learners – 34%). The students with the intermediate level of prior knowledge, namely communicative learners, demonstrate the same tendency (72% in comparison with cognitive-linguistic learners – 64%).

Learners with high domain knowledge, and a cognitive-linguistic type, they are characterized by a profound level of prior knowledge. They prefer to focus their attention mainly on the verbal information alone and feel uncomfortable under the conditions of making use of multimedia information (97% in comparison to communicative learners - 28%). Additionally, high-level prior knowledge students with right-brain dominance factually tend to the extensive use of multimedia information and learning (96%).

In conclusion we would like to sum up the major results of our longitudinal research on continuing education and neurological, psychological as well as methodological considerations in multimedia environments. The data presented suggests that the two models of a student's individual neuro-psycho-methodological profile make it possible to differentiate between individual differences of students while organizing multimedia information and learning. Due to this fact the most effective, portable system of their diagnostics proved to be the neuro-psycho-methodological approach of defining the neurological profile, which is characterised by the simplicity, validity and time-saving technique. Consequently, the knowledge of peculiar dominant left and-right- brain characteristics of a learner essentially improves the process of continuing bilingual learning and training, informs the student on the most optimal strategies in information processing.

Taking into account these findings obtained in modern neuropsychology there are several good reasons for further study of the interaction between the mind and machine:

What way will the brain cope with and abundance of information increasing from day to day? Will it cope with the help of a machine or not?

Judging by our experimental data obtained from 693 subjects (390 men and 303 women) of various age groups: 5-6, 13-15, 17-23 years old, we have received the profound evidence that "technology ....can contribute substantially to the active, experiential learning that Dewey advocated decades ago " [39, p.1].

The next string of questions which raised in the course of our longitudinal study was:

Is there any danger that the brain will not manage with the complexity itself? What changes will take place in a human brain under information overloading? Are there any self-defence and self-control mechanisms in the brain preventing it from overloading?

The experiment proved that the adequate choice of either left-brain or right-brain strategies (or their combination) by a student is one of the powerful components of an active learning environment, with the use of appropriate technology as one more vital component of effective teaching in bilingual education which might both regulate and prevent information overload and provides self-defence and self-control mechanisms to work efficiently.

One more block of questions under the current research which are still unanswered:

What brain systems and structures are the most vulnerable?

Is it necessary to suppress emotions and open the way to intellect?

As is evident from the subjects' reports, left-handed students (72%) exercised negative emotions on aural comprehension tasks followed by the usage of communications technologies accompanied by writing with a word processor or distant classes through telecommunications. They also found inconvenient for them to work with technologies that encourage learners to work collaboratively to create group texts or with technologies that engage students in dialogue with another class: long-distance team-teaching partnerships or the implementation of case-method technologies. More than 60% of right-handed subjects coined their emotional state under these technologies as being very close to the burnout syndrome. And what mattered much under those circumstances was the desire not to suppress emotions but "...to reveal them as deep as possible."

And finally, we strongly believe that the assumptions and procedures used in studies on the questions mentioned above are still the matter of controversy and form a relatively new but extremely fertile field of cognitive research in the 21-st century.

References

1. Ovando, C.J.: Bilingual and ESL Classrooms: Teaching in Multilingual Contexts. McGraw-Hill (1998) 79

2. Furedy, J.: A Pre-Socratic Biobehavioral Approach to the Experimental Analysis of Real Cognitive Functions Versus Computer Information – Processing Metaphors: Back to the Future. In: International Conference on Psychology.: (2000) 95
3. Fabbro, F.: The Neurolinguistics of Bilingualism, London: Psychology Press (1999) 14.
4. Lefrancois, Guy R.: Psychology for Teaching: A Bear Is Politically Correct. 9th Ed.(1997) 426
5. Brown, K.: Balancing the Tools of Technology with Our Own Humanity: The Use of Technology in Building Partnerships and Communities. In J.V.Tinajero and A.F.Ada (Eds.), The Power of Two Languages: Literacy and Biliteracy for Spanish-speaking Students. MacMillan/McGraw-Hill (1993) 178-198
6. Ovando, C.J.: Bilingual and ESL Classrooms: Teaching in Multilingual Contexts. McGraw-Hill (1998) 119
7. Krauss, M.: Extending Enquiry Beyond the Classroom: Electronic Conversations with ESL Students. CAELL Journal, 5(1), (1994) 2-11
8. Gaer, S ., Ferenz,K.: Telecommunications and Interactive Writing Projects. CAELL Journal, 4(2), (1993) 2-5
9. Willis, J., Stephens, E.C., Matthew, K.I.: Technology, Reading, and Language Arts. Boston: Allyn and Bacon (1996)
10. Hunt, N.: A Review of Advanced Technologies for L2 Learning. TESOL Journal, 3(1), (1993) 8-9
11. O'Malley, J.M., Chamot, A.U.: Learning Strategies in Second Language Acquisition. Cambridge: Cambridge University Press (1990)
12. Derkach, L.N.: A New Approach to Evaluating Brain Functioning under Bilingualism. Abstracts of the 24-th International Congress of Applied Psychology. San Francisco (1998) 437
13. Derkach, L.N.: Designing Teacher's Motivation and Job Commitment for the 21-st Century. Abstracts to the 6-th European Congress of Psychology. Rome (1999) 132-133
14. Derkach, L.N.: New Ideas on Cognitive Development in Adolescents under Bilingualism. 2nd Brazilian Congress on Psychology of Development. Gramado, Brazil (1998) 16-17
15. Derkach, L.N.: Management Education and Brain: New Directions in Theory and Research. Proceedings of the 4th CBE Annual Conference for Administrative Sciences. United Arab Emirates University. Al Ain (2000) 57-70
16. Derkach, L.N.: Continuing Education and Brain: Future Perspectives. Proceedings of the International Conference on Millennium Dawn in Training and Continuing Education. College of Engineering University of Bahrain. Bahrain (2001) 349-357
17. Annet, M.: A Classification of Hand Preference by Association Analysis. In: British Journal of Psychology. 61 (1970) 303-321
18. Luria, A.R.: Major Problems of Neurolinguistics. Moscow State University Publishing Press (1995) 388
19. Derkach, L.N., Kovalenko, J.V.: Brain Functioning and Foreign Language Learning. In: 2-nd National TESOL Ukraine Conference "The Art of Science of TESOL". Vinnitsia Pedagogical Institute Press (1997) 125-126
20. Derkach, L.N., Egorova, L.E.: Associative Test and Verbal Memory in Schoolchildren. In: Tashkent Pedagogical Institute Press (1982) 50-59
21. Self-Assessment Test "Brain Works". In: Derkach, L.N., Kovalenko, J.V., Marchenko, A.V., Erokhina, I.V. English Textbook for Geography University Students. Dnepropetrovsk National University Press (1999) 100
22. Hector, J.H.: Teaching and Learning Via Internet: Experiences with Introductory Statistics. International Journal of Psychology, Sweden (2000) 136
23. Retschitzki, J., Baumann, F., et al.: New Technologies and Learning Strategies in Higher Education. International Journal of Psychology, Stockholm, Sweden (2000) 135
24. McVough, W.: Learning Style and Type of Knowledge Learned in Distance WEB Based Versus Traditional College Classes. International Journal of Psychology, Sweden (2000) 247

25. Issing, L.: Instructional Design for Media Technologies Abstracts of the 6-th European Congress of Psychology, Rome (1999) 215
26. Loarer, E.: Psychological Expertise of Educational Multimedia. Abstracts of the 6-th European Congress of Psychology, Rome, Italy (1999) 267
27. Mayer, R.: Aids to Multimedia Learning. Abstracts of the 6-th European European Congress of Psychology, Rome, Italy (1999) 288
28. Jausovec, N., Gerlic, I.: Multimedia Differences in Cognitive Processes Observed with EEG. Abstracts of the 6-th European Congress of Psychology, Rome, Italy (1999) 217
29. Chidambaram, L.: Relational Development in Computer-Supported Groups. MIS Quartely, Vol.20, 2 (1996) 143-166
30. Vreede, H.J., Briggs, R.O., Santanen, E.L.: Group Support Systems for Innovative Information Science Education. Journal of Informatics Education and Research, Vol.1, 1 (1999) 1-11
31. Kaas, D.: Internet Councelling Benefits Vs Pitfalls. International Conference on Psychology, Haifa, Israel (2000) 150
32. Dewey, J.: Democracy and Education. McMillan, New York (1916)
33. Chiarello, C. Interpretation of Word Meanings by the Cerebral Hemispheres: One Is Not Enough. In: Schwanenflugel (eds.): The Psychology of Word Meanings. Hillsdale: Earlbaum (1999) 251-278
34. Josif, G.: Student Modelling for Intelligent Tutoring Systems. Abstracts to 6-th European Congress of Psychologists, Rome (1999) 214
35. Cem Tanova: The Cognitive Styles and Learning Preferences of Undergraduate Business and Executive MBA Students. Proceedings of the 4th CBE Annual Conference for Administrative Sciences. United Arab Emirates University, Al Ain (2000) 89-92
36. Makin, P., Cooper, C., Cox, Ch.: Organisations and Psychological Contract. The British Psychological Society, London (1999) 61
37. Bryden, M.P.: Laterality: Functional Asymmetry in the Intact Brain. Academic Press, New York (1982)
38. Makin, P., Cooper, C., Cox, Ch.: Organisations and Psychological Contract. The British Psychological Society, London (1999) 74
39. O'Neil, J.: Using Technology to Support "Authentic Learning". ASCD Update, 35(8), 1(1993) 4-5

Digital Image Creation and Analysis as a Means to Examine Learning and Cognition

Brad Hokanson

Department of Design, Housing, and Apparel
University of Minnesota, U.S.A.
bhokanson@che.umn.edu

Abstract. This paper presents research on the development of digital collages and investigations into the cognitive efforts of graphic designers. The research venue was an upper level undergraduate course in computer graphics. Work in the course focused on the use of digital imaging software (Adobe Photoshop) and the creation, manipulation, use and understanding of images. Graphic design students manipulated images to address a series of complex topics. This study began with a hypothesis: computers may extend our cognitive abilities, inspired at least in part by the work of McLuhan.

1 Introduction

The research questions remain essential: How is thought amplified or extended through the use of computers? How do computers affect the way we think and learn? How has our method or process of cognition changed through the use of various media; and how has our thought process been structured through the use of symbol systems such as writing, drawing, or speech? These questions are focused through an examination of images and language.

How we solve problems and create is tied to how we think. We use technology to solve problems. Examining the process of problem solving sheds light on cognition; whether the process involves use of the computer or symbol systems such as drawing, writing, or mathematics.

Changing the process of generation, of creation, changes the process of how we think. If we generate ideas using different symbol systems or media, we are using different processes to think. Subsequently, different creative processes mean the thinking process, cognition itself, will be different. And, as cognition is different, so too will be learning.

The true goal of learning is not the repetition of facts, but rather an improvement in cognitive processes.

M. Beynon, C.L. Nehaniv, and K. Dautenhahn (Eds.): CT 2001, LNAI 2117, pp. 226–232, 2001.

2 Overview

Subjects were 17 undergraduates studying graphic design; three males and 14 females ranging in age from 20 to 24. All were graphic design students and all were considered advanced students in the program. The course was required of graphic design majors and centered on the use of digital imagery through the use of Adobe Photoshop and other graphics software.

The study focused specifically on a group of learners with specific characteristics; a highly developed level of skill in visual material due to choice and education, a highly educated group, and highly computerized group due to their profession and education. Examination of the use of a different symbol system than that used by mainstream society required this selection, and also allowed a more focused examination of the concept of symbol systems themselves. A more 'well-rounded' group would not have had the interest in developing an understanding of a highly distilled and rarefied symbol system or computer software.

This paper summarizes the study by examining how the findings support the original hypotheses and answer the initial research questions.

Six methods, including interviews, standardized tests, and image analysis were used to gather data for the study; meaningful findings will be discussed. How the findings relate to the body of research literature in the field will be examined, and the extension of the findings to other areas will be noted. Ideas for future research will follow, with some concluding remarks.

3 Observations

This investigation began with an understanding of the use of symbol systems, the use of media in support of thought. Linguistic research extending from Whorf [15] to recent work by Hunt and Agnoli [3] supports the concept of different 'languages', different symbol systems directing our thoughts in varied directions. This study, while not proving their hypotheses on cognition and language, remains a part of that tradition.

Extending Salomon's [12,13] concepts of internalization, it had been anticipated that a professionally oriented participant group with a high orientation toward visual or other divergent, non-text symbols would most readily present evidence of cognitive activity involving the use of these symbol systems. Extensive and focused use of the symbol systems was also desired of the study activity. Both of these conditions were achieved and remain present in the field in general.

The strongest indications of support for the original hypotheses, that computers and alternative symbol systems advance and assist cognition, are evident in the interviews and journals maintained by the participants.

Interviews proved most effective in the development of an understanding of the participants and their thought processes. The interviews presented parallel sets of ideas in the mixed use of images and words, of integrating words and pictures into processes. Participants uniformly linked the creation of images to the use and manipulation of words.

It appeared that learning software was the educational focus of the students as opposed to any other educational activity. Interviews later noted that skills were developing in communication and in the use of images, and "along the way, we learned Photoshop." (student interview). This is a critical difference, in that this recognizes that learning of specific procedural knowledge is required to successfully examine or present other information or ideas. In other words, you need to know how to paint, how to write, or how to manipulate digital images before one can communicate or develop ideas.

The comments made by students in their journals illustrated their development in the use of a new software program and subsequent use of images. The electronic journals, while limited in their clarity, provided a series of minor observations that supported other observations.

Portions of the study, specifically the surveys and the image analysis provide secondary support of the hypothesis. Surveys provided interesting observations. As would be expected in any successful educational activity, perceptions of skill level changed in the subject domain; students believed they learned the software. Additionally, a large but not significant change in the participant's perceived writing ability was recorded. It is hypothesized that this unexpected finding is due to extensive practice in the use of a variety of symbols for communication, and that that use has transferred to their skill with other media, particularly writing. This clearly is a venue for further research.

Evaluation of the images was limited to the more mechanical aspects of the images, and aesthetic or content based concerns were not included as part of this study. The images, evolved and got more complex over the course of the study. It was hypothesized that increased complexity of image corresponds to an increased complexity of thought on the given topic.

The final research methods, the Watson/Glazer Critical Thinking Analysis (WGCTA) and semantic webs were unsupportive of the initial hypothesis. The outside metric used in the study, the Watson/Glazer Critical Thinking Analysis, did not record significant change. Semantic mapping did not show an improvement in student cognition or understanding.

4 Discussion

First, two different theoretical interpretations are presented with the goal of unifying the empirical information and background research into a cohesive whole. Each deals with how we use language, how we use images, and how we think. First is a discussion of the interplay or interaction that occurs between image and word. Second, there is included a discussion of the nature of intelligence and media. Woven through these is the phenomena of efficiency, how the participants communicate and solve problems through the use of various symbols.

4.1 Words and Images Connected

Central to this study is the interaction and intersection between word and image. The study began with word based information which was translated to visual images by the participants. The requirement of translation, of conversion of the written concepts to visual images is an activity at the boundary between two symbol systems, words and images.

The measurements and observations of the study illustrated a complex series of processes that varied within and between individuals. Each participant appeared to have their own method of working, of creating images.

The most important generalization was the use of an inclusive as opposed to an exclusive process in the creating of images. Simply put, participants would use words *and* images in the creation of new images. As illustrated in the interviews, words acted as a shorthand in the development of new images. Words helped create images. It appears that in an 'image only' palette or symbol system, where the use of text is removed from the primary means of expression, words still play an important part in the development of images, and by extension, ideas.

While most students used simple words or linear lists to summarize or trigger ideas for image development, the semantic maps originally used as a measurement instrument were adopted by some students and integrated into their methodology. Semantic maps were viewed as a two dimensional listing of where information should be. In any case, words remained a strong element in the design process.

What was discovered was a pragmatic, iterative, series of methods chosen for expediency. Through interviews and journals, students noted using tools that matched their experience, selecting symbol systems as to which was the most expedient. Speed and pragmatism became the most important elements in choosing the working symbol system; images were used when most appropriate. Translation back and forth between symbol systems, between text and image, was a common occurrence.

This pragmatic awareness of efficiency, a form of meta-awareness, is consistent with published research on ability and computer use in learning. For example, writing students chose the structures that allowed them the highest probability of success with the easiest or most expeditious method (ironically, often with counter productive results for learning) [1]. One must remember that the goal of educational technology is to make learning more effective, not easier.

The magnitude of efficiency allowed by the computer can make a difference in the ease and acceptance of use of alternative symbol systems. When learners are comfortable with the computer, and find that creating on the computer is as good and more expedient than other methods, it will be used [11]. When it is not perceived as more efficacious, it is not used. In this study, words were used as a fast shorthand; the images never got fast enough, in spite of their acceleration by the computer. The principles of internalization, and creation within a given symbol system may not have been achieved.

Humans focus on expedient methods of resolving work. For example, the initial substantial investment in learning to read and write is paid back in the continued ability to gather information and to think. A similar set of phenomena occurs with the use of imagery and other non-word symbol systems. It takes effort to learn, but after learning a cognitive medium, the capability is expansive and accelerates cognition.

4.2 The Nature of Intelligence

Olson's [8,9] concept, that "intelligence is skill in a media" was, from the beginning, a central issue of the study. Recent research in divergent learning "styles" lends some support to the use of various systems. The central finding of the study, the intense, bi-directional interaction between words and images advances this aphorism, leading, perhaps to a redefinition of intelligence.

Initially, one can see through previous published research that various people understand and learn information through different media. There are claimed to be various learning styles, e.g. Kolb [7], Gardner's multiple intelligences in divergent media [4], varied receptivity toward different languages as noted by Whorf [15], *et al.*, and the symbol systems of Salomon [12]. Intelligence has been primarily evaluated through words, and specifically through writing through the past few millennia. Similarly, a word-based test of critical thinking, the WGCTA, found no development as part of this study. Recognizing the range of symbol systems to communicate and think, we may find that intelligence may be skill in *any* medium.

This study leads one to hypothesize that there may be a possible correlation between writing and imaging skills requiring additional investigation. This correlation or interaction may imply some transference of skills between different media and different symbol systems. The ability to decode a sequence of symbols may be applied to a different symbol system, reading words may lead to reading pictures; reading numbers may lead to (or require) reading graphical displays.

The ability to change and shift media or symbol systems may be a critical element in the use of computers and multi-media of the future. Intelligence may be skill in multiple media...using multiple sources of exploration, both perception and thought, using the symbol system or media that causes the most appropriate benefit.

In the investigation of advancement of cognitive processes, "thinking" more or faster through the use of different symbol systems, and correspondingly the use of a different perceptual method, a wider range of symbol systems will need to be employed. Supplemented by a greater access to computing capability, translation to and between various symbol systems may rise in importance. This interaction, developed as a skill, transferable, also implies that developing skill in one media may improve skill in others as well.

Intelligence is skill in the use and transferability of symbol systems; it is the use of multiple systems, highly refined systems for specialized work, and the transfer between various symbol systems in the pursuit of thought.

5 Extension

As with any study, the extension of the study to a larger group or a different situation will temper the findings. Expanding circles centering on the common student in graphic design, involved in digital imagery can be constructed. Through that focus on one aspect of human computer use and its application to thought and communication, certain generalizations may be made, and which require additional research.

The study involved a small class in a specialized field of study. They developed a high level of skill in using computer software to manipulate images, a skill that built on years of study in the visual arts. The results can be considered narrowly generalizable to others in their field, or as part of an understanding of the nature of humans using symbol systems divergent from the mainstream of text.

Findings of this study could be extended to others of similar populations; specifically those highly skilled in graphic media and familiar with computers. The value of the study in application to other populations comes from the focused nature of the personal interaction with the computer and not the specific software or symbol system. By this extension, the findings could shed light on others working intensively with other software applications. It may also illuminate the use of images and the nature of problem solving.

While most studies are in some way limited, each begins with a goal of using the research to understand the larger picture based on a small study. The pragmatic requirements of empirical research must be balanced with the applicability to other and larger situations. Specifically, this study may increase our knowledge of cognition, the use of symbol systems, and the use of the computer as an aid to thought.

6 Directions for Further Research

The nature of research is to develop further questions, not just answers. Hence, what follows is a series of directions for further research examining the border between image and word, examining the nature of computer use, and continuing to attempt to understand the nature of intelligence and media.

Some of the research methods employed uncovered interesting information, and will direct future research. Specific beginning points include a perceived improvement in writing and a substantial interaction between words and images. Additional study focusing on these two areas may increase our understanding of the use of varied symbol systems.

The two major observations that integrated results from the varied research methods also suggest future research directions. The first observation, that the use of words and images was interactive and iterative, is important to the understanding of images, to visual education, and education in general. It's extension, that intelligence is tied to our use of media, translation between media is similarly important.

Clearly, research into the efficacy of these switching of symbol systems would be beneficial in understanding the use of images as a means to thought. Extension of theoretical understanding on *translation* of symbol systems can also provide some understanding: Summarizing, for example, remains important as a tool to understand information. How does this correlate between translation between symbol systems or *visually* summarizing information or concepts? Additional research is needed into manipulating symbols as a means to stimulate higher order thought.

Other venues that attempt to isolate the use of symbol systems as manipulated, changed, or applied via computers should also be examined. For example, an investigation as to the changed capabilities of a writer with and without the use of computer may illuminate differences in ability. Another possible venue is an examination of the use of and translation between different symbol systems.

7 Conclusion

The computer exists as a meta-medium [6], one with access and ability for a variety of symbol systems and media. It allows the use of a wider range of symbols and symbol systems, and represents many alternative media in a single cognitive venue. Speed and efficiency is of great importance in the use of any media, and that speed of use has a major impact on the users' and learners' selection of symbol systems...and on their ability to think. What began with an examination of the use of computers in accelerating thought is ending with a greater recognition of the diversity of thought.

The study began its investigation with the goal of investigating the assistance one received through the use of computer based symbols in cognitive processes. The study focused on the specific use of a given symbol system and the access to that symbol system through computers. Literature reviewed presented an understanding of symbols in thought, and on the manipulation of symbols. A series of methods were outlined for use in the study and applied. The findings presented a series of mixed results that partially illuminate the use of the computer and digital imagery.

Words were found to be useful in planning and processing the ideas to be developed. A bridge, cited in journals and interviews, between word, information, and image was found. It appears that the participants were skilled in the use of words in the development of ideas; even those students that professed a lack of skill in the use of words still employed words to some degree in the development of images.

The interaction between symbol systems, between words and elsewhere remains a valid and interesting area for research and study.

References

1. Cochran-Smith, Marilyn, Paris C.L., and Kahn, J.L, *Learning to Write Differently*, Norwood, N.J., Ablex, (1992).
2. Eisner, E. (1997, January). Cognition and representation, *The Phi Delta Kappan*, pp. 349-353.
3. Gardner, H. (1993). *Frames of Mind: The Theory of Multiple Intelligences*. New York: Basic Books.
4. Hunt, E. & Angoli, F. (1991) The Whorfian Hypothesis: A cognitive psychology perspective, *Psychological Review,* 98(3):377-389.
5. Innis, H. A. (1954). *The Bias of Communication*. Toronto: University of Toronto.
6. Kay, A. (1984). Computer Software, *Scientific American*, 254 (3) 53-59.
7. Kolb, D. (1978). Learning Style Inventory Technical Manual. Boston: McBer.
8. McLuhan, M. (1964) *Understanding media, the extensions of man.* Cambridge, MA: MIT Press.
9. Olson, D.R., (1966). On Cognitive Strategies, in Bruner, J. S., Olver, R.R., and Greenfield, P.M. et. al., *Studies in Cognitive Growth* , New York: Wiley and Sons.
10. Olson, David R., (1974). Introduction, *Yearbook of the National Society for the Study of Education* ,73d, Chicago: University of Chicago Press.
11. Perkins, D. N., (1985). The fingertip effect: How information-processing technology shapes thinking, *Educational Researcher*, 14(6) p. 11-17.
12. Salomon, G. (1979). *Interaction of Media, Cognition and Learning: An Exploration of How Symbolic Forms Cultivate Mental Skills and Affect Knowledge Acquisition.* San Francisco: Jossey-Bass.
13. Salomon, G. (1997). Of mind and media; How culture's symbolic forms affect learning and thinking. *Phi Delta Kappan*, January 1997.
14. Von Bertalanffy, L. (1965). On the definition of the symbol, in J. R. Royce (Ed.), *Psychology and the Symbol* (pp. 26-72). New York: Random House.
15. Whorf, B. L. (1956). *Language, Thought, and Reality: Selected Writings of Benjamin Lee Whorf.* New York: Wiley.

Woven Stories as a Cognitive Tool

Petri Gerdt[1], Piet Kommers[2], Chee-Kit Looi[3], and Erkki Sutinen[1]

[1] Dept. of Computer Science, University of Joensuu
{pgerdt,sutinen}@cs.joensuu.fi
[2] Dept. of Educational Technology, Univ. Twente, The Netherlands
P.A.M.Kommers@edte.utwente.nl
[3] Institute of Systems Science, National University of Singapore
cheekit@iss.nus.edu.sg

Abstract. Woven Stories is a web-based application that allows users to compose their stories, and link appropriate story sections with pre-existing sections authored by someone else. As a co-authoring environment, Woven Stories supports the user not only as his or her individual cognitive tool, but also as a shared platform to reflect ideas and thought processes of other users with related interests. Thus, a group of users can apply Woven Stories to tasks such as creative problem solving.

1 Introduction

Motivation. Within the current trends of distance, virtual, and mobile environments, the emphasis on creating contents and content-related services seems to overshadow the need for simpler, but more generic cognitive tools. Since the concept of a story forms a versatile starting point for representing various kinds of information and knowledge in different contexts, it is important to find out how the available technology could contribute not only to browsing existing stories, but as an activating platform for users to create, reflect and co-author their own stories. We are interested in how the cognitive process of composing a story, even like this paper, can benefit from appropriate computerized tools. More specifically, we are looking into the area from the perspective of having several co-authors working on the same – *woven* – story.

The skills of storytelling and understanding basic story structures and meanings are essential to cognitive activities such as casual conversation, understanding of literature, and successful communication in general. Learning, schooling, and education form an area where the need for cognitive processing is most apparent; at the same time, it is also a field for stories. Children must be taught the essential storytelling skills and they can be brought up with the help of stories. Adults can use stories to achieve more complicated goals, such as structuring and reorganizing knowledge. Stories serve as flexible and generic cognitive tools: when listening to a story, one can easily use imagination to interpret a personal experience from the perspective of the story.

Telling stories is a powerful way of actively constructing knowledge. A story can be viewed as a set of concepts linked together by a narrative. An author has

M. Beynon, C.L. Nehaniv, and K. Dautenhahn (Eds.): CT 2001, LNAI 2117, pp. 233–247, 2001.

to resolve the relationships of the concepts included in order to fit them into a greater scheme, that of the story line. This process forces the author to reflect on her own experiences, to relate the information to existing structures, maybe forming new ones.

The Concept of Woven Stories. Woven stories can be regarded as a new conceptual tool. In brief, woven stories allow several authors to write stories in a shared story space. The epistemic representations provided by the related software allow authors to build upon and link to already made stories and even to superimpose new stories on top of the templates of existing story lines. Thus, an individual author could build on the beginning of a story, authored by someone else. Or, the author could include only some discrete sections, perhaps written by several authors, into his or her story.

The concept of woven stories was presented in [10], along with its first implementation. According to the original idea, authoring stories collaboratively can be viewed as a dialogue between the participating authors in a socially constructivist setting. Authors express their views by writing story sections and linking them to other sections. Different authors might want a different outcome to a certain section and thus link alternative sections to that section.

The previous implementation of woven stories [10], based on web technology, is intended for people of all ages. The goal of the present paper is to elaborate on the idea of woven stories, and to supply a novel implementation.

Applications of Woven Stories. Superficially, one might identify woven stories with other text-based tools, like text processors or concept mappers for individual users, or news groups or chatting software for collaborative groups. However, the difference is clear: computer-assisted storytelling systems, or story processors for short, are tools designed to help in the very task of composing a story, with related features, whereas general-purpose text software serves a larger audience, with tools not specifically targeted to a story author's needs, like that of maintaining an interesting thread together with meaningful explanations.

The particular goal of the woven stories concept is to facilitate learning by authoring stories collaboratively. It has many different applications, for example collaborative learning, creative problem solving, and thought processing in a certain problem domain.

Structure of This Paper. Section 2 focuses on computerized stories. It highlights different designs and implementations, and shows how technology extends the use of stories to a broad spectrum of applications. Section 3 considers one of these designs, namely woven stories, as a cognitive tool. A broader perspective, taking account of conceptual awareness, reinforces the relevance of woven stories as cognitive tools (Sect. 4). An implementation of woven stories is presented in Sect. 5.

2 Computerized Stories

In this section, we illustrate the broad application of stories in computer systems through a survey of related research. Many projects deal with teaching children how to tell stories. Others realize how fundamental stories are in human communication and support composition of stories to facilitate communication. The human mind uses stories to understand and memorize new experiences and objects [28]; thus, the use of stories in educational designs is well-grounded.

Storytelling Tools for Children. Storytelling systems form a category of learning environments that are often intended for children. The way in which these systems facilitate the storytelling process varies a great deal. The Graphic StoryWriter [31] system automatically generates a story as the user, a child aged 4-7, manipulates graphic elements. The system guides the young author to include all basic story elements in the story. However, the types of stories that one can compose with the Graphic StoryWriter are limited.

Recent research has introduced innovative ways of learning and storytelling by alternative ways of interacting with computer systems. StoryMat [27] is a play environment where a child can play with a stuffed animal on a mat. The child's movements and voice are recorded to inspire other children to play. PETS (Personal Electronic Teller of Stories) [7] is a system where the users first construct a robotic pet from building bricks and then compose a story, to be acted out by the robotic pet. SAGE (Storyteller Agent Generation Environment) [32] is a system where the stories composed by children are told by a programmable stuffed animal.

KidPad and Klump are storytelling tools that allow children to author stories collaboratively [2]. This is achieved by shoulder-to-shoulder collaboration which means that the children use several input devices, such as more than one mouse connected to the same workstation, simultaneously.

In the NIMIS approach (Networked Interactive Multimedia In Schools) [13], young students work in a new kind of computer integrated classroom which offers tools for multimedia creation and use. The pedagogical goals of the NIMIS project refer to skills like reading, writing, and narration. The enhanced classroom facilitates different computer aided activities like reading through writing, story creation and story writing with cartoons.

Storytelling Tools for Special Needs. Computers can support individuals with communications disabilities to tell stories and to develop their personal identity. In a study, non-speaking persons used a computer system that facilitates the use of pre-stored written material in conversations. The system enhanced the communication skills of the users [23].

In another experiment, an identity construction environment [4] was used by young patients who were confined to their beds by their medical conditions. The environment, called Zora [3], was used by the patients in order to understand their identity and role in a community through storytelling and interaction in a virtual city. The results of the experiment with the patients reported in [4] are encouraging, and justify the need to develop tools like Zora.

Structuring with Stories and Narratives. Laurillard et al. [18] state that it is imperative to include a clear story line, or a narrative, in educational material. People tend to memorize narratives more easily than isolated concepts. Furthermore, people realize narratives and information by narrative construction. Narrative construction is an active process of meaning-making, stimulated by the information provided by the environment and the personal knowledge of the user [25]. It is far more beneficial for a learner to construct a narrative than merely follow one.

A narrative can be used as an alternative to the traditional spatial or semantic organization of digital data, for example when navigating through a given environment [24]. Persons who have problems in dealing with spatially or semantically organized data may benefit especially from arranging data in a certain domain with the help of a narrative. The narrativization [24] or narrative construction [25] of data may be an affective experience that promotes learning and remembering of data.

One example of using a story for structuring and facilitating learning in a CS2 course is reported in [34]. The participants of the course implement a multimedia project as group work, based on a story given by the lecturers. The story functions as the focal point of the project. Instead of a narrative, the students construct a multimedia application around the story. Thus, they learn not only software engineering but also story telling.

3 Woven Stories as a Cognitive Tool: Beyond the Hypertext Metaphor

Weaving stories is not just an exercise aimed at developing students' skills in constructing a narrative from story fragments contributed by co-students. The main asset of this study technique is to build upon the ideas of those who have written from a different perspective. Since learning can be regarded primarily as a cognitive activity, it is clearly not sufficient to coach and support students at the level of concrete tasks; cognitive tools for learning should address the underlying mental processes like perspective taking, abstraction, analogy, imagination and conceptual awareness. The significance of conceptual awareness may at first seem obscure, but the need to offer support in this respect is now becoming obvious and urgent as more and more study techniques make use of 'concept mapping'. The general opinion nowadays is that concept mapping is in the core of meta cognition: becoming aware of what you know and what you still do not know. Also the consciousness on the structure of our knowledge is a vital notion before we can learn effectively. The best context to think about concept mapping is the area of cognitive tools like simulations, expert systems, decision support systems and Computer Mediated Communication (CMC) tools. Cognitive tools are meant to overcome the limitations of the human mind in thinking, learning and problem solving. Many of them mainly assist the user by offering visualizations like advanced Computer Aided Design (CAD). However, as concept mapping aims to give support in conceptual domains, it is a non-trivial problem how to choose the right symbols, operations and metaphors.

Bevilacqua [5] defined hypertext as an organizing principle, like the 15th-century invention of alphabetical order or the Platonic invention of dialectical argument. Ted Nelson, who coined the word in the 1960s, defines "hyper" as "extended, generalized, and multidimensional" [20]. Michael Heim writes, "text derives originally from the Latin word for weaving and for interwoven material, and it comes to have an extraordinary accuracy of meaning in the case of word processing" [12], p. 160-161. This image of a multidimensional fabric of knowledge linked with all its intellectual antecedents is one that is familiar to librarians, teachers and finally also to students. In a sense, we've been advocates of hypertext all along; encyclopedias, card catalogs, citation indexes, and abstracts all make up this invisible web of knowledge. We are used to such organization; in fact, it remains central to our way of teaching others how the library works. However, the electronic hypertext document has few of the built-in frustrations of the paper system.

The essence of hypertext is a dynamic linking of concepts allowing the student to follow preferences instantaneously and to be in control. The scope of a topic is no longer defined by the editor or author and is limited only by the initiative of the student. As Heim explains, "instead of searching for a footnote or going to find another document referred to, the dynamic footnote, or link, can automatically bring the appended or referenced material to the screen. The referenced material could be a paragraph or an article or an entire book. A return key brings the student back to the point in the original text where the link symbol appeared" [12], p. 162. However, the student may also choose not to follow diversions, but to continue through a particular document without interruption. It is this interactivity with the database that is the key to hypertext systems; pictures, sound, and text can be instantly retrieved according to the student's needs or whims.

Currently there are two types of hypertext: static and dynamic, [6], p. 250. Static hypertext does not permit changes to the database, but it is interactively browsable. In dynamic hypertext the student may add or subtract data and links. An important aspect of many dynamic hypertext systems is the ability to maintain multiple versions of a document as it changes over time. This allows the writer to track the history of a document and weigh up alternative versions simultaneously. In a multi-user environment, this allows the original writer to maintain the first version of a document even after others have changed it. As hypertext systems have progressed over the past 20 years, several problems have surfaced. Among the most vexing issues facing hypertext developers are orientation to the database, cognitive overload, and compatibility.

It is feared that students who are used to finding their way through books with the aid of tables of contents, indexes, footnotes, and marginalia might become lost within hypertext systems. However, new visual cues are integrated into most hypertext systems to lessen feelings of disorientation. As databases grow, navigational tools such as the global map of links and documents and the history of paths taken, though complex themselves, become necessary. And, as hypertext documents develop standards, students will develop "pattern recognition" of those standards, much as they do with city bus maps. Over time and use, hypertext will probably change our way of thinking; perhaps, as we learn how to move non-sequentially in texts, the feeling of not knowing where we are will no longer be an issue.

Another criticism of hypertext is that students are presented with so much meta information for navigation and control that the targeted learning processes may suffer from cognitive overload. While reading through a document, choices must constantly be made about which links to follow and which to ignore. Following several paths at once may lead to the navigation problem described above. Although this problem is not new with hypertext, computerized access does somethimes add an overwhelming dimension to it. The issues of standards and compatibility have yet to be addressed. Some may argue that imposing standards while hypertext is still in an experimental stage will dampen creativity, but the reality is that currently we are developing what Van Dam calls "docu-islands" [33] of knowledge that are incompatible with one another. Just when it seems that compatibility problems of microcomputers have eased somewhat, new, more complex hypermedia document systems will make all those interconnections obsolete. It is not too soon to press for standards and compatibility to ensure not only connectivity but also ease of use.

4 Conceptual Awareness

Before tools like concept mapping and associated knowledge representation tools can successfully be adopted in learning situations, the nature of human thinking and creativity should be taken into account. Learning is all too often identified with "the quickest road between prior and desired knowledge." One dominant paradigm to make this identification is "instruction": giving the right information at the right moment so that the learner can precisely fit the new on top of the known information. Even if this instructional mechanism is highly sophisticated in terms of pacing and feedback, it still assumes that learning is essentially a constellation of information in the human mind. Norman's theory on linear versus web teaching is an early reference to how fragile the (linear) information linkage process can be. Linear teaching means that students get new information that 'fits' with their prior knowledge. The more lessons follow each other, the more brittle does this grafting mechanism become. The alternative here is in the web-teaching sequence; the student is stimulated to articulate his/her prior ideas and make them as explicit as possible before any new information is given. Moreover there will be a linkage between the most crucial concepts in the students' prior knowledge and the new concepts to be learned. The really new information is then interconnected to the new anchoring concepts; [22]. The idea of web teaching has now received a new dimension as the web (WWW) is becoming the default target knowledge structure. Norman's idea of web teaching also fits quite precisely with Ausubel's sequencing method of 'Advance Organizer'. Rumelhart & Norman [26] proposed the idea of a more comprehensive portfolio comprising various learning mechanisms: accretion, restructuring and tuning. This highlights the fact that after having access to new facts, learners need to restructure their knowledge framework; it also recognizes that effective and sustainable knowledge needs an organic structure in order to become coherent with the dominant entities and relations in the new knowledge domain.

Concept Maps. Concept mapping is an activity derived from psychological research meant to depict one's knowledge, ideas, convictions and beliefs. It can be used to make ones ideas more explicit, and find related ideas that would otherwise stay hidden if one only thinks about it. 'Concept mapping' is presented here as a private knowledge assessment tool for the student. Giving more attention to the autonomy of students in their learning, and acknowledging the constructionistic mechanisms of one's cognition, concept mapping may act as a continuous monitoring device for the student's progress and deficits; [29]. A good way to organize and especially to re-organize information in a learning- and problem solving context is to map your ideas and associations into two-dimensional space, to create a structure known as a scheme or a "concept map". Constructing concept maps stimulates us to externalize, articulate and pull together information we already know about a subject and understand new information as we learn. An essential quality of effective concept mapping tools is to elicit the appropriate level of complexity and detail in the students' explorations. Both available concept entities and relational operators in the mapping tool should prompt those semantic associations that are essential to reconcile known and unknown. Concept mapping as a procedure is iterative. It stimulates a learner to determine the contours of his/her knowledge. Concept mapping in educational contexts can serve in four roles:

1. As a design method to be used as a structural scaffolding technique before and during the development of hypermedia products.
2. As a navigation device for students who need orientation while they explore wide information domains like hypermedia documents on CD-ROMs or WWW.
3. As a knowledge elicitation technique to be used by students as they try to articulate and synthesize their actual state of knowledge in the various stages of the learning process. As a knowledge elicitation technique stimulating retrospection and encouraging the user to be reactive, the concept mapping activity might be an essential first step to improving the students navigation skills in hypermedia browsing.
4. As an authentic knowledge assessment tool to enable students to diagnose their own level of understanding and to detect misconceptions.

The Nature of Concepts. It is important to distinguish the way we see a concept in our imagination from computer-driven ideas such as the information entities used to support its implementation or the abstract classes used in an object-oriented specification. Though an object-oriented representation might be exactly tuned to a certain stage of learning, it may not be taken as prescriptive. Conceptual learning tools need generic entities and relations so that they are versatile enough to address the many possible manifestations of a single concept in real settings. Rather than seeing concept entities as information packages, it would be more appropriate to see them as vivid personalities able to reflect on themselves, to make contact with other concepts, to arrange contact between other concepts, and even able to change themselves. Picturing prior knowledge then resembles a genetic approach rather than an attempt to find unique and prescriptive representations of truth. Its only goal is to depict views on a relatively unfamiliar topic that may stimulate the student to become receptive to new ideas from the teacher and fellow students.

Concept Mapping as a Technique to Regulate Cognitive Processes. Concept mapping is a technique to represent mental schemata and the structure of information. Concept mapping may be appropriate for

- orienting students (like Ausubels Advance Organizer; [1])
- articulating prior and final knowledge
- exchanging views and ideas among students at a distance
- transfering learned knowledge between different topics and domains
- diagnosing misconceptions

Making mental schemata and information structures explicit in such maps allows evaluation of these structures. Making comparisons, calculating measures and inferring logical consequences are facilitated by explicit concept maps. Concept mapping can be categorized as a cognitive tool that may be implemented in computer software. Cognitive tools are based upon a *constructionistic* epistemology. The goals and design of constructionism are not well supported by previous technological innovations.

Traditional technologies such as programmed instruction and techniques such as instructional design are *objectivistic*. Cognitive tools are constructionistic as they actively engage learners in creation of knowledge that reflects their comprehension and conception of the information rather than focusing on the presentation of objective knowledge. Cognitive tools are learner-controlled, not teacher- or technology-driven. Cognitive tools are not designed to reduce information processing, that is, make a task easier, as has been the goal of instructional design and most instructional technologies. Learning is mediated by thinking (mental processes). Thinking is activated by learning activities, and learning activities are mediated by instructional interventions, including those mediated by technologies. One role of delivery technologies should be to display thinking tools: tools that facilitate thinking processes. Deeper information processing results from activating appropriate schemata, using them to interpret new information, assimilating new information back into the schemata, reorganizing them in light of the newly interpreted information, and then using those newly aggrandized schemata to explain, interpret, or infer new knowledge. The learner needs to perform such operations on the newly acquired information that will relate this to prior knowledge. Prior knowledge may be incomplete or even contradictory to the information to be learned (misconceptions). In this case conflicts will occur between the old and the new information. Media have become more integral to both learning and work settings. Teaching problem-solving skills, co-operation, design and the integration of knowledge is part of the learning process itself, and should be an intrinsic element in a learning task. With the coming of massive and yet flexible information resources such as hypermedia and Internet-based connections in learning settings, greater demands are made on the student's initiative and learning management skills. Constructionism is an attempt to let students create their own mental concepts and let them construct agglomerations of concepts from prior knowledge. However entropic mechanisms in the student's mind may lead to early misconceptions and fragmented abstractions. To meet the challenge of an increasing student autonomy whilst reducing the negative effects of the constructionistic approach, it becomes opportune to define procedures for planning, problem-solving and inter-student negotiation about conceptual structures.

Computer-Based Tools for Cognitive Collaboration in Design. Tomorrow's companies, schools and private houses will have access to virtually all other personal workplaces in the world; not only by oral conversations (telephone), but also by sharing written documents, video fragments, databases, schematic drafts, planning charts, outlines etc. Besides the gain in functionality for working and learning, this also brings along the need for new skills, attitudes and willingness to communicate about premature ideas which are far from ready for expression in formal documents or distribution in one's job environment. Mind tools such as are promoted in the constructionistic approach [16] advocate concept-based activities like using simulations, building small scale knowledge systems and creating concept maps. However this approach still lacks the means to integrate mind tools with a concrete design problem.

Baseline Tools for Concept Mapping. In media design, it is to be exptected that most effort is expended in the refinement process of information analysis, working out a scenario, video recording and mixing, and finally compiling and shaping technical documentation. In order that students learn basic principles, and become flexible in their problem-solving approach, it is crucial to allow them to explore different strategies and perspectives associated with a given goal. Schematic representations like concept maps promise to be effective in negotiation about primitive notions amongst students. Schemes are defined as sets of actions that constitute part of a way of acting on the world, or a partial way of looking at the world arising out of actions. Schematic conceptual representations are generalized characterizations of schemes that allow students to transfer knowledge from one type of problem to another. Planning, monitoring and controlling one's own learning process should not reduce the flexibility to change one's conceptual perception of the problem space. This is why special attention will be given in our proposed project to flexible concept representations that allow conceptual shift, but at the same time stimulate the student to develop meta cognition and a readiness to communicate intermediate stages with other students.

The function of the concept map representation is to stimulate the student to take a more global viewpoint on their chosen approach to controlling the learning process, and to make it easier for students to benefit from alternative problem approaches. We expect that depending on the design phase, the level of prior knowledge, and the stage in communication, there is a need for a specific concept mapping tool, with its own entities, symbols and procedures. Some concept mapping tools to accompany learning are:

- SemNet ([8])
- Learning Tool ([17])
- Textvision (2D and 3D) ([14], [15])
- Inspiration ([11])

Further experiments will be carried out to establish the most appropriate way to apply concept mapping as proposed in this section to the technique of woven stories.

5 The Woven Stories 2 Prototype

The Woven Stories 2 (WS2) prototype system is a further development of the woven stories concept and prototype system presented in [10]. The prototype presented in [10] is a system that uses a WWW browser to facilitate co-authoring over the Internet on a limited scale. It is possible to form tree like structures of story sections authored by different users in it.

The WS2 is a completely new system. It is developed as a computer-based collaborative writing environment, that supports both synchronous and asynchronous work. WS2 takes the realization of the woven stories concept further by removing the limitations of the first prototype. In WS2 the number of sections inserted by the authors is not limited; the sections can be linked arbitrarily. WS2 also offers improved user control and support for multiple documents. The WS2 system supports the authoring of hyperdocuments which consist of sections that may be linked together. A section's content is a piece of text (a story section), a paragraph. Together the sections and the links between them can form trees or even graphs where (in traditional computer science terminology) the nodes are story sections and the edges are represented by the links in between.

Co-authoring over the Internet with the WS2 system makes participation of authors from all over the world possible by removing geographical barriers, by making distances irrelevant. Any number of authors (users) can contribute sections to a WS2 hyperdocument. The authors have total control over their own sections: they may modify, or delete them as they please. The authors are allowed to link their own sections to sections authored by others. The WS2 system supports an unlimited number of hyperdocuments which the clients can access one hyperdocument at a time.

The WS2 client has a structural view of the story that shows the story sections as rectangles and their relations as arrows between them. The structural view is intended to be used as a overview of the hyperdocument, through which new elements are added or existing ones modified. A section can be viewed by clicking on it with the mouse, new sections and links between them are added similarly with mouseclicks producing content sensitive menus in a CAD-like manner. The client includes a chat window for communication between the authors currently online. The user interface of the WS2 client is based on the

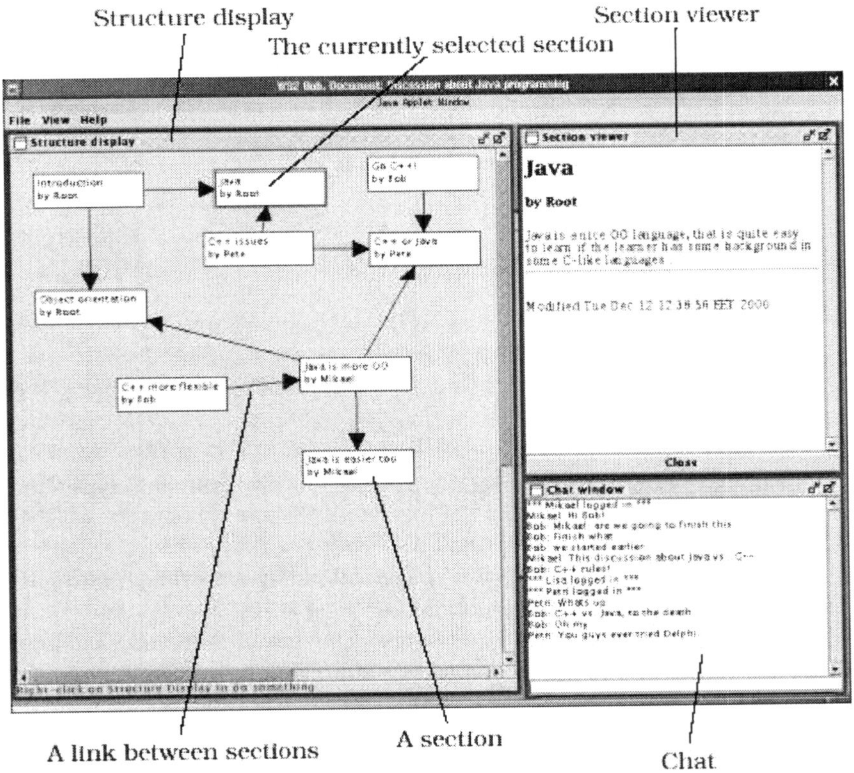

Fig. 1. The layout of the WS2 client interface

relaxed WYSIWIS (What You See Is What I See) concept [30]. The authors share the same structural view of the story, but they can browse the sections independently (see Fig. 1).

The WS2 system controls the access to the system by requesting a password and login that must be specified by an administrator of the WS2 system. Furthermore users are divided into administrators, lesser administrators and users. The administrators may add users and modify all existing hyperdocuments, even deleting whole documents along with all their sections and links. Lesser administrators can add new empty hyperdocuments and specify which users can add sections and links to them. User can add sections and link them to other sections to documents to which they have been given access.

WS2 is based on a client-server architecture, where the server resides on a WWW-server and manages data and mediates communications between the clients. The clients are separate programs that the users of the system can download from a web site. The clients contact the server over the WWW and the number of clients interacting with the server at the same time is not limited.

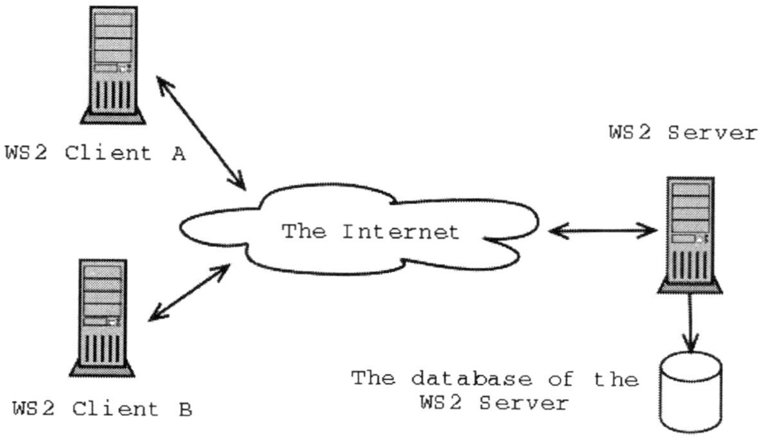

Fig. 2. Overview of the general architecture of the WS2 prototype

The WS2 system works on a real time basis: all changes to the stored material are immediately notified to all clients online (see Fig. 2).

The WS2 system is completely coded with the Java 2 Software Development Kit (SDK), version 1.3, enabling the server and client to run on any computer that has the Java 2 Runtime Environment (J2RE). An important feature in the WS2 system is the communication between the client(s) and the server which is implemented with HTTP tunneling. In HTTP tunneling the messages sent between the client and server are embedded into HTTP requests. This is done in order to make communication possible through firewalls that enable normal HTTP-transactions. HTTP tunneling makes it possible to use the WW2 system from virtually everywhere if Internet access and J2RE is available.

The aims of the Woven Stories 2 prototype are to test how co-authoring works and what is required of a tool that facilitates co-authoring and to get experience in developing a distributed authoring system. The system is still being developed and many computer supported collaborative writing issues like group awareness [9], proper version control [19] and better support for co-authoring in general [21] will be addressed. As the prototype evolves the focus of the testing will be shifted into the use of the system and the concept of co-authoring as a tool of collaborative learning, creative problem solving, and the processing of ideas. The prototype presented in this paper will be adapted into specific application areas of the woven stories concept as the special needs of those areas become clearer.

6 Discussion

As any meaningful computer application should, a system using the woven stories concept enhances a user's capabilities in the area for which the software was designed. In the case of story telling, the crucial issue is supporting an individual author in his or her task of composing a story for a given context. The possible contexts for composition cover range widely; they include a child's fantasy, a designer's problem-solving process and an elderly citizen's biography. In any case, the writing aid should enrich and deepen the process of authoring the story. The process itself is a highly cognitive one, requiring instruments for reflection, thinking, and creativity.

One of the key foci in woven stories is that of group processing. A story is seldom an independent, abstract piece of work, with no relevant environment. It is rarely restricted to an individual's perspective on the surrounding world, whether physical or mental. Rather, stories are born out of coincidental interactions with other people on almost random occasions. In a way, woven stories emulates this social network, or in fact, provides its users with a real-time www-based context where new stories emerge out of encounters with ideas or written paragraphs of which the authors were previously unaware of.

The role of computer in enhancing the authoring process is obvious in the concept of woven stories. For example, woven stories can deepen and intensify the narrative, suggest novel approaches that take advantage of virtually unrelated aspects in an open-ended manner, and fortify creativity in problem solving. In the last application area, a common method is to broaden one's mind by trying to apply apparently unrelated ideas to the problem in question. In a woven story, this could be implemented by having the system look for story portions very distant from an author's current text paragraph. Thus, a system weaves stories by itself; these automatically created links could easily stimulate a problem solver to look into his or her stale ideas with fresh eyes.

As a simple metaphor, woven stories presents designers with several opportunities to develop its expressive power. For instance, the original multi-layer character of classic stories actually indicates that the story itself is a generic one and can be applied, say, for multiple ages in different but still relevant ways. From the computing point of view, this means that a generic story can be parameterized for an individual author. This, again, implies that an author of a single story could have the computer show how another subplot, written by another author, could be applied to his or her situation. This is a mechanism to emulate a transfer for which the context might be (amongst other possibilities) educational – *do what you expect from others in your situation* – and therapeutic – *what would I do in my neighbour's situation?*.

References

1. Ausubel, D.: *The Psychology of Meaningful Verbal Learning.* New York, Grune & Stratton, 1963.
2. Benford, S., Bederson, B. B., Akesson, K., Bayon, V., Druin, A., Hansson, P., Hourcade, J. P., Ingram, R., Neale, H., O'Malley, C., Simsarian, K. T., Stanton, D., Sundblad, Y., Taxén, G.: Designing storytelling technologies to encouraging collaboration between young children. In *Proceedings of the CHI 2000 conference on Human factors in computing systems*, 556–563, 2000.

3. Bers, U. M.: Zora: a Graphical Multi-user Environment to Share Stories about the Self. In C. Hoadley and J. Roschelle (Eds.): *Proceedings of the Computer Support for Collaborative Learning (CSCL) 1999* Mahwah, Lawrence Erlbaum Associates, 1999.
4. Bers, M. U., Gonzalez-Heydrich, J., DeMaso, D. R.: Identity construction environments: supporting a virtual therapeutic community of pediatric patients undergoing dialysis. In *Proceedings of the SIG-CHI on Human factors in computing systems*, 380–387, 2001.
5. Bevilacqua, A. F.: Hypertext: Behind the Hype. *American Libraries* 20(2), 158–162, 1989.
6. Byers, T. J.: Built by association. *PC World*, 5, 244-251, 1987.
7. Druin, A., Montemayor, J., Hendler, J., et al.: Designing PETS: A Personal Electric Teller of Stories. In *Proceedings of the CHI 99 conference on human factors in computing systems: the CHI is the limit*, Pittsburgh, 326–329, 1999.
8. Fisher, K. M.: SemNet: A tool for Personal Knowledge Construction. In: Kommers, P. A. M., Jonassen, D. H., Mayes, J. T.: *Cognitive Tools for Learning*, NATO ASI Series Vol 81, Berlin, Springer Verlag, 1991.
9. Gutwin, C., Greenberg, S.: Effects of awareness support on groupware usability. In *Conference proceedings on Human factors in computing*, Los Angeles, 511–518, 1998.
10. Harviainen, T., Hassinen, M., Kommers, P., Sutinen, E.: Woven stories: collaboratively authoring microworlds via the Internet. *International Journal of Continuing Engineering Education and Life-long Learning*, 9 (3/4), 328–340, 1999.
11. Helfgott, D., Helfgott, M., Hoof, B.: *Inspiration, The visual way to quickly develop and communicate ideas*, Inspiration Software Inc, 1993.
12. Heim, M.: *Electronic Language: A Philosophical Study of Word Processing*, New Haven, Yale University Press, 1987.
13. Hoppe, U., Lingnau, A., Machado, I., Paiva, A., Prada, R., Tewissen, F.: Supporting collaborative activities in computer integrated classrooms-the NIMIS approach. *Groupware, 2000. CRIWG 2000. Proceedings. Sixth International Workshop on* , 94–101, 2000.
14. Kommers, P. A. M.: Virtual Structures in Hypermedia Resources. In *Proceedings of the HCI'91 International Conference*, Berlin, Springer Verlag, 1343–1351, 1990.
15. Kommers, P. A. M., Vries, de, S.: TextVision and the Visualization of Knowledge: School-based Evaluation of its Acceptanceat two Levels of Schooling. In Kommers, P. A. M., Jonassen, D. H., Mayes, T. (Eds): *Mind Tools: Cognitive Technologies for Modelling Knowledge*, Springer Verlag, Berlin, 1991.
16. Kommers, P. A. M., Jonassen, D. H., Mayes, J. T. (Eds): *Cognitive Tools for Learning*. NATO ASI Series. Series F: Computer and Systems Sciences; 81, Berlin/Heidelberg, Springer, 1992.
17. Kozma, R. B.: The impact of computer-based tools and rhetorical prompts on writing processes and products. *Cognition and Instruction*, 8, 1–27, 1991.
18. Laurillard, D., Stratfold, M., Luckin, R., Plowman, L., Taylor, J.: Affordances for Learning in a Non-Linear Narrative Medium. *Journal of Interactive Media in Education*, (2), 2000.
19. Lee, B. G., Chang, K. H., Narayanan, N. H.: An integrated approach to version control management in computer supported collaborative writing. In *Proceedings of the 36th annual conference on Southeast regional conference*, 34–43, 1998.
20. Nelson, T.: A Conceptual Framework for Man-Machine Everything. In *Proceedings of the AFIPS National Joint Computer Conference*, 1973.

21. Neuwirth, C. M., Kaufer, D. S., Chandhok, R., Morris, J. H.: Issues in the design of computer support for co-authoring and commenting. In *Proceedings of the conference on Computer-supported cooperative work*, 183–195, 1990.
22. Norman, D. A., Rumelhart, D. E.,: Memory and knowledge. In Norman, D. A., Rumelhart, D. E., The LNR Research Group (Eds): *Explorations in cognition*, San Francisco, Freeman, 1975.
23. O'Mara, D. A., Waller, A., Tait, L., Hood, H., Booth, L., Brophy-Arnott, B.: Developing personal identity through story telling. *Speech and Language Processing for Disabled and Elderly People (Ref. No. 2000/025), IEE Seminar on*, 9/1–9/4, 2000.
24. Persson, P.: Supporting Navigation in Digital Environments: A Narrative Approach, In *Exploring Navigation: Towards a Framework for Design and Evaluation in Electronic Spaces*, SICS Technical Report T98:01, SICS, Stockholm.
25. Plowman, L., Luckin, R. Laurillard, D., Stratfold, M., Taylor, J.: Designing Multimedia for Learning: Narrative Guidance and Narrative Construction. In *CHI 1999*, 310–317, 1999.
26. Rumelhart, D. E., Norman, D. A.: Accretion, tuning and restructuring: Three modes of learning. In Cotton, J. W., Klatzky, R. (Eds): *Semantic factors in cognition*, Hillsdale, NJ, Erlbaum, 1978.
27. Ryokai, K., Cassell, J.: StoryMat: A Play Space with Narrative Memories. In *Proceedings of the 1999 International conference on Intelligent user interfaces*, Rodendo Beach, 1999.
28. Schank, R. C., Abelson, R. P.: Knowledge and Memory: The Real Story. In Robert S. Wyer, Jr (Ed) *Knowledge and Memory: The Real Story*, Hillsdale, NJ, Lawrence Erlbaum Associates, 1995.
29. Shavelson, R. J., Lang, H., Lewin, B.: *On concept maps as potential "authentic" assessments in science* (CSE Technical report No. 388). Los Angeles, CA, National Center for Research on Evaluation, Standards, and Student Testing (CRESST), UCLA, 1994.
30. Stefik, M., Bobrow, D. G., Foster, G., Lanning, S., Tatar, D.: WYSIWIS revised: early experiences with multiuser interfaces. *ACM Transactions on Information Systems*, 5(2), 147–167, 1987.
31. Steiner, K. S., Moher, T. G.: Graphic StoryWriter: An Interactive Environment for Emergent Storytelling. *Conference proceedings on Human factors in computing systems* (CHI 92), Monterey, 357–364, 1992.
32. Umaschi, M., Cassell, J.: Storytelling systems: constructing the innerface of the interface. In proceedings of the *Second International Conference on Cognitive Technology, 1997. Humanizing the Information Age*, 98–10, 1997.
33. Van Dam, A.: Hypertext '87 keynote address. *Communications of the ACM*, 31, 887-895, 1988.
34. Wolz, U., Domen D., McAucliffe, M.: Multi-media integrated into CS 2: An interactive children's story as a unifying class project. In *Proceedings of ItiCSE '97*, Uppsala, 103–110, 1997.

The Narrative Intelligence Hypothesis: In Search of the Transactional Format of Narratives in Humans and Other Social Animals

Kerstin Dautenhahn

Adaptive Systems Research Group
University of Hertfordshire
Department of Computer Science
College Lane, Hatfield, Hertfordshire, AL10 9AB, UK
K.Dautenhahn@herts.ac.uk

Abstract. This article discusses narrative intelligence in the context of the evolution of primate (social) intelligence, and with respect to the particular cognitive limits that constrain the development of human social networks and societies. The Narrative Intelligence Hypothesis suggests that the evolutionary origin of communicating in a narrative format co-evolved with increasingly complex social dynamics among our human ancestors. This article gives examples of social interactions in non-human primates and how these interactions can be interpreted in terms of nonverbal narratives. The particular format of preverbal narrative that infants learn through transactions with others is important for the development of communication and social skills. A possible impairment of the construction of narrative formats in children with autism is discussed. Implications of the Narrative Intelligence Hypothesis for research into communication and social interactions in animals and robots are outlined. The article concludes by discussing implications for humane technology development.

1 Introduction: The Social Animals

Humans are primates, and share fundamental cognitive and behavioral characteristics with other primates, in particular apes (orangutan, gorilla, chimpanzee, bonobo). Although it is widely accepted that humans and other apes have a common ancestor and that human behavior and cognition is grounded in evolutionary 'older' characteristics, many people believe that human intelligence and human culture are 'unique' and qualitatively different from most (if not all) other non-human animals. Human language often serves as an example of a 'unique' characteristic. With a few exceptions [Read & Miller 95], most discussions on the 'narrative mind' neglect the evolutionary origins of narrative. Therefore, it is not surprising that most research on narrative focuses almost exclusively on language in humans (see e.g. [Turner 96]). The work that is presented in this paper attempts to complement these works: instead of focussing on differences between human and other animal societies, we point out similarities and evolutionary shared histories of primates with specific regard to the origins and the transactional format of narratives (cf. [Dautenhahn 99], [Dautenhahn, to appear]).

M. Beynon, C.L. Nehaniv, and K. Dautenhahn (Eds.): CT 2001, LNAI 2117, pp. 248-266, 2001.

The article sets off by reviewing the main arguments of a debate that is currently discussed intensively in primatology and anthropology, namely that the characteristic properties of human 'minds' and human culture are grounded in the human's capacity to use language, and that the primary function of language was that it affords to cope with increasingly complex social dynamics. Based on this framework on the social origin of human intelligence we discuss the *Narrative Intelligence Hypothesis* (NIH), first suggested in [Dautenhahn 99a], that points out the intertwined relationship between the evolution of narrative and the evolution of social complexity in primate societies. The underlying assumptions and arguments are discussed in more detail. The NIH as referred to in this paper consists of the following line of arguments:

a) *individualized societies* are a necessary (but possibly not sufficient) 'substrate' for the evolution of narratives. In such societies members know each other and relate to each other on an individual level (animals with 'personalities', 'minds'), and interact with each other through transactional processes,
b) the specific *narrative format* of such transactions serves an *important communicative function* among primates, and possibly independently in other groups of species that live in individualized societies,
c) narrative co-evolved along and in order to cope with increasingly complex dynamics in the primate social field,
d) the evolution of communication in terms of narrative language (story-telling) was an important factor in human evolution that has shaped the evolution of human cognition, societies and human culture. The use of language in a narrative format provided an efficient means of 'social grooming'
e) human cultures which are fundamentally 'narrative' in nature provide an environment that young human primates are immersed in and facilitates not only the development of a skilled story-teller and communicator, but the development of an autobiographical self.

The NIH is speculative and part of ongoing research. The particular contribution of this article is that it analyses in more detail the *structure* and *canonical format* of narrative that can be found in different verbal and non-verbal social interactions among primates, and in preverbal communication of human infants. The relationships between narrative and culture and autobiography are touched upon but discussed in more detail elsewhere ([Dautenhahn 99a,c], [Dautenhahn, to appear], [Dautenhahn, submitted]).

Implications of the NIH are on the one hand a better understanding of the origins of narrative intelligence in humans and other animals, on the other hand such an understanding can point out issues that are relevant in the design of *narrative technology* that meets the social and cognitive needs of human users of such technology.

2 The Social Brain Hypothesis

Primate societies belong to individualized societies with complex kinds of social interaction and the development of various forms of social relationships and networks. In individualized societies group members individually recognize each other and interact with each other based on a history of interactions as part of a social network. Many mammal species (such as primates, elephants, cetaceans) live in highly individualized societies, so do bird species such as corvids and parrots. Preserving social coherence and managing cooperation and competition with group members are important aspects of living in individualized societies. Dealing with such a complex social field often requires sophisticated means of interaction and communication which are important for the Narrative Intelligence Hypothesis that is discussed in this article.

In the context of human (or generally primate) intelligence the *Social Intelligence Hypothesis* (SIH), sometimes also called Machiavellian Intelligence Hypothesis or *Social Brain Hypothesis*, suggests that the primate brain and primate intelligence evolved in adaptation to the need to operate in large groups where structure and cohesion of the group required a detailed understanding of group members, cf. [Byrne & Whiten 88], [Whiten & Byrne 97], [Byrne 97]. It is assumed that social complexity that required the evolution of social skills (which allow to interpret, predict and manipulate conspecifics) has been a prominent selective factor accelerating primate brain evolution, given that maintaining a large brain is very costly. Identifying friends and allies, predicting others' behavior, knowing how to form alliances, manipulating group members, making war, love and peace, are important ingredients of primate politics [de Waal 82]. Thus, there are two interesting aspects to human sociality: it served as an evolutionary constraint which led to an increase of brain size in primates, which in return led to an increased capacity to further develop social complexity.

Research in primatology that studies and compares cognitive and behavioral complexity in and among primate species can give exciting hints on the origins of human cultures and societies. Particularly relevant for the theme of this article are the potential relationships between social complexity and brain evolution. A detailed analysis by Dunbar and his collaborators gives evidence (e.g. [Dunbar 92,93,98] and other publications) that the size of a cohesive social group in primates is a function of relative neocortical volume (volume of neocortex divided by volume of the rest of the brain). This evidence supports the argument that social complexity played a causal role in primate brain evolution, namely that in order to manage larger groups, bigger brains are needed to provide the required 'information processing capacity'. Note, that group size as such is not the only indicator of social complexity: other researchers have found supporting evidence, e.g. that primate species with relatively larger neocortices exhibit more *complex social strategies* than species with smaller neocortices [Pawlowski et al 98].

What specifically characterizes social complexity? Here, a definition that applies to many systems from machines to animal societies might be useful. According to [Philips & Austad 96] complexity is a function of: 1) the number of functionally distinct elements (parts, jobs, roles), 2) the number of ways in which these elements can interact to perpetuate the system or to promote its goals (or, if it is an artifact, the goals of its users), 3) the number of different elements (parts, jobs, roles) any individual within the system can assume at different times or at a given time, and 4) the capacity of the system to transform itself to meet new contingencies (i.e. the capacity of the system to produce new elements or new relations between elements). According to Philips and Austad conditions 1 and 2 can be applied to many systems from machines to societies. Conditions 3 and 4 are particularly suited to social organization. Condition 3 refers to the number of different roles an animal can play in a social network, and how these roles and relations to other animals dynamically change over time (condition 4).

How are social networks and relations established and maintained? Judging from our own experience as a member of human society, communicating via *language* seems to be the dominant mechanism for this purpose. Non-human primates in the wild do not seem to use a human-like language. Here, social cohesion is maintained through time by social grooming. Social grooming patterns generally reflect social relationships, they are used as a means to establish coalition bonds, for reconciliation and consolation and other important aspects of primate politics. Social grooming is a one-to-one behavior extended over time that poses particular constraints on the amount of time an animal can spend on it, given other needs such as feeding, sleeping etc. Also, cognitive constraints limit the complexity of social dynamics that primates can cope with, as discussed in the following paragraph.

Given the neocortical size of modern humans, Dunbar (1993) extrapolated from the non-human primate regression (relative neocortical volume vs. group size) and predicted a group size of 150 for human societies. This number limits the number of relationships that an individual human can monitor simultaneously, it is the upper group size limit which still allows social contacts that can be regularly maintained, supporting effective coordination of tasks and information-flow via direct person-to-person contacts. The number 150 is supported by evidence from analyzing contemporary and historical human societies. But how do humans preserve cohesion in groups of 150 individuals, a function that (physical) social grooming serves in non-human primate societies? In terms of survival needs (resting, feeding etc.) primates can only afford to spend around 20 % of their time on social interactions and social grooming, much less than a group size of 150 requires. It was therefore suggested by Dunbar (1993) that in order to preserve stability and coherence in human societies, human language has evolved as an efficient mechanism of social bonding, replacing social grooming mechanisms in non-human primate societies with direct physical contact (allowing only much smaller groups). Following this argument, language allowed an increase in group size while still preserving stability and cohesion within the group. The next section will elaborate this argument further by analyzing what the particular features of communication via language are that makes it an efficient 'social glue' in human societies.

3 The Narrative Intelligence Hypothesis

In the context of the evolution of human intelligence, the Social Intelligence Hypothesis offers little explanation for the evolution of specific ape and human kinds of intelligence (e.g. involving mental representations): clear evidence for a systematic monkey-ape difference in neocortex ratio is lacking. Great apes do not form systematically larger groups than monkeys do, which draws attention to physical rather than social factors (e.g. tool use, processing plant food etc.) that drove the evolution of mental representations in apes and humans. Why have in particular human apes evolved sophisticated representational and mental skills, are there any candidate factors that could have accelerated the evolution of human intelligence? If the evolution of language played an important role, as suggested by others (e.g. [Dunbar 93], [Donald 93]), what are the particular characteristics of language that matter?

We discussed previously [Dautenhahn 99a] that a closer look at the ontogeny of language and narrative, i.e. the role of language in the development of children could provide an important hint: Evidence shows that narratives play a crucial role in how young human primates become socially skilled individuals with an autobiography, being able to effectively communicate with others [Nelson 93; Engel 95].

Narrative psychology suggests that stories are the most efficient and natural human way to communicate, in particular to communicate about others [Bruner 87; 90; 91]. As Reader and Miller suggest "stories are universally basic to conversation and meaning making", and as developmental and cross-cultural studies suggest "humans appear to have a readiness, from the beginning of life, to hear and understand stories" [Read & Miller 95: p. 143]. The *Narrative Intelligence Hypothesis* [Dautenhahn 99a] interprets such evidence from the ontogeny of human language in the context of primate evolution: it proposes that the evolutionary origin of communicating in stories co-evolved with increasing social dynamics among our human ancestors, in particular the necessity to communicate about third-party relationships (which in humans seems to reach the highest degree of sophistication among all apes, cf. gossip and manipulation, [Sinderman 82]). According to the NIH human narrative intelligence might have evolved because the structure and format of narrative is particularly suited to communicate about the social world.

Looking at human evolution, we can observe an evolutionary trend from physical contact (non-human primates) to vocal communication and language (hominids) to communicating in stories (highly 'enculturated' humans living in complex societies) correlated with an increase in complexity and sophistication of social interaction and 'mindreading'. This trend demonstrates the evolution of increasingly efficient mechanisms for time-sharing the processes of social bonding. While physical grooming is generally a dyadic activity, language can be used in a variety of ways extending the dyadic use in dialogues to e.g. one-to-many communication as it is today used extensively in the mass media (television, books, email etc.). It can be estimated [Dunbar 93] that the human bonding mechanism of language is about 2.8 times as efficient as social grooming (the non-human primate bonding mechanism). Indeed, evidence suggests that conversational groups usually consist of one speaker plus two or three listeners. Of course larger groups can be formed easily, but in terms of actively participating and following different arguments within the group 1+2(3) seem to be the upper limit for avoiding information processing overload in the primate social brain. Also, language because of its representational nature affords documentation, preservation in storage media and transmission of (social) knowledge to the next generation, as well as communication between geographically separated locations [Donald 93].

Discussions in the social domain (e.g. on social relationships and feelings of group members) are fundamentally about personal meaning, different from e.g. discussions in the technical domain (e.g. about how to operate a tool or where to find food). *Narrative might be the 'natural' format for encoding and transmitting meaningful, socially relevant information (e.g. emotions and intentions of group members).* Humans use language to learn about other people and third-party relationships, to manipulate people, to bond with people, to break up or reinforce relationships. Studies show that people spend about 60 % of conversations on gossiping about relationships and personal experiences [Dunbar 93]. Thus, a primary role of language might have been to communicate about social issues, to get to know other group members, to synchronize group behavior, to preserve group cohesion.

Although humans use gestures, facial expressions, body language and other nonverbal means to convey (social) meaning, human communication is dominated by verbal communication, which is serial in nature (although in face-to-face interaction accompanied by non-verbal cues). Thus, given the serial communication channel of human language, what is the best means to communicate social issues - so important for primates as argued above - namely, learning about the who, what, and why? Physical social grooming, the main group cohesion mechanism in non-human primates is 'holistic', parallel, spatial, sensual, meaningful. How can a stream of symbols that are in themselves meaningless convey meaning such as bodily grooming does? I argue that narrative structure and format seems to be particularly suited: usually a narrative gives a certain introduction of the characters (making contact between individuals, actors, listener and speaker), develops a plot, namely a sequence of actions that convey meaning (value, pleasurable, unpleasurable), usually with a high point and a resolution (reinforcement or break-up of relationships), and focuses on unusual events rather than stereotypical events. In this way, stories seem to give language a structure which resembles (and goes beyond) physical grooming, namely replacing physical presence and actions by the creation of a mental picture of physical actions, providing the stage, actors, intentions and a storyline. Story-telling also gives more flexibility than social grooming as to the actors and content of the stories: stories can include people that are part of the current audience, as well as absent persons, historical characters, fictional characters, etc. Stories that are told by a skilled story-teller (e.g. using appropriate body language, exploiting prosody, and possessing a rich repertoire of verbal expressions) can give very good examples of 'the power of words'. The format of a story can provide sensual, emotional, and meaningful aspects to otherwise 'factual' information, e.g. people often clearly remember works of literature that elicited strong emotional responses and were influential at a particular time during their lives.

To summarize, the following strategies of coping with a complex social field in primate societies were outlined in the preceding sections:

a) non-verbal, physical social grooming as a means of preserving group cohesion, limited to one-to-one interaction
b) communicating about social matters and relating to others in the narrative format of transactions with non-verbal 'enacted' stories
c) using language and verbal narratives in order to cope with social life

The Narrative Intelligence Hypothesis suggests that the evolution of human societies might have gone through these different stages, not replacing preceding stages, but adding additional strategies that extend an individual's repertoire of social interaction, ranging from physical contact (e.g. in families and very close relationships), to preverbal 'narrative' communication in transactions with others (let alone the subtleties of body language and nonverbal social cues, not necessarily conscious, cf. [Hall 68], [Farnell 99], [Gill et al 99]), to developing into a skilled story-teller within the first years of life and refining these skills throughout one's life. The next section gives a few examples of where we might find narratives in the behavior of humans, other animals, and possibly even artifacts. To begin with, we need to have a closer look that the specific canonical format of narrative.

4 In Search for Narratives

4.1 What Exactly Are Narratives?

Many definitions and theories of narrative and narrative intelligence exist in the literature. This paper follows a particular theory formulated by Jerome Bruner and discussed in his publications, e.g [Bruner 87,90,91].

Bruner's account of narrative needs to be placed in the context of his distinctions of two complimentary modes of thought and understanding the world Bruner (1990):

The *paradigmatic mode* or logico-scientific one is based on the idea of a formal, mathematical system of descriptions and explanations. Discourse in this mode requires consistency and noncontradiction. Tools such as logic, symbol-systems, mathematics, sciences, and automata have been developed in order to experience and learn about the 'truth' in the *physical world.*

The second mode of thought according to Bruner is *the narrative mode* that deals with human intentions and that we use to understand the *social and cultural world* through stories. These stories remain stories whether they are true or not, and whether they are based on facts or fiction.

Thus, stories are primarily dealing with people and their intentions, they are about the social and cultural domain rather than the domain of the physical word. Narratives are often centered towards subjective and personal experience. According to Bruner (1991) narrative is a conventional form that is culturally transmitted and constrained. Children are not born as skilled story-tellers, the grow up immersed in a culture of story-tellers (parents, peers) who help them develop and shape their narrative skills and autobiographical selves [Nelson 93; Engel 95]. Narrative is not just a way of representing or communicating about reality, it is constituting and understanding (social) reality.

More specifically, Bruner (1991) discusses different characteristics of narrative, properties that distinguish an utterance or any part of language from a story:

1. Narratives describe sequences of events, in 'human time' rather than 'clock time'
2. Narratives are about 'unusual events', 'things worth telling' that can nevertheless be embedded in generic scripts
3. Narratives describe people or other agents, endowed with intentional states, acting in a setting in a way that is relevant to their beliefs, desires, theories, values etc.
4. Narratives must have a plot that conveys meaning, and a high point

The complete list of characteristics is described in [Bruner 91]. A detailed discussions of these criteria and implications for agent design are discussed in [Sengers 2000].

Particularly important for the theme of this paper are characteristics 2 and 3: different from episodic memory that can for example be represented in scripts [Schank & Abelson 77], such as the famous 'restaurant script', narratives are about breaches and violations to routine behavior. Also, stories are about the social field, about people as intentional and mental agents, and how they relate to each other.

Narrative capacities (understanding and producing stories) are capacities shaped by society, but developing in an *individual* being (cf. [Nehaniv 97], [Dautenhahn & Coles 01]). Also, stories have an important meaning for the individual agent, e.g. stories that children tell to *themselves* play an important part of a child's abilities to make meaning of events (cf. [Nelson 89], [Engel 95]). Similarly, a human professional story-teller might rehearse relevant material in solitude, but in the actual performance takes into account the specific audience, its reactions, and other indications on how the audience 'might think and feel', so that the actual story can be adapted appropriately. Thus, stories, at least for fundamentally social animals such as humans, are most effective in communication in a social context:

> "We converse in order to understand the world, exchange information, persuade, cooperate, deal with problems, and plan for the future. Other human beings are a central focus on each of these domains: We wish to understand other people and their social interactions; we need to deal with problems involving others; and other people are at the heart of many of our plans for the future." [Read & Miller 95: p. 147].

Bruner's above mentioned criteria of narrative structure and format do not only apply to stories that are told and written, but equally well to other formats such as comics [McCloud 94]. Human culture has developed various means of artistic expression (sequential visual arts, dance, pantomine, literature etc.) which are fundamentally 'narrative' in nature, conveying meaning about people and how people relate to the world. Children who are immersed in human culture, exposed to those narratives, develop as skilled story-tellers, as is shown in the following story of an 11-year old when asked to write a story about a robot. Note, that the story fits very well Bruner's criteria:

> "In America there was a professor called Peter Brainared and in 1978 he created a robot called Weebo. Weebo could do all sorts of things: she could create holograms, have a data bank of what the professor was going to do, show cartoon strips of what she was feeling like by having a television screen on top of her head which could open and close when she wanted to tell Peter how she felt. And she could record what she saw on television or what people said to her. Weebo looked like a flying saucer about as big as an eleven year old's head also she could fly. Peter Brainared had a girlfriend called Sarah and they were going to get married but he didn't turn up for the wedding because he was too busy with his experiments so she arranged for another one and another one but he still didn't turn up, so she broke off the engagement and when he heard this he told Weebo how much he loved her and she recorded it, went round to Sarah's house and showed her the clip on her television screen to show Sarah how much he loved her and it brought Sarah and Peter back together." (study described in [Bumby & Dautenhahn 99])

Please note that although a central protagonist in the above story is a robot, it is depicted as an intentional agent embedded in a social context [Dennett 87].

4.2 Narratives and Autism

Traditionally Jerome Bruner, Katherine Nelson, Susan Engel (see references above) and other psychologists interested in the nature and development of narratives have a particular viewpoint of narratives in terms of human verbal story-telling. Interestingly, Bruner and Feldman (1993) proposed the *narrative deficit hypothesis of autism*, a theory of autism that is based on a failure of infants to participate in narrative construction through preverbal transactional formats. Children with autism generally have difficulty in communication and social interaction with other people [Jordan 99], and this theory suggest that these deficits could be explained in terms of a deficit in narrative communication skills. As it is discussed later in this article, this work gives important hints on the transactional structure of narratives, a structure that we believe is of wider importance, not limited to the specific context of autism.

What is a narrative transactional format? Bruner and Feldman distinguish different stages. They suggest that the first transactional process is about reciprocal attribution of intentionality, of agency. The characteristic format of preverbal transactions is according to Bruner and Feldman a narrative one, consisting of four stages:

1) canonical steady state
2) precipitating event
3) a restoration
4) a coda marking the end.

An example is the peek-a-book game where 1) mutual eye gaze is established between infant and caretaker, 2) the caretaker hides her face behind an object, 3) the object is removed revealing the face again, and 4) "Boo", marking the end of the game.

Bruner and Feldman suggest that problems of people with autism in the social domain are due to an inability early in their lives to get engaged in 'appropriate' transactions with other people. These transactions normally enable a child to develop a narrative encoding of experiences that allows to represent *culturally canonical forms of human action and interaction.* Normally this leads a child at 2-3 years of age to rework experiences in terms of stories, until she ultimately develops into a skilled story-teller [Engel 95].

As research by Meltzoff, Gopnik, Moore and others suggest, transactional formats play a crucial role very early in a child's life when she takes the first steps of becoming a 'mindreader' and socially skilled individual: reciprocal imitation games are a format of interaction that contributes to the mutual attribution of agency [Meltzoff & Gopnik 93], [Meltzoff & Moore 99]. Immediate imitation creates intersubjective experience [Nadel et al 99]. By mastering interpersonal timing and sharing of topics in such dyadic interactions children's transition from primary to pragmatic communication is supported. It seems that imitation games with caretakers play an important part in a child's development of the concept of 'person' and [Meltzoff & Gopnik 93; Meltzoff & Moore 99], a major milestone in the development of social cognition in humans.

Data by Bruner and Feldman (1993) and others indicates that children with autism seem to have difficulty in organizing their experiences in a narrative format, as well as a difficulty in understanding the narrative format that people usually use to regulate their interactions. People with autism tend to *describe* rather than to *narrate*, lacking the specific causal, temporal, and intentional pragmatic markers needed for story-making. A preliminary study reported by Bruner and Feldman (1993) with high-functioning children with autism indicated that although they understood stories (gave appropriate answers when asked questions during the reading of the story), they showed great difficult in retelling the story, i.e. *composing* a story based on what they know. The stories they told preserved many events and the correct sequence, but lacked the proper emphasis on *important and meaningful events*, events that motivated the plot and the actors. The stories lacked the narrative bent and did not conform to the canonical cultural expectations that people expect in ordinary social interaction. Such a lack of meaning-making makes conversations in ordinary life extremely difficult, although, as Bruner and Feldman note, people with autism show a strong desire to engage in conversations [Bruner & Feldman 93].

Generally, there are different aspects to narrative in communication: expressing or telling stories, recognizing stories (understanding narrative in other agents) and experiencing the world through narratives (being an autobiographic agent) [Nehaniv 99a].

The following example of narratives in robots concentrates on the mechanisms of how a single agent can *express* episodic memory and what could make them narratives.

4.3 Narratives in Robots

The project *Memory-Based Interaction in Autonomous Social Robots* takes an Artificial Life perspective on stories and narratives [Dautenhahn & Nehaniv 98], [Nehaniv & Dautenhahn 98]. The minimal working definition of stories used in the project is as follows: "Stories are sequences of actions, expressed by an autonomous agent (including movements as well as 'speech acts'), which can be related to previous situations in the agent's autobiographical memory" [Dautenhahn & Coles 01]. Note, that in the context of this particular project we use the term 'story' in terms of episodical events. Such stories clearly do not fulfill Bruner's list of criteria for narratives and might therefore be better called *pre-narratives.* A computational framework was developed which supports systematic experimental studies of story-telling in autonomous behavior-based robotic agents (simulated and physical robots).

A particular goal in this project is to study *minimal experimental conditions* of how story-telling might emerge from episodic memory [Coles & Dautenhahn 00], [Dautenhahn & Coles 01]. An initial experimental study [Dautenhahn & Coles 01] investigated memory-based controllers and computational mechanisms of 'story-telling' for robotic agents. We showed that 'story-telling' (i.e. using episodic memory) can be beneficial even to a single agent (cf. [Nehaniv 97]) since it increases the behavioral variability of the reactive agent. Thus, from the point of view of developing Artificial Life *post-reactive* agents [Nehaniv et al 99], we can speculate that minimal mechanisms in the 'first story-telling animal' (not necessarily social) might have survived because the animal was better adapted to a dynamic environment. Later, this capacity could have been used and further developed in a social, communicative context. Note, that for our pre-narratives we do not presuppose any knowledge of the *meaning* or *interpretation* of stories. We did not want to impose meaning by a human designer since meaning implies *meaning for a particular agent*, evaluated from its own (historical) perspective [Nehaniv 99a]. Similarly, *understanding* in this framework means that the agent's stories are grounded in its own experiences rather than imposed by a human designer.

Such kind of research with an experimental computational and robotic test-bed demonstrates a bottom-up approach towards studying narrative and how it can arise and evolve from pre-narrative formats (e.g. episodic memory abilities and formats that are necessary but not sufficient for narratives, as discussed in previous sections) in agents and agent societies. Also, it can provide a means to design and study narrative robots with 'meaningful' narratives that are grounded in the robot's own experiences and means of interacting with the world and other agents (including robots), so as to contribute to the robot's agenda to survive. This approach is different from the common approach to building robots with a body language where non-verbal or scripted narrative behavior is imposed onto the robot purely by design so as to make the robots believable and entertaining for human observers (cf. [Bruce et al 00]).

4.4 Narratives in Animal Behavior?

Stories have an extended temporal horizon, they relate to past and future, they are created depending on the (social) context. Do animals use narrative formats in transactions? Studies e.g. with bonobos, Grey parrots and dolphins on animal language capacities teach the animals a language (using gestures, icons or imitating human sounds), and test the animal's language capacities primarily in interactions with humans [Savage-Rumbaugh et al 86; Pepperberg 99; Herman 01]. In the wild, the extent to which animals use a communication system as complex as human language is still controversial, e.g. dolphins and whales are good candidates for sophisticated communicators. However, looking for verbal and acoustic channels of communication might disguise the *nonverbal*, transactional nature of narratives, as shown in *preverbal* precursor of narratives in the developing child, and possibly evolutionary precursors of (*non-verbal*) narrative that can be found in non-human animals. Michael Arbib (2001) proposes an evolutionary origin of human language in non-verbal communication and body language that can be found in many social species (e.g. mammals, birds). He suggests that imitation (and the primate mirror neuron system [Gallese et al 96]) provided the major mechanisms that facilitated the transition from body language and non-verbal imitation to verbal communication. This work supports the arguments as presented in this paper, namely proposing a) the existence of a strong link between non-verbal, preverbal and verbal communication, and b) the important role of dynamic formats of interactions, such as imitative games, in the development of social communication.

With this focus on interactional structure and non-verbal narratives, how can stories in non-human species look like?

Let us consider Frans de Waal's description of an event of reconciliation in chimpanzees.

> "On this occasion Nikkie, the leader of the group, has slapped Hennie during a passiing charge. Hennie, a young adult female of nine years, sits apart for a while feeling with her hand the spot on her back where Nikkie hit her. Then she seems to forget the incident; she lies down in the grass, staring in the distance. More than fifteen minutes later Hennie slowly gets up and walks straight to a group that includes Nikkie and the oldest female, Mama. Hennie approaches Nikkie, greeting him with soft pant grunts. Then she stretches out her arm to offer Nikkie the back of her hand for a kiss. Nikkie's hand kiss consists of taking Hennie's whole hand rather unceremoniously into his mouth. This contact is followed by a mouth-to-mouth kiss. Then Hennie walks over to Mama with a nervous grin. Mama places a hand on Hennie's back and gently pats her until the grin disappears". ([de Waal 89], pp 39,42)

This example shows that the agent (Hennie) is interacting with an eye to future relationships, considering past and very recent experiences. Hennie, Nikkie and Mama have histories, autobiographic histories as individual agents [Dautenhahn 96], as well as a history of relationships among each other and as members of a larger group. Although the event might be interpreted purely on the basis of behavioristic stimulus-response rules, for many primatologists the interpretation of the event in terms of intentional agents and social relationships is the most plausible explanation.

Interestingly, Hennie's interaction with Nikkie shows the canonical format of narrative transactions among intentional agents described in section 4.2:

1) canonical state: greeting: soft pant grunts
2) precipitating event: Hennie reaches out to Nikkie (attempt at reconciling relationship)
3) restoration: kissing (relationship is restored)
4) end: Hennie is comforted by Mama

The second example we discuss is a different type of primate social interaction, namely tactical deception whereby the agent shifts the target's attention to part of its own body. In this particular case the agent (female Olive baboon) distracts the target (male) with intimate behavior.

> "One of the female baboons at Gilgil grew particularly fond of meat, although the males do most of the hunting. A male, one who does not willingly share, caught an antelope. The female edged up to him and groomed him until he lolled back under her attentions. She then snatched the antelope carcass and ran". (cited in [Whiten & Byrne 88]).

Here, the analysis in terms of transactional narrative formats looks as follows:

1) canonical state: male brings antelope, female waits
2) precipitating event: distraction by grooming
3) restoration: female snatches food and runs away (resolution, female achieves goal)
4) end: female eats meat (not described)

Episodes of animal behavior as described above are different from other instances of animal behavior that possess a certain structure and appear in sequences, such as the chase-trip-bite hunting behavior of cheetahs. Also, the alarm calls of vervet monkeys [Cheney & Seyfarth 90], although serving an important communicative function in a social group and having a component of social learning, are not narrative in nature. It is not the short length of such calls that makes it difficult to interpret them in terms of narrative, it is the fact that their primary function is to change the *behavior* of others as a response to a non-social stimulus, i.e. the sight of a predator, causing an appropriate behavior such as running to the trees after hearing a leopard alarm. The narrative format in animal behavior on the other hand refers to *communicative* and *transactional* contexts where communication is *about* the social field, e.g. group members, their experiences and relationships among them. Narratives are *constructed* based on the current context and the social context (communicator/speaker plus recipients/audience). The primate protagonists described above apparently interacted with respect to the social context, i.e. considering the social network and relationships among group members, with the purpose of influencing and manipulating others. Thus, such kind of non-verbal narratives are fundamentally social in nature.

For a more detailed analysis of narrative formats in animal behavior a lot more work is necessary. For example, the characteristics of the transactional format that Bruner and Feldman suggested (1993) need to be elaborated, possibly revised or replaced, and might need to be adapted to specific constraints of the primate social field [Tomasello & Call 97], so our interpretation can only give a first hint on what aspects one might be looking for when searching for narrative formats in animal communication.

5 The Narrative Intelligence Hypothesis Revisited

If human language and narrative intelligence, rooted in nonverbal narrative intelligence in non-human primates, has evolved to deal with an increasing need to communicate in more and more complex societies, what predictions can be made based on this hypothesis? How could the Narrative Intelligence Hypothesis be tested? What are important research directions based on the importance of narrative in animals and artifacts?

Let us first consider how the NIH might be confirmed or dismissed. As with other hypotheses on the origin of primate/human intelligence and language, animal behavior and communicative abilities are not directly documented in the fossil record, they can only be inferred indirectly from anatomical features (e.g. the vocal system that is necessary to produce a human-like language) and remains that indicate social structures (e.g. remains of nests or resting places, or groups of animals that died together). However, recent primate species that could serve as models of ancestors of the human species might give clues on what groups of primate species one might analyze if wanting to trace the origins of human narrative intelligence. Possible narrative structures confirmed in primate behavior might then be correlated with the complexity of the social field in these species.

With respect to the evolution of human societies, Russel [Russell 93] discusses four levels of social organization which might serve as models for the evolution of human societies: a) the 'shrew'-type pre-primates: solitary, many offspring, insectivores, e.g. *Purgatorius*, a 70-million-year-old fossil, b) the 'mouse-lemur'-type primates: bush-living, nocturnal, strong mother-daughter bonding (stable matrilines), social learning (offspring learns from mother), solitary males and social groups of mothers and daughters, e.g. the 50-million-year-old fossil *Shoshonius cooperi*, c) the 'Lemur catta'-type diurnal lemurs: appearing about 54 million years ago, social groups (troops), dominant females, submissive males, stable matrilines, occasionally consort bonds between single male and female, e.g. *Adapidae*, d) the 'chimpanzee'-type lemur-ape: appearing about 24 million years ago, groups of dominant males and submissive females, stable families of mothers and their offspring, male power coalitions, e.g. *Dryopithecus*. The social organization of recent species of apes shows variations of this pattern: of harem-structures (gorilla), solitary lifestyle (orangutan). Russel discusses how human societies can be interpreted and discussed as variations of such primate social patterns.

The Narrative Intelligence Hypothesis would predict that comparative studies of communicative, and in particular narrative formats of interactions across primates species with different social organizations, e.g. as described above, can identify a correlation between the complexity of the narrative format and an increasing complexity of the primate social field. Such an increase of social complexity need not be limited to group size, but could also cover all other aspects of social complexity that we discussed previously, such as an increasing number of different types of interactions and roles of group member, and the dynamics of how the social network can change and adapt to changes. Such stages of social organization can be related to behavioral as well as cognitive and mental capacities of primates. The NIH suggests a search for the narrative format in interactions, a format that is so efficiently suited to communicate and deal with the complexity of social life.

What kind of research directions and research methods could the NIH inspire?

Testing with Robotic and Computational Models:
As indicated in section 4.3 artifacts could provide scientific tools to explore and experimentally test the design space of narrative intelligence. Narratives in this sense need to have a 'meaning' for an (intentional) agent. The approach of using artifacts as experimental test beds has been used successfully for many years in the areas of adaptive behavior and artificial life, yielding many interesting results that a) help understanding animal behavior, b) help designing life-like artifacts, in this case artifacts with narrative skills.

Study and Analysis of Animal Narrative Capacities:
Since the Narrative Intelligence Hypothesis does not assume any 'novel' development in the transition from nonverbal (through evolution) or preverbal (development) to verbal narrative intelligence, a detailed study and analysis of the structure and format of animal narrative communication is required in order to develop a proper framework. Many vertebrate species are highly social (e.g. non-human primates, dolphins, whales, elephants, bird species such as crows and parrots) and use non-verbal means of body language in interaction and communication. Narrative intelligence has on the one hand a communicative function (as a means of discourse and dialogue), but it has also an individual dimension (understanding and thinking in terms of narrative). Revealing narrative structure in animal communication might therefore further our understanding about meaningful events in the lives of these animals.

Interesting open research questions (this is not an exhaustive list):

- Relationship between preverbal and verbal narrative intelligence in humans (ontogeny)
- Relationship between nonverbal narrative intelligence in non-human animals and narrative intelligence in humans (phylogeny)
- The format of nonverbal narrative intelligence in animals (species specific? Specific to social organization of animal societies?)
- Can we identify narrative 'modes of thought' in different animal species?

6 Implications for Human Society and Technology

There are many implications of the Social Brain Hypothesis and the Narrative Intelligence Hypothesis for technology development. Human cognitive and narrative capacities are constrained by human evolution. Even technological extensions and enhancements (new media, new means of communication, new interfaces and implants) need to operate within the boundaries set out by biology.

Firstly, for people whose real social networks are smaller than 150, the roles of friends and social partners might be filled by other 'partners', either human beings e.g. actors in movies and soap operas, news presenters or presenters of daily chat shows, fictional characters such as Captain Kirk in Star Trek or Homer Simpson, including computer game characters such as Lara Croft in Tomb Raider. Although many of such 'social relationships' are rather uni-directional (we bond with them but they do not bond with us and do not even know of our existence), they might serve a similar role than real human networks [Dunbar 96]. The boundaries between real and artificial are often nebulous or ambiguous, cf. interactions with chatbots or MUD robots in multi-user on-line environments [Foner 00], or a new generation of embodied conversational agents, e.g. software agents that might serve as real estate agents [Bickmore & Cassell 00]. Today, new interactive game software can create believable illusions that agents truly bond with their users, e.g. the Norns in the computer game Creatures, or robotic pets such as Furbies or Aibo's that are extending such acquaintances even to the physical level. However, this extension of real social networks (see [Turkle 95]) is not without limitations, constrained by the cognitive group size limit of 150 that characterizes human primates social networks. As Dunbar argues (1996) modern information technology might change a number of characteristics of how and with whom and with what speed we communicate, but not influence the size of social networks, nor the necessities of direct personal contact that need to provide trust and credibility to social relationships. Note, that the cognitive group size refers to individually knowing somebody: humans have developed varies means of coping with very large group sizes, e.g. military ranks, castes, stereotypes, (possibly prejudices), etc. Although we might 'know' the names of thousands of people (e.g. as entries in a database) such 'knowing' is not based on individual knowledge.

Language is a dominant means of communication in modern societies human that can do remarkable things seemingly without limits. "Yet underlying it all are minds that are not infinitely flexible, whose cognitive predispositions are designed to handle the kinds of small-scale societies that have characterized all but the last minutes of our evolutionary history." [Dunbar 96].

Building narrative technology, in particular interactive environments is a growing area, ranging from applications in education and therapy to entertainment. In the project AURORA we are developing a robotic agent as a therapeutic tool for children with autism. Here, giving the robot story-telling skills could address issues that are relevant to Bruner and Feldman's narrative deficit hypothesis (see section 4.2). At present the robot we use in the AURORA project is not historically grounded, it reacts based on the here and now ([Dautenhahn 99b], [Dautenhahn & Werry 00], [Werry et al, this volume], [Dautenhahn, in press]), but architectures such as the one we studied for the robotic story-teller (section 4.3) might be applied and studied with respect to their therapeutic effect.

Generally, we can expect that empowering human skills of forming and maintaining social networks might be advanced by supporting the development of narrative skills in children and adults. As we discussed in this article, narratives are not only entertaining and fun, they serve an important cognitive function in the development of social cognition and a sense of self [Dennett 89]. However, as discussed in [Nehaniv

99b], humane technology needs to respect human narrative grounding in order to avoid undesirable and unforeseen effects.

The narratives of the future might reflect our ability to preserve coherence and structure in human societies that consist of increasingly fragmented, temporally and geographically distributed social networks. In shaping this development it is important to investigate the evolutionary heritage of our narrative capacities and the natural boundaries it provides. Also, appreciating the stories other non-human animals tell will allow us to put our familiar stories-as-we-know-them into the broader perspective of stories-as-they-could-be.

Acknowledgements. The project *Memory-Based Interaction in Autonomous Social Robots* is supported by a grant from The Nuffield Foundation, NUF-NAL 00. The AURORA project is supported by EPSRC (GR/M62648), *Applied AI Systems Inc.*, and the National Autistic Society (NAS). Thanks to Phoebe Sengers, Michael Mateas and Chrystopher Nehaniv for fruitful discussions on narrative over the past few years. The ideas expressed in this paper are nevertheless my own.

References

Arbib, M. (2001). The mirror system, imitation, and the evolution of language. In K. Dautenhahn and C. L. Nehaniv (Eds.) *Imitation in Animals and Artifacts*, MIT Press.

Bickmore, T. & J. Cassell (2000). In K. Dautenhahn, (Ed.), *Proc. Socially Intelligent Agents - The Human in the Loop*, AAAI Fall Symposium 2000, AAAI Press, Technical Report FS-00-04 (4-8).

Bruce, A., J. Knight, S. Listopad, B. Magerko & I. R. Nourbakhsh (2000). Robot Improv: Using Drama to Create Believable Agents. *Workshop on Interactive Robotics and Entertainment (WIRE-2000)*, Pittsburgh, April 2000.

Byrne, R. W. & A. Whiten, (Eds.), (1988). *Machiavellian Intelligence*. Clarendon Press.

Byrne, R. W. (1997). Machiavellian intelligence. *Evolutionary Anthropology,* 5, 172-180.

Bruner, J. (1987). *Actual Minds, Possible Worlds*. Cambridge: Harvard University Press.

Bruner, J. (1990). *Acts of Meaning*. Cambridge: Harvard University Press.

Bruner, J. (1991). The Narrative Construction of Reality. *Critical Inquiry,* 18(1), 1-21.

Bruner, J. & C. Feldman (1993). Theories of mind and the problem of autism. In S. Baron-Cohen et al. (Eds.), *Understanding other Minds: Perspectives from Autism.* Oxford University Press, Oxford.

Bumby, K. & K. Dautenhahn (1999). Investigating Children's Attitudes Towards Robots: A Case Study. *Proc. CT99, The Third International Cognitive Technology Conference*, August, San Francisco, 359-374.

Coles, S. & K. Dautenhahn (2000). A robotic story-teller.In Proc. *SIRS2000, 8th Symposium on Intelligent Robotic Systems*, The University of Reading, England, 18-20 July 2000.

Cheney, D. L. & R. M. Seyfarth (1990). *How Monkeys See the World.* University of Chicago Press.

Dautenhahn, K. (1996). Embodiment in animals and artifacts. In *Proc. Embodied Cognition and Action*, pages 27-32, AAAI Press, Technical report FS-96-02, 1996.

Dautenhahn, K. (1999a). The lemur's tale - Story-telling in primates and other socially intelligent agents. In M. Mateas & P. Sengers, (Eds.), *Proc. Narrative Intelligence*, AAAI Fall Symposium 1999, AAAI Press, Technical Report FS-99-01 (59-66).

Dautenhahn, K. (1999b). Robots as social actors: AURORA and the case of autism. *Proc. CT99, The Third International Cognitive Technology Conference*, August, San Francisco, 359-374, URL: http://www.added.com.au/cogtech/CT99/Dautenhahn.htm (last accessed 30/10/2000).
Dautenhahn, K. (1999c). Embodiment and interaction in socially intelligent life-like agents. In C. L. Nehaniv, (Ed.), *Computation for Metaphors, Analogy and Agents*, 102-142. Springer Lecture Notes in Artificial Intelligence, Volume 1562.
Dautenhahn, K. (to appear). Stories of Lemurs and Robots - The Social Origin of Story-Telling. To appear in *Narrative Intelligence*, edited by Michael Mateas and Phoebe Sengers, John Benjamins Publishing Company.
Dautenhahn, K. (in press). Design Issues on Interactive Environments for Children with Autism. *CyberPsychology and Behavior*, in press.
Dautenhahn, K. (submitted). Socially Intelligent Agents - Towards a Science of Social Minds. Submitted to *Minds and Machines*, January 2001.
Dautenhahn, K. & C. L. Nehaniv (1998). Artificial life and natural stories. In *Proc. Third International Symposium on Artificial Life and Robotics* (AROB III'98 - January 19-21, 1998, Beppu, Japan), volume 2, 435-439.
Dautenhahn, K. & I. Werry (2000). Issues of robot-human interaction dynamics in the rehabilitation of children with autism. *Proc. From Animals to Animats, The Sixth International Conference on the Simulation of Adaptive Behavior (SAB2000)* (519-528), 11 - 15 September 2000, Paris, France, 2000.
Dautenhahn, K. & Steven Coles (2001). Narrative Intelligence from the bottom up: A computational framework for the study of story-telling in autonomous agents. To appear in Special Issue of the *Journal of Artificial Societies and Social Simulation (JASSS)* on Starting from Society: The application of social analogies to computational systems.
Dennett, D. C. (1987) *The intentional stance*. MIT Press.
Dennett, D. C. (1989/91). The origins of selves. *Cogito,* 3, 163-73, Autumn 1989. Reprinted in Daniel Kolak and R. Martin, eds., (1991), *Self & Identity: Contemporary Philosophical Issues*, Macmillan.
de Waal, F. (1982). *Chimpanzee Politics: Power and sex among apes*. Jonathan Cape, London.
de Waal, F. (1989). *Peacemaking among Primates*. Harvard University Press.
Donald, M. (1993). Precis of *Origins of the modern mind: Three stages in the evolution of culture and cognition. Behavioral and Brain Sciences,* 16, 737-791.
Dunbar, R. I. M. (1992). Neocortex size as a constraint on group size in primates. *Journal of Human Evolution*, 20, 469-493.
Dunbar, R. I. M. (1993). Coevolution of neocortical size, group size and language in humans. *Behavioral and Brain Sciences,* 16, 681-735.
Dunbar, R. I. M. (1996). *Grooming, Gossip and the Evolution of Language*. Faber and Faber Limited.
Dunbar, R. I. M. (1998). The social brain hypothesis. *Evolutionary Anthropology*, 6, 178-190.
Engel, S. (1995/99). *The Stories Children Tell: Making Sense of the Narratives of Childhood.* W. H. Freeman and Company.
Farnell, B. (1999). Moving Bodies, Acting Selves. *Annu. Rev. Anthropol.*, 28, 341-373.
Gallese, V., L. Fadiga, L. Fogassi & G. Rizzolatti (1996). Action recognition in the premotor cortex. *Brain*, 119, 593-609.
Foner, L. N. (2000). Are we having fun yet? Using social agents in social domains.In K. Dautenhahn (Ed.) *Human Cognition and Social Agent Technology*, John Benjamins Publishing Company, chapter 12(323-348).
Gill, S. P., M. Kawamori, Y. Katagiri & A. Shimojima (1999). Pragmatics of body moves. *Proc. CT99, The Third International Cognitive Technology Conference*, August, San Francisco, 359-374.
Hall, E. T. (1968). Proxemics. *Current Anthropology*, 9(2-3), 83-95.
Herman, L. M. (2001). Vocal, social, and self imitation by bottlenosed dolphins. In K. Dautenhahn & C. L. Nehaniv (Eds.) Imitation in Animals and Artifacts, MIT Press.

Jordon, R. (1999). *Autistic Spectrum Disorders: An introductory handbook for practitioners.* David Fulton Publishers.

Meltzoff, A. N. & A. Gopnik (1993). The role of imitation in understanding persons and developing a theory of mind. In S. Baron-Cohen, H. Tager-Flusberg & D. J. Cohen (Eds.), *Understanding other minds* (335-366), Oxford University Press.

Meltzoff, A. N. & M. K. Moore (1999). Persons and representation: why infant imitation is important for theories of human development. In J. Nadel & G. Butterworth (Eds.), *Imitation in Infancy* (9-35), Cambridge University Press.

McCloud, S. (1994) *Understanding Comics - The Invisible Art.* HarperPerennial.

Nadel, J., C. Guerini, A. Peze, & C. Rivet (1999). The evolving nature of imitation as a format of communication. In J. Nadel & G. Butterworth (Eds.), *Imitation in Infancy* (209-234). Cambridge University Press.

Nehaniv, C. L. (1997). What's Your Story? - Irreversibility, Algebra, Autobiographic Agents". In: K. Dautenhahn, ed., Socially Intelligent Agents: Papers from the 1997 AAAI Fall Symposium (November 1997, MIT, Cambridge, Massachusetts) FS-97-02, American Association for Artificial Intelligence Press, 150-153.

Nehaniv, C. L. & K. Dautenhahn (1998). Embodiment and Memories - Algebras of Time and History for Autobiographic Agents. In R. Trappel (Ed.) *Proc. 14th European Meeting on Cybernetics and Systems Research* (651-656).

Nehaniv, C. L., K. Dautenhahn, & M. J. Loomes (1999). Constructive Biology and Approaches to Temporal Grounding in Post-Reactive Robotics. In G. T. McKee & P. Schenker (Eds.) *Sensor Fusion and Decentralized Control in Robotics Systems II* (September 19-20, 1999, Boston, Massachusetts), Proc. of *SPIE* Vol. 3839(156-167).

Nehaniv, C. L. (1999a). Narrative for artifacts: Transcending context and self. In M. Mateas & P. Sengers, (Eds.), *Proc. Narrative Intelligence*, AAAI Fall Symposium 1999, AAAI Press, Technical Report FS-99-01 (101-104).

Nehaniv, C. L. (1999b). Story-Telling and Emotion: Cognitive Technology Considerations in Networking Temporally and Affectively Grounded Minds. In Proc. Third International Conference on Cognitive Technology: Networked Minds (CT'99), Aug. 11-14, 1999 San Francisco/Silicon Valley, USA, 313-322.

Nelson, K., Ed., (1989). *Narratives from the crib.* Harvard University Press.

Nelson, K. (1993). The psychological and social origins of autobiographical memory, *Psychological Science,* 4(1), 7-14.

Philips, M. & S. N. Austad (1996). Animal communication and social evolution. In M. Bekoff & D. Jamieson, (Eds.), *Readings in Animal Cognition* (257-267). MIT Press.

Pawlowski, B., C. B. Lowen, & R. I. M. Dunbar (1998). Neocortex size, social skills and mating success in primates. *Behaviour*, 135, 357-368.

Pepperberg, I. M. (1999). *The Alex Studies. Cognitive and Communicative Abilities of Grey Parrots.* Harvard University Press, Cambridge, MA.

Read, S. J. & L. C. Miller (1995). Stories are fundamental to meaning and memory: For social creatures, could it be otherwise? In R. S. Wyer, (Ed.), *Knowledge and Memory: the Real Story*, chapter 7 (139-152). Lawrence Erlbaum Associates, Hillsdale, New Jersey.

Russell, R. J. (1993). *The Lemurs' Legacy. The Evolution of Power, Sex, and Love.* G.P. Putnam's Sons, New York.

Savage-Rumbaugh, E. S., K. McDonald, R. A. Sevcik, W. D. Hopkins, & E. Rubert (1986). Spontaneous symbol acquisition and communicative use by pygmy chimpanzees (*Pan paniscus*). *Journal of Experimental Psychology: General,* 115, 211-235.

Schank, R. C. & R. P. Abelson (1977). *Scripts, Plans, Goals and Understanding: An Inquiry into Human Knowledge Structures.* Erlbaum, Hillsdale, NJ.

Sengers, P. (2000). Narrative Intelligence. In K. Dautenhahn, (Ed.), *Human Cognition and Social Agent Technology*, chapter 1 (1-26), John Benjamins Publishing Company.

Sindermann, C. J., (1982). *Winning the Games Scientists Play.* Plenum Press, New York, London.

Tomasello, M. & J. Call (1997). *Primate Cognition.* Oxford University Press.

Turkle, S. (1995). *Life on the screen, identity in the age of the internet*, Simon and Schuster.
Turner, M. (1996). *The Literary Mind.* Oxford University Press.
Werry, I., K. Dautenhahn, B. Ogden & W. Harwin (2001). Can Social Interaction Skills Be Taught by a Social Agent? The Role of a Robotic Mediator in Autism Therapy. This volume.
Whiten, A., & R. W. Byrne, (Eds.), (1997). *Machiavellian Intelligence II: Extensions and Evaluations,* Cambridge University Press.
Whiten, A. & R. W. Byrne (1988). The manipulation of attention in primate tactical deception. In Byrne & Whiten, Eds, *Machiavellian Intelligence*. Clarendon Press, chapter 16(211-237).

Building Rules

Ronnie Goldstein[1], Ivan Kalas[2], Richard Noss[3], and Dave Pratt[4]

[1] Open University, Milton Keynes, UK
[2] Comenius University, Bratislava, Slovakia
[3] Institute of Education, University of London, UK
[4] Institute of Education, University of Warwick, UK

Abstract. This paper reports on aspects of the Playground project in which young children (age 6 to 8) are writing and sharing their own computer video-games. We discuss how structures in the kernel language influenced the design of one of the project's playgrounds and in turn children's thinking and use of rules. One feature of the paper is the range of children's responses to the task of translating their ideas for games into formal rules; kernel features, such as object orientation and the use of events, at times support and at other times constrain those responses.

1 Introduction

In the Playground Project[1], a team of researchers and developers from several countries is exploring how young children (6 to 8 years) construct computer-based games. Children build their games in a *playground*, a space containing game-building tools, games already built by children or developers, and sub-elements of a game, such as game objects, rules, parts of rules and scenery. At the broadest level, we are interested in how they think about the rules that underlie such games including how their meanings for such rules evolve and are expressed through the use of the virtual tools that we provide. This perspective of the children's meaning-making recognises the dialectic relationship between the design of the tools themselves and the evolution of thinking about the sorts of rules that might define computer-based games.

Our main theoretical construct is that of *webbing* [1]. The notion of webbing is strongly influenced by the paradigm of situated cognition [2] in which meaning, as constructed in everyday experience, has little to do with generalised theory and much to do with strategy constructed *in situ*, and based heavily on the structuring resources available in that setting. Noss and Hoyles propose a network of links, encompassing not only individual meanings but also resources external to that person.

The idea of webbing emphasises that learners come to construct new knowledge by forging and reforging internal connections through the interaction of internal and external resources during activity and in reflection upon it, and so reinforces the dependence that such knowledge inevitably has upon the particular attributes of those tools as understood by children [see 3 for a seminal analysis of how new representational forms may contribute fundamentally new kinds of literacies]. Representational systems are constitutive of knowledge construction, and new kinds of systems might

[1] EU ESE Project #29329: see http://www.ioe.ac.uk/playground.

M. Beynon, C.L. Nehaniv, and K. Dautenhahn (Eds.): CT 2001, LNAI 2117, pp. 267–281, 2001.

at the very least, make available classes of knowledge that are unreachable without it [see 4].

A key facet of design for digitally-based systems is the affordance to inspect and manipulate the mechanisms beneath the surface — this is, incidentally, genuinely unique to digital technologies, and does potentially change the situation with respect to learning. The challenge is to build systems, which support the construction of objects and ideas (together) by the learner (this is a restatement of the 'constructionist' programme eloquently stated by Seymour Papert and some others). Giving learners a sense that they can build things for themselves, and thus think about and reflect on what they are building, has been a central tenet of the learner-programming movement of the last two or three decades.

Yet the idea of children programming – constructing and manipulating executable representations of objects and relationships – has not always been straightforward. (For a history of children's programming, see Noss and Hoyles [1].) One reason for this has been the essential identification between the act of programming, and the manipulation of lines of text. Despite some genuinely novel and powerful attempts to make this situation as palatable as possible (Logo is the most obvious example), these have not always been entirely successful. This situation has now changed, and there is a sufficiently large set of examples of programming systems where the primary mode of interaction is no longer restricted to textual interface. We are trying, therefore, to design a system that is at once open and accessible, that allows the novice to operate within new worlds, yet simultaneously allows her to see what works and how it works. While we want the learner to be fluent *within* the medium, we also want her to reflect on the structures of the medium itself [see 5 for an interesting view of this problem].

The Playground Project integrates three strands of work that might conventionally have been seen as distinct stages of development. One strand (the *kernel*) is the development of programming environments that provide the infrastructure for playgrounds. In the project, there are two kernels, ToonTalk [6] and Imagine [7]. For the purposes of this paper, we shall focus on the latter, a powerful object-oriented member of the Logo family but we wish to acknowledge the influence of elements of ToonTalk on the design of the Imagine playground. Imagine was published in UK by Logotron in January 2001. A second strand (the *playground*) has been the building of an environment, written in Imagine, which young children can use to construct their own games. This new environment is called *Pathways*, an interactive, open, visual, multimedia environment. The third strand (*meaning-making*) has been the observation of children using this environment. Unusually, all three strands have been implemented iteratively and in parallel so that each informs the ongoing development of the others. We see this integrated iterative development as a natural manifestation of our interest on the intertwined relationship between tools and learning.

In this paper, we continue the theme of interlinked strands (kernel, playground and meaning-making) by discussing how each is contributing to our understanding of three closely connected arenas, the Game Formal, the Game Outside and the Game Inside, each representing a different type of activity as will be described below. It turns out the main issues to be raised about how children think about rules emerge from activity across these three arenas. The aim of this paper is to articulate these issues by describing activity on that interface. Our approach in this paper is to illustrate activity on this interface by reporting on excerpts from sessions involving two researchers (in the transcribed excerpts, R refers to either one of them) and usually two children. The two researchers acted as participant observers in a clinical interviews spread over between 3 and 6 sessions with each pair of children. The children had not previously used the software. All had occasional access to computers in school and used computers at home. The sessions took place outside the classroom. The activity and discussions of the children with each other and with the researchers were video-taped, transcribed and interpreted. The work is ongoing. Here we report on emergent ideas apparent at this relatively early stage of the analysis.

2 The Kernel: The Imagine Programming Language

The kernel for the Pathways playground, Imagine, is an innovative modern programming environment especially designed for children to explore and develop. It is a powerful tool for designing open visual applications for education and entertainment.

Imagine is a new 32-bit version of Logo. [See 7 for more details.] We report here only those features of the kernel, which have directly supported and influenced the process of the design of Pathways.

- Substantially strong direct manipulation tools are provided in Imagine, which supports each direct step of the manipulation by an unambiguous language construct. This approach has turned out to be a productive choice for Pathways because the entire playground is a language of direct manipulations – through simple visual structures (rules) children express the behaviour of each object-actor of the game. Each rule is internally represented by a Logo language construct in a relatively straightforward way.
- The object-oriented metaphor is built into Imagine in a non-intrusive way. Imagine contains a set of primitive classes. The user can create either instances of those classes, i.e. individual objects, or define his or her own classes by further specifying already existing classes and creating instances of new classes. Moreover, the user can create an instance, then develop its behaviour, then use this individual object as a prototype and create instances based on that particular instance. This provides for easy and intuitive multiplication of objects. Later on we will present how this concept is implemented in Pathways through the magic wand tool.
- Event-driven programming considerably simplifies specifying reactions to events such as clicking an object by the mouse (any button), pressing a key, dragging an object, colliding with another object, or controlling an object by a joystick. There is a set of standard events recognised by Imagine. However, creating user-defined events can extend that set.
- Parallel processing is inherent. An Imagine project usually consists of several objects (actors), which follow their own behaviours. Although they may interact (e.g. through events) significantly, each of them may be controlled by its own independent process running in parallel with others.
- Imagine includes the facilities for text to speech and speech to text (if a speech engine is installed in the computer). To realise the voice output, the `say` com-

mand with any text input is used. Imagine, in co-operation with the speech engine, can also analyse the voice input, use the user-defined voice-menu to transform it into the associated Logo instruction and run it.

3 The Game Formal: The Pathways Playground

Pathways is a formal iconic language, which aims to make the process of building games accessible to very young children. The rules created by children in Pathways define the game that has been built. We refer to this definition as the Game Formal.

The software allows children to choose from many different background screens and to place a number of objects on the screen (see Fig. 1). The objects can be given various shapes and they can be made to move automatically or to be dependent on an input device. An underpinning hypothesis is that young children can create complex behaviours by teaching objects a relatively small number of simple rules expressed in an iconic language. An object can hold several rules. For example, the pathways in Figure 6 apply to an object that has been given the shape of a tiger. The two rules combine so that the tiger is controlled by the joystick and when button 1 of the joystick is pressed, the tiger jumps upwards on the screen (see the translation in the third column of Figure 6). Children move from defining rules to playing the game by clicking a switch that turns the game on but they can return at any time to making rules by switching the game off again.

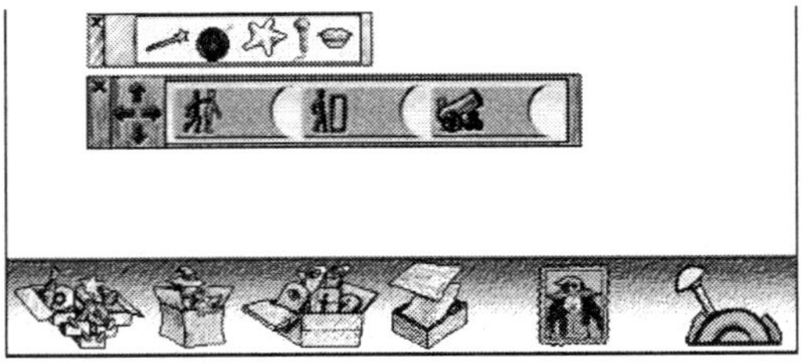

Fig. 1. This section of the Pathways screen shows two boxes that were opened from the icons in the control bar at the foot of the screen. The Toolbox at the top of the figure contains the following icons: a wand for copying; a bomb for deleting; a star for changing various settings of an object including its shape; a microphone for giving comments to objects and the mouth for speaking comments. The opened Stonesbox (below the Toolbox) is currently revealing some of the when stones. The when stones represent conditions such as "when I am touching". The Stonesbox also contains do stones that represent actions such as "I jump" and behaviours that allow an object to be controlled by the mouse, joystick or keyboard. There are further icons in the control box to open a stored game, change the background, to create a new object in a game and to switch a game on or off.

4 The Game Outside

The games to which we refer in this paper were created by children of age 7 with some support as appropriate from the researchers. We refer to such a game as the

Game Outside to distinguish it from its formal definition, the Game Formal, and the rather less precise imagined game that takes place in the child's head (discussed in the next section). In Figs. 2, 3, and 4, we briefly describe the games referred to in this paper.

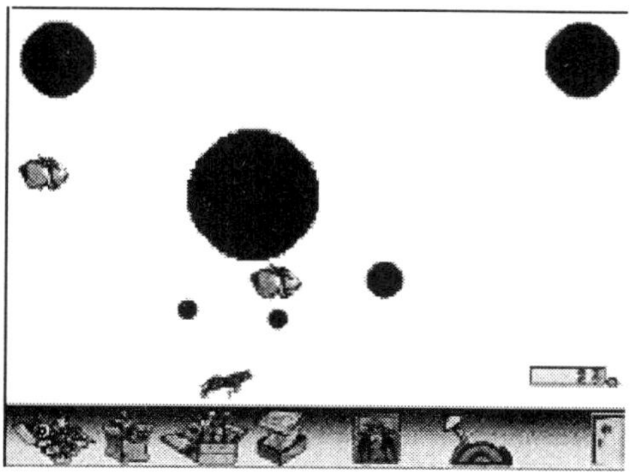

Fig. 2. Matthew and Hannah built this game, Dodgerz. The player controls the tiger with a joystick and he has to catch the rabbits as quickly as possible. The rabbits move automatically around the screen, bouncing off the walls. (Matthew and Hannah referred to the circles as walls.) When the tiger touches a rabbit, the rabbit disappears and a sound is made. If the tiger touches a wall, one of the rabbits that has already been "eaten" by the tiger, may reappear. (We have removed the grassy background in the game in order to improve the clarity of the picture.)

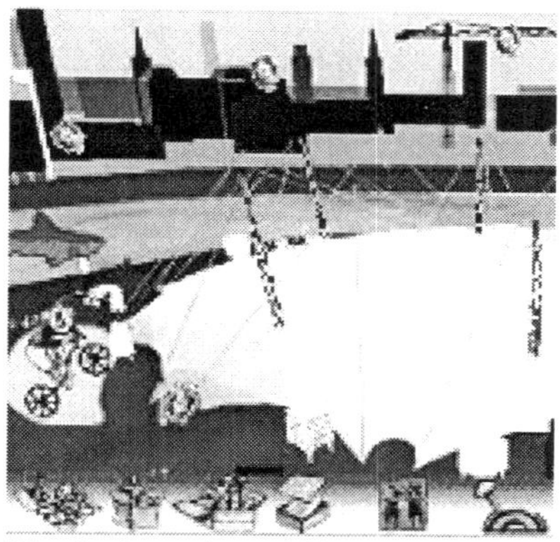

Fig. 3. A section of Harry and Laura's Dome game is shown. One player controls the clown with a joystick while a second player controls the shark with the mouse. The turtles move automatically and randomly around the screen. The players compete to collect the six spikes (only four are shown here) from the roof of the Millennium Dome before the turtles explode them.

5 The Game Inside

When young children work with Pathways, the ideas that they have in their heads about the games they want to create are very important. They are important to the children because their work belongs to them and they are not simply regurgitating what adults have told them to do. They are important to their teachers because of the motivation that this ownership stimulates. And, not insignificantly, they are important to us, the researchers, because we want to learn about the children's thoughts as they engage with the entire process of creating a game.

Before children have started to use the computer we tell them that they are about to design their own games and their imaginations run riot. Of course their thoughts are constrained by their experience and the games they want to design are usually based on recent experiences or on video-games they have played before. When Laura and Harry were about to start Laura said she would like to create a Millennium Dome

Fig. 4. A section of Eleanor's game is shown. This game is more like a story in which the cat pushes the meat (with its bottom!) towards the dog. When the dog and the meat touch, the meat disappears, a sound is made, and the cat returns.

game. It became clear after further discussion that Laura had recently visited the Dome. She could imagine a game in which the aim was to collect the spikes from the top of the Dome and transfer them to certain zones. The imagined game went further into fantasy as they suggested that their characters (i.e. the ones they controlled) might be a dog for Laura and a fish for Harry.

As they progressed, Laura and Harry obviously learned many things about Pathways and what they might be able to achieve using the software. We had written a simple game to start them off and they had begun by changing the rules for the dog. This bottom-up approach is not inappropriate but, even after working with some of the rules for more than one session, their fantasised game still had a lot of meaning. In the excerpt below Harry had been immersed in a discussion with Laura about certain rules and which objects should own those rules. Suddenly he surprises us:

1. H: Because that big spike … I wonder what would happen if we get all the spikes. I wonder if the game will actually finish.

In fact Harry repeated that the game might have automatically stopped when all the spikes had exploded despite no rule having being written to that effect. The imagined game often seemed to interpose with the game being built. An appropriate ending for the Dome game could have been when all the spikes were exploded and Harry talked as if this might have happened automatically. Was Harry just articulating this imagined ending, was he assuming that the initial game he and Laura had been given contained this ending or did he really think this could have happened without his instruction?

At a still later stage their thinking about the game seemed to be entirely governed by the rules that were to be given to the objects of the game. Harry and Laura were discussing how their game might progress and the phrases they were using were close in their form to the rules as they would have been expressed by the mouth in Pathways. They seemed to have a very clear appreciation of the relationship between the Game Inside and the Game Formal.

2. H: It could be a two-player game. When, whoever gets more, the most, spikes in there, wins the game.
3. H: The turtle could be one and the shark could be the other.
4. L wants to use the dog and the shark from last time.
5. L: When the shark touches the spikes, it could explode. We could tell the shark or dog to hide when it gets touched and then to show again.
6. R(esearcher): Say that again slowly.
7. L: When the shark touches the spikes, it could explode.

But the facility to talk in terms of the formal rules of the software does not develop so fast in all young children and there does not always seem to be a clear-cut distinction between what they want to happen and what will happen. At some points the children's thinking is dominated by their fantasies. At other times they are constrained by the computer and what they can achieve using the tools of Pathways. It may be that some children who are not so comfortable with formal rules prefer to spend more of their time in their fantasies. We have seen children who have given a few rules to some objects, play their 'computer games' with many rules that are still only inside their heads. We do, however, expect a transition so that, as time progresses, the computer becomes more dominant and there are many more Game Formal episodes. In this respect, however, as in most other respects, learning is not a simple linear process.

6 On The Interface

From our point of view, the most interesting activities concern the movements between the Game Inside, Game Formal and Game Outside. In this section we focus on the negotiation that takes place when children try to implement their ideas and when they find that the game they have built is not quite what they had imagined. We wish to focus on two structures in the kernel that were exploited when designing Pathways and which, it turns out, have implications for the activities that take place.

6.1 Working with Objects

When children write instructions in Pathways they choose an object on the screen and put the appropriate stones into one of the pathways for that object. Rules are not written in some general rule-factory area and given to the computer; they are written for particular objects. This is emphasised by the mouth tool in the software which, when placed on a stone or a rule, will speak in the first person singular: 'I move forwards' or 'when I am touching the dog, I hide'. The software's object orientation and the parallel processing that goes along with it have many consequences for children who are thinking about their game and learning to write rules formally.

Where is the Rule? We have instances in our data where children have written a rule that they subsequently decide to alter but they are unable to locate it without opening the pathways for each of their objects and inspecting them. In the following excerpt Harry and Laura decided to create more zones in their game by copying the one they already had. They covered the spikes with the zones to hide them. They played the game and found that when the shark picked up a spike and carried it to the zone, the spike blew up. They began to look for a rule that caused the explosion. They looked in one of the zones. There was no rule.

8. L: Maybe it was that one. Go to the other one.
9. H open up the box of the next zone. There are no rules again.
10. H: It hasn't got any rules.
11. H opens up another box belonging to a different zone. There are still no rules.
12. H: I don't think they have got any.
13. They check the remaining zone. Still no rule.
14. R: Where is this rule that you're thinking of that makes the spike go off?
15. There is no response.

Harry and Laura wanted to remove the offending rule but failed to find its location. They had forgotten, or were not able to reconstruct, that the rule belonged to the spike itself. This issue is a clear example of how sometimes the object orientation of the software can be problematic.

Where Does the Rule Belong? We have seen several instances of children having to work hard at where a new rule should be written. Often it is clear and straight-forward which of the objects in the game needs to be given a certain rule. For instance, the rule may only concern one object. But rules may also involve more than one object and this makes it more difficult to decide where the rule belongs.

The excerpt below took place shortly after the previous one. Harry and Laura were discussing how to insert their own rule to explode the shark. The current rule, "when I am touching the shark, I blow up", belonged to the spike.

16. H: When the shark touches the spike it blows up
17. L: When the shark touches the ...
18. H: touches the spike
19. L: Yeah, when the shark touches the spike. (L picks up the "when I am touching stone", presumably in preparation to give it to the shark.)
20. L: When I touch...the shark isn't it. So I don't particularly need that (L puts the stone back into the stones box). I just need to change that (pointing to the input to the stone in the spike's pathway).
21. L clicks on the stone and changes the input to a shark.
22. L: When I touch the shark ...
23. H: I blow up. The shark blows up.
24. L: When I touch the shark, I bomb.
25. H: Yes, but that will mean that bombs, because we're on the ...
26. R: It will mean what bombs?
27. H: The spike.
28. R: Because the spike is "I" at the moment.
29. L: The spike would get bombed. The shark wouldn't. I think we need those rules in the shark.

Harry and Laura stated their rule in many different ways. Their language alternated between the spike being the object that owned the rule and the shark being the object. This oscillation reflected a lack of clarity about where the rule should go. They started by putting the rule into the spike's rule box but in the end Laura expressed the thought that it would have been better to have given it to the shark since it was the shark that they wanted to explode.

The Same Rule for Several Objects or Several Different Rules for One Object. There are occasions when there is a choice to be made: whether to give an identical rule to each of several objects or to give a number of similar rules to one object. Matthew and Hannah had been working together building a maze game but in the excerpt below Matthew was on his own. He had several walls on the screen and each of them was a different object. He wanted the game to be stopped when the tiger (controlled by the joystick) touched any of the walls.

30. R: What other penalties could you have when you touch the wall?

31. M: I've just thought of the worst one ever. The game stops... Oh no. Which would I have to change the rule on. The tiger or ...(Long pause) It's these (pointing to the wall). I need to do it there. So it's only with one thing... I need to do it to each of the walls because if the tiger will only go into one wall it will happen. If it goes into a different one, nothing will happen.
32. R: So if the tiger's rule was "When I go into a wall", say again what you think the problem would be.
33. M: Uhm, if it was the tiger that said if I touch the wall, the game stops, it can only do one wall. So I need to do it with each of them (the walls) the tiger.
34. Matthew opens up a piece of wall, erases the bounce stone, and places the 'stop the game' stone after the 'when' stone (see Figure 5).

He then copied the rule from the one piece of wall to all the others.

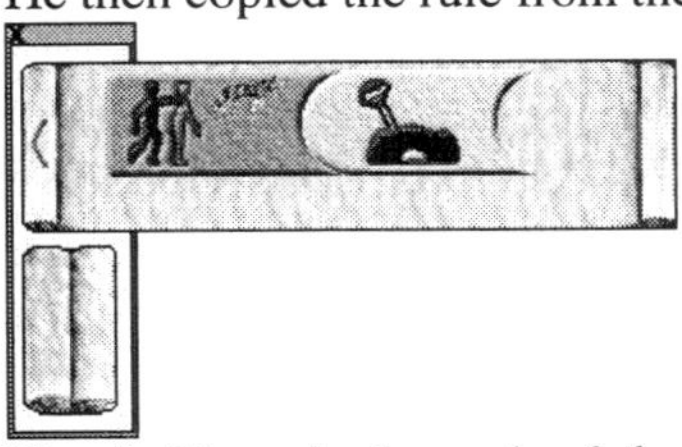

Fig. 5. The rule for each of the walls

This excerpt is quite extraordinary. Matthew appreciated that he could give the rule to either the tiger or to all of the walls. He determined that giving the rules to the walls was more efficient since he could copy the same rule for each wall; if the tiger had been given all the rules, each of them would have needed to be slightly different (one for each section of wall).

In the specification for a future version of Pathways it will be possible to identify objects with a particular shape as the input to stones like "when I am touching". A great advantage then will be that in situations like the one described above, only one rule will be required for the tiger: "When I am touching any wall stop the game".

Confounded Rules. When children are writing a number of different rules for several different objects some of them may be contradictory or have some unexpected influence on other rules in their program. At this stage in our research we cannot definitely say that inconsistencies are encouraged but we do have examples to suggest that this may be the case. Hannah and Matthew built the following rules for the tiger in their game (see Fig. 6). The two pathways in the tiger's rules are, in a sense, contradictory: pressing the button of the joystick has no visible effect on the screen since the path of the tiger is, in any case, controlled by the joystick.

Object	Iconic image of rules	Translation
 Tiger	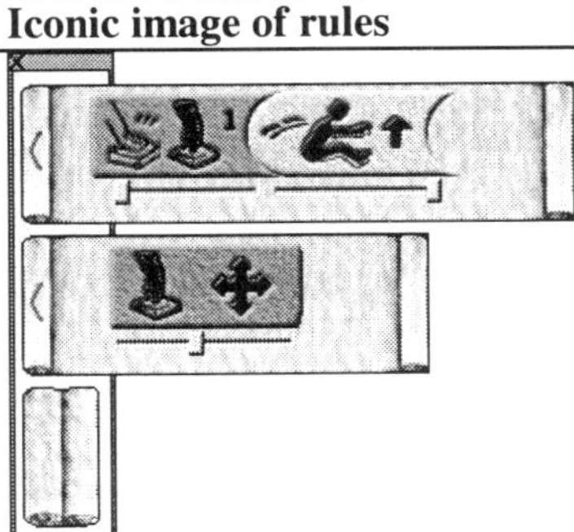	When joystick button 1 is pressed, I jump up 30. Follow the joystick in all directions

Fig. 6. Matthew and Hannah taught the tiger to jump whenever the joystick button was pressed and to follow the movement of the joystick.

6.2 The Language of Rules

We have discussed how the kernel provides structures for identifying the occurrence of certain events such as when a joystick button is pressed. The availability of events in the kernel was one of the inspirations for the particular implementation of the way rules are expressed in the Pathways language. Consider in more detail the example above. The command, `OnJoystickButton1Down [setycor ycor + 30]` in Imagine is represented in Pathways as in the first rule in Figure 6.

Although Pathways does not contain a large collection of when stones, we realised as we worked with children that the sort they needed for their video-games did not match the events provided within Imagine. In some cases it was possible to create new events but often it was necessary to simulate events in Pathways. For example, "When I am near" has no corresponding event in the kernel and so it has been simulated. In order to achieve this it was necessary to create separate modes for playing the game and for editing it. Thus the child now needs to signal that she wishes to leave the editing mode and enter the play mode of Pathways. When she switches the game on, the rules, including the simulated events, are executed. In fact, it turned out that there were benefits in separating edit and play modes since entering the play mode signals an intention to move from the Game Formal to the Game Outside. This is a step that is usually associated with a desire to test the changes made or it might simply be to have a break from the demanding aspects of focussing on the rules. Below we discuss two main issues relating to the language of rules.

Modelling Natural Language. The design of Pathways was intended to model natural language, though obviously the latter is infinitely more flexible. It was felt that by using an iconic language that corresponded to one way of phrasing rules in natural language, children might find it easier to talk to each other about the rules and the transition from Game Inside to Game Formal might be facilitated.

As with natural language there is a grammar in the language of Pathways. There are certain features that were included in the design of Pathways to highlight particular aspects of the grammar. For instance, conditionals and actions are separated into when stones and do stones, and this separation is marked by differing colours and shapes. The shape of the stones is such that a do stone can fit to the right of a when stone or to the right of another do stone; a when stones does not fit to the right of a do stone. As we suggest below, these helpful features are not sufficient for all children to appreciate the grammar of the rules in Pathways – indeed it seems that there are much deeper psychological

Object	Iconic image of rules	Translation
 Spike	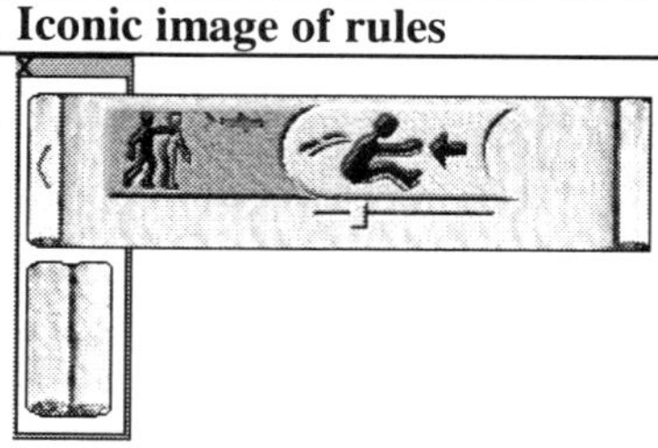	When I am touching the shark, jump left

Fig. 7. The children wanted the spike to jump whenever it touched the shark.

pro-cesses involved – but we have evidence of children using this grammar to help them formalise their ideas.

In the following excerpt, Harry and Laura were building their Dome game. The player's character was a shark. At this stage in their activities, they wanted their shark to jump whenever it touched a particular spike and they were considering a rule to be given to that spike.

35. L: When I touch the shark, I jump.
36. Harry begins to put the jump stone in first. Laura stops him.
37. L: No, don't do that first, because we ...
38. Harry reacts by putting the jump stone back and looking for the "when I am touching" stone. They place the stone in the blank pathway for the new spike. They alter the input so that their when stone refers to the shark.
39. R: What does that bit of the rule say so far?
40. L: When I touch the shark, I ...jump.
41. R: You want it to jump.
42. They find the jump stone and place it alongside the conditional when stone (see Fig. 7).

Laura prevented Harry from inserting the do stone (line 37) because she recognised that the grammar of a rule had to be of the form 'when' followed by 'do'. Harry seemed to share this way of thinking about rules and he appreciated the reason for Laura's intervention without it having been articulated.

The next excerpt shows how Matthew was aware of the grammar of a rule. Matthew and Hannah were working on the rules for each piece of wall in their maze. They wanted the wall to send a message to the score. Later they realised that this should happen when the player's character, a tiger, touched the wall but at this stage they were focussing on the message rather than the touching.

43. M finds the when stone with the envelope on it.
44. H: No, not that one. That's receive a message, silly.
45. H tells M to go back to the do stone with the envelope on it.
46. M: So we are having ... what stones are they? Aren't they do stones?
47. R: Yes.
48. M: So we are having a do stone before a when stone?

Hannah understood that the wall should send a message rather than receive it (line 44). Matthew on the other hand knew that he should insert a when stone first (lines 46 and 48). The important issue for us is that Matthew, though using the wrong stone, was appreciating the grammar of the rules in Pathways.

The kernel includes the facility for changing text to speech (as well as speech input) and this was exploited in Pathways through the provision of a mouth tool. This tool can be picked up by the child and clicked over any rule or stone so that a message is spoken by that rule or stone in the first person singular. This facility was used by the children on frequent occasions and we conjecture that, along with the above support, the mouth tool provides a means for children to model the language they use on that used by the system.

Given the iconic nature of the Pathways language and its closeness to natural language, it would be easy to assume that children find the experience of working with this language quite comfortable. In fact, our experience of working with children has raised a particularly important issue about the transition from Game Inside/Outside to Game Formal.

Moving from Narrative to Conditional. The When/Do structure that Pathways uses for rules is a particular type of conditional statement. Suppose you are observing a hammer moving across a screen. You notice that it moves on top of a gong and a sound is played. As an observer the above account might be a reasonable story of what happened. Even as a player where you control the hammer with a mouse, a similar account seems valid. You might relate the story as "I moved the hammer across the screen then it touched the gong and then there was a sound." We describe this type of account as narrative. Now suppose you are programming the computer to display this effect in Pathways. The simple linear narrative of how the hammer moved towards, then over the gong and a sound was made could easily have been written in a conventional language such as Logo. However, when the story becomes a game, a narrative approach to programming the game will not suffice. To implement a player controlled game, you need to use events. In Pathways, you need to teach the hammer a rule such as "When I am touching the gong, I make a sound." The perspective of the programmer of games is quite different from that of the observer or player of games.

Children's experience of games is usually as an observer or player. Thus one focus of our research is to observe whether such young children are able to take the programmer's perspective and manipulate When/Do type structures. What do we mean by 'manipulate' here? An essential ability for the programming of games is to transform ideas belonging to the Game Inside into the formal rules of the Game Formal that instantiate the Game Outside. However, this transformation works in both directions, so as children work, they are inspired, perhaps by the discovery of new stones, or by the inevitable constraints of the Pathways language, to renegotiate the Game Inside.

It turns out that some children adopt the programmer's perspective remarkably easily. This excerpt took place at the beginning of Matthew and Hannah's second session with Pathways. They had decided that their character, a tiger, should jump around their maze.

49. H: How do you make the thing jump?
50. M: That's what we need to think about.
51. H: When you press …
52. M: If you press …
53. H: When you press a button, it jumps or something.
54. M: When you press left button, I jump.

Matthew and Hannah discussed the rule needed to make the tiger jump in a language closely corresponding to that of Pathways, even though they had not at that point instantiated the rule using the icons. Matthew and Hannah are able children who appear to think in terms of the formal language they need to use to write rules. Any transition that might be happening between Game Inside and Game Formal is very fluent and it is not easy to find times when a conditional account is appropriate and they make any use of narrative language.

Object	Iconic image of rules	Translation
 Spike	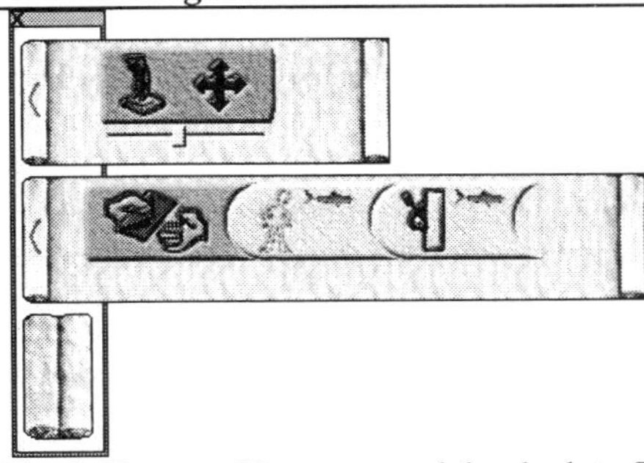	Follow the joystick in all directions When I receive the red message, hide the shark and show the shark.

Fig. 8. Harry and Laura wanted the shark to flash.

In the following excerpt, Harry and Laura demonstrate how they were able to distinguish between the use of narrative when in the role of player and conditional language when acting as programmers. At this point in the development of their Dome game, they had just programmed one of the spikes to make the shark flash when it received a message (see Fig. 8).

55. L: But it goes (she makes a funny on and off sound here) doesn't it?
56. H: It keeps on flashing.
57. R: Can someone explain to me, because I really haven't a clue, what this second rule means.
58. H: When the spike sends the message, I hide and then I …
59. L: It hides the shark then it shows the shark.
60. R: In the game what will it do?
61. L: It will hide, then it will show it, then it will hide, then it will show it (L makes the funny on off sound).

In line 58, Harry responded in conditional language to the researcher's question about the rule by taking on the programmer's perspective, whereas in line 61 Laura used the narrative to describe what happened in the game.

Other children found the adoption of the conditional language problematic even though they understood that the expectation was for them to build a game of their own. Eleanor is a child who needed much more guidance in order to begin using the conditional language. In our third session with Eleanor, we instigated a teaching experiment in which we tried to help Eleanor to appreciate the need for when stones in rules. In the fourth session she had begun to implement an idea in which a cat would carry meat to a dog. In Eleanor's Game Inside, the dog should eat the meat. We expected that after the teaching experiment, Eleanor would recognise a solution such as giving the dog the rule: "When I touch the meat, I hide the meat". We found, however, that Eleanor continued to use a narrative account.

62. R: Tell me what you want the rule to say, in your own words.
63. E: I want it to say that the cat goes over to the meat before the dog, and the cat picks up the meat and pushes it along to the dog and then dog eats it and then they just run round the street.

Later, Eleanor was still working on the same problem.

64. E: I'm looking for a stone in here that I can fit inside that one, that will tell it to touch something.
65. R: What's this rule doing, this stone? Do you want another one or do you want the same one as this now?
66. E: I want a different one to actually tell it to touch the meat and then make it go back ... It's not bringing back the meat because it's supposed to, well what I was thinking it would do was touch the meat, turn around and bring it back.

Eleanor is using narrative language without any conditional statements even though she recognises that the meat must touch the dog before there is any further action. A few minutes later, she does pick up the appropriate when stone from the Stonesbox and she solves her problem. We can not claim that Eleanor henceforth was able to move smoothly between narrative and conditional modes, as could the previous children, but there was evidence of increasing flexibility. Thus, towards the end of her fifth session, Eleanor wanted her cat to move and she suggested:

67. E: Maybe we could have When I am touching an object, I move forward 30.

Although we have only worked with a small group of children so far, it is already clear that the modes of articulation about rules vary dramatically. Matthew appeared to be firmly rooted in the type of conditional language used by programmers and he was so at ease with this mode of articulation that he rarely expressed any narrative thinking. Harry and Laura were sufficiently flexible to take on either the programmer's perspective using conditional language, or the observer/player's perspective using a narrative account. Eleanor tended to adopt the narrative account whatever the context but with sufficient guidance and support appeared able to begin using the conditional.

7 Discussion

First we summarise the main issues raised in the reporting of activity on the interfaces between Game Formal, Game Inside and Game Outside.

- Children do not always find it easy to recognise which object from their Game Inside should own a particular rule emerging in the Game Formal. Children also have occasional problems locating a rule since the program is distributed across many objects and it is often not obvious to which object certain rules should be given. It is an issue sometimes whether to give one rule to several objects or to give several different rules to one object. We have also noticed incidences of rules that (unintentionally) impact on the execution of other rules.
- The use of events is essential in the programming of games and this has led to the when/do structure of rules in Pathways. Children react very differently to the demands of translating their Game Inside into a Game Formal, which must be defined in terms of when/do rules. Some children appear to move smoothly between a narrative account of their game given from the player's perspective and a conditional account that a programmer has to utilise. Other children of this age find the conditional account much less natural than a narrative account. Such children need much more support than is given in the software alone. We are still

exploring what type of extra support might help children with the transition from narrative to conditional language.

The story of how Eleanor, Hannah, Harry, Laura and Matthew were able to create such interesting games is essentially a story of webbing, in which the combination of the external resources in Pathways and the inputs by the researchers are juxtaposed with the internal resources of the children. The stories highlight the contrasting nature of those internal resources. In Matthew's case they are powerful and they allow him to tune [REF] very quickly into the external resources available in Pathways. Harry and Laura's internal resources also enabled them to employ the formal language of Pathways in a creative way. In contrast, Eleanor's resources seemed to limit her to narrative accounts. Eleanor was eventually successful in beginning to use conditional language effectively. In webbing terms this indicates a reforging of connections. We expect, given the nature of our theoretical framework, that this new knowledge about conditionals is situated within the Pathways context. We also recognise that Eleanor's new situated knowledge remains potentially available as an internal resource for future meaning-making. However, for this potential to be released so that she might draw on these experiences in non-Pathways contexts, further support would need to be offered through the external resources of that new situation. We see the role of the researcher or teacher as critical in the webbing both within and beyond Pathways. Our focus with children such as Eleanor is to gain a better understanding of how their thinking about rules might be reforged in the light of such support.

The insights gained from children's activities sometimes cause us to reconsider the design of the playground and, to accommodate such changes, the kernel itself may need to be modified occasionally. The management of such deeply embedded iterative design has been difficult and at times painful. Nevertheless, there have been and continue to be tremendous gains in terms of the creation of tools that are cognitive in nature, in the sense that they are designed to support the activity that takes place on the interface between the Game Inside and the Game Formal.

References

1. Noss, R. and Hoyles, C.: Windows on Mathematical Meanings: Learning Cultures and Computers. Kluwer Academic Publishers, London (1996)
2. Lave, J.: Cognition in Practice. Cambridge University Press, Cambridge (1988)
3. diSessa, A.: Changing Minds: Computers, Learning, and Literacy. MIT Press, (2000)
4. Kaput, J., Hoyles, C., and Noss, R.: Developing New Notations for a Learnable Mathematics in the Computational Era. In L. English (Ed.), Handbook of International Research in Mathematics Education Lawrence Erlbaum(in press)
5. Hancock, C.. The medium and the curriculum: reflections on transparent tools and tacit mathematics. In A. diSessa, C. Hoyles, and R. Noss (Eds.), Computers for Exploratory Learning Springer-Verlag. Berlin. (1995) 221-240
6. Kahn, K.: ToonTalk - An Animated Programming Environment for Children. Journal of Visual Languages and Computing. June (1996)
7. Kalas, I. and Blaho, A.: Imagine... New Generation of Logo: Programmable Pictures. In Proceedings of WCC2000. Beijing. (2000)

Virtual Mental Space: Interacting with the Characters of Works of Literature

Boris Galitsky

iAskWeb, Inc.
261 Main Str #1B Waltham MA 02451 USA
boris@iaskweb.com

Abstract. We are building the software implementation of a virtual mental space: its inhabitants interact with the literature characters by asking questions in natural language. The players are encouraged to inquiry about intentions and desires, knowledge and beliefs, pretending and deceiving of the literature characters. Works of literature are identified based on the patterns of mental interaction between the characters. Reasoning about mental states and natural language processing with the semantic focus are the techniques required to implement the literature search component of the virtual mental world.

1 Introduction

Interactive forms of entertainment and education are usually based on such modalities as video, sound, speech, etc. Combinations of various modalities in computer entertainment and educational systems are important to achieve the overall impression of being close to the "real world" and being "smart". We believe that the latter component should involve, in particular, an interactive mental environment. The phenomena of "mental virtual world", its cognitive features and sociological consequences have been subject to rather limited exploration in comparison with the "physical virtual reality", traditionally approached from the viewpoint of simulation, visualization and perception. In spite of the recent rise in educational technology, there is still an insufficient number of computational tools, specifically designed to encourage exploration of the inner world, identity, self-reflection and communication issues.

The software implementation of a mental world by means of a game is an important component of cognitive technology, accompanying perception of physical reality. Stories are one of the many ways in which a person is presented to others and herself [1,2]. Training of mental reasoning is an important way of developing emotional and intellectual capabilities of children and adults with various mental disorders [6,9].

In recent years, there was a lot of attention to the formal background of reasoning about mental states and actions [3,4], in particular, the BDI (belief-desire-intention) multiagent architecture [8]. However, reasoning about mental states of software and human agents has just started to find a variety of applications. Planning and scheduling, multiagent management and prediction of investors' mental states are some ex-

M. Beynon, C.L. Nehaniv, and K. Dautenhahn (Eds.): CT 2001, LNAI 2117, pp. 282-288, 2001.

amples of implementation of reasoning about beliefs [6,7]. Customers of these systems are usually aware that there is some kind of implementation of reasoning about their beliefs and intentions.

In this study, we focus on the design and implementation of an interactive entertainment environment based on the mental states of the characters of works of literature (WOL). The game players are involved in the literature world by means of asking question about the mental states of the abstract characters. The system is capable of understanding these natural language (NL) questions and providing the introduction to the adequate WOL scenario. In addition, the system is capable of accepting the scenarios from the players in NL, extracting the expressions about mental attributes. In contrast to behavior simulation and storytelling systems [1,2,7], we focus on just a single interaction modality: question-answering and knowledge sharing in NL, excluding speech interaction. Though our system is implemented as an information retrieval one, the users perceive it as a game, being impressed with the intellectual challenge, associated with WOL querying, and emergent literature plots.

The Internet entertainment tool is developed, which allows a player to "intelligently" access books without explicit mentioning of its title or its author's name. Our system introduces to customers a new way to exchange ideas about the literature plots, using the belief states of characters. NL querying of the literature library is implemented as well as its extension: a player can input a brief description and essential mental states (conflict) of a WOL such that the other players can access this WOL via the search (www.dimacs.rutgers.edu/~galitsky/AU).

2 Mental States of Literature Characters

What is the role of mental states of literature characters in the classification and schematization of the works of literature? We have built the library (database) of WOLs, which includes the manually extracted mental states of their characters. We collected as many WOLs as it was necessary to represent the totality of mental states, encoded by logical formulas of the certain complexity [3]. Note that a mental state is represented by a set of mental formulas, consisting from the mental predicates *know, want, inform, pretend,* etc. The mental formulas are expressed in a second-order predicate language, extended by the metapredicates of mental states and actions [6]. Analysis of our database allows us to make the following conclusions:

1. As a rule, the main plot of a WOL deals with the development of human emotions, expressible via the basic (want-know-believe) and derived (pretend, deceive, etc.) mental predicates. A single mental state expresses the very essence of a particular WOL for the small forms (a verse, a story, a sketch, etc.). When one considers a novel, a poem, a drama, etc., which has a more complex nature, then a set of individual plots can be revealed. Each of these plots is depicting its own structure of mental states that is not necessarily unique. Taken all together, they have the highly complex forms, appropriate to identify the WOL.
2. Extraction of the mental states from a WOL allows us to clarify psychological, social and philosophical problems, encoded by this work. The mental compo-

nents, in contrast to the "physical" ones are frequently expressed implicitly and contain some forms of ambiguity.

3. The same mental formula may be a part of different WOLs, written by the distinguishing authors. Therefore, it is impossible to identify a certain WOL or author when we take into consideration just a single mental formula. However, the frequency of repetition of certain mental formulas shows us the importance of the problem raised by a WOL.
4. The sets of mental formulas are sufficient to identify a WOL. The possibility to recognize a certain author according to a collection of mental states of his or her WOLs is beyond our current considerations.

3 Design of the Literature Search System

The following problems have to be resolved to implement the literature search based on the mental state of characters:

1) NL understanding of a query or statement (see [5] for details).
2) Domain representation in the form of semantic headers, where mental formulas are assigned to the textual representation (abstract) of WOLs.
3) Synthesis of all well-written mental formulas in the given vocabulary.
4) Matching of the translated NL query against the semantic headers. We use the approximate match in case of failure of the direct match.
5) Synthesis of canonical NL sentence based on mental formula.
6) Keyword search initiation in case of failure of mental formula search.

Figure 1 presents the interaction between the respective components 1)-6) of the WOL search system. The flexible system architecture allows functioning in two modes: WOL search and WOL database extension. Similar NL processing (component 1) and mental formula analysis (components 3-5) features are employed in both modes, but the component interaction is different.

Component 3) is required, because the traditional axioms for knowledge and belief (see, for example, [3,4]) are insufficient to handle the totality of all mental formulas, representing the real-life situations. We developed the algorithm to extract the realizable mental formula from the totality of all well-written mental formulas, represented via metapredicates. In addition, introduction of the classes of equality of mental formulas are required for the approximate match of mental formulas 4), which is also inconsistent with traditional axiomatic of reasoning about knowledge and belief. NL synthesis of mental expression is necessary for the verification of the system's deduction. A player needs this component to verify that he/she was understood by the system correctly before starting to evaluate the answer. NL synthesis in such strictly limited domain as mental expression is straightforward and does not require special considerations. We mention that the components 1) and 2) are developed as a commercial product for advising in financial domains [5]. However, semantic rules for the analysis of mental formulas require specific machinery for complex embedded expressions and metapredicate substitutions.

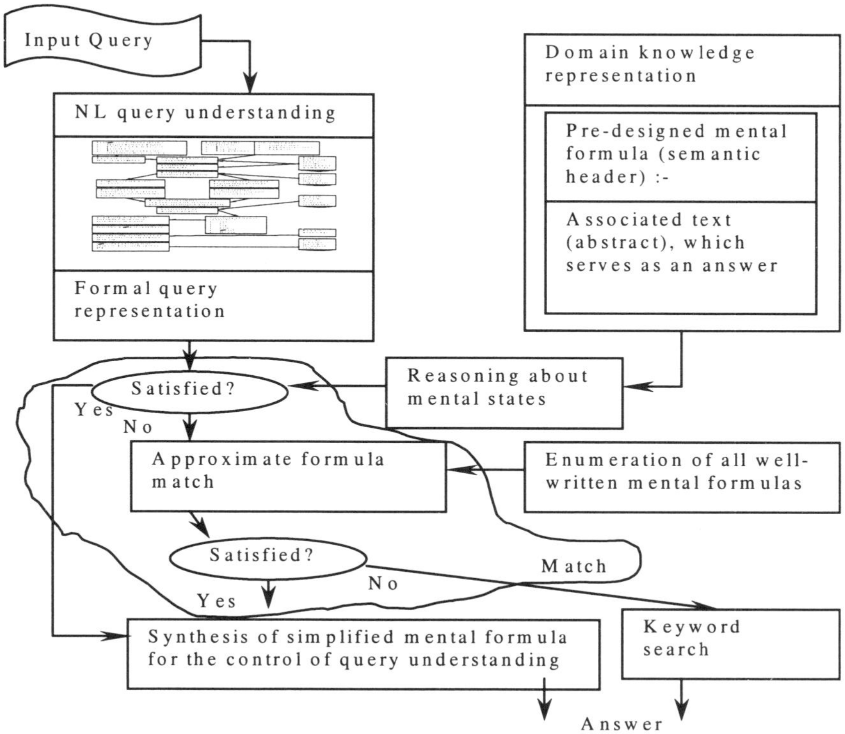

Fig. 1. The scheme of the literature search system. In the search mode, NL query, containing the mental states and action of WOL characters, is converted into mental formula. If there is no semantic header in the domain knowledge component, which satisfies the resultant translation formula for a query, the approximate match is initiated. Using the enumeration of all well-formed mental formulas, the system builds the approximation and deductive link between the translated query and existing semantic header. If the approximate match fails, the system generates the list of keywords to search through the WOL abstracts. The same components are employed when a customer inputs a new WOL. It is possible to explicitly input the corresponding semantic header via NL or initiate the system to generate it independently given the WOL abstract.

The special question-answering technique for the weakly structured domains has been developed to link the formal representation of a question with the formal expression of the essential idea of an answer (Tables 1, 2). These expressions, enumerating the key mental states and actions of the WOL characters, are called the *semantic headers* of answers [5]. The mode of automatic annotation, where a customer introduces an abstract of a plot and the system prepares it as an answer for the other customers, takes advantage of the flexibility properties of the semantic header technique.

Based on the same technique, the version for autistic children has been developed, where question – answering is focused on the participants of a scene (Fig. 2).

4 The Literature Search Demo Page

Table 1. Introductory page to the virtual mental space system with the sample questions and statements. Note that the subjects of search are identified using the mental entities only.

Please demonstrate your knowledge of classical literature, from medieval to modern, asking questions about the mental states of the characters and compare the system results with your own imagination. How do you typically look for a book (on the Internet)? Using a search engine, you type the title of this book or the author. Sometimes, knowledge on a general topic will help. Have you ever dreamed about searching for a book, which matches your current spiritual state, mentioning the mental interactions between the persons involved in a work of literature?
When does a person want another person not to pretend that he does not know anything?
What happens when a person pretends that he/she wants to know something?
Why does a person pretend that he does not understand that the other person does not want something?
When would a person believe that another person does not pretend that somebody wants something?
In what circumstances would a person believe that another person pretends about some intention?
Does she pretend that she believes that he does not pretend that he is not a murderer of her husband?
A lover believes that the husband does not want the lover to be in love with his wife.

Table 2. Formal representation for the first query in Table 1 above and the resultant answer

*want(person, not pretend(other_person,_, not know(other_person, Smth))):- outputAsAnAnswer(**The Heaven and the Hell** by P. Merimee*
An inquisitor wants the girl to tell him about her lover's heretical ideas. The girl gets scared and tries to shield her lover from the Inquisition. She assures the inquisitor that he did not participate in composing the libel on the Pope.
The inquisitor gets surprised at hearing about the libel as he never mentioned it during the conversation. The girl gets even more frightened and does her best to assure the inquisitor that it was he who introduced the subject and mentioned it. She says that she has never heard of anything of that sort. The inquisitor interrupts her and tells her to stop pretending. However, the girl keeps insisting on him being innocent and the inquisitor lets her go.
Later the inquisitor meets the girl once again. This time he produces her lover's love letter to another woman. Poor girl turns pale with offence and indignation and betrays her lover. She says that he is the person who composed the libel on the Pope.

5 Conclusions

Interaction with the virtual space of the literature characters is demonstrated to be a novel entertainment area, appealing to adults as well as to children, interacting with the characters of the scenes in NL. Since the players are suggested to both ask questions and share the literature knowledge, the system encourages the cooperation among the members of the players' community. In the demo we have built, the system only recognizes the questions and statements, involving the terms for mental states and actions. This way we encourage the players to stay within a "pure" mental space and to increase the complexity of queries and statements we expect the system to handle properly. Observing the game players, we discovered that they frequently try to obtain the exhaustive list of WOLs, memorize the querying results and enjoy sharing WOL plots with the others.

Does Mike see that the dog is eating the sausages?
Does Nick know what is happening with Mike and the dog?
Which way does Nick express his emotions?
Does Fred know whether Peter knows what is happening with sausages?
What would Fred do if he wants to let Peter know what is happening?

Fig. 2. Interacting with the virtual mental space of the scene characters. NL system answers the questions about mental states of Mike, Peter, Fred, Nick and the dog (on the left). Examples of questions the children may ask the system about, the characters of the scene are on the right. Involving more and more complex mental states helps the playing children to develop creativity and imagination of thinking, as well as the communication skills of understanding other's mental states.

References

1. Umaschi, M. and Cassell, J. Storytelling systems: constructing the Innerface of the Interface, in *Second Intl.Cognitive Technology Conf. (CT'97),* IEEE Computer Science Press (1997).
2. Hayes-Roth, B., van Gent, R. Story-Making with Improvisational Puppets, in *International Conference on Autonomous Agents'97* (1997).
3. Fagin, R., Halpern, J. Y., Moses, Y., Vardi, M. Y. *Reasoning about Knowledge.* MIT Press, Cambridge, MA, London, England (1995).
4. Konolige, K. *A Deduction Model of Belief.* Morgan Kaufmann Publ. (1986).
5. Galitsky, B. Technique of semantic headers: a manual for knowledge engineers. *DIMACS Tech. Report #2000-29* Rutgers University (2000).

6. Galitsky, B. The formal scenario and metalanguage support means to reason with it. *DI-MACS Tech. Report #98-27*, Rutgers University (1998).
7. Galitsky, B. Automatic generation of the multiagent story for the Internet advertisement. *AAAI Spring Symposium Series Workshop on the Intelligent Agents*, Stanford CA (1999).
8. Cohen P.R., Levesque H.J., Intention is choice with commitment *Artificial Intelligence* 42:213-261 (1990).
9. Hadwin, J., Baron-Cohen, S., Howlin, P., Hill, K Does teaching theory of mind have an effect on the ability to develop conversation in children with autism? *J. Autism Development Disorder* 27(5):519-37.

The Plausibility Problem: An Initial Analysis

Benedict du Boulay and Rosemary Luckin

School of Cognitive and Computing Sciences,
University of Sussex,
Brighton BN1 9QH.
{bend,rosel}@cogs.susx.ac.uk

Abstract. Many interactive systems in everyday use carry out roles that are also performed – or have previously been performed – by human beings. Our expectations of how such systems will and, more importantly, should, behave is tempered both by our experience of how humans normally perform in those roles and by our experience and beliefs about what it is possible and reasonable for machines to do. So, an important factor underpinning the acceptability of such systems is the plausibility with which the role they are performing is viewed by their users.
We identify three kinds of potential plausibility issue, depending on whether (i) the system is seen by its users to be a machine acting in its own right, or (ii) the machine is seen to be a proxy, either acting on behalf of a human or providing a channel of communication to a human, or (iii) the status of the machine is unclear between the first two cases.

1 Introduction

Many interactive systems in everyday use carry out roles that are also performed – or have previously been performed – by human beings. Good examples of such systems can be found in computer-supported training. Here users perform some task and their performance is commented on by the system. However, as information and communication technologies are used in the lives of a greater number and variety of people, so the number of human-like roles these systems perform or mediate increases. The internet has brought new forms of interaction into people's homes, work and leisure environments. For example, One2One's 'Ask Yasmin' interactive customer service assistant can help people find out about mobile phone service options; the search engine 'Ask Jeeves' answers users questions in order to help them search for information on the world wide web and Amazon.com offers its users suggestions about the types of book they might like to read. Our expectations of how such systems will and, more importantly, should, behave is tempered both by our experience of how humans normally perform in those roles and by our experience and beliefs about what it is possible and reasonable for machines to do. So, an important factor underpinning the acceptability of such systems is the plausibility with which the role they are performing is viewed by their users. With respect to training systems, Lepper et al. [13] define the issue as follows:

M. Beynon, C.L. Nehaniv, and K. Dautenhahn (Eds.): CT 2001, LNAI 2117, pp. 289–300, 2001.

> "Even if the computer could accurately diagnose the student's affective state and even if the computer could respond to that state (in combination with its diagnosis of the learner's cognitive state) exactly as a human tutor would, there remains one final potential difficulty: the plausibility, or perhaps the acceptability, problem. The issue here is whether the same actions and the same statements that human tutors use will have the same effect if delivered instead by a computer, even a computer with a virtually human voice." [13] (page 102)

The notion of plausibility is closely related to the notion of credibility [6]. Credibility is bound up with such concepts as believability, trustworthiness and expertise. Plausibility is more subtle and is concerned with effectiveness and acceptability within a role and relies on our sense of the differential social roles that humans and computers may be expected to play. So plausibility is one way of talking about a whole species of interactive system design issues where the designer is attempting to mobilise inter-subjectivity as a persuasive, seductive or supportive interactional device.

The design challenge raised by the plausibility problem is, first, to identify the situations in which the plausibility of a system becomes an "issue" for its users and, second, to establish whether and when it actually becomes a "problem" [5]. The debate surrounding Eliza and its more specific variants such as Parry, indicate that there are circumstances where users can suspend belief (as if watching a film or a play) and not be concerned over the status of their conversational partner as a machine. Eliza also reminds us that there may be circumstances where direct human-human interaction might be unwelcome and that any such inhibition may be usefully reduced by a machine acting as a conversational partner. Re-exploring people's reactions to Eliza-like systems is extremely timely with the advent of virtual representatives being used to host web sites and offer advice (see for example `http://www.one2one.co.uk` and `http://www.axcess.com`). In some circumstances, implausibility might be counter-productive, causing users to distrust and then fail to make the best use of some system. On other occasions it may not be a problem at all: the provision of an ironic interchange that serves to amuse, perhaps even to motivate, for example.

Our research question is not whether individuals respond to systems in ways that are similar to their response to other humans. Nor is it simply whether giving such systems surface human-like characteristics (such as voice output or displaying an animated face) makes a difference. The central notion is, in training systems or in other areas such as information provision, healthcare, e-commerce or leisure systems, whether copying the tactics normally employed by humans playing roles in those areas (trainer, salesman, advice-giver for example) works when the role player is a machine. In some ways detecting implausibility is easier than detecting plausibility. For instance, one indication that a system is behaving implausibly might be when it evokes irritation in the person using it (e.g. through the jaunty friendliness that some systems adopt). Another reaction could be reduced engagement with the task in hand. We are certainly interested in affective responses (such as irritation) that might accompany implausible machine exchanges. However, we are especially interested in what further effects on task performance might follow from this. Such effects could include failure to answer subsequent questions in the exchange, provision of partial or incorrect information, adopting a frivolous mode of response, becoming distracted, or simply abandoning the session that is underway. Reactions will vary according to the circumstances in which the system is being used. When using an e-commerce site users may well simply 'vote with their feet' and abandon the interaction, whereas users of training systems may well not have this latitude and so persist.

Our investigation of the factors which contribute to systems being regarded as plausible and those which undermine takes into account the system purpose and its characteristics. We may suppose that a dumb system playing a limited role, whose role is expected to be limited and which, in fact, acts in a limited way will be perfectly plausible, whilst a similar system that moves outside reasonable parameters for its role (e.g. an advice system pretending to be sympathetic) may, for that very reason, appear implausible. In developing teaching and training systems, we have encountered various manifestations of the plausibility problem. For example, systems withholding help deliberately [4], or systems apparently forgetting what has been taught by the human learner in a learning companion system [19], or issues concerned with users' lack of belief about the capability of the system to deliver what they need, e.g. help of appropriate quality [16].

This paper examines the nature of the Plausibility Problem as a particular example of situations in which an attempt to simulate inter-subjective understanding is made by or through an interactive system (and issues of plausibility thereby arise). The roles explored are taken from educational contexts and include helping and advising as well as evaluation.

We identify three kinds of potential plausibility issue, depending on whether (i) the system is seen by its users to be a machine acting in its own right, or (ii) the machine is seen to be a proxy, either acting on behalf of a human or providing a channel of communication to a human, or (iii) the status of the machine is unclear between the first two cases.

In the first case plausibility is bound up with issues as to whether a machine, as a machine, is acting outside the bounds of what the user, in that context, thinks is reasonable. The second case is much less of an issue for us in that the system is seen as a proxy for a human and therefore any plausibility issue will tend to be associated with the person for whom (or to whom) the machine is a proxy. Of course, there may be issues of the effectiveness of its role as a proxy or of its facilitating communication, but these are not really plausibility issues.

The third case does raise plausibility issues, especially where the user cannot judge whether the machine is acting in its own right or not. In that case, if the user thinks that it is so acting in its own right, when in fact it is just a proxy, the user may regard some behaviour of the system as implausible which might have been regarded as plausible had the behaviour come from a human (or her proxy). Likewise, if the system is thought of as a proxy for a human but is in fact acting in its own right, an implausibility judgement may be made about the way that the supposed human is acting.

In the days where computers were largely stand-alone, their status as self-contained *vs* being proxy or communication channels was perhaps more clearcut. With the ubiquity of networking, the issue of whether (or to what extent) a system is a proxy is much more complex. This blurring is accentuated by systems which attempt to simulate human face-to-face interactions through the use of animated pedagogical agents, see e.g. [9]. With the rapid improvement in graphical and audio technology these systems can now bring a wider range of more human-like interaction tactics to bear such as a change of facial expression, or a change of verbal emphasis.

This paper is divided into two main sections. The next section provides examples of implausibility judgements where the system is regarded as a machine acting in its own right, case (i) above. The second section looks briefly at examples where plausibility judgements are bound up with uncertainty about the status of the machine, case (iii) above.

2 It's Just a Machine – and Machines Should Not Do That

2.1 Human Teachers Can Say That, But Not Machine Teachers

Del Soldato [3,4] implemented various of the motivational tactics, e.g. derived by [10,11,12,13] in a prototype tutor to teach rudimentary debugging of Prolog programs. Included in her system was a set of (motivational) rules intended to maintain the students' sense of confidence and control. These rules might suggest easy problems to a student who needed a boost in confidence, or might be rather 'firmer' with students who had not exhibited much effort and also seemed self-confident.

The system (MORE) was evaluated by comparing a version with the motivational rules switched on with one where they were disabled. The version using motivational rules was generally liked by students but two negative reactions from students are noteworthy. One of the rules in the system was designed to prevent the student prematurely abandoning a problem and moving on to the next one, if the system believed that the student was not exhibiting enough "effort", as measured by the number of actions the student had taken in the partial solution.

> "One subject was showing signs of boredom from the start of the interaction. ... After a little effort trying to solve a problem, the subject gave up and the tutor encouraged him to continue and offered help. The subject kept working, grumbling that the tutor was not letting him leave. When comparing the two versions of the tutor he recalled precisely this event, complaining that he had not been allowed to quit the interaction." [3](page 77)

Further rules were concerned with deciding how specific a help message should be delivered in response to a help request – not dissimilar to the rules in Sherlock, see e.g. [14], or indeed to the Contingent Teaching strategy [21]. However in some circumstances the help system refused to offer any help at all in response to a request from the student, in the belief that such students needed to build up their sense of control and that they were becoming too dependent on the system.

> "The subjects who were refused a requested hint, on the contrary, reacted strongly against the tutor's decision to skip helping (ironically exclaiming "Thank you" was a common reaction). Two subjects tried the giving-up option immediately after having had their help requests not satisfied. One case resulted in the desired help delivery (the confidence model value was low), but the other subject, who happened to be very confident and skilled, was offered another problem to solve, and later commented that he was actually seeking help."
>
> "One of the subjects annoyed by having his help request rejected by the tutor commented: "I want to feel I am in control of the machine, and if I ask for help I want the machine to give me help". When asked whether human teachers can skip help, the answer was: "But a human teacher knows when to skip help. I interact with the human teacher but I want to be in control of the machine". It is interesting to note that the subject used to work as a system manager." [3](pages 76–77)

In both these cases the student was surprised that the system behaved in the way that it did – not we believe because the system's response was thought to be educationally unwarranted, but because it was "merely" a machine and it was not for it, *as a machine*, to frustrate the human learner's wishes.

2.2 Human Students Would Do That, But Not Machine Students

There is increasing interest in the development of learner companion systems of various kinds, see e.g. [1]. Here the idea is that the human learner has access to a (more or less) experienced, computer-based fellow learner who can either provide help, act as a learning role model, or through its mistakes act as a reflective device for the human learner. For instance, [19] describes a system where the human learner teaches a weaker companion system boolean algebra in order to better understand the topic herself. The learning companion (LC) was not an 'embodied' agent, but essentially an unseen entity communicated with via a simple text and push-button interface.

Some care was taken to make the weaker companion act in a realistic way. In particular, it did not always "understand" what the human student tried to teach it, it did not always follow the advice offered by the human student, and it sometimes forgot what it had been taught. Ramirez Uresti notes that some students were "very annoyed to observe that the LC did not 'learn' all the concepts that had been so carefully taught to it". Moreover, this judgement about plausibility had knock-on effects for later in the interaction:

> "However, after some teaching incidents, students started to diminish the quality of their teaching until just the rule needed for the current step was taught to the LC. ... Once students noticed that the LC was not learning quickly they started to teach only one rule instead of a complete heuristic. This combination of teaching all the strategy and then having to teach it again and again may have been detrimental to the perception of the week LC and of the teaching process. It may also explain why the weak LC was described in the post-test as not very exciting and annoying." [19] (page 110–111)

3 Human Teachers Can Do That, But Not Machines

Learners' expectations are an important factor of the plausibility problem. Increasingly learners are exposed to computers in their learning and in other aspects of their lives. They absorb the cultural computation conventions and facilities for giving help. These build up expectations of the degree of focussed assistance that they might reasonably expect.

In the next example, see Sect. 3.1 below, the plausibility problem may be responsible for results which confounded expectations. There are a number of differences between this system and those of del Soldato and Ramirez Uresti, described above. It was aimed at school children, specifically designed to be similar to other educational systems they had used and was evaluated in the children's everyday class. It also explored a topic – simple ecology – that the children were learning at school and, in the versions that decided how helpful to be, was designed to ensure that the child succeeded as far as possible, even if this meant that the system did most of the work.

3.1 A System That 'Wants' to Help

Three versions of a tutorial assistant which aimed to help learners aged 10-11 years explore food webs and chains were implemented within a simulated microworld called the Ecolab [15]. The system was developed to explore the way in which Vygotsky's Zone of Proximal Development might be used to inform software design. The child can add different organisms to her simulated Ecolab world and the complexity of the feeding relationships and the abstractness of the terminology presented to the learner can be varied. The simulated Ecolab world can be viewed differently, for example in the style of a food web diagram, as a bar chart of each organism's energy level or as a picture of the organisms in their simulated habitat. The activities the learner was required to complete could be "differentiated" (i.e. made easier) if necessary and different levels (i.e. qualities) of help were available.

One version of the system – VIS – maintained a sophisticated learner model and took control of almost all decisions for the learner. It selected the nature and content of the activity, the level of complexity, level of terminology abstraction, differentiation of the activity and the level of help. The only option left within the learner's control was the choice of which view to use to look at her Ecolab. A second version of the assistant – WIS – offered learners suggestions about activities and differentiation levels. They were offered help, the level of which was decided on a contingently calculated basis [21]. They could choose to reject the help offered or select the "more help" option. The third system variation was called NIS. It offered 2 levels of help to learners as they tried to complete a particular task. The first level consisted of feedback and an offer of further help. The second level, which was made available if the child accepted this offer, involved the assisting computer completing the task in which the child was currently embroiled. Of the three systems NIS offered the smallest number of different levels of help and allowed the greatest freedom of choice to the child. She could select what she wanted to learn about, what sort of activity she wanted to try, how difficult she wanted it to be and then accept help if she wanted it. The choices were completely up to the individual child, with not even a suggestion of what might be tried being offered by the system.

Three groups of 10 children (matched for ability) worked with the three systems. Outcomes were evaluated both through pre/post-test scores on a test of understanding of various aspects of food webs and chains, and via an analysis of what activities the children engaged in and how much help they sought and received. Pre/post-test comparisons showed that VIS produced greater learning gains than WIS and NIS, see [15,18] for details. Our focus here is not on the learning gains but on the help seeking behaviour of the students.

3.2 Children Who Don't Ask for Help

It is clear from the records logged by the systems of each child's interactions that none of the NIS users accepted the option of seeking more help when offered feedback. There is a clear and typical pattern within the interactions of NIS users: actions are attempted, feedback is given with the offer of help, help is not accepted. The action is re-attempted and once completed successfully it is repeated, interspersed with view changes and further organism additions at differing rates of frequency. Only one of the NIS users asked for a differentiated activity and only two attempted to interact at anything other than the simplest level of complexity or terminology abstraction. The child who tried the differentiated activities chose the highest level of differentiation and once the activities were done he returned to the typical NIS pattern. The help seeking or lack of it is particularly marked in the two children who opted to try the most advanced level of interaction. Both made errors in their initial attempts at completing the food web building action selected, but neither opted to take more help when offered. Few activities were attempted and those that were chosen were accessed with the lowest level of differentiation. The same food web building activity was repeated in both sessions of computer use and in both sessions errors were made. The presence of these errors and the apparent desire to tackle more complex concepts would suggest that the children were willing to move beyond what they already understood. However, the lack of collaborative support restricted their opportunities for success and their progress was limited. What could have been a challenging interaction became a repetitive experience of limited scope.

Unlike the NIS users, all the WIS users accepted help above the basic level and the majority used help of the highest level and then remained at this level. A typical WIS approach would be to try an action, take as much help as needed to succeed with this action and then repeat it before trying another different action. Activities were requested with differentiation. In the majority of cases this differentiation was at the highest level. Without question the WIS users were more willing to attempt actions with which they were going to need help. There were members of this group who progressed through the curriculum both in terms of complexity and terminology abstraction. This is a direct contrast to the NIS user group.

3.3 Why Do Some Children Seek Help and Others Not?

The clear difference between one group's willingness to use help over and above simple feedback (WIS) and the other group's complete lack of help seeking is interesting. The help instances for the NIS users were either simple feedback or a demonstration of the particular action being attempted: equivalent to the highest level of help in WIS or VIS. All but one of the NIS users made mistakes and were given feedback, but none of them accepted the offer of further help. It is difficult to explain this startling lack of help seeking behaviour and any attempts are clearly speculative.

The only difference between the WIS and NIS system with regard to differentiation or the presentation of help is in the way that WIS suggests that the user try a particular level of differentiation for an activity or ask for help. This policy of offering suggestions was not universally successful. WIS users received suggestions about which activities they should try. These were however accepted less often than the suggestions about the differentiation of an activity. If a suggestion was enough to allow the child to accept an easier activity then it seems reasonable to consider the possibility that without the suggestions, the NIS users viewed choosing a more difficult activity as being somehow better and therefore what they should be attempting.

As part of the design of the experiment, note was taken of the computer programs the children had experienced previously. One tentative explanation of the different behaviours is that children did not believe that either asking for more help or for an easier activity would be successful. The WIS users received suggestions and once the higher levels of help were experienced they were taken up and used prolifically. In this sense the WIS system demonstrated its plausibility as a useful source of assistance in a way that the children never gave the NIS system a chance to show.

A further factor which is consistent with this help seeking behaviour is found in the observation that none of the children accessed the system help menu or system help buttons. These were available to explain the purpose of the various interface buttons and the way that action command dialogues could be completed. The children had all used a demo of the system, which allowed them to determine the nature of the interface and none reported problems at the post-test interview. However, when observing the children using the system it was clear that there were occasions when they were unsure about a button or a box and yet they did not use the help button provided. This may well be an interface issue which needs attention in any further implementations of VIS. However, it may also be part of the same plausibility problem.

3.4 Turning to a Wizard for Help

In order to further explore children's perceptions about the type of help that computing technology can afford we have subsequently conducted a series of small empirical investigations. Working with children can be difficult: they are less willing and able to express their thoughts and ideas. We therefore used an adaptation of the 'Wizard of Oz' technique: previously used to simulate human computer interfaces with the human 'wizard's' existence being unknown to the user [2]. However, in this case the user and the wizard were working on the same apparatus: a paper-based computer and were able to view each other's interactions continuously. Pairs of children used the paper-based version of the Ecolab software, the one playing the role of the computer; the other the role of the learner. In this way we hoped to elicit information about children's perceptions of the types of help that computers could and should provide for them when using the software to learn about ecology [8,7].

Early results indicate that children can accept the possibility that a computer might be more helpful on some occasions than on others and that this lack of consistency in the 'behaviour' of the technology is not viewed as unacceptable or implausible. Sometimes the children tried to help the 'user' as best they could, on other occasions they chose to make it difficult. For example, one child, when playing the role of the computer preferred to make his learner manage with little help; he explained his selection: "It is the hardest ... and computers are really mean". However, we have yet to see whether or not the replacement of the child 'wizard' with a software implementation will yield the same results. This will raise questions about the 'location' of the implausibility: does it arise from the interface or the wider context in which the interactions occur?

4 Is It a Machine or a Person?

The nature of a network of computers further clouds the plausibility landscape and blurs the boundaries between when users are interacting with technology and when they may be interacting with other human beings. In contrast to the current HCI impetus for increasing usability through hiding how applications work, there is increasing evidence to suggest that people have a poor understanding of how networked technologies, and in particular the Internet actually work [20]. The Internet is still a relatively new phenomenon that allows data exchange between networks of computers connected via national and international telecommunications systems to other connected networks that wish to communicate. Thanks to agreed transfer protocols and address standardisation these networks appear seamless to users who can read and download files from remote machines, publish to those using remote machines, communicate via multi-media or use their personal computers as terminals. Whilst this seamlessness has clear benefits, it creates the illusion of a faultless network of connections which is far from the truth. The Internet is unstable, unpredictable and inherently unreliable.

In order to try and ascertain the implications of networked technologies and people's conceptions and misconceptions, we conducted an empirical study with 9-10 year old children. The use of children in this study offered us the opportunity to tackle early understandings and hopefully even pin down when misconceptions and potential plausibility issues might occur. During a series of studies with a class of 9-11 year old children over a two year period we talked to children about their expectations of what the internet would and could offer [17]. The children in this study produced simple representations of the Internet that often focused upon the sort of computer that they were familiar with. There were however many instances in which they included references to the sort of activities that the Internet enables. The most popular facilities children envisaged to be available as a result of the Internet were communication, research or information retrieval using the WWW and – to a lesser, though increasing, extent – the publication of work. Despite the common occurrence of interpersonal communication however, humans were not frequently seen as integral to children's representations of the Internet. However, some children did talk about the internet as an animate object that "knows" things. And yet, when asked about their feelings about publishing their own work on the internet the concerns they raised were only ever couched in terms of their worries about what other people would think about them and their work. Would the spelling and grammar be good enough, for example?

5 Conclusions

We have started to map out some examples of the plausibility issue, and tried to show why it is more than simply about designing for a smooth and agreeable interaction. Our examples are taken from education but future work will examine other areas such as advice-giving and e-commerce where similar issues are likely to arise.

This early work does not as yet allow us to draw firm conclusions about when and where the plausibility problem occurs with any precision. It does however, indicate the complexity of the issue and suggest that people's perceptions about what networked technologies can and should do are not consistent, nor are they identical to those that prevail for stand-alone systems. The plausibility problem is a changing and moving target that is not going to disappear as the sophistication and ubiquity of the technology increases.

References

1. T.-W. Chan. Learning companion systems, social learning systems, and the global learning club. *Journal of Artificial Intelligence in Education*, 7(2):125–159, 1996.
2. N. Dahlback, A. Jonsson, and L. Ahrenberg. Wizard of Oz studies – why and how. In M. T. Maybury and W. Wahlster, editors, *Readings in Intelligent User Interfaces*. Morgan Kaufmann, San Francisco, 1998.
3. T. del Soldato. Motivation in tutoring systems. Technical Report CSRP 303, School of Cognitive and Computing Sciences, University of Sussex, 1994.
4. T. del Soldato and B. du Boulay. Implementation of motivational tactics in tutoring systems. *Journal of Artificial Intelligence in Education*, 6(4):337–378, 1996.
5. B. du Boulay, R. Luckin, and T. del Soldato. The plausibility problem: Human teaching tactics in the 'hands' of a machine. In S. P. Lajoie and M. Vivet, editors, *Artificial Intelligence in Education: Proceedings of the International Conference of the AI-ED Society on Artificial Intelligence and Education, Le Mans France*, pages 225–232. IOS Press, 1999.
6. B. Fogg and H. Tseng. The elements of computer credibility. In *Proceedings of CHI'99*, pages 80–87, Pittsburgh, 1999.
7. L. Hammerton and R. Luckin. Children and the internet: a study of 9 - 11 year olds. *Paper to be presented in a workshop at AIED 2001*, San Antonio, Texas, 2001.
8. L. Hammerton and R. Luckin. How to help? investigating children's opinions on help. *Poster to be presented at AIED 2001*, San Antonio, Texas, 2001.
9. W. L. Johnson, J. W. Rickel, and J. C. Lester. Animated pedagogical agents: Face-to-face interaction in interactive learning environments. *International Journal of Artificial Intelligence in Education*, 11(1):47–78, 2000.
10. J. M. Keller. Motivational design of instruction. In C. M. Reigeluth, editor, *Instructional-design Theories and Models: An Overview of their Current Status*. Lawrence Erlbaum, 1983.
11. M. R. Lepper. Motivational considerations in the study of instruction. *Cognition and Instruction*, 5(4):289–309, 1988.
12. M. R. Lepper and R. Chabay. Socializing the intelligent tutor: Bringing empathy to computer tutors. In H. Mandl and A. Lesgold, editors, *Learning Issues for Intelligent Tutoring Systems*, pages 242–257. Springer-Verlag, New York, 1988.
13. M. R. Lepper, M. Woolverton, D. L. Mumme, and J.-L. Gurtner. Motivational techniques of expert human tutors: Lessons for the design of computer-based tutors. In S. P. Lajoie and S. J. Derry, editors, *Computers as Cognitive Tools*, pages 75–105. Lawrence Erlbaum, Hillsdale, New Jersey, 1993.
14. A. Lesgold, S. Lajoie, M. Bunzo, and G. Eggan. Sherlock: A coached practice environment for an electronics troubleshooting job. In J. H. Larkin and R. W. Chabay, editors, *Computer-Assisted Instruction and Intelligent Tutoring Systems*, pages 289–317. Lawrence Erlbaum Associates, Hillsdale, New Jersey, 1992.
15. R. Luckin. 'ECOLAB': Explorations in the Zone of Proximal Development. Technical Report CSRP 386, School of Cognitive and Computing Sciences, University of Sussex, 1998.
16. R. Luckin and B. du Boulay. Ecolab: The development and evaluation of a vygotskian design framework. *International Journal of Artificial Intelligence in Education*, 10(2):198–220, 1999.
17. R. Luckin and J. Rimmer. Children and the internet: a study of 9–11 year olds perceptions of networked technologies. Technical report, School of Cognitive and Computing Sciences, University of Sussex. In preparation.

18. S. Puntambekar and B. du Boulay. Design and development of MIST – a system to help students develop metacognition. *Journal of Educational Computing Research*, 16(1):1–35, 1997.
19. J. A. Ramirez Uresti. Should I teach my computer peer? some issues in teaching a learning companion. In G. Gauthier, C. Frasson, and K. VanLehn, editors, *Intelligent Tutoring Systems: 5th International Conference, ITS 2000, Montreal*, number 1839 in Lecture Notes in Computer Science, pages 103–112. Springer, 2000.
20. L. Sheeran, M. Sasse, J. Rimmer, and I. Wakeman. Back to basics: Is a better understanding of the internet a precursor for effective use of the web? In *NordiCHI,Stockholm*, 2000.
21. D. J. Wood and D. J. Middleton. A study of assisted problem solving. *British Journal of Psychology*, 66:181–191, 1975.

Computer Interfaces: From Communication to Mind-Prosthesis Metaphor

Georgi Stojanov and Kire Stojanoski

Computer Science Department, Electrical Engineering Faculty,
SS Cyril and Methodius University in Skopje, Macedonia
geos@cerera.etf.ukim.edu.mk, kires@ekumena.org.mk

Abstract. This paper explores underlying metaphors about the process of working with computers and computer interfaces. First, we identify the most widespread implicit metaphor, which we claim to be the *conduit metaphor*. While interacting with a computer, users are implicitly put in a *conversational situation,* unlike the situation they find themselves in when interacting with most other artifacts. We advance arguments for this thesis taken from the history of computers, computer interfaces, and their current design. Nowadays we are witnessing a shift from this, all pervasive metaphor towards another emerging metaphor where computers are beginning gradually to be perceived as an augmentation, or prosthesis for the perceptive and cognitive capabilities. In this, transition phase, we can see people advocating views where the two metaphors are mixed. Then, we put forward the claim that the prosthesis metaphor is far more fruitful, productive, and explicative and we indicate some of the practical implications of adopting this metaphor.

1 Introduction: On Being Interactive

We want to draw attention to an important terminological issue concerning *interactivity*. Usually, when talking about human-computer interaction (HCI) people implicitly assume *linguistic* interactivity, without necessarily being aware of that. This may lead to confusions and misunderstandings, as we will point out later in the text. To make it clear, we can contrast the interactions that we have with our cars and the interactions we have with computers. The claim that we make in this section is that since they first appeared, computers have been construed as *conversational partners*, by extrapolation we can describe HCI in terms of Reddy's conduit metaphor [17]. On the other hand, we could hardly say that we think of our cars as conversational partners. This is inherently a non-linguistic interaction, and people don't expect their car to talk to them. Indeed, they are not even comfortable with that concept. In our opinion, it is for this reason that, although the technology is available, cars that talk (to warn you to fasten your seat belt, for example) are uncommon. Instead, various visual or auditory cues for that purpose are provided.

Why do we think that the conduit metaphor was (and maybe still is) appropriate to describe the implicit understanding of the human-computer interaction? We can safely say that, in the beginning, the frame for HCI construal was the following one: the human was "giving commands" to the machine, which was supposed to "under

M. Beynon, C.L. Nehaniv, and K. Dautenhahn (Eds.): CT 2001, LNAI 2117, pp. 301-310, 2001.

stand (extract/unpack the meaning from them) and, eventually, obey and give a response" (Fig. 1).

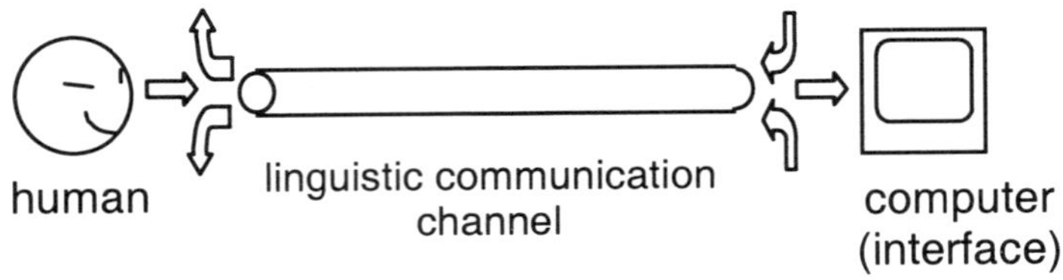

Fig. 1. Reddy's conduit metaphor applied to human-computer (interface) interaction

Whatever role was played by technological limitations, it is not by chance that the very first computer interfaces were based on the command line concept, where users would type in human language-like commands. The computer was construed as some kind of interlocutor, with whom the user communicated via the command prompt. The keyboard, which was and still is the main input device, was a kind of *invitation for conversation* [7]. It is clear that designers treated computers and computer interfaces in this way. People were hearing of "programming languages", "electronic brains" and other anthropomorphic metaphors, which basically have shaped the source domain in terms of which they were to understand the new entity [25]. Biocca in [2] points out to some research work that has revealed how widespread this metaphor is. So, the expectations were raised, and it came naturally to address computers as conversational partners. In the command line interface, users couldn't see the internal structure, although they were talking about it, imagining it, and manipulating it, via the commands issued at the prompt (see Fig. 2 for schematic representation). The delays between the time issuing the command and seeing the effects of it only reinforced the conversational metaphor.

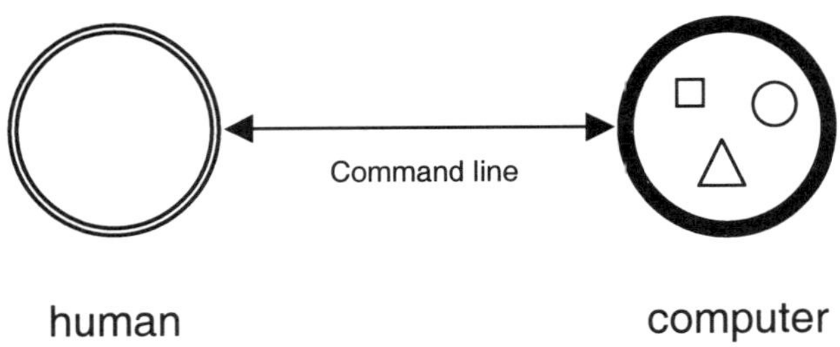

Fig. 2. Schematic representation of human and computer as conversational partners

That this was and is still so, we can tell from the huge body of examples of how people talk about this interaction. For example, Huhtamo [7] says "Thus, it has been asserted that interactive systems position us in a 'conversational' situation... But with whom or with what this conversation takes place (e.g. the machine, the software, the maker 'behind' the software, oneself, other people or 'avatars', non-human but human-like entities etc.) is a much more complex question." In other words, that the human-computer interaction IS a conversational situation is not even questioned: it is assumed. Huhtamo wants only to discover WHO the other participant is. Because of this situation, projects like Weizenbaum's ELIZA [31], or Colby's PARRY [3] were possible. In these programs, Weizenbaum and Colby play with the assumptions and expectations that the user has in a specific conversational situation.

According to the conduit metaphor, the partners in the conversation are supposed to share the meaning, which is packed into the words. From this stems the frustration of users when computers don't respond in the expected way.

2 On Being *There*

The next phase in interface evolution came with the introduction of desktop and similar physical metaphors. In these GUIs, user's knowledge of the physical world was used in order to speed-up the learning process. Here, the system offered some representations of the objects in it (files, folders...) and direct manipulation with them with pointing devices (mouse, light-pen) became possible.

In this phase the UI is becoming spatial and it gives to the user a possibility to get a glimpse inside the artificial interlocutor. Users now can use the variety of their audio-visual capacities instead of just language oriented. Such a UI uses spatial metaphors and allows its users a glimpse inside the artificial interlocutor. Because they have explicit representations of the computer system's constitutive parts, they can manipulate them directly (see Fig. 3). Moreover, since the duration of the feedback loop of these actions is negligible, these spatial interactions can be easily considered as being executed inside of single, symbiotic system consisted of the user and the user interface. (We can say that fast feedback gives us a feeling of owning the items that participate in the interaction loop.) In contrast, the command line interface lacks such immediacy: there is packing, transmitting, unpacking - which reinforces the feeling that we are communicating with another entity.

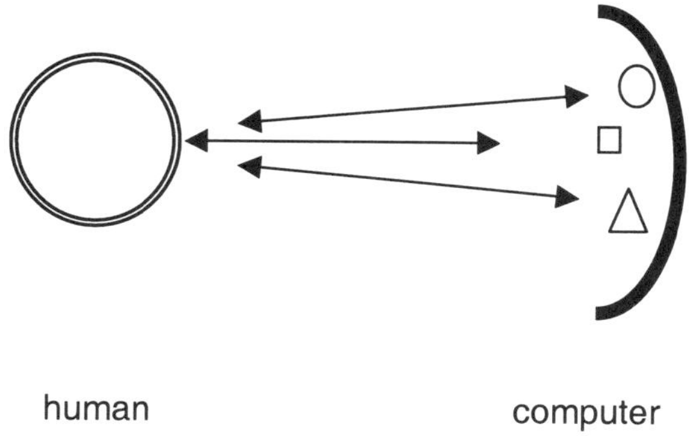

Fig. 3. Introduction of GUIs and spatial metaphors enabled users to see some representations of the internal structures and to directly manipulate them

Even in a GUI, the conversational metaphor is still present because the user can be interrupted in their current work by splash screen linguistic messages that notofy errors or give other information. One can claim that the conduit metaphor is still applicable, and we do agree that in a sense there is an isomorphic mapping among the commands and the actions. We assert that, although applicable, it is not productive enough for the complexity of the underlying systems. On one hand, the users are invited to enter the simulated physical reality with the pointing devices being extensions of their limbs (rudimentary prosthesis and immersion), while on the other, there is still the sense of conversing with the computer (interface) entity (Fig. 4).

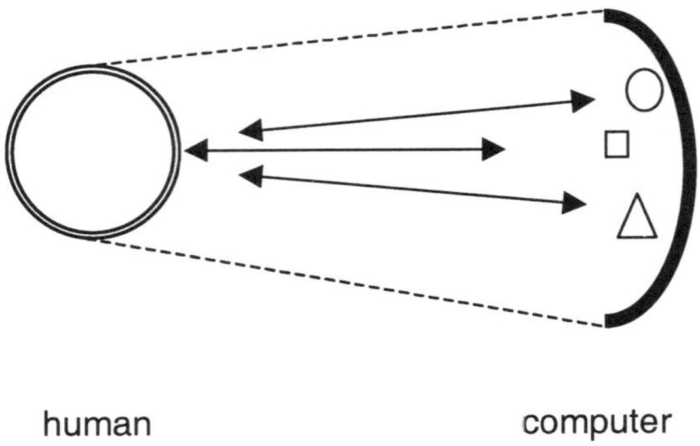

Fig. 4. Mixing the "Being *there*" and "communicating with *it*" metaphors

This situation is a bit weird, and Janney, in [8], which has the gist of a warning against computer-induced schizophrenia, gives a nice analysis of it. Whilst still using the term 'prosthesis' to describe the computer, he introduces the slightly obscure notion of 'interacting with the prosthesis' Making analogy with prosthesis which replace body parts (artificial hip, denture, wooden leg…) he discusses the problems that arise at the contact points between the body (mind) and the prosthesis. This motivates the question: "[W]hat happens to the human *mind* at the interface with the mechanical prosthesis, where the two have to communicate with each other in order for a prosthetic extension to take place[?]" (emphasis original). This is a good example of a blend of meanings: prosthesis as partner. Consistent understanding would treat the interface as *either* prosthesis *or* as partner. Naturally, mixing those two understandings (implicit metaphors) introduces problems. Brenda Laurel in [10] is consistent in using the computer as conversational partner metaphor, and this consequently drives her to the conclusion that the user and the computer should come to the point where they share 'mutual knowledge, mutual beliefs, and mutual assumptions'. As research in AI tells us, we are far from being at that point, and from this we conclude that conversational partner metaphor is not fruitful. There are further evidences of this in the many not very successful (to say the least) attempts to introduce anthropomorphic design in user interfaces (e.g. Microsoft Bob, Microsoft Office Assistants, etc.). The problems that arise in this special case illustrate a general difficulty with physical metaphors, which are always problematic, since there are many unfulfilled expectations as well as hidden features.

The other option, which we're advocating in this paper and which is reinforced by the appearance of the Internet, is to apply the mind-prosthesis metaphor in a consistent way.

3 On Being

"...[E]ach technology needs time to develop as a medium that enhances the experience of people. In my family, my grandparents used the telephone for emergencies and my parents used it to do routine business and make plans. I used it as an extension of my social an emotional communication."

Sherry Turkle in an interview,
www.well.com/user/hlr/texts/mindtomind/turkleint3.html

Additional complexity came with the appearance and the allpervasiveness of the Internet. The Internet had to be conceptualized in some way, and there was no clear way to extend/apply the conduit metaphor here. Emerging implicit understanding is that the user with the computer is an agent behaving in the Internet environment (like a person surfing with their surfing board).

In this paper we make this new metaphor explicit, and view the interface as a mind-prosthesis for the acting agent, or as its augmentation. This shift in perspective makes profound impact on the process of interface design.

At this point we feel it is worth commenting on other construal of the notion of augmentation. In [9], Kari Kuutti and Victor Kaptelinin, discuss some problems with augmentation, commenting on the work of Douglas Engelbart and others. Some of their observations about the problems of augmentation are not amenable to treatment using the user interface approach. The issues of difference between 'individual' and 'internal' in human cognition, processes of assimilation and adaptation of cognitive capabilities, asymmetry between human beings and 'artifacts as information processing units' are naturally resolved when these 'artifacts' are to be considered as prostheses. We agree with Kuutti and Kaptelinin's thesis but for superficial differences that stem from a different use of the word "augmentation". What we mean by augmentation (prosthesis) can be interpreted as a set of functional organs in the sense of Vygotsky [30] and Leont'ev Activity Theory [11]; as adding diverse possibilities for structural couplings [12], [29], between the user and their environment; or adding diverse possibilities for repetitive interactivity [1].

In our thesis, augmentation is of secondary importance only. As Marshal McLuhan points out [13] extensions and self-amputations go together. Augmentation of one's ability comes as a result of an amputation of another ability, which becomes less important. (A classical illustration for that is, again, the car. As it increases the magnitude of locomotion on one hand, it reduces the physical exercising (its biological function) of the human legs, on the other). Mind prosthesis is necessary to the user who needs a way to operate in the new environment. Augmentation is needed for the augmented world.

Conceptualization of the new environments is closely related to the appearance of additional dimensions within the UI. We now have haptic, olfactory, kinetic, and other dimensions. These new nonverbal interactions add to the user immersiveness in the interface, widening the human bandwidth in processing the information [2], [26].

We give here some common examples. Writing this paper on the screen, our attention to the spelling is not as great as it would be if there were no spellchecker acting as a functional organ that we never even think of explicitly invoking because, as part of our system, it is always present. Or, consider the use of a search engine to look for a specific piece of information. Because of reliability and repeatability, many users do not use bookmarks anymore; instead they just remember the keywords. In a way, if you know the keywords, you have the information. So you can easily perceive the search engine to be somehow a part of yourself, just like your hand. This is because users have internalized this search with a search engine and keywords only trigger the internalized searching activity.

On the spatial side, we argue that zoom enabled UIs are much more productive than the schizophrenia inducing windows metaphors (to paraphrase Janney). Here, the user can use their spatial orientation intelligence and topological intuition to navigate through the content (as for example in Jef Raskin's ZoomWorld [16], or the pad++ from the HCI lab at the University of New Mexico). Items referenced by their location, rather than by name, are more visible, and can be more easily manipulated and associated with other items and their environment.

When we adopt the mind prosthesis metaphor the process of interface design should be seen as constructing functional organs for the mind, creating new structural couplings, or adding new possibilities for interactivity. We are designing external organizational structures, external memory, and external rules. These organs should be adapted to the limitations and bottlenecks of the human awareness and, when possible, imitate the conscious processes that are used by humans to overcome these limitations. Considering the possibilities of today's computers, there is a very large space for the adaptation of the prosthesis to the individual mind. This actually is personalization, which is not just like making lists of interests of the user and grading of content, but it also about considering individual psychological and psychophysical characteristics [4],[20],[23],[24]. When all these principles are met, and the person has internalized the functional organs, the outcome is a new entity with extended perception of itself. There are two (symmetrical?) points of view of this process. The person has assimilated the possibilities offered by interface or they have accommodated to those possibilities. (Fig. 5)

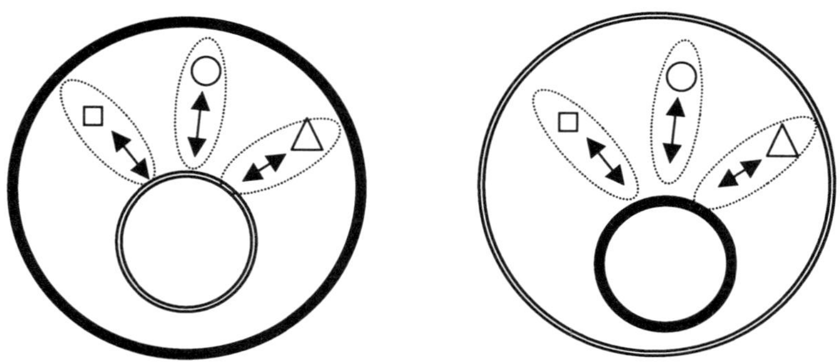

Fig. 5. Two different perspectives on what happens when the human user has sucsessfully developed various *functional organs* (see text for explanation)

A criterion for a good HCI model is how successfully it manages to maintain the user in a state of *flow* [4] which is defined as a proper balance between user skills and the degree of challenge of the task (Fig 6). That is, the user is engaged in interactions that occupy most of their attention without information overload (what you do not need is not displayed), without interruptions by the UI (no error splash screens), and the temporal sensitivity of the user is matched by the interface (real-time feedback).

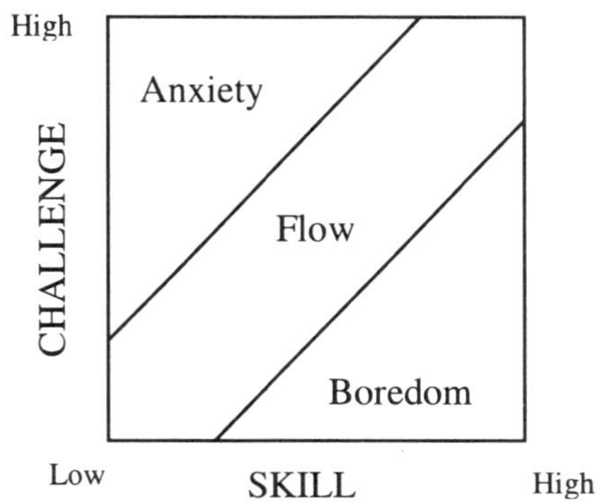

Fig. 6. Flow region defined by skill and challenge balance

The user interface should present only the most relevant information to the user, but there should be clearly defined pathways leading to the relevant details. This will prevent autoamputation [8], [13], induced by information overload.

The communication of the symbiotic (human + prosthesis) system to the outer world is still understood by the conduit metaphor (Fig. 7). The information that goes through that channel is basically twofold: one is the contact to the other *symbiots* (ICQing, IRCing, MUDing) and the other is the link to the expanding world of the *electronic noosphere* (free surfing, searching). Electronic noosphere can be viewed as a kind of implementation of the more general concept of noosphere, the "thinking layer of the earth," as introduced in [5]. There is emerging field of research in the social sciences focusing on computer-mediated human communication [27], [28], [6], [18], [19].

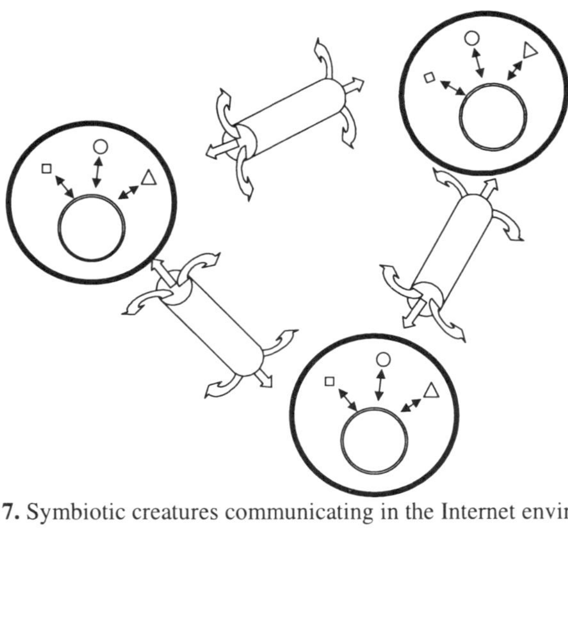

Fig. 7. Symbiotic creatures communicating in the Internet environment

4 Concluding Remarks

The new metaphor can be regarded as a unifier of the research work in interface design, and conceptual and philosophical research in augmented reality. In both areas, agents are symbiotic creatures. In emerging practice there is a variety of new research threads in data visualization, sonification, user interfaces specialized for handicapped persons, artificially induced synesthesia and so on Even though all of these fields may appear to be divergent and to deal with different subjects, the mind-prosthesis metaphor comes as a natural umbrella. It is all about enhancing the capabilities of symbiotic creatures behaving in diverse environments.

Now we can extract some practical guidelines for HCI design that reinforce the understanding of interfaces as mind-prosthesis and facilitate the process of entering the flow experience. Those features include:

zooming facility,

possibility of building different organizational patterns (for example, additional visual representations of the same internal elements),

continuous navigation (the user should have clear orientation within the abstract space topology - the user could thus make use of their spatial intelligence in dealing with the interface),

tracking the history of the user actions (as much as is possible in detail on a system level),

redundancy in the audio-visual cues for the actions and responses,

screening out the unnecessary data,

complying to the capacity of the human short term memory,

avoiding elements that interrupt the on-going user activity.

New input-output devices (i.e. haptic, olfactory, gaze tracking devices) already make it possible to imagine interfaces where, instead of showing a window containing some error message, the interface can induce a physical pressure on some part of user's body. Finding a way to relieve the pressure, would translate to removing the cause for the error in the system.

Freed from the real world physical metaphors, designers should think of more innovative abstract conceptualizations of digital environments, leading to interfaces that expand the cognitive reach of their users.

Coda

Actually we can see that some of these features are already recognized and recommended by certain professionals in HCI: Nielsen is advocating simplicity in web interface design [14]; Norman says the computer should be "quiet, invisible, unobtrusive" [15]; Raskin takes the zooming facility to be essential for the interface [16].

In implementing two utility programs we have tried to incorporate mind prosthesis reinforcing features for their interfaces. FF Read is an experimental program for fast reading on the screen. Unlike other speed reading programs, it does not use the standard Windows interface, but it borrows the concepts and the controls of the gaming world instead. The words of the text are flowing on the screen, changing their position, size and color, but in a way that provides a natural focus for the user attention. The user can customize the speed and different modes of flow.

M3Dutility is a utility program, that is used to define interactive sequences with 3D objects generated with programs like Maya and 3DMax Studio. In contrast to similar programs where the screen is cluttered with numerous windows dispersing the user's attention, on launching M3DUtility the user faces a calming almost blank screen. When some 3D scene is imported, the hierarchy of the scene is drawn, and the user can zoom in and out on particular objects to see the details. With simple operations, he or she can group the objects involved in a particular interaction sequence into ensembles called views. What has been achieved can be seen by activating the (single) preview window. The prototype is available for download at www.multi3D.com.

References

1. Bickhard, M.H., Representational Content in Humans and Machines, *Journal of Experimental and Theoretical Artificial Intelligence*, 5, pp 285-333, (1993).
2. Biocca, F., Human-Bandwidth and the Design of Internet2 Interfaces:Human Factors and Psychosocial Challenge, *Internet2 Sociotechnical Summit*, September 13-15, The University of Michigan Media Union, Ann Arbor, Michigan (1999).
3. Colby, K.M., *Artificial Paranoia. Artificial Intelligence, Volume 2*, pp 1-25, (1971).
4. Csikszemtmihalyi, M. Flow: *The Psychology of Optimal Experience*. New York: Harper and Row (1990).
5. De Chardin, P. T., *Le Phenomene Humain*, Edition de Seuil, Paris, (1955).
6. Ellsworth, J. H., Using Computer-Mediated Communication in Teaching University Courses, In Zane L. Berge and Mauri P. Collins (Eds.), *Computer Mediated Communication and the Online Classroom*, Volume 1: Overview and Perspectives, pp 29-36, Hampton Press (1995).
7. Huhtamo, E., Seeking Deeper Contact. In: Eckhard Diesing and Ralf Sausmikat (eds.) catalogue of European Media Art Festival Osnabruck, pp 251-272, (1993).
8. Janney, R. W. The Prosthesis as Partner: Pragmatics and the Human-Computer Interface. *Proceedings of CT97 Conference Aizu*, pp 82-87, IEEE Computer Society (1997).
9. Kuutti, K., Kaptelinin, V., Rethinking Cognitive Tools: From Augmentation to Mediation (Extended Abstract), *Proceedings of CT97 Conference* Aizu pp 31-32, IEEE Computer Society (1997).
10. Laurel, B. (ed.) *The Art of Human-Computer Interface Design*, Addison-Wesley (1990).
11. Leont'ev, A.N., *Activity, Consciousness, and Personality*. Englewood Cliffs, NJ: Prentice-Hall (1978).
12. Maturana, H. R.&Varela, F. J., *The Tree of Knowledge, The Biological Roots of Human Understanding*, New Science Library (1987).
13. McLuhan, M., *Understanding Media: The Extensions of Man*, 3rd Edition, McGraw-Hill, New York (1964).
14. Nielsen, J., *Designing Web Usability: The Practice of Simplicity*, New Riders Publishing, Indianapolis (1999).
15. Norman, D. A., The Invisible Computer. Cambridge, MA: MIT Press (1998)
16. Raskin, J., *The Humane Interface: New Directions for Designing Interactive Systems*, Adison-Wesley (2000).
17. Reddy, M., The conduit metaphor: A case of frame conflict in our language about language, in A. Ortony (ed.), *Metaphor and Thought*. pp 284-324, Cambridge University Press (1993).

18. Reid, F. J. M., Malinek, V., Stott, C. J. T., and Evans, J. St. B. T., The Messaging Threshold in Computer-Mediated Communication, *Ergonomics*, Vol. 39, No. 8, pp 1017-1037, (1996).
19. Santoro, G. M.. What is Computer Mediated Education?, In Zane L. Berge and Mauri P. Collins (Eds.), *Computer Mediated Communication and the Online Classroom*, volume 1: Overview and Perspectives, pp 11-27, Hampton Press (1995).
20. Stojanov, G. *Expectancy Theory and Interpretation of EXG curves in the Context of Machine and Biological Intelligence*, PhD Thesis, University Sts. Cyril and Methodius in Skopje, Macedonia (1997).
21. Stojanov, G., Bozinovski, S., Trajkovski G. Interactionist-Expectative View on Agency and Learning, *IMACS Journal on Mathematics and Computers in Simulation*, 44, pp 295-310, (1997).
22. Stojanov, G., Embodiment as Metaphor: Metaphorizing-in The Environment, in C. Nehaniv (ed.) *Computation for Metaphors, Analogy and Agents,* Lecture Notes in Artificial Intelligence, Vol. 1562, pp 88-101, Springer-Verlag (1999).
23. Stojanov, G., Gerbino, W., Investigating Perception in Humans Inhabiting Simple Virtual Environments, poster presented at European Conference on Visual Perception; abstract published in supplement of Perception, Vol. 28, pp 78, Trieste (1999).
24. Stojanov, G., Kulakov, A., Trajkovski, G., Investigating Perception in Humans Inhabiting Simple Virtual Environments: An Enactivist View, poster presented at Cognitive Science Conference on Perception, Consciousness and Art, Vrije Universitaet in Bruxelles (1999).
25. Stojanov, G., Programming Languages, Their Reality and Origins, and What Can That Tell us About Natural Language Origins, published abstract of the 12th LOS Meeting, pp 24, Baltimore, July 1996, USA (1996).
26. Stojanov, G., The Fourth Dialogue Between Hylas and Phylonus about the End of a Science or George Berkeley the Virtual Realist, Vindicated, in S. Bozinovski (ed.) *The Artificial Intelligence* (in Macedonian), GOCMAR, (English version at www.soros.org.mk/image/geos/providenceFACE.ps) (1994).
27. Turkle, S., Constructions and Reconstructions of Self in Virtual Reality: Playing in the MUDs. *Mind, Culture, and Activity*, 1, (3) pp 72-84, (1994).
28. Turkle, S., *Life on the Screen: Identity in the Age of the Internet,* New York: Simon and Schuster, 1995. Paperback edition, New York: Touchstone (1997).
29. Varela, J.F., Thompson, E., Rosch, E., *The Embodied Mind*, MIT Press (1991).
30. Vygotsky, L.S., *Mind in Society.* Cambridge, MA. Harward University Press (1978).
31. Weizenbaum, J., ELIZA - A Computer Program For the Study of Natural Language Communication Between Man and Machine, *Communication of the ACM* Volume 9, Number 1, pp 36-45, (1966).

Meaning and Relevance

Reinhard Riedl

Department of Computer Science, University of Zurich, Winterthurerstr. 190
CH-8057 Zurich, Switzerland
riedl@ifi.unizh.ch
http://www.ifi.unizh.ch/~riedl

Abstract. We discuss, how information brokering through virtual spaces for information exchange fails, and what can be done in order to improve the success of asynchronous information publishing in virtual spaces.

1 Introduction

Literacy skills play an ever increasingly important role in business and in private life. Pieces of art or scientific projects are often perceived as descriptive texts only, i.e. they are no longer perceived in their natural context as objects per se, but they appear as objects embedded into a virtual space for publishing texts and pictures. This changes reality. In the following, we shall investigate how virtual spaces for information exchange work, where information is exchanged by publishing it at a given address.

Some experts on teleportation have put the thesis forward, that reality depends on what we could know about it rather than on what we do know, or what our mind constructs as a reality, or what mathematical models describe. Although this thesis originally resulted from the attempt to explain the superposition of the living and the dead cat in Schroedinger's gedankenexperiment, it also provides a useful basis for the investigation of the use of virtual spaces for communication exchange. If we can only know the representation of an information object in a virtual space, then we may only construct such a reality which provides us with exactly that information, which is contained in the representation.

There is a strong tendency in Western culture to violate that principle by assuming that human intelligence can do better. Smart users of the Internet try to figure out

- where some information or a particular information object could be found
- what data mean, which are provided by an information object
- how relevant the information provided by an information object is

Smart engineers try to

- find the experts with the appropriate domain knowledge

M. Beynon, C.L. Nehaniv, and K. Dautenhahn (Eds.): CT 2001, LNAI 2117, pp. 311–324, 2001.

- figure out what the problem descriptions provided by the customers, the management, and the experts mean
- identify the relevance of the various pieces of information, which they deduced analytically

Smart managers try to

- nurture the convergence of communities of practice [6] by selecting the right people for interdisciplinary cooperations
- understand the information provided by project managers and project partners
- find out, which risks are relevant and which are not

On the one hand, other than computers, humans have the ability of putting information into a tacit context, on the other hand, experience also tells us that information technology often fails because processes are defined such that only the smart user, engineers, and managers can succeed. Indeed, on the one hand it is true that what we can know is more than what is available as data in a virtual information space, while on the other hand there is a lot of what we definitely cannot know, when we try to obtain information from a virtual space for information publishing. As information systems are built for a large class of users, there is a considerable twilight zone between these domains of possible and impossible knowledge, and there rather fuzzy than clear cut boundaries.

Furthermore, there is some domain of a posteriori context knowledge, which is void upon the act of publishing, and which only emerges from the act of accessing and interpreting the published data, that is knowledge which results from opening the door to the box with the cat. Exposing data to the context of a new usage context redesigns what we can know about its meaning, and thus it creates a new meaning a posteriori. For example, it is exactly this emergent knowledge, which interdisciplinary research work tries to achieve.

The design of a virtual space for information publishing and exchange ought to support the publishing and retrieval processes in such a way that information can indeed be exchanged and that at the same time new meaning can be invented. As long as users feel the freedom of interpretation, both goals do not contradict each other in practice, and thus they can be pursued jointly. Unfortunately, investigations indicate some significant loss of analytic capabilities of structural perception among young people occurred during the last decades, which hinders the inter-networking of knowledge domains and thus on the one hand it reduces the domain of usable context knowledge and on the other hand, it prevents the emergence of new knowledge.

Part of this loss seems to be due to an increased speed of life and to a reduction of boundaries between the private and the public. Our life is becoming faster due to an increased rate of context switches and due to an increased speed of context evolution. That creates stress and thus disables content networking to some extent. Events determine contemporary life rather than plans and cultural cycles, and deadline scheduling directs the rhythm of activities. Thus. some form of macro-flow is created, which replaces self-criticism and reflection with interaction. And we are observing an imperialist capturing of public space by private life. In fact, people complain about loosing their privacy as the borders to public life get lost, and public space is invaded by people's privacy. As a consequence, on the one hand general boundaries of problem areas are lost and on the other hand small cultural niches are redefined by local codes of thinking and behavior. That change puts a lot of pressure on young people who try to develop a holistic understanding of the world until it might eventually extinguishes the intellectual abilities for creative perception of unfamiliar content, while at the same time many people adapt their thinking to knowledge niches with well-defined pseudo-logical codes.

Further, new scientific and technological trends like 'the semantic web' strengthen that development. Contrary to attempts of building expert systems in the 80's, they focus on very restricted domains of practice, for which they create ontologies. Due to the principles of complexity, such attempts are much more promising, the more since we now have carrier technologies like XML and RDF/RDFS, which are nearly universally accepted as a basis for the application of ontologies in semantic annotations of documents. However, while that supports the cooperation of experts in one domain, it does not change the principal situation for the challenge of interdisciplinary networking of thoughts. Rather, it tightens the coupling of experts to their own ontologies, which may hinder cross-over thinking. Thereby, one of the main problems is the idea, that ontologies should be used to enable communication in such a way, that it guarantees that the same word refers to the same object. In many practical communication scenarios, such an object does not exist, but rather affordance have to be translated from one domain of practice to another. An illustrative example is provided by the technical term marriage and affiliated terms, which describe quite different concepts in different countries, but still there is the need to deal with this concept in the everyday situation that people migrate from one country to another.

According to traditional Western concepts of philosophy, and by abuse of philosophical terminology, intellect refers to some ability of acting upon data input interpreted as information based on internal images of the mind, which are created from that information and which are used for the deduction of actions to be taken. These images do not necessarily take the form of visual information, nor does their actual visualization necessarily reflect physical realty. However, in the recent past intellect has become more and more involved with multimedia and realistic representations of physical realty. As a consequence the meaning of meaning is changing its nature. Although the results are not clear yet, it seems that this will reduce people's inclination for universal abstraction and it will increase the social binding to local codes of interpretation.

How can information technology cope with these changes and with a most complex variety of language niches? First, it is easier to simulate intelligence in a social or business niche with a strong coupling between language and meaning and with few or no mental images beyond language than it is in an environment, where a general purpose language forms the basis of an interdisciplinary, cross-cultural communication. Second, it is becoming a much more challenging task to support interdisciplinary communication with information technology, if ecological niches have to be connected, which lack the experience of cross-cultural communication based on a high standard of common purpose language. In the following we shall collect preliminary research results from various areas where either virtual or imaginary spaces for information exchange are used. We shall derive some conclusions from our observations, both for practical improvements and for future research work, but in most cases we shall only be able to name the problems on the surface and to indicate promising directions of future research.

2 Background

The research results presented in this section are primarily based on interviews and on behavior monitoring, which is complemented with the results of statistical analyses of log files. We have analyzed the limitations for information technology in large Intra-nets, we have analyzed the limitations of project management procedures for interdisciplinary projects, and we studied the interrelationship between theater science and e-business. In particular, we have performed a case study on the Intra-net of a major international provider of financial services, we have performed a case study of an interdisciplinary engineering project, we have studied, how web-marketing works at another major international provider of financial services, and we have tried to apply the theory of communication tools from theater and communication sciences for the design of SW-agents. Hereby, our main purpose was to achieve a better understanding how virtual spaces for information exchange work in practice, and how new functionality could improve their success. We first summarize the results and then we discuss them in more detail. We hope that they will not be considered as the end of a research story, but rather as an intriguing starting point for more in depth empirical research.

First, our case studied strongly indicated that the information management in the company which runs the Intra-Web should take much more care of the change of communication patterns caused by the introduction of Web-technology. Web-based communication leads to some implicit form of publish/subscribe communication, which often fails as implicit expectations are not met. The management ought to be aware of the frustrations about the limitations of information technology and it ought to understand the impact of information technology on organizational and social structures. In particular, information management is recommended to focus its activities on the existing search problems and on the relevance problem. (See also [5].)

Second, our study of the success of the web-marketing activities indicated, that they should be based on clear formulations of goals and on monitoring whether these goals are achieved, and how the marketing could be improved. The analysis of log-files based on state-of-the-art market research techniques tailored for virtual markets can provide the information needed for monitoring the success of the pursuit of marketing goals through web-marketing. The management ought to supervise the monitoring and use the side-results on user-profiling for a user-focused strategic orientation in the market. (For a more detailed discussion see [10].)

Third, from our case study of an interdisciplinary research project and its comparison with other interdisciplinary research projects, we have concluded that interdisciplinary research and development projects should focus on the convergence of language and concepts rather than hurry for quick results, since exchange of knowledge resulting from convergence is an asset by itself. The management has to be aware that equal partnership and a single, common project goal cannot both be achieved in short projects. Bundling of projects from multiple disciplines, each heading for its own joint goal of interdisciplinary cooperation, and each lead by one team, might provide a working alternative.

Fourth, we could show that information agents can be modeled based on the context/subtext paradigm from theater science, which further supports an exchange and trading of market knowledge. Information search has some structural similarities with the usage of language in written texts, as it is formulated by Zipf's laws and by Heaps' law (see [8]). The usage of virtual characters for the communication of meaning and relevance similar to classical theater provides an option which would deserve more attention. Hamlet's problem of having no private language is inverted in contemporary society and business by the fact that protected public space is diminishing. Information technology should provide protected virtual communication spaces with access control and it should provide guidelines for the interaction with these spaces. (Such rules of conduct should not be confused with netiquette rules, which focuses on politeness. Instead, they are supposed to nurture knowledge sharing and to protect those willing to share their knowledge.)

Part of the knowledge management in the Intra-net of the international company, where we performed our first case study, is genuinely embedded into the company's Intra-Web. Interviews with employees show that the Intra-Web is used as a virtual space for information exchange with the implicit assumption that it would implement some form of asynchronous and even rather synchronous publish/subscribe communication. Hereby, imaginary channels are defined by common interest or by the position of the receiver in the organizational hierarchy. The traditional push principle for communication between the management and the employees is replaced by the pull principle, which also dominates the knowledge management both on the company level and in internal projects. However, the technology and the knowledge management processes do not properly support these communication models. People spend hours with searching for information, which they never find, although it exists and although they have been informed about its existence. In project meetings, they employees often get only vague descriptions where to find the information, which are either not precise enough to help them find the information object, or they are no longer valid as the URI of the information object has been changed, since it was spotted by recommending person.

If people happen to find the desired information object nevertheless, it is often difficult for them to decide on its relevance. Although there is usually given enough context description that some more or less appropriate meaning can be assigned to the data, it is often hard to guess how relevant that information is. Since there is nearly no garbage collection and there are no strict guidelines defining where directions or guidelines have to be published on the Intra-Web, it happens quite often that the Web-page found presents outdated information, which cannot be deduced from the Web page itself. In other cases, the organizational relevance of information is unclear because the role of the publisher is not defined. The Web page may present his own view, or it may present the view of a group, or it might even be guideline from the management. As a result, most users of the Intra-Web are very frustrated about the Intra-Web.

We have experimented with the search tools and we have analyzed the log-files of one search engine. This showed that even experienced web-users could not locate information with the tools available although they knew about its existence. It revealed that search sessions are very short, and very few users apply thesaurus like variations of search terms or use complex search terms. In some cases, this seems to be the result of frustration, in other cases it clearly results from the fact that people have never been given an appropriate training how to use the search infrastructure. The misfortunate situation is prolonged by the fact that projects which headed for tools supporting the characteristic push version of the publish/subscribe paradigm, e.g. parsing agents, were stopped by the management, which decided to improve the situation with content management tools creating XML-files. We expect that these tools will indeed somewhat improve the situation, but there is no reason to assume that the situation will change fundamentally, as the basic search problem is not tackled.

We suggest, that a better training is provided for the employees how to use available tools, that a relevance annotation regime is enforced, which allows only signed information to be published and which tracks responsibilities for signed information. Further, we suggest that advanced monitoring and profiling instruments are installed in order to find out about user acceptance of processes and tools, and that some form of Boehm's spiral [1] for permanent Intra-net re-engineering is installed, based on experimental introduction of tool prototypes and relying on complimentary, cyclic system consolidation. Clearly, all results must be anonymized. Furthermore, we would like to emphasize that in the scenario investigated the two main problems are search and relevance, while the problem of deducing meaning from data exists primarily in the heads of those managers who want to prevent company-wide knowledge sharing and who argue that interdisciplinary knowledge exchange is impossible.

Our experience with Web-marketing confirms the findings in the case study. The concepts of communication were even less reflected by the designers of the Web-site in the other company. They had only one question about customer behavior, namely whether their site was accessed throughout the whole day or not. The main problem seemed to be that marketing and management people were not aware of the possibilities to carry out market research on log data (although the management kindly support a project spending about 0.75 person years on a feasibility study plus the a prototypical realization of tool for data preparation). The Web was rather considered as a publish only medium assuming that this publishing would work similar to the distribution of brochures. Thus a lot of care was taken to communicate an intended meaning and to clarify the relevance of the information, but the capabilities of the technology to find out about how people dealt with the information were not used. In other words, similar as in the Intra-net of the first company, no verification and validation of communication activities took place.

The second case study on interdisciplinary research and development was carried out in an 18 months R&D-project with a head-count between 55 and 60 participants from 15 teams, which spent slightly less than 200 000 Euro per month. In this project it took between 11 and 15 months until a partial common agreement was achieved on what the project was all about. A first agreement was achieved at a workshop in month 11, where 8 teams participated. Afterwards, that agreement was not properly communicated to the other 7 partners and thus it was questioned by these partners plus three partners participating in the workshop. A second, more or less identical, agreement was achieved at a workshop at then end of month 15, where 13 partners participated. Even then it was not clear, whether this agreement was understood and accepted by all present partners. At that time it became clear that various fundamentals terms used in the project were understood differently by different key players in the project, however it was much to late to make significant changes.

The project was based on the exchange of partial results through deliverables at fixed dates, which created clear dependencies between the various partners. None of these exchanges worked. In the original project plan, the technical partners had suggested the use of visual prototypes as boundary objects for the communication between the technical and the non-technical partners. They did not use that term, though, but they described them as a tool for the investigation of conceptual system requirements. This turned out to be one of the really bad ideas of the project. First, some of the non-technical partners misspelled the visual prototypes as virtual prototypes and after this error had been corrected formally, they later interpreted them as graphical user interfaces. That eventually hindered the cooperation in the project as the technical partners insisted on the role of boundary objects for the visual prototypes. Matters were made worse by the fact that the term requirements specification was understood differently by the technical and the non-technical partners. While the technical partners expected conceptual UML-specifications, the responsible non-technical partners provided wish lists and they argued that GUIs cannot be specified in UML. Both groups could provide arguments for their expectations, and their behavior, respectively. On the one hand, the delivery of conceptual UML specifications had been promised by the non-technical partners. On the other hand, all partners had agreed that the project was a feasibility study based on explorative prototyping, while the project was organized as an application prototyping project according to the waterfall project plan. The team which understood the visual prototypes as GUIs argued that an explorative GUI design required every new month a new visual prototype, while the technical partners spent their time resources in obtaining conceptual requirements specifications themselves. That ontological chaos impeded a fast convergence of the various opinions about the ultimate goal of the project, which finally provided rather disjoint results from the various teams of experts involved. The inherent contradictions in project concepts were not even resolved when the disastrous consequences were visible to all project partners. Instead, the management sticked to the principle that a joint project result had to be delivered instead of the achieved results in the various disciplines.

However, language problems do not fully explain what happened and why it happened. There were five different types of goals present in the project. People wanted to solve real world problems, people wanted to gain system engineering experience, people wanted to gain technical experience, people wanted to perform empiric, scientific research, and people wanted to improve their personal contacts. At the end, there were two groups of people, those primarily pursuing one or various of the first four goals and those primarily pursuing the last goal. This does not imply that people in the first group were not interested in good relations. It rather relates to the individual perception of personal benefit. In both groups technical and non-technical people were present, some teams had members in both groups, and both groups still communicated with each other, but cooperation failed. It ought to be stressed, that while differences in goals played an important negative role in the project, in most case where cooperation worked between experts from different disciplines, this was due to a convergence of language, namely i]t was initiated by the emergence of a common wording.

We suggest that the project management in complex, interdisciplinary projects is split into administrative management and scientific leadership. The latter should be carried out by someone who understands the goals and capabilities of the various project partners and who is able to monitor and to supervise the relationships between the various partners. At the beginning of the project, about 9 months should be provided for convergence issues. That abstract task ought to be 'embodied' with specific experimental tasks, which have to be performed under strict time constraints. Boundary objects should be used in this process, but their role and meaning has to be clarified for all partners in the project before the project starts. Successful convergence means language sharing and reciprocal understanding of context, defined by goals, knowledge, tools, and evaluation standards. The project must state and supervise the pursuit of this goal. In the scenario investigated the two main problems were obtaining usable input from other disciplines and associating meaning to information provided by other disciplines, while relevance was only a second order problem.

In [7] we have shown how context/subtext models from theater sciences can be exploited for the design of information agents and for the exchange of market knowledge in e-commerce. Analyses in the case study depicted above show, that it is possible to implement such information agents in real world scenarios, but they also indicate that the success will depend on the 'load' of the scenario, i.e. on the structural patterns of that load. In the following we shall discuss various structures and problems, which are shared by the performing arts and by information technology.

3 Meaning and Relevance in Theater

Theater has a long experience in the communication of meaning and relevance. Its traditional idea is the assignment of text to characters. Characters provide a context for text and they indicate the relevance of statements. Originally characters were 'defined' by masks. Later on, there were fixed character types, until eventually a free interplay between the character guided interpretation of text and text-based interpretation of characters evolved. In the recent past, the American director Robert Wilson has reintroduced icon-like gestures to define basic relations between characters, that is to visualized the subtext of a character. And various dramatists, for example Werner Schwab, have forced directors to return to more synthetic play in order to master the extremes of his language. Both concepts partially destroy the individual nature of characters, although they proceed into quite different directions. Lots of similar attempts may be observed, but only few of these really question the whole principle of characters as carriers of context for text, which enables the audience to decide on the meaning and the relevance of the text. Such a questioning takes place, however, when the ideas of the author or director are allocated to a homogeneous set of characters or to one monolithic character, and when the traditional story is replaced by a virtual timeliness with the text only aiming at the characterization of a generic type of character. There has been coined the terminology of post-dramatic theater (compare [3] for an extensive analysis which describes a theater with deconstructed role structures, dematerialized figuration, and without traditional narrative elements. For example, Elfriede Jelinek has illuminatingly described her work as putting language surfaces in opposition thus exhibiting the new paradigm. In a way, theater thus reflects the demateralization of media on an abstract level (although it still opposes zeitgeist by its persisting principle to bring people together and to let them physically share the perception of materialized communication). More important, post-dramatic theater destroys and reinvents viewing habits and expertises, and it provides completely new cognitive challenges for the audience. We are observing an emancipation of meaning and relevance from characters, but the historic relevance of this emancipation is not clear at all. Information technology can benefit from the experiences collected, but empiric studies of the change of cognitive abilities are needed. On the other hand, experiments on perception in theater can be supported by information technology, and they can be simulated in virtual information exchange spaces. We conjecture, that empiric research will provide a lot of insight on the phenomenon of flow. In [11] the importance of the flow construct for web marketing has been described. Yet, the prerequisites for its appearance still remain unclear. Flow usually involves a merging of actions and awareness with most intense concentration, but in theater a comparable phenomenon may be observed, while people are staying passive. The success or failure of putting up of language surfaces in theater might rely on similar mechanisms of mental interaction like the flow of a web-surfer, whose only choices are to select the next piece of information in each step. However, contrary to the results of our case studies this is so far only philosophical speculation.

In order to understand the interrelations between theater and human-to-system interaction in computer science. the problems of meaning and perception have to examined in more depth. Gibson [2] suggested that perception of the world is based on affordances, that is the perception of opportunities for action. Applying this concept to text objects, meaning would point from an abstract representation of information to an affordance for interaction with the environment (and for exerting control on the environment). Thus the question how to design a virtual space for information exchange reads, how to design this space such that objects may be put into the space which clearly point to some selected affordance. However, we have seen above that meaning and relevance constitute different problems because their importance is different for different scenarios. Thus the design tasks reads: How can we design a virtual space for information exchange that the intended pointer from an object to an affordance is understood and the relevance of this pointer is understood by those spotting the object. Hereby, relevance relates to both objective and subjective factors, as it describes the feasibility and the attractivity of the exploitation by the receiver of the information represented by the referring object.

The design question indicated above concerns the factoring of objects, putting them into the space, and re/perceiving them from/in the space. The 'pointer'-metaphor fits very well with many forms of asynchronous communication through the Internet and synchronous communication on stage. In dramatic theater, textual information objects are carried by characters. This is essential for the process of perception and the deduction of affordances, which is partially carried out synchronously and partially carried out asynchronously (as we think about what we saw on stage). And the reference to the affordances is also deduced from the text itself. Thus, there results an interplay between the meta-meaning and the meta-relationships. It is a characteristic feature of the traditional theater, that there is no real world verification and falsification allowed. It is only imaginary reality, which we are observing on stage. This increases the freedom to change the references with directed randomness suggested by what is going on on stage. Theater thus facilitates self-exploration of the audience without the need for immediate testing of the ideas. This is true both for actresses and actors, and for the audience. Information technology has taken this one step forward with the invention of chat rooms where everyone may perform as an actor if she finds partners for virtual play. These observations suggest that we should try to draw up virtual stages with a limited set of actions, well understood affordances, and monitoring and evaluation facilities in order to investigate how meaning may change during interaction.

Market research in virtual markets [9] considers web-sites as stages with web-surfers as actors and it provides computable informations about actions taken on this stage. The underlying commercial concept is that web-marketing ought to provide customers with affordances, whose exploitation by the customer provides benefit for providers. When these affordances are referred to by information objects published, we can trace how the content of referring information influences the access to affordances, which enables us to conclude on the meaning represented by information objects. Knowing the meaning of words is essential to design successful marketing-sites, and therefore, there is a commercial interest in that type of research. Indeed, market research on log-files can provide us with information, how successful these pointers are, which gives a hint on what kind of meaning plus relevance customers associate with a particular pointer. From another perspective, in a second step of investigation, we could monitor the construction of virtual pointers in the mind in the future. Beyond that, really good marketing sites involve customer interaction, and integrate customer-to-site/customer-to-company as well as customer-to-customer interaction into the site's architectural design, for example by supporting discussion forums. The benefit of this is that the owners of a site can listen to real people talking about real problems with their products and real affordances provided their products.

From these observations we could proceed in two directions: exploiting the character principle for the management of meaning and relevance for general communication support in virtual spaces for information exchange, such as use-groups, and introducing artistic character play in virtual environments. Humans have different emotional artifacts/tools for communication and observations as well as a wide variety of research results indicate that the use of these tools may be modeled as a dynamic system with bifurcations. At some point, some tool obtains dominance over the other tools. This may result from the structure of interaction such as rhythm and challenge, e.g. when flow emerges, but it also may result from reflection, or it may be triggered by the perception of key gestures or key words, in which case meaning is responsible for the bifurcation, i.e. for the selection of a primary communication tool. Right now, we are preparing experiments in order to demonstrate that inference empirically. Thereby, role play will be used to stimulate a selection of a dominating emotional communication tool, and we shall compare surfing behavior with respect to different such tool selections.

As there is a wide area of unsettled questions, we suggest further research to focus on the issue of how 'characters' can be used for putting information objects into virtual spaces for information exchange such that the perception of intended meaning and actual relevance is improved. This will be of particular relevance to virtual marketing, but it might also be applied in virtual teaching. In collaborative filtering and situated user guidance, there have been developed various tools, which support implicit collaboration by the visualization of access of users to information (in case that the user has authorized the system to provide a colleague with tracing data on his access behavior). Compare [4]). Some e-commerce-sites provide their customers on rather explicit information what other groups of customers have bought. Characters could be used that users can classify themselves thus leading an automatic self-clustering of users with a character representation of a cluster. Further, characters could be chosen as guides through a system, providing recommendations and specific information. Contrary to the situation in theater, the reactions of users can be implicitly monitored by analyzing the log-files of the web-Server.

In order to improve our understanding of how web-publishing and experiments on channeling of emotions can be brought together, we have analyzed the log-files of a theater magazine on the Web with about 10 000 page accesses per month. In particular, we have filtered out the context of web-accesses if it was provided by the browser. It turned out that an automatic classification was impossible to achieve, as knowledge of what is on the accessed pages is a conditio sine qua non for the classification of accesses, and in particular for the interpretation of the context provided by the browsers. However, human classification provided some remarkable results. For example, every sixth visitor entering the site upon the suggestion of a search engine seemed to be looking for pornography, The overall amount of visitors who had searched for something else than contained in the theater magazine was close to 40 %. One third of these misdirections was "created" by words in the full text. Although these visitors could have concluded from the summary presented by the search engines, they chose to visit the site, and some enjoyed quite long sessions. Other classifications showed that about two third of the visitors seemed to know very precisely what they were looking for while one third seemed to surf around without a special goal or ambition. It was obvious that motives were strongly differing as was the access behavior. One of the reasons was clearly the different nature and quality of texts - ranging from light poetry to 'heavy' German philosophy, ranging from mature texts to rather immature ones - and the different types of events, to which the texts referred - ranging from children's theater, performed readings, and contemporary dance to classical ballet, provocative actionist performances, Shakespeare and Beckett. Interviews showed that first readers had serious problems in understanding what it was all about, and classification of the contents of the site varied strongly from "fast writing" to "complex philosophical thoughts". Sites similar to the one investigated would be ideal test-beds for research on how people interested in the fine arts interact with hyper-media. The chaotic nature of the site and the multiple meanings either repel visitors or they strongly attract them, which may destroy some stereotypes in surfing and which leave others in place. We shall continue our monitoring of the behavior of readers of the virtual theater magazine in order to find out, how changes over time occur.

4 Conclusions

We may distinguish three different tasks for publishing in virtual spaces for information exchange.

1. Providing meta-information with respect to the information content, which enable tool-based support of the human retrieval tasks, i.e. which points to some class of affordances in a way, which can be understood by a machine

2. Providing meta-information with respect to the relevance, which enable both tool-based support of the human retrieval tasks and human decisions on the relevance of information
3. Providing information, whose meaning is clear for the human user, i.e. which clearly and precisely points to some particular affordance

According to the results of our case studies discussed above, the first task is really the hard task as the dimension of any general ontological space is very high. So far, it seems that solutions only exist for restricted domains of practice and not for information brokering between different such domains. The second task is considerably easier, as we may link relevance with individuals and roles. Information objects can be signed by human publisher with a well defined organizational role or with a social role known to the consumers of information. The first type of information can be processed by expert systems. The second type of information may be compared with tracings of user behavior by a recommender system performing statistical analyses. Depending on the scenario, the results may or may not support automatic recommendations. Further, both types of information help human users to decide on the relevance of an information object. Moreover, users may share information about their interaction behavior with the virtual information space with other users. That may facilitates human retrieval of information based on the social relevance of information. The main difference between the second and the first task is the considerably smaller essential complexity of the problem space for the second task. In addition, time-stamping of information objects may successfully support decisions on the actuality of an information object. Finally, the third task is also less of a problem in various information scenarios, as the content of an information object may be linked with further explanatory information provided for the whole information exchange space. However, the third task is the core problem for any type of virtual information exchange space in interdisciplinary cooperation, or if we head for publishing of information, which can be understood globally by machines. Anyway, the key to an improvement of the success of a virtual space for information exchange is the annotation of information with context, which must be jointly supported by tools as well as by processes and by communication culture.

References

1. B.W. Boehm, A spiral model of software development and enhancement, in ACM Sigsoft Software Engineering Noptes 11/4, 1986
2. J.J. Gibson, The Ecological Approach to Visual Perception, Haughton Mifflin Boston, 1979
3. H. T. Lehmann, Postrdramatisches Theater, Verlag der Autoren, 1999
4. C. Lueg, Supporting Situated Information Seeking: Communication, Interaction, and Collaboration, PhD Thesis, Faculty of Sciences, University of Zurich, Switerland, 1999
5. C Lueg and R. Riedl, How Information Technology Could Benefit from Modern Approaches to Knowledge Management, Proceedings of the 3rd Conference on Practical Applications of Knowledge Management – PAKM 2000, Basel 2000

6. Pawlowski S. D. Pawlowski, S.D. Robey, and A. Raven, Supporting Shared Information Systems: Boundary Objects, Communities, and Brokering, Proceedings ICIS 2001
7. R.Riedl, Agent Views of Electronic Markets, Proceedings SCI'2000, Florida 2000
8. R. Riedl: Need for Trace Benchmarks, in "Performance Evaluations with Realistic Applications", edited by R. Eigenmann (SPEC), MIT-Press 2001
9. R. Riedl, Customer-Centered Models for Web-sites and Intra-nets, Proceedings of HPCN Europe 2001, Amsterdam, 2001
10. Report-Based Re-engineering of Web-Sites and Intranets, Proceedings of e-business and e-work, Venice 2001 (in preparation)
11. J. Sterne, World Wide Web Marketing: Integrating the Web into Your Marketing Strategy, 2nd Edition, Wiley and Sons, 1999
12. E. Wenger, Communities of Practice: Learning, Meaning, and Identity, Cambridge University Press, 1998

Cognitive Dimensions of Notations: Design Tools for Cognitive Technology

A.F. Blackwell[1], C. Britton[2], A. Cox[2], T.R.G. Green [3], C. Gurr[4], G. Kadoda[5], M.S. Kutar[2], M. Loomes[2], C.L. Nehaniv[2], M. Petre[6], C. Roast[7], C. Roe[8], A. Wong[8], and R.M. Young[2]

[1] University of Cambridge, [2] University of Hertfordshire, [3] University of Leeds, [4] University of Edinburgh, [5] University of Bournemouth, [6] Open University, [7] Sheffield Hallam University, [8] University of Warwick, U.K.
Alan.Blackwell@cl.cam.ac.uk

Abstract. The Cognitive Dimensions of Notations framework has been created to assist the designers of notational systems and information artifacts to evaluate their designs with respect to the impact that they will have on the users of those designs. The framework emphasizes the design choices available to such designers, including characterization of the user's activity, and the inevitable tradeoffs that will occur between potential design options. The resulting framework has been under development for over 10 years, and now has an active community of researchers devoted to it. This paper first introduces Cognitive Dimensions. It then summarizes the current activity, especially the results of a one-day workshop devoted to Cognitive Dimensions in December 2000, and reviews the ways in which it applies to the field of Cognitive Technology.

1 Introduction

The title of this meeting is *Cognitive Technology: Instruments of Mind.* In this paper, we try to characterize the ways that the instruments of our minds are compromised by the restrictions that our bodies and physical environment place on them. This can be regarded as a proposed approach to the study and practice of cognitive ergonomics. Moreover, it also represents as an approach toward meeting the goals of Cognitive Technology by developing methodological tools with which to describe, analyze, and predict the cognitive impact that existing artifacts and artifacts under design will have on their human users. Let us consider a (trivially) simple example to start with. Any cognitive technology transfers information from our heads to our physical environment so that we can "offload" it from short-term memory, and also so we can interact with it. A piece of paper with visible marks on it is one of the simplest such technologies. A very large piece of paper with many small marks can carry a great deal of information, and represent complex structures. But there are limits imposed on this information and its complexity. They are not imposed by the piece of paper (which can be made arbitrarily large) but by our bodies. There is a limit on the ability of our eyes to see far away, and especially on their ability to resolve small marks that are far away. These limitations have predictable effects on the value of this particular cognitive technology: Where we might want to gain a visual overview of the whole infor-

M. Beynon, C.L. Nehaniv, and K. Dautenhahn (Eds.): CT 2001, LNAI 2117, pp. 325-341, 2001.

mation structure, we cannot do so because we can't see all of it at once. When we need to refer to some specific component of the information, we must search for it by scanning the paper a section at a time.

These observations may seem trivial in the case of large pieces of paper, which are limited (to say the least) as representatives of cognitive technology. But the limitations can be even more severe in more advanced cognitive technologies. Digital technologies can record far more information than single sheets of paper, and they can describe far more complex information structures, not limited by a two dimensional surface. But despite the promise of ubiquitous computing, wall-sized displays and intelligent paper, we generally find that the computer screen only offers a restricted window onto these large and complex information structures. This means that the problem of visibility – initially only a physical restriction imposed by our eyes – also becomes a problem of how to control mechanisms for scrolling and zooming. If we use our arms, hands and fingers to operate them, the simple problem of reading an information display becomes compromised by our bodily limits: manual dexterity, reaction times, positional stability and other factors.

Thus far we have only considered the question of *visibility*, and we have assumed that the user of this information artifact is simply reading information off the display. In fact most of the interesting applications of cognitive technology involve more complex activities – creating information structures, modifying them, adding information to them, or exploring possible design options for completely new information structures. Visibility is an important consideration for almost all of these activities, but many of them place additional constraints on the user beyond simple physical perception and interaction. Examples include *viscosity* – the difficulty of making small changes to the information structure; *provisionality* – ways in which the user can express parts of the structure that are not yet precisely defined; and many others. We call these attributes of information artifacts the *Cognitive Dimensions of Notations* (CDs). In the same way that visibility has a predictable relationship to important aspects of the cognitive activity of reading (above we observed ability to see overall structure, efficiency of searching for specific components), so the other CDs can be used to predict the consequences of using an information artifact for other types of activity.

Who needs this kind of analysis? It is clear that we are not saying anything profound about human cognition. Neither are we saying anything new about sophisticated information structures, algorithms or tools. The reason that the CDs framework has been developed is that people who are designing new information artifacts – the developers of cognitive technologies – often find themselves encountering the same problems over and over again when designing different systems.

Expert designers of cognitive technologies learn by experience, and eventually (with luck) produce well-designed information artifacts that are appropriate to the user's activity. Unfortunately many developers of new cognitive technologies are not expert at anticipating and providing for the user's needs. They are computer scientists or engineers who understand the technical problems they are addressing far better than they understand the problems of the user. We believe that this problem is best addressed by providing a vocabulary for discussing the design problems that might arise – a vocabulary informed by research in cognitive psychology, but oriented toward the understanding of a system developer. The Cognitive Dimensions of Notations are such a vocabulary. There are other techniques for analyzing the usability of computer systems, but these often focus on the finest details of interaction – key-press times, visual recognition or memory retrieval. Instead the CDs framework attempts to describe the most relevant aspects of our interaction with information artifacts at a broad-brush level, intended to be useful as a discussion tool for designers.

2 The Role of Cognitive Dimensions

Cognitive Dimensions of Notations (CDs) is a framework for describing the usability of notational systems (e.g. word processors, computer-aided design tools, or music notation) and information artifacts (e.g. watches, radios or central heating controllers). The CDs framework aims to do this by providing a vocabulary that can be used by designers when investigating the cognitive implications of their design decisions. Designers of notational systems do realize that their decisions have an impact on usability, and that the usability problems with notations have cognitive implications for the user. But many designers only know those things in an intuitive way. This makes it difficult for them to discuss usability issues, especially as they seldom have any formal education in cognitive psychology.

This situation becomes more serious in cases where the design process involves making decisions about design tradeoffs. Perhaps the design can be improved in one respect, but only at the expense of making it worse in some other respect. Or perhaps it can be made more appropriate for a particular user group (e.g. the elderly), but only at the expense of becoming less usable for some other user group (e.g. those who have very little time). Or more insidiously, perhaps the design can be altered so that it is suitable for users when they are carrying out a certain task, but then becomes unusable for another important task. As an example, consider a notation that expresses some complex procedure on a screen in flow diagram form. Flow diagrams make the possible interactions between different events a lot clearer, but they take up more room on the screen than a simple textual list. And if the user is actually modifying the diagrams, all the connecting lines make it more difficult to change the diagram because they have to be moved around and tidied up after changes. These are generic properties of notational systems, which CDs describe by names like *hidden dependencies* (the visibility of relationships), *diffuseness* (the amount of space that the notation takes up), and *viscosity* (the amount of effort required to make small changes to the notation).

None of these is necessarily a problem; that depends on what the user wants to do – e.g. viscosity is not a problem if the user doesn't need to make any changes. So the framework considers dimensions in the context of user activities.

The CDs framework has been designed for situations where the designer is making choices about notations or representations, and where usability tradeoffs are a factor in the design. It is particularly difficult to design new notational systems and information artifacts. CDs describe some common properties of notations that allow the designer to anticipate the effect of design changes, and make more conscious choices about tradeoffs without having actually to build and evaluate prototypes.

The development of the CDs framework was initiated by Thomas Green in a 1989 publication (Green 1989). Since then over 50 research papers have been published on topics related to the CDs, including a longer description applying CDs to the domain of visual programming languages (Green and Petre 1996) and a tutorial aimed at professional designers (Green and Blackwell 1998). This paper reports the results of a meeting held in December 2000 at the University of Hertfordshire of researchers who are currently pursuing projects related to CDs. It describes how the state of the art in CDs research can contribute to the overall objectives of cognitive technology.

3 Summary of the CDs Framework

As mentioned above, we describe CDs as providing, not only a vocabulary for use by designers, but a framework for thinking about the nature of notational systems and the way that people interact with them. This framework provides a structure in which to understand the vocabulary itself, but also includes a number of theoretical activities that extend beyond the demands of many designers applying the vocabulary in more restricted design contexts.

The framework includes definitions of notations and notational systems, characterization of the human activities involving notational systems, a description of the ways that multiple notations can interact within a single system, and a minimal process for applying the resulting insights in a design context for use in evaluating and improving a design. More recently, as larger numbers of researchers have adopted the CDs framework as a research tool, the framework has also developed some reflective components applicable to extending and refining the framework itself. The later parts of this paper, and the workshop from which it has been derived, deal with this latter aspect.

However it is first necessary to review the established parts of the framework. We start with the definitions of notational systems. A notation consists of marks (often visible, though possibly sensed by some other means) made on some medium. Examples include ink on paper, patterns of light on a video screen, and many others. It is possible for several notations to be mixed within a single medium: a computer screen may display multiple windows, each running a different application with its own notation. Even within a window, there may be multiple notations – the main notation of the application, but also generic sub-notations such as menu bars, dialogs etc. A notational system contains both a notation and an environment (such as an editor) for manipulating that notation. CDs describe usability properties of the system, not just the notation. Where the system includes sub-notations, users generally interact with them through sub-devices, which have their own cognitive dimensions. We describe some self-contained notational systems as “information artifacts”. These include things such as telephones, central heating controls, and many ubiquitous automated systems beyond the range of typical computer applications. In all these cases, the notation expresses some structure, more or less complex.

It is important to note that none of the cognitive dimensions are necessarily good or bad by themselves. The usability profile of a system or artifact depends on what kind of activity the user will be engaging in, and on the structure of the information contained in the notation. The activities that are least demanding in terms of usability profile are simply *searching* for a single piece of information (such as looking up a name in a telephone book) and *incrementally understanding* the content of the information structure expressed by a notation (such as reading a textbook). The more interesting activities are those that involve extending the notation: *incrementing* an existing structure by adding new information, *transcribing* information from one notational form to another, *modifying* the structure, or exploring possible new information structures in *exploratory design*.

These are the main theoretical foundations of the framework – at this point we will give a brief review of the set of dimensions, with thumbnail definitions of each. These descriptions are very brief – note that they are more fully described, with illustrative examples and explanation, in many other publications, including a tutorial that is available online (Green and Blackwell 1998).

Review of Dimensions

Viscosity: Resistance to Change.
A viscous system needs many user actions to accomplish one goal. Changing all headings to upper-case may need one action per heading. (Environments containing suitable abstractions can reduce viscosity.) We distinguish repetition viscosity, many actions of the same type, from knock-on viscosity, where further actions are required to restore consistency.

Visibility: Ability to View Components Easily.
Systems that bury information in encapsulations reduce visibility. Since examples are important for problem-solving, such systems are to be deprecated for exploratory activities; likewise, if consistency of transcription is to be maintained, high visibility may be needed.

Premature Commitment: Constraints on the Order of Doing Things.
Self-explanatory. Examples: being forced to declare identifiers too soon; choosing a search path down a decision tree; having to select your cutlery before you choose your food.

Hidden Dependencies: Important Links between Entities Are Not Visible.
If one entity cites another entity, which in turn cites a third, changing the value of the third entity may have unexpected repercussions. Examples: cells of spreadsheets; style definitions in Word; complex class hierarchies; HTML links. There are sometimes actions that cause dependencies to get frozen – e.g. soft figure numbering can be frozen when changing platforms; these interactions with changes over time are still problematic in the framework.

Role-Expressiveness: The Purpose of an Entity Is Readily Inferred.
Role-expressive notations make it easy to discover why the programmer or composer has built the structure in a particular way; in other notations each entity looks much the same and discovering their relationships is difficult. Assessing role-expressiveness requires a reasonable conjecture about cognitive representations (see the Prolog analysis below) but does not require the analyst to develop his/her own cognitive model or analysis.

Error-Proneness: The Notation Invites Mistakes and the System Gives Little Protection.
Enough is known about the cognitive psychology of slips and errors to predict that certain notations will invite them. Prevention (e.g. check digits, declarations of identifiers, etc) can redeem the problem.

Abstraction: Types and Availability of Abstraction Mechanisms.
Abstractions (redefinitions) change the underlying notation. Macros, data structures, global find-and-replace commands, quick-dial telephone codes, and word-processor styles are all abstractions. Some are persistent, some are transient.

Abstractions, if the user is allowed to modify them, always require an abstraction manager -- a redefinition sub-device. It will sometimes have its own notation and environment (e.g. the Word style sheet manager) but not always (for example, a class hierarchy can be built in a conventional text editor).

Systems that allow many abstractions are potentially difficult to learn.

Secondary Notation: Extra Information in Means Other Than Formal Syntax.
Users often need to record things that have not been anticipated by the notation designer. Rather than anticipating every possible user requirement, many systems support secondary notations that can be used however the user likes. One example is comments in a programming language, another is the use of colors or format choices to indicate information additional to the content of text.

Closeness of Mapping: Closeness of Representation to Domain.
How closely related is the notation to the entities it is describing?

Consistency: Similar Semantics Are Expressed in Similar Syntactic Forms.
Users often infer the structure of information artifacts from patterns in notation. If similar information is obscured by presenting it in different ways, usability is compromised.

Diffuseness: Verbosity of Language.
Some notations can be annoyingly long-winded, or occupy too much valuable "real-estate" within a display area. Big icons and long words reduce the available working area.

Hard Mental Operations: High Demand on Cognitive Resources.
A notation can make things complex or difficult to work out in your head, by making inordinate demands on working memory, or requiring deeply nested goal structures.

Provisionality: Degree of Commitment to Actions or Marks.
Premature commitment refers to hard constraints on the order of doing things, but whether or not hard constraints exist, it can be useful to make provisional actions – recording potential design options, sketching, or playing "what-if" games.

Progressive Evaluation: Work-to-Date Can Be Checked at Any Time.
Evaluation is an important part of the design process, and notational systems can facilitate evaluation by allowing users to stop in the middle to check work so far, find out how much progress has been made, or check what stage in the work they are up to. A major advantage of interpreted programming environments such as BASIC is that users can try out partially-completed versions of the product program, perhaps leaving type information or declarations incomplete.

Application

In a design context, the dimensions would be applied after identifying a "main" notation to be analysed. In the course of the analysis, sub-devices might be identified, offering separate notations for purposes such as extending the main notation (an abstraction manager sub-device). The designer would assess usability with respect to some activity profile describing the activities that the user is likely to carry out. The dimensional characteristics of the notational system can then have their implications assessed with respect to that profile. Where problems are identified, the framework offers design manoeuvres by which those problems might be addressed although they potentially involve tradeoffs, in which changing the design of the notational system on one dimension may result in additional changes of the system properties on another dimension.

4 Current Frontiers in CDs Research

This section summarizes the presentation, some discussion, and the results from the December workshop described above.

Activities and Profiles

Profiles are where users' activities mesh with the cognitive dimensions of the notation: a profile specifies what is needed to support an activity. No dimension is evaluative on its own - one can't know whether it is relevant until one knows what activity is to be supported. There have been several attempts to define a broadly useful set of generic activities. Hendry and Green (1994) defined three different types of activity using notational structures: incremental growth, transcription and presentation. This list has been refined in various ways. The original CDs tutorial defined four activities in constructing notations: *transcription*, *incrementation*, *modification* and *exploratory design*. Soon afterward the CDs questionnaire for user evaluation added a fifth: *search*.

The December workshop also considered the newly proposed *exploratory understanding*, which is relevant both to notational tools such as software visualization systems, and to distributed notations such as the world-wide web. We expect that this will offer new insights from related analysis techniques such as information foraging theories. There may be further activities related to other areas of human activity that have not yet been addressed by CDs to date. The workshop offered some possible new activities including *play*, *competition*, and *community building*. But these are dangerous – the addition of new activities introduces credibility obstacles for the framework to a greater extent than the addition of new dimensions does.

We also feel that the activities are currently formulated in too abstract a way, despite the fact that they are critical to the evaluative use of CDs. We have taken great pain that every dimension should be described with illustrative examples, case studies, and associated advice for designers. Activities, on the other hand, are described at a rather abstract level in terms of the structure of information and constraints on the notational environment. This makes it unlikely that usability profiles will be exploited effectively. The workshop concluded that the activities must be paraphrased in everyday language to make them as accessible to designers as the dimensions themselves. These descriptions will be supplemented by examples of relevant tasks, some of which may be juxtaposed within the context of a specific class of information artifact: this is currently being pursued through a series of simulated central heating controllers, implemented in JavaScript, and available on the web through the CDs archive site.

Britton, Kutar, and Jones have studied the creation of a CDs profile for a specific task: the validation of a requirements specification, and reported on this work at the workshop. They wished to evaluate the comprehensibility of different specification languages for non-specialist readers. This profile therefore measured the *intelligibility* of specifications, characterized by the user activities of a) extracting information from the representation and b) checking the correspondence of the represented information with existing knowledge. These activities are not externally observable, but form the basis for user activities that can be observed.

Selecting a limited set of dimensions resulted in a more streamlined profile and allowed them to concentrate on those dimensions that were of particular interest. These were then used to compare two specifications of the temporal aspects of an interactive system. One was written in the logic language $TRIO_{\neq}$, the other in a version of extended to make temporal properties easier to understand. The conclusion was that prior selection of a subset of CDs may be unhelpful. Using the full set of dimensions can produce some unexpected, but useful results and should be done in order to discover as much information as possible. This suggests that profiles should describe the weighting of dimensions for different activities, rather than attempting to eliminate dimensions. (See also the related paper by Kutar, Nehaniv, and Britton in this volume, which discusses the cognitive impacts of various design choices for notations used to specify temporal properties of interactive.)

The evaluation of some notational system should always be conducted according to a defined profile of use – we suggest that this might be called a profile *instance*, as opposed to more generic sets of dimensions with associated consequences and trade-offs that would be called a profile *class*. The result of assessment for a specific profile instance is a CDs *assessment*. CDs assessment can be achieved by relatively untrained users of CDs, while the creation of new profile classes is more difficult, potentially requiring the assistance of CDs researchers acting as consultants. This effort might be reduced by creating profile *clusters* that describe a group of related profiles. The process of assessment itself will be facilitated by having a better-constructed set of standard questions, such as: what is the notation of the main device; how do the dimensions apply to it; what abstractions are available; are there abstraction managers; and are the abstractions transient or persistent?

Trade-Offs

Trade-offs are frequently-observed patterns in CDs analyses – they are situations in which one source of difficulty is fixed at the expense of creating another type of difficulty. At present too little is known about what trade-offs occur in real life, but some observations will be reported. Questions arising are: are these tradeoffs correctly identified and specified? Are they always correct, or only in certain situations? Can we find more examples? Is there a methodology we can use to account for and correct them? How do we (CDs researchers) communicate the ideas for use by designers?

One way to communicate is by looking for everyday language; see for instance the questionnaire developed by Blackwell and Green. This questionnaire, along with other resources, is available from the CDs archive site – a URL is included in the bibliography. Another attempt at communication is to present working examples for consideration. All the examples need to present alternative solutions, in a minimalist form, in order to emphasize the tradeoffs. Some examples can be seen at the following URL:
http://www.ndirect.co.uk/~thomas.green/workStuff/devices/controllers/HeatingA2.html

Formalization

Several current research projects are investigating approaches to formalization of CDs. At its most basic level such a theory would be expected to be descriptively adequate - replicating examples of cognitive dimensions. A more mature theory would be expected to predict instances of dimensions and provide general theorems regarding cognitive dimensions. Clearly, the eventual goal is a theory which is valid within recognized boundaries and which is capable of directly contributing to our understanding.

To aid the process of validation Roast et. al. have developed a tool for modeling formal interpretations of the dimensions (called CiDa). The tool is designed to support theory validation through enabling the consequences of posited CD definitions to be examined. CiDa analysis requires that the target system is modeled in terms of simple non-deterministic state based machine and that states of this machine are associated with potential user goals. The objective of this work is to develop CDs theory through an example-driven approach, where it is the artifact that is modeled rather than the cognitive processes. The ideal is that it should be possible to observe the artifact, model it, and then validate the model. CiDa creates formal models of a variety of tasks, rather than being restricted to tasks that have been selected to illustrate specific CDs.

The Empirical Modelling (EM) research group based at the University of Warwick aim to analyze artifacts by focusing on identifying patterns of agency and dependency through observation and experiment, and embodying these patterns in computer models called Interactive Situation Models (ISMs). An artifact comprises many different aspects of state. The explicit state is the visible state of the artifact. The internal state is all the physical states of the information artifact. The mental state is the state that users project upon the artifact when considering expectations about possible next state/interpretation of current state. The situational state is knowledge of the real world context to which the artifact refers. The EM group suggest that CDs relate to the way in which the above aspects of state interact in trying to make appropriate use of an information artifact. Their current research indicates that the construction of an ISM of an artifact may give a modeller a better understanding of the CDs of that artifact. This work is reported in detail elsewhere in this meeting. For more information see: http://www.dcs.warwick.ac.uk/modelling .

Operationalization

An alternative approach to formalization is operationalization: identifying practical questions and activities that help designers of information artifacts to reason about cognitive consequences of making a particular collection of design choices. This work starts from the perspective that cognitive dimensions lay out a design space, and that they provide a 'broad brush' framework supporting reasoning about how those choices place the design in the space. Again, cognitive dimensions are not binary, but descriptive, establishing where in a space of inter-related factors and choices a design lies.

As demonstrated in the Green and Petre (1996) paper, this approach identifies pragmatic 'yardsticks' and 'straw tests'. These are not canonical or definitive tests, but simply a set of practical questions used to fuel a cognitive dimensions analysis. They are cast in operational terms: they enquire how the effects of the dimension translate into work required. They are meant both to make the evaluation concrete and to provide a basis for comparison between designs or design choices. For example, regarding 'Hidden dependencies': Is every dependency overtly indicated in both directions? Is the indication perceptual or only symbolic? Or regarding 'Imposed Look-Ahead': Are there order constraints? Are there internal dependencies?

For some dimensions, we also apply some 'straw' tests: simple tests based on typical activities (modification, examination, comparison) and chosen to measure 'work done' in terms of the dimension. For example, timing typical modifications in order to evaluate 'viscosity'.

The value in this approach is its immediacy; the usage is pragmatic and accessible, making a cognitive dimensions analysis a low-cost tool to add to a design repertoire. Putting CDs readily into use is the best way to demonstrate their relevance to practice. But the process of operationalization itself is informative and feeds back into cognitive dimensions theory, giving perspective on definitions and concepts, exposing interrelationships among design choices, reflecting on the impact of tasks and environments, and so on.

5 Extending the Framework with New Dimensions

The core of the Cognitive Dimensions of Notations framework is the list of dimensions itself. This list has been gradually expanding – Thomas Green's early publications (Green 1989,1990,1991) described only a few selected dimensions, as did other researchers in early publications (Gilmore 1991). By the time the Green and Petre (1996) paper was published, 13 dimensions were listed. Green and Petre did not claim that the set of dimensions was then complete. On the contrary, they have continued to encourage discussion of new additions. As it turns out, the process of defining new dimensions has slowed down. This may partly be because the existence of a definitive publication made the initial step of defining one more dimension a daunting one.

More importantly, few researchers have seen the addition of new dimensions as an important end in itself. The 1996 paper, under the heading of "Future progress in cognitive dimensions", observed that the framework was incomplete – but not in the sense that more dimensions were urgently needed. Rather it emphasized the need for formalization and applicability.

Nevertheless, new dimensions do get proposed from time to time. Some of these proposals have been published, but more of them exist only in the form of informal conversations with Green and other central researchers. But it is neither necessary nor desirable for the development of the framework to depend on any individual acting as a gatekeeper / coordinator for new additions. The December workshop therefore considered possible future approaches to the process of identifying and defining new Cognitive Dimensions.

Some Examples

Some examples of a few candidate dimensions, taken from informal sources, are included here. Some of these have been published before, but most are appropriated from other research fields (in the sense that they are inspired by authors who did not consider themselves to be working on cognitive dimensions). None of them should be considered at this stage to have canonical status – in fact the question of how to assemble the canon is the main topic of discussion.

Creative Ambiguity
The extent to which a notation encourages or enables the user to see something different when looking at it a second time (based on work by Hewson (1991), by Goldschmidt (1991), and by Fish and Scrivener (1990))

Specificity
The notation uses elements that have a limited number of potential meanings (irrespective of their defined meaning in this notation), rather than a wide range of conventional uses (based on work by Stenning and Oberlander 1995)

Detail in Context
It is possible to see how elements relate to others within the same notational layer (rather than to elements in other layers, which is role expressiveness), and it is possible to move between them with sensible transitions, such as Fisheye views (based on work by Furnas (1986) and by Carpendale, Cowperthwaite and Fracchia (1995))

Indexing
The notation includes elements to help the user find specific parts.

Synopsie
(originally "grokkiness") The notation provides an understanding of the whole when you "stand back and look". This was described as "Gestalt view" by some of the respondents in the survey by Whitley and Blackwell (1997).

Free Rides
New information is generated as a result of following the notational rules (based on work by Cheng (1998) and by Shimojima (1996))

Useful Awkwardness
It's not always good to be able to do things easily. Awkward interfaces can force the user to reflect on the task, with an overall gain in efficiency (based on discussions with Marian Petre, and work by O'Hara & Payne (1999))

Unevenness
Because things are easy to do, the system pushes your ideas in a certain direction (based on work by Stacey (1995))

Lability
The notation changes shape easily

Permissiveness
The notation allows several different ways of doing things (based on work by Thimbleby, not yet published).

Where Do They Come From?

As is apparent from the above list, most candidates for new dimensions come from other research, whether or not the author is aware of the CDs framework. This is a good thing. One objective of CDs is that they should be credibly derived from psychological or cognitive science research. This is largely what gives them authority among notation designers (and the implication is intentional, through the use of the word "cognitive").

This suggests that an immediate point of good practice would be to encourage the participation of the original researchers in the process of defining new dimensions. This would obviously include due credit via citation of the author's original work, as well as the opportunity for the original author to review the dimension derived from his or her work – both our characterization of the dimension itself, and the way that it is related to the rest of the framework through profiles, tradeoffs, dependencies and design manœuvres.

Criteria for Acceptance

What are the criteria that define a good (or even an acceptable) new cognitive dimension of notations? The process by which the current set were derived has been the subject of reflection, but not thorough documentation.

As the number of dimensions grows, it is also becoming crucial to identify a useful subset for new users (including undergraduate courses). Commercial users are already impatient with the size of the set that exists now. We could perhaps create a CDs-lite for commercial friends – perhaps with 7 plus or minus 2 dimensions. These might be selected as the most important, or possibly the easiest to understand. We might possibly adopt Jack Carroll's minimal documentation approach to presentation, so that people only have to deal with the dimensions that they need.

Orthogonality

Most important, the term "dimension" was chosen to imply that these are mutually orthogonal – they all describe different directions within the design space. Furthermore, it is hoped that the trade-off relationships between them might be similar to those of the Ideal Gas Law – so that it is probably not possible to design a notation system that achieves specific values on any two dimensions, without having the value of a third imposed by necessary constraints. But these notions of orthogonality are intuitive rather than exact, and they are described in this way mainly so that designers recognize the nature of the constraints on their design. There is ongoing work on formalization of dimensions that should allow more precise statements to be made regarding orthogonality and trade-offs for a few dimensions, but such analysis cannot yet be required when proposing new dimensions.

Instead, mutual orthogonality can only really be tested at present via a qualitative approach – going through all current dimensions, and checking to see whether any of them might describe the same phenomenon as that described by the proposed new dimension. This checking ought to be done by more than one person. It is so common for individual researchers to misunderstand the nature of one or two of the dimensions, that it is highly likely a proposed new dimension will simply be a rediscovery of an existing dimension (which the researcher had understood to refer to something else). It is also necessary to be aware that the new dimension might simply be the obverse case of an existing dimension.

Granularity

The CDs seem to describe activities at a reasonably consistent level of granularity. They should probably continue to describe phenomena at a similar scale. They do not directly describe large cognitive tasks (design a system, write a play), but the structural constituents of those tasks. They also tend not to describe low-level perceptual processes. Some things that are too low a level of granularity might include Gestalt phenomena, or observations related to individual motions (e.g. selection target size, as analyzed by Fitts' law). If they were to be characterized using GOMS analysis, we might say that CDs do not apply either to leaf nodes in the goal tree, or to the whole tree, but to sub-trees.

Object of Description

There is an outstanding question regarding what it is that the dimensions are supposed to describe. Some possible options for suitable objects of description (no doubt not a complete list) are:

(i) structural properties of the information within the notation/device
(ii) the external representation of that structure
(iii) the semantics of that information
(iv) the relationship between the notated information and domain-level concepts – some of which are inevitably not notated

Depending on which of these are chosen, the CDs field gets bigger or smaller. Useful awkwardness and permissiveness are both defined partly by domain-level concepts, so they might not be members of the CDs list, if we restrict objects of description to (say) (i) & (ii).

Effect of Manipulation
It ought to be possible to consider each dimension and say 'if you change the design in the following way, you will move its value on this dimension'. This is a criterion of understanding how the dimension works, as well as the basis for design manœuvres. When we define a new dimension, we should be able to say something about how to manipulate it.

Applicability
One of the desirable properties of a CD is that it should make sense to talk about it in a wide range of different situations. This has not always been achieved with the current set of dimensions.

Polarity
As CDs are not supposed to be either good or bad (more on this below), they should have interesting properties in both directions – i.e. both when present and absent. Error-proneness is not a very good dimension when considered from this perspective.

Choosing Names
It is hard to find good names for new dimensions. "Grokkiness" (which persisted for almost a year) shows just how hard it is! Some of the criteria for good names include:

Length of Name
It seems like one or two words should be enough (Closeness of Mapping is really on the limit).

Vernacularity
CDs should sound both technical and approachable at the same time. They must sound sufficiently technical that they don't get confused with everyday meanings, and that they can be accorded some respect by notation designers. In an effort to get something sufficiently technical, we have sometimes had mixed results, either by resorting to neologism (grokkiness) or archaism (synopsie).

There is also a problem of cultural specificity. It turns out that the term "knock-on viscosity" is unintelligible to Americans (recently reported by Margaret Burnett, and confirmed by several other delegates at VL2000). Some Americans guess correctly, but others think that it might have something to do with door knockers. They have suggested "domino" or "consequent" viscosity – is either of these too technical, or too approachable?

Polarity
It gives a false impression of the CDs framework if readers treat the dimensions as representing "usability problems" rather than trade-offs. But this constantly happens, especially if the audience is already familiar with Nielsen's heuristic analysis of usability. We have partly caused the problem ourselves, because most of the names do imply negative consequences "Hidden dependencies" rather than "Visible dependencies", for example. There are several options for addressing this problem:

- Choose neutral names (desirable, but hard to achieve).
- Purposely choose names with alternating obverse polarities.
- Choose positive names if at all possible (to avoid the usability problem assumption).
- Provide dual definitions for all dimensions, illustrating positive and negative aspects.

With regard to polarity, it is also important to remember that dimensions only become evaluative when applied to some specific activity. For this reason, it should be possible to describe the characteristics of a dimension without any evaluative emphasis – evaluative observations should ideally be localized within the profile.

Supporting Apparatus
A cognitive dimension is more than just a name and a definition. All of the current dimensions are supported by a range of documentary and tutorial apparatus.

Examples
Each dimension is supported by examples of situations in which it can occur, with the consequences of that occurrence. There should be one "killer example" that immediately reveals to the reader the essence of the dimension. Ideally, examples should be drawn both from programming and other user interface domains.

Pictorial Examples
In future, it would be very useful for every "killer example" to be supported by a pictorial illustration that can be incorporated in published papers referring to and citing the dimension. There is no real harm in repeating the same illustration, and a nicely illustrated example would help to promulgate CDs as a whole. We hope to add some examples of such reusable illustrations to the Cognitive Dimensions archive site.

Impact
Different dimensions have different impacts on various activity types and profiles. Some kind of characterization should be attempted.

Trade-Offs
Should be noted. But if there is a specific trade-off that invariably occurs, that might be a sign that this dimension is only the obverse case of an existing dimension, rather than an orthogonal dimension.

Sources
Research sources should be cited, both as supporting evidence, and also to give appropriate credit to previous researchers.

Manœuvres and Workarounds
It is valuable to have some observations regarding design manœuvres and also the ways that users might try to work around the effects of the dimension.

6 Conclusion

Many of the usability evaluation methods that have been applied to cognitive technologies in the past were derived from models of machine ergonomics, stressing manual efficiency rather than appropriateness to the user. A reaction to this has now led to an alternative emphasis on anecdotal transfer of trade skills and aesthetic criteria (as, for example, in Tufte, 1983). The current usability criteria for activities such as Information Architecture for Web design combine these cognition-free accounts of design criteria with an idealized view of the contributions offered by technological innovation.

The CDs framework offers an account of information artifacts that respects the value of the user's activity, seeking to recognize the cognitive constraints that the artifact places on that activity. This is very much in accordance with the overall goals of the Cognitive Technology field.

The CDs framework has, over the last 10 years, developed into a useful tool. But it is not complete, and further work remains to be done. This paper has presented a "state of the nation view" from active researchers in the field, and also offered a joint agenda for ongoing research. Within the context of Cognitive Technology, this has served two purposes. First, the ultimate goals of the CDs framework are closely aligned with those of Cognitive Technology, and we wish to see further cross-fertilization in future. Second, we have offered in this paper an insight into the process of developing and maintaining a theoretical framework as it is transferred into the wider research community and also to industrial practitioners. We believe that this process of "rubbing up against" a broader community of users and collaborators has enriched the CDs framework. This is an experience that we recommend to other researchers developing theoretical models for Cognitive Technology.

References

Note that many of these publications are available online from the Cognitive Dimensions archive site: http://www.cl.cam.ac.uk/~afb21/CognitiveDimensions/

Blackwell, A.F. & Green, T.R.G. (2000). A Cognitive Dimensions questionnaire optimised for users. In A.F. Blackwell & E. Bilotta (Eds.) *Proceedings of the Twelfth Annual Meeting of the Psychology of Programming Interest Group*, 137-152.

Carpendale, M.S.T., Cowperthwaite D.J. and Fracchia, F. D. (1995). 3-Dimensional pliable surfaces for the effective presentation of visual information information navigation. *Proceedings of the ACM Symposium on User Interface Software and Technology* p.217-226.

Cheng, P.C. (1998). AVOW diagrams: A novel representational system for understanding electricity. In *Proceedings Thinking with Diagrams 98: Is there a science of diagrams?* pp. 86-93.

Fish, J. & Scrivener, S. (1990). Amplifying the mind's eye: Sketching and visual cognition *Leonardo*, 23(1), 117-126.

Furnas, G.W. (1986). Generalized fisheye views visualizing complex information spaces. *Proceedings of ACM CHI'86 Conference on Human Factors in Computing Systems* p.16-23.

Gilmore, D. J. (1991) Visibility: a dimensional analysis. In D. Diaper and N. V. Hammond (Eds.) *People and Computers VI*. Cambridge University Press.

Goldschmidt, G. (1991). The dialectics of sketching. *Creativity Research Journal*, 4(2), 123-143.

Green, T. R. G. & Petre, M. (1996) Usability analysis of visual programming environments: a 'cognitive dimensions' framework. *Journal of Visual Languages and Computing*, 7, 131-174.

Green, T. R. G. (1989). Cognitive dimensions of notations. In *People and Computers V*, A Sutcliffe and L Macaulay (Ed.) Cambridge University Press: Cambridge., pp. 443-460.

Green, T. R. G. (1990) The cognitive dimension of viscosity: a sticky problem for HCI. In D. Diaper, D. Gilmore, G. Cockton and B. Shackel (Eds.) *Human-Computer Interaction — INTERACT '90*. Elsevier.

Green, T. R. G. (1991) Describing information artefacts with cognitive dimensions and structure maps. In D. Diaper and N. V. Hammond (Eds.) Proceedings of "*HCI'91: Usability Now*", Annual Conference of BCS Human-Computer Interaction Group. Cambridge University Press.

Green, T.R.G. & Blackwell, A.F. (1998). Design for usability using Cognitive Dimensions. Tutorial presented at *British Computer Society conference on Human Computer Interaction HCI'98*. Available online from the Cognitive Dimensions archive site http://www.cl.cam.ac.uk/~afb21/CognitiveDimensions/

Hendry, D. G. and Green, T. R. G. (1994) Creating, comprehending, and explaining spreadsheets: a cognitive interpretation of what discretionary users think of the spreadsheet model. Int. J. Human-Computer Studies, 40(6), 1033-1065.

Hewson, R. (1991). Deciding through doing: The role of sketching in typographic design. *ACM SIGCHI Bulletin*, 23(4), 39-40.

O'Hara K.P., and Payne, S.J. (1999). Planning and the user interface: The effects of lockout time and error recovery cost *International Journal of Human-Computer Studies* 50(1), 41-59.

Shimojima, A. (1996). Operational constraints in diagrammatic reasoning. In G. Allwein & J. Barwise (Eds) *Logical reasoning with diagrams*. Oxford: Oxford University Press, pp. 27-48.

Simos, M. & Blackwell, A.F. (1998). Pruning the tree of trees: The evaluation of notations for domain modeling. In J. Domingue & P. Mulholland (Eds.), *Proceedings of the 10th Annual Meeting of the Psychology of Programming Interest Group*, pp. 92-99.

Stacey, M. K. (1995) Distorting design: unevenness as a cognitive dimension of design tools. In G. Allen, J. Wilkinson & P. Wright (eds.), *Adjunct Proceedings of HCI'95*. Huddersfield: University of Huddersfield School of Computing and Mathematics.

Stenning, K. & Oberlander, J. (1995). A cognitive theory of graphical and linguistic reasoning: Logic and implementation. *Cognitive Science*, 19(1), 97-140.

Tufte, E. (1983). *The visual display of quantitative information*. Graphics Press, Cheshire, Connecticut.

Whitley, K.N. and Blackwell, A.F. (1997). Visual programming: the outlook from academia and industry. In S. Wiedenbeck & J. Scholtz (Eds.), *Proceedings of the 7th Workshop on Empirical Studies of Programmers*, pp. 180-208.

The Cognitive Dimensions of an Artifact vis-à-vis Individual Human Users: Studies with Notations for the Temporal Specification of Interactive Systems

Maria S. Kutar, Chrystopher L. Nehaniv, Carol Britton, and Sara Jones

Interactive Systems Engineering Research Group
University of Hertfordshire
Hatfield, Herts AL10 9AB
United Kingdom
`http://homepages.feis.herts.ac.uk/~comqcln/ISE.html`

Abstract. Cognitive Technology explores ways in which the cognitive fit between people and technology may be optimized. If this goal is to be achieved we will require methods of assessing tools and information arftifacts in order that we may properly examine the interplay between human cognition and technologies. Examination of this relationship neccessitates recogniton of the fact that it will be shaped by the cognitive and embodiment characteristics of the user, as well as the activity being carried out and the nature of the technology or artifact itself. The Cognitive Dimensions (CDs) framework provides a set of tools and measures which may contribute to Cognitive Technology's aims. CDs provide a pragmatic approach to the assessment of information artifacts, highlighting properties which affect cognition. We argue that not only may CDs be of benefit to Cognitive Technology, but that Cognitive Technology provides a broader context for understanding the importance and impact of CDs. Greater awareness of the importance of particular characteristics of users may serve to inform the application of CDs. In this paper we provide a general introduction to the CD framework, and show how CDs have been used in the evaluation and improvement of a temporal specification notation for interactive systems. We then discuss the ways in which user characteristics may be taken into account in applications of the CD framework. We illustrate the discussion with examples showing the differing impact of a temporal specification notation's properties on experienced and novice users.

1 Motivation

Cognitive Technology seeks ways to optimize the cognitive fit between people and technology. This goal requires methods for assessing the tools and information artifacts for how well they fit particular populations of users, carrying out particular activities using the technologies. Cognitive Dimensions (Green 1989, Green 1991) can provide one set of tools and measures for achieving some of Cognitive

M. Beynon, C.L. Nehaniv, and K. Dautenhahn (Eds.): CT 2001, LNAI 2117, pp. 342–355, 2001.

Technology's aims. Cognitive Dimensions (CDs) afford a pragmatic approach to assessing information artifacts; they may be applied to any information structure (including, for example, those of notations or interfaces), and they provide a ready-made vocabulary for evaluation. In addition, the dimensions have already been used successfully in a number of cases, both by Green and collaborators (Green 1991, Modugno, Green & Myers 1994, Green & Blackwell 1996), as well as by other authors (Britton & Jones 1999, Cox 1999, Kadoda 1999, Roast 1995, Yang et al. 1997, Shum 1991).

Conversely, we argue that Cognitive Technology provides a broader context for understanding the importance and impact of the Cognitive Dimensions framework, and that it can serve to inform their application. We argue that cognitive and embodiment issues of users, as well as the activities that these different users perform, matter in the construction of cognitive dimensions profiles for artifacts. That is, different cognitive dimensions profile values are likely to arise for different users of the same artifact, or the same user performing a different activity with the same artifact. There is not a single, Platonic, unchanging user sitting somewhere in an abstract realm. Recognizing the diversity of users is a basic premise of appropriate design (see, for example (Shneiderman 1997)). Moreover, we must recognize users' learning about artifact affordances over time and the artifact's changes, if any, in response to the history of interaction as factors. Users change their modes of interaction as they become familiar with an artifact (Thomas 1998), or as the user's cognition and skills change. The context of the activity as well as the cognitive, social or physical characteristics of the user may all play a role in the assessment of the artifact. The design of the artifact itself might also be changing or it may be used in a changing context, and these changes may well affect a cognitive dimensions profile.

Several of the identified cognitive dimensions clearly recognize the role of the user as a cognitive agent. The dimension of "error-proneness" cannot be decoupled completely from the user making the error. Some notations or information structures might be more or less error-prone depending on the cognitive characteristics and capabilities of the user. The profile value in the dimension of "hard mental operations" for example, certainly depends on "for whom". Dealing with $2 + 3$ might be hard for a small child, but not for educated adults. Using quantifiers (at least single, non-nested ones) is of no difficulty to a mathematician or an experienced competent formal specifier. Different users have different capacities and needs (blind users, mathematically trained users, users that prefer visual vs. textual presentations, etc.)

Cognitive dimensions articulate concepts that are important and relevant in the context of specification languages, allowing a better understanding of those concepts and the inter-relationships between them. Specification notations have been evaluated for users with an "untrained eye" (Britton & Jones 1999). It was appropriate to restrict the class of users when creating a cognitive dimensions profile. The untrained eye is only one of many kinds of eyes, and its owner needs to be considered in analyzing interaction with the notation or information artifact. Work on modelling different classes of users suggests that they respond well to interfaces that take their cognitive styles into account (e.g. (Barker, Jones, Britton & Messer 2000)). Finally, the dimensions are helpful to a researcher who is a computer scientist, rather than a cognitive psychologist, in that they provide an accessible route into the cognitive psychology literature, and a framework and vocabulary in which discussion can take place. In the following section we provide a brief introduction to the cognitive dimensions framework.

2 Cognitive Dimensions Overview

The aim of the cognitive dimensions framework (Green 1989, Green 1991) is to provide tools which may be used to evaluate the usability of information structures. The dimensions are 'thinking tools' rather than strict guidelines, with a focus on usability aspects which make learning and doing hard for mental, rather than physical reasons. Cognitive dimensions are aimed at the non-HCI specialist and comprise a broad-brush approach rather than detailed guidelines.

The cognitive dimensions framework may be applied to both interactive artifacts such as word processors, and non-interactive artifacts such as music notation, and programming or specification languages. An artifact may be analyzed and an evaluation derived which can assist in determining the artifact's suitability for a particular task. It should be noted that the artifact is considered in conjunction with the environment in which it is to be used. We may think of the combination of an artifact and its environment as a 'system', and it is this combination, the 'system', to which the analysis is applied. Consequently, a single specification language, for example, may be considered in a number of different environments, each 'system' resulting in a different usability profile. This is a key feature of cognitive dimensions as, rather than providing a generalized analysis, they may be used to evaluate an artifact's suitability for a particular task or purpose. Considering an information structure in the context in which it is used greatly enhances the practical applicability of the dimensions. Table 1 shows the different dimensions and their short descriptions.

In order to enable the cognitive dimensions framework to be used to evaluate the usability of an information artifact for a specific activity, a cognitive dimensions profile is required. The profile shows the extent to which the dimensions are considered to be desirable for that activity. For example, the dimension of viscosity refers to the resistance to change of an artifact. Clearly, the desirability of viscosity will be dependent upon the activity in which a user is engaged: it is quite acceptable if that activity is transcription, but will be harmful if the user is attempting to modify the artifact. A number of different user activities have been identified, and profiles for transcription, incrementation, modification and exploration are given in (Green 1991). Further profiles may be derived for users of artifacts and notations as required, providing the opportunity to take into account not only different activities, but also the characteristics of various categories of users themselves; this issue is explored in detail in Sect. 4 below.

Table 1.

Dimension	Description
Abstraction	Types and availability of structuring and abstraction mechanisms
Secondary Notation	Extra information carried by means other than the official syntax
Diffuseness	Verbosity of the language
Hidden Dependencies	Important links between entities are not visible
Visibility	Ability to view components easily
Consistency	Similar semantics are expressed in similar syntactic forms
Closeness of Mapping	Closeness of representation to domain
Role Expressiveness	The purpose of a component is readily inferred
Premature Commitment	Constraints on the order of doing things
Provisionality	Degree of commitment to actions or marks
Viscosity	Resistance to change
Progressive Evaluation	Work-to-date can be checked at any time
Error Proneness	Notation invites mistakes
Hard Mental Operations	High demand on cognitive resources

In the following section we use our previous experience of using the cognitive dimensions framework to illustrate their relevance to the goals of Cognitive Technology.

3 Cognitive Dimensions Illustration

We have used cognitive dimensions to evaluate the real-time temporal logic $TRIO_{\neq}$ (Corsetti, Montanari & Ratto 1991), a specification notation designed to allow representation of time at multiple granularities (Kutar, Britton & Nehaniv 2000). In this case our user is one with experience of formal modelling, who is familiar with everyday concepts of time. In certain circumstances $TRIO_{\neq}$ forces use of a single time granularity, which may cause difficulties to users of the notation.[1] Scaife and Rogers (Scaife & Rogers 1996) identify the characteristic of rerepresentation which is particularly relevant to the use of unnatural time granularities in a representation. *Re-representation* refers to the way in which different representations which have the same abstract structure influence problem solving. For example Zhang and Norman (Zhang & Norman 1994) describe the difference between carrying out the same multiplication using roman or arabic numerals. The same formal structure is represented by LXVIII times X and 68 times 10, but the former is much more difficult to manipulate to find the solution, for someone who is used to working with arabic numerals. (The roman system

[1] The notion of *time granularity* refers to the hierarchical relations between units of temporal measurement of differing lengths.

is also more difficult for humans in general because of its lack of a positional, regular hierarchical form (see for example (Nehaniv 1997)). This is reflective of the way in which different representations of time influence the creation and understanding of temporal representations. For example, if a week is represented as 10,080 minutes then although the temporal periods represented are identical, the user's approach to understanding and manipulating the representation are changed. Cognitive dimensions analysis highlights the problem of rerepresentation through a number of different dimensions: *abstraction, closeness of mapping* (closeness of representation to the domain), *error-proneness* (whether the notation invites mistake), and *hard mental operations* (level of demand on cognitive resources).

The cognitive dimensions analysis of $\text{TRIO}_{\neq}$ confirmed that the representation of time at unnatural granularities affects usability of the notation in a number of ways. We used factors highlighted by the cognitive dimensions analysis to make a number of changes to the way in which time is represented in the notation. The introduction of further *abstraction mechanisms* to the notation allows time to be represented in a manner which reflects everyday *natural usage of time granularity*, so that for example, a week is represented as a week-long interval, rather than in terms of minutes or hours. This also supports reference to regularly paced or re-occuring events (such as meetings on the first Monday of each month). Moreover, the modified notation, Natural Granularity $\text{TRIO}_{\neq}$ (NGT), also supports *relativization of temporal reference*, e.g. it enables one to easily refer to "11 a.m. on Thursday of last week", rather than forcing the user of the notation to calculate the number of seconds or minutes from the present moment to that particular time. (See (Kutar, Nehaniv & Britton in press) for more details). This means that the notation mirrors everyday usage of the domain more closely (*closeness of mapping*). This eliminates the need to translate between different time granularities, and reduces both the *error-proneness* of the notation and the demand on cognitive resources (*hard mental operations*).

A cognitive dimensions analysis of the altered notation indicated that by eliminating the need for rerepresentation of time, usability of the notation is increased. A further advantage of the use of cognitive dimensions is that there are a number of known trade-offs between the different dimensions. Therefore when changes were being made to the notation we were aware of the possibility that changes in one direction may have an affect on other aspects. For example the introduction of abstraction mechanisms may impact upon the dimensions of viscosity, visibility and hidden dependencies. Use of the cognitive dimensions framework highlighted these relationships and allowed us to examine whether the changes we had made had impacted on such areas in an adverse manner.

In our work, we have identified issues of rerepresentation (Scaife & Rogers 1996), closeness of mapping (Norman 1988), granularity, and pace (Dix 1992) as key in the shortcomings of a particular specification language for temporal properties of interactive systems. The explanation of the difference between our improved version of this notation and the original one lies in the particular cognitive and apparently universal characteristics of possible human users of this notation. These characteristics can to a certain extent be decoupled from individual human users (but could not from, say, automated users (e.g., user models or software agents), or members of a hypothetical intelligent species with radically differing cognitive and cultural characteristics from humans.)

The improved NGT notation supports human agents doing temporal specification since it takes into account the cognitive characteristics of these human users (specifiers) and also of the human users of the systems being specified (end-users). Several affordances existing in human cognition are closely matched by the improved notation. These include:

1. *the coordinated use of different granularities,*
2. *an agent-centered perspective on events in a temporal stream,*
3. *the ability to handle naturally multiple granularities in understanding regular – or paced – events,*
4. *the easy expressibility of regular or recurring events in life,*
5. *chunking and zooming in a hierarchy.*

Supporting these human capacities also naturally leads to changes in the cognitive dimensions of error-proneness and closeness of mapping.

Use of the cognitive dimensions framework facilitated our work with temporal specification notations in a number of ways. CDs articulate concepts that are important and relevant in the context of specification notations, allowing a better understanding of those concepts and the interrelationships between them. CDs are not prescriptive, nor should they be considered as categorizing 'good' and 'bad' properties. Instead CDs may be viewed as a tool which uncovers aspects of a notation which may influence its utility. Recognition is given to the fact that certain properties may be more desirable in some situations, and they provide a tool with which assessment of the usability of the notation may be made. In the following section we examine further how the framework may be of relevance to the interaction between cognition and the design of particular technologies.

4 Cognitive Dimensions for Cognitive Technology

We believe that the solely task-oriented nature of the cognitive dimensions framework frequently has unduly restricted its applicability. The use of profiles to determine the desirability of the task-neutral individual dimensions for particular activities, illustrates that the utility of the framework lies in its ability to be applied in a wide range of areas and to a variety of artifacts. The focus on cognitive and embodiment characteristics of users engendered by cognitive technology suggests that there is an additional variable, that of *user characteristics*, which is relevant to analysis of any task, and taking this variable into account should be understood as crucial for the cognitive dimensions framework. As already mentioned, examination of usability characteristics of an artifact in isolation from the user who carries out activities in conjunction with that artifact, provides only a single level of insight into the artifact. Whilst we do not discount the information that this may yield about characteristics of the artifact which influence its utility in relation to activity, we believe that consideration of the characteristics of the user of the artifact should also be viewed as a primary consideration. Indeed, it should be clear that the possibility of valid analysis by treating all users as interchangable when studying an information artifact will be the exceptional case rather than the rule in applying CDs. When it is possible, it will surely represent a great simplification for researchers and designers, but we certainly cannot expect this to be the case in general.

The cognitive dimensions framework may be seen to accommodate cognitive and embodiment characteristics of individual (and categories of) users via two routes. The first of these is through the dimensions themselves. Although notionally task- and user-neutral the consideration of the user cannot be entirely decoupled from all of the dimensions. In particular, the dimensions of *hard mental operations*, and *role expressiveness* require us to assume some notional user. Cognitive characteristics of different users inevitably affect any analysis of an artifact under these dimensions. In a similar manner the dimension of *visibility* requires that *embodiment* and *experience* characteristics of the user are considered. The nature of these dimensions means that they cannot be applied to an artifact in a manner that enables evaluation of that artifact for all users. This must be recognized as compromising the effectiveness of the CDs framework where it is claimed to be "user-neutral" or implicitly treated as such. Certainly the intention has been that the dimensions *alone* are user-neutral and indeed task-neutral, and it is only the development of a profile which *alters* this (cf. (Green & Blackwell 1998)). Indeed, our point is that development of a profile requires taking individual users and tasks into account and it would be methodologically questionable (at best) and outright wrong (in most cases) to suppose that user- and task-neutrality could be maintained when a profile is to be developed. If the concept of the cognitive dimensions profile is extended to take into account user and task characteristics, we recognize that these dimensions identify characteristics which are of great importance when evaluating the usability of artifacts.

The second way in which the importance of user characteristics may be recognized by the cognitive dimensions framework, is through the cognitive dimensions profile. Currently the concept of the profile may be defined as *the extent to which the properties expressed by the cognitive dimensions are considered to be desirable for a particular activity.* The user is not recognized as an influencing factor in the determination of a profile for a particular activity. Indeed the profiles which are provided in the cognitive dimensions tutorial (Green & Blackwell 1998) (*incrementation*, *transcription*, *modification* and *exploratory design*) are all user-neutral. However, it is clear that this set of profiles is merely illustrative. We believe that development of profiles which take into account characteristics of individual, or categories of users enable the cognitive dimensions framework to be applied in a far more productive manner. This approach enables us to consider not simply an artifact in relation to a notional generic user, but to consider the usability of that artifact from the viewpoint of any individual user.

Furthermore, a new *temporal aspect* to the dimensions may be uncovered as it becomes possible to consider the changing relationship between the artifact and the user over time. This enables us to more clearly understand the impact of experience in a user's interaction with an artifact, and to view the relationship between the user's cognition and embodiment with the artifact as this relationship develops over time. Thus the extension of the concept of the cognitive dimensions profile to incorporate the user enables the applicability of the cognitive dimensions framework to be extended in a temporal dimension.

Previous work in our research group (Britton & Jones 1999, Britton & Kutar 2001) has focused on the development of a cognitive dimensions profile which may be used to examine the intelligibility of languages used in the specification of software for untrained users. In this work the context of interest was the validation of a requirements specification, and the ultimate aim was to evaluate different specification languages in terms of how easy it would be for readers who are not computer professionals to understand a specification written in the language (Britton & Jones 1999). Untrained users were considered both in the context of the tasks that they would have to carry out, and in terms of the characteristics that would have a bearing on how they were able to perform in the tasks. The task considered was that of requirements validation, where a specification of requirements for a system is read, discussed and checked to ensure that it records accurately what the clients and users want. Certain characteristics of the untrained user were considered to be particularly relevant in carrying out this task. A typical user: – would have a sound knowledge of the subject matter of the specification (the problem domain), would not have a mathematical background, would not be familiar with languages used in the specification of software, and would not be familiar with the process of validating a software specification.

The next stage in the work was to determine which of the cognitive dimensions related to this type of user carrying out this particular task. The dimensions that were found to be particularly useful in this context were closeness of mapping, role expressiveness, visibility, secondary notation, hard mental operations, hidden dependencies, consistency and abstraction gradient. As an example, the cognitive dimension of abstraction gradient helped to focus on the structuring mechanisms of specification languages. This was important in the context of the untrained user, since a clear structure is one of the main ways in which a specification can be made more intelligible. The dimensions of visibility and secondary notation highlighted the need to provide extra help (in terms of layout or use of color, for example) for untrained users in understanding the specification. A further example relates to the dimensions of closeness of mapping and role expressiveness, which highlight the need for a close relationship between elements of the representation and elements in the problem domain. One of the dimensions that was not included in this work was viscosity. This was because it was felt that the readers (who were not computer professionals) would suggest changes and make annotations to the specification, but that the actual changes to the notation itself would be carried out by others (developers or requirements engineers). It was therefore not necessary to consider viscosity as part of the evaluation of specification languages for this type of user. The final cognitive dimensions profile used in this research consisted of eight dimensions that were found to be of relevance in this particular context. Consideration of both task and user resulted in a streamlined cognitive dimensions profile and allowed the researchers to concentrate on those dimensions that were of particular interest. However, further research (Britton & Kutar 2001)) suggests that use of a subset of dimensions may not always be desirable as it raises the possibility that some factors may be overlooked.

5 Accommodating the User in Cognitive Dimensions

In this section we use two contrasting profiles to illustrate the manner in which accommodating characteristics of the user allows the cognitive dimensions framework to be used to greater effect. We show two profiles which may be used to evaluate notations used in the temporal specification of interactive systems, and then examine how the profiles may be used in a cognitive dimensions analysis of such a notation.

The activity under consideration in these profiles is the specification of temporal properties of interactive systems where those temporal properties occur at a range of time granularities. The notation that we will consider in our discussion is a real-time temporal logic, $\text{TRIO}_{\neq}$ (Corsetti et al. 1991) which has been developed for this purpose. (This is the same notation which was analysed and discussed in Sect. 3 above.) This notation is formal in nature and we consider profiles for two contrasting categories of user. The first of these is an experienced formal specifier, who is familiar with formal notations. The second perspective that we will consider is that of a novice user, who may have some mathematical background, but who has no experience of formal methods; this user could be, for example, a first year computer science undergraduate. For the purposes of this exercise, we will assume that other user characteristics, such as cognitive capabilities and embodiment, are identical for both experienced and novice users.

Taking these characteristics of the users into account, in addition to the activities under consideration, results in differing profiles: were activity to be the only consideration a single profile would be expected to be applicable to both of our categories of user. Table 2 shows the two profiles. The desirability of each dimension is taken from the following set: very useful, useful, neutral, harmful, very harmful.

It can be seen that by taking into account the characteristics of the user when creating a cognitive dimensions profile a different profile is derived. For some dimensions the desirability of the property expressed by the dimension is identical, for others it is slightly different, and for others there is substantial change. In addition to this, we must also take into account the fact that some dimensions are inherently user dependent, as discussed above. We discuss each of these categories below.

Table 2.

Dimension	Experienced	Novice
Abstraction	very useful	neutral
Secondary Notation	useful	harmful
Diffuseness	neutral	useful
Hidden Dependencies	harmful	harmful
Visibility	useful	very useful
Consistency	useful	very useful
Closeness of Mapping	useful	very useful
Role Expressiveness	useful	very useful
Premature Commitment	very harmful	harmful
Provisionality	harmful	very harmful
Viscosity	harmful	very harmful
Progressive Evaluation	useful	very useful
Error Proneness	harmful	harmful
Hard Mental Operations	harmful	harmful

Identical Desirability

For three of the dimensions the desirability remains unchanged across the different categories of user. For example an error-prone notation is recognised to be harmful for both the novice and the experienced user. Whilst an experienced user is arguably in a position where they are better able to work round such a difficulty, the fact remains that a notation which itself invites mistake is undesirable. Therefore we do not distinguish between categories of user for this dimension. In a similar manner the profile shows the dimensions of hidden dependencies and hard mental operations to be equally undesirable for our two categories of user (although see Sect. 5 below for further discussion of the user dependent dimensions including hard mental operations).

Differing Desirability

Substantially Different Desirability. For two of the dimensions, *abstraction* and *secondary notation*, it can be seen that the desirability is strongly dependent upon user experience. Our profiles show that abstraction mechanisms are very useful for the experienced specifier, but neutral for the novice. In our notation we can identify two different types of abstraction mechanism; those relating to time, and all remaining abstraction mechanisms. For both categories of user temporal abstractions, such as those which represent time at different granularities, are a useful feature. They eliminate the need to represent time at 'unnatural' granularities. However, other forms of abstraction mechanism may be harmful to the novice user. The cognitive dimensions framework recognizes different ways in which abstraction may relate to the novice user. The term *abstraction barrier* is used to describe the need for a user to master a number of new abstractions in order to master the system. *Abstraction hunger* is used to describe systems which require user-defined abstractions. For our notation, both of these are of importance. Without training in formal methods, a user may have difficulty in mastering the abstractions contained within formal notations, such as how the logical connectives and existential qualifiers may be applied to real systems (abstraction barrier). In addition, real-time temporal logics such as $\mathrm{TRIO}_{\neq}$ require the user of the notation to create their own abstractions to specify a system; the notation itself provides no mechanisms which assist in achieving this (abstraction hunger). This is a feature which may be beneficial to the experienced specifier, who is not restricted by the notation in his abstraction of the system being specified. However, for the novice unfamiliar with the process an abstraction hungry system can be harmful. Taking into account the potential benefits of the temporal abstractions, the overall desirability for the novice specifier is neutral. In contrast, as we have discussed, for the experienced specifier abstraction is desirable overall.

The dimension of secondary notation again has a differing desirability. For the experienced specifier, secondary notation enables the structuring of a specification in a manner which can aid understanding, and this dimensions may be viewed as an extra tool which may be used in the creation and presentation of the specification. For the novice specifier however, the addition of non-syntactic methods of adding information to a specification may be seen as detrimental. The novice user may, for example, use secondary notation in place of features of the notation where those features are less familiar to the user than the nuances of secondary notation. This may impact upon the user's employment of the notation in the longer term as such usage may be learned and become habitual. Alternatively, if the notation *requires* usage of secondary notation for effective specification, this may create a barrier to the user learning how the notation is most effectively employed. Therefore, although the potential for secondary notation may be helpful to the experienced user, the converse is true for the novice.

Slightly Different Desirability. For a number of the dimensions there is only a slight difference in desirability for the two categories of user. These are diffuseness, visibility, consistency, closeness of mapping, role expressiveness, premature commitment, provisionality, viscosity and progressive evaluation. These differ in our profiles as we are now able to show that while these dimensions are desirable or harmful for our activity, the effect may be magnified, or have differing important for the novice user. For example, the diffuseness of our notation is seen to be neutral for the experienced user. However, for the novice, excessive terseness is recognized as being potentially harmful (Green & Blackwell 1998), and so the dimension of diffuseness is desirable for the novice user.

The dimensions of visibility, consistency, closeness of mapping, role expressiveness and progressive evaluation are considered to be desirable in a notation for the activity of temporally specifying interactive systems. However, these are all properties of a notation which, whilst desirable for the experienced user, are very important for the novice user. Recognizing user characteristics in a cognitive dimensions profile enables this difference to be recognized.

User-Dependent Dimensions. As we noted in Sect. 4 above, three of the dimensions are clearly affected by cognitive or embodiment characteristics of the user: hard mental operations, role expressiveness and visibility. We have made the assumption that both categories of user share identical embodiment characteristics, and can therefore discount the user dependency of visibility from the profile. For the remaining dimensions of hard mental operations and role expressiveness the cognitive characteristics of the user will continue to be of relevance. Even within our category of experienced formal specifiers, the dimensions of hard mental operations is likely to vary according to individual users, their previously acquired skills and experience, and their abilities and perceptions. Therefore although we note that it may be harmful for any category of user if the cognitive load is increased through properties of the notation, we argue that this dimension should be excluded from a user-centered profile. The dimension of role-expressiveness is also partially user-dependent. However, we believe that as this dimension may also be influenced by the artifact, and it is possible to distinguish its desirability for various categories of users, it may be included in a user-centered profile.

User-Centered Profile Summary

We have illustrated that the cognitive dimensions framework may be extended to take into account characteristics of various categories of user for a common activity. The consideration of user experience results in a changed cognitive dimensions profile. Although for many of the dimensions the difference in profiles is small it has enabled recognition that a dimension which is, for example, generally desirable, for an activity, may become very desirable for novice users. This enables a cognitive dimensions analysis of an artifact to illustrate that the artifact poses particular difficulties for novice users. We believe that the approach will be similarly applicable to other cognitive and embodiment characteristics of users.

6 Discussion

The cognitive dimensions framework articulates concepts which may influence the quality of interaction with information artifacts. However the framework has been developed in a manner which is user-neutral. This unduly restricts the applicability of a useful evaluative framework. We have shown how the cognitive dimensions framework may be used in a manner which takes into account cognitive and embodiment characteristics of different categories of user engaged in a particular activity. Used together with regard for the embodied and situated nature of the cognition of particular humans, performing particular activities, in a particular context with a particular artifact, cognitive dimensions provide good in-roads to evaluation of the artifact and discussions on how it could be improved. Consideration of the agent-centered interaction with the artifact, taking into account particularities of the agent's embodiment, cognitive capacities, and experience, can suggest the appropriate generalisation of one cognitive dimensions analysis to other agents and to similar artifacts.

Tools change our perceptions and embodiments. Tools, including notational systems, can extend cognitive capabilities. Such an understanding can yield not only technological solutions to real world problems but also, and mainly, tools designed to be sensitive to the cognitive capabilities, affective characteristics, and temporal embeddedness of their users (Nehaniv 1997, Nehaniv 1999*b*, Nehaniv 1999*a*). Cognitive dimensions can be used as a tool to help achieve some of these goals. We have used them as a tool to help us derive a more humane, cognitively friendly notation for specifying the temporal properties of interactive systems. Taking histories of interaction into account together with the embodied, cognitive and situated aspects of interaction as users learn, and awareness of changing artifacts may lead to applications in areas where an extended cognitive dimensions approach can help us achieve even more of these goals.

References

Barker, T., Jones, S., Britton, C. & Messer, D. J. (2000), Individual Cognitive Style and Performance in a Multimedia Learning Application, *in* 'EURO EDUCATION 2000 Conference, Aalborg, Denmark, 8-10 February, 2000'.

Britton, C. & Jones, S. (1999), 'The Untrained Eye: How Languages for Software Specification Support Understanding in Untrained Users', *Human Computer Interaction* **14**, 191–244.

Britton, C. & Kutar, M. (2001), Cognitive Dimensions Profiles: A Cautionary Tale, *in* G. Kadoda, ed., 'Proceedings of the Thirteenth Annual Meeting of The Psychology of Programming Interest Group'.

Corsetti, E., Montanari, A. & Ratto, E. (1991), 'Dealing with Different Time Granularities in Formal specifications of Real-time Systems', *The Journal of Real-Time Systems* **3**, 191–215.

Cox, K. (1999), Cognitive Dimensions of Use Cases - Feedback from a Student Questionaire, *in* A. Blackwell & E. Bilotta, eds, 'Proceedings of the Twelth Annual Meeting of The Psychology of Programming Interest Group', Memoria, Cosenza, Italy.

Dix, A. (1992), Pace and Interaction, *in* Monk, Diaper & Harrison, eds, 'People and Computers VII', Cambridge University Press.

Green, T. (1989), Cognitive Dimensions of Notations, *in* A. Sutcliffe & L. Macaulay, eds, 'People and Computers V, Proceedings of HCI'89', Cambridge University Press.

Green, T. (1991), Describing Information Artefacts with Cognitive Dimensions and Structure Maps, *in* D. Diaper & N. Hammond, eds, 'People and Computers VI, Proceedings of HCI'91', Cambridge University Press.

Green, T. & Blackwell, A. (1996), 'Thinking about Visual Programs', In Thinking with diagrams (IEE Colloquium Digest No: 96/010). Institute for Electronic Engineers, London.

Green, T. & Blackwell, A. (1998), 'A Tutorial on Cognitive Dimensions', Available online at: http://www.ndirect.co.uk/ thomas.green/workStuff/Papers/index.html.

Kadoda, G. (1999), A Cognitive Dimensions View of the Differences Between Designers and Users of Theorem Proving Assistants, *in* A. Blackwell & E. Bilotta, eds, 'Proceedings of the Twelth Annual Meeting of The Psychology of Programming Interest Group', Memoria, Cosenza, Italy.

Kutar, M., Britton, C. & Nehaniv, C. (2000), Specifying Multiple Time Granularities in Interactive Systems, *in* P. Palanque & F. Paternó, eds, 'Interactive Systems: Design, Specification and Verification', Springer, pp. 51–63. Lecture Notes in Computer Science, Vol. 1946.

Kutar, M., Nehaniv, C. & Britton, C. (in press), NGT: Natural Specification of Temporal Properties of Interactive Systems with Multiple Time Granularities, *in* 'Design, Specification, and Verification of Interactive Systems: 2001 (8th International Workshop DSV-IS, Glasgow, Scotland, 13-15 June 2001)', Springer Lecture Notes in Computer Science.

Modugno, F., Green, T. & Myers, B. (1994), Visual Programming in a Visual Domain: A Case Study of Cognitive Dimensions, *in* 'People and Computers IX, Proceedings of HCI'94', Cambridge University Press.

Nehaniv, C. L. (1997), Formal Models for Understanding: Coordinate Systems and Cognitive Empowerment, *in* 'Proceedings of the Second International Conference on Cognitive Technology', IEEE Computer Society, pp. 147–162.

Nehaniv, C. L. (1999*a*), Narrative for Artifacts: Transcending Context and Self, *in* 'Narrative Intelligence: Papers from the 1999 AAAI Fall Symposium, (5-7 November 1999 - North Falmouth, Massachusetts)', Vol. FS-99-01, American Association for Artificial Intelligence, pp. 101–104.

Nehaniv, C. L. (1999*b*), Story-Telling and Emotion: Cognitive Technology Considerations in Networking Temporally and Affectively Grounded Minds, *in* 'Third International Conference on Cognitive Technology: Networked Minds (CT'99), Aug. 11-14, 1999 San Francisco/Silicon Valley, USA', pp. 313–322.

Norman, D. (1988), *The Psychology of Everyday Things*, Harper Collins.

Roast, C. (1995), Modelling Temporal Requirements for Interactive Behaviour, *in* 'Proceedings of the International Symposium on Human Factors in Telecommunications'.

Scaife, M. & Rogers, Y. (1996), 'External Cognition: How do Graphical Representations Work?', *International Journal of Human-Computer Studies* **45**, 185–213.

Shneiderman, B. (1997), *Designing the User Interface: Strategies for Effective Human-Computer Interaction*, 3rd edn, Addison-Wesley.

Shum, S. (1991), Cognitive Dimensions of Design Rationale, *in* D. Diaper & N. Hammond, eds, 'People and Computers VI, Proceedings of HCI'91', Cambridge University Press.

Thomas, R. (1998), *Long Term Human-Computer Interaction: An Exploratory Perspective*, Springer Verlag.

Yang, S. et al. (1997), 'Representation Design Benchmarks: A Design-Time Aid for VPL Navigable Static Representations', *Journal of Visual Languages and Computing* **8**, 563–599.

Zhang, J. & Norman, D. (1994), 'An Account of how Readers Search for Information in Diagrams', *Cognitive Science* **18**, 87–122.

Interactive Situation Models for Cognitive Aspects of User-Artefact Interaction

Meurig Beynon, Chris Roe, Ashley Ward, and Allan Wong

The Empirical Modelling Research Group, Department of Computer Science,
University of Warwick, Coventry CV4 7AL, U.K.
`http://www.dcs.warwick.ac.uk/modelling/`

Abstract. Cognitive aspects of human interaction with artefacts is a central concern for Cognitive Technology. Techniques to investigate them will gain greater significance as new products and technologies more closely customised to specific users are introduced. The study of Cognitive Dimensions is a well-established technique that can be used to support and direct empirical investigation of cognitive aspects of artefact use. This paper proposes a complementary technique, based on constructing 'interactive situation models', that applies to the study of specific user-artefact interactions. It interprets the cognitive activities of the user through interrelating situational, explicit, mental and internal aspects of state. The application of this approach in analysing, recording and classifying such activities is illustrated with reference to a simple case study based on modelling the use of an actual digital watch. The paper concludes with a brief discussion of possible connections with Cognitive Dimensions and implications for 'invisible computing'.

1 Introduction

A central concern of Cognitive Technology (CT) is the impact that the use of artefacts can have upon the mind of the user, and its broader implications for users in their social, cultural or administrative context. The study of CT demands analyses and techniques that can take full account of the interplay between human cognition and technological products. As computer-based technology advances, and new modes of human-computer interaction are being developed, cognitive aspects of human-computer interaction acquire ever greater significance.

In current practice in designing and implementing artefacts, the activities that relate most strongly to the agenda of CT are arguably the empirical studies undertaken by interface designers in developing an *information artefact* (IA) [8]. These involve monitoring the way in which potential users interact with an IA, and observing the problems they encounter. In this context, the experimenter is not necessarily explicitly concerned with what goes on in a user's mind, but sees the consequences of common mistakes and misconceptions, and explores practical steps that can be taken to eliminate them through redesign. Such empirical activity is intimately – if not necessarily directly – concerned with both human cognition and technological development.

M. Beynon, C.L. Nehaniv, and K. Dautenhahn (Eds.): CT 2001, LNAI 2117, pp. 356–372, 2001.

The informal nature of the testing activity and the nature of the insights that are gained through experiments with users pose challenges for representation. The intuitions of gifted and experienced designers typically play a crucial role. By drawing on their experience, good designers become astute at interpreting user difficulties and relating them to problems in the design of the artefact and its interface. The tacit acquisition and application of knowledge may serve the purpose in some design contexts, but requires skill and judgement that is not easy to develop and to share. This motivates the search for supporting techniques and frameworks that can make the analysis of cognitive issues in artefact and interface design more systematic, and results of the analysis more accessible for recording, exploring and communicating.

The study of cognitive dimensions (CDs), as introduced by Thomas Green in [7], is one approach to this issue. CDs provide a generic framework to guide the empirical study of IAs. This involves the identification of generic activities that are represented in user interaction with IAs, and the investigation of the cognitive demands made on the user in carrying them out. The study of CDs can inform the design of IAs, but it also serves a broader function of framing the agenda for discussion of their qualities and deficiencies from the perspective of a typical user. Knowledge about trade-offs between dimensions, for instance, is the same kind of knowledge that an experienced designer draws on when evaluating a redesign.

Human interaction with information artefacts can be conceived and viewed from two perspectives. In closed user-artefact interaction, the roles of the human user and responses of the artefact are enacted within a stable well-established context, where all instances of use are precisely identified and characterised in the user manual. The appropriate model for such use of the information artefact is as depicted in Fig. 1. The archetype for closed user-artefact interaction is provided by standard use of a device such as a digital watch that has been specifically designed to perform particular functions in appropriate situations according to preconceived conventions for interpretation.

The concept of closed user-artefact interaction implicitly imposes a stereotype upon the user. Modern developments in technology motivate a different perspective on human interaction with artefacts. As Cooper points out in [5], sophisticated computer-based artefacts take on the characteristics of the computer. This means that their behaviour can be customised, reprogrammed and reinterpreted by the user; their responses can be adapted to the user and the situation. Because individual users can directly shape the artefact, their personal experience, their competence, their knowledge and conception of the intended function of the artefact become crucially significant. Even the concept of specific uses of the artefact may be suspect, and the role of the person involved in the interaction is more aptly characterised as *user-designer* rather than mere user. Both use and experiment then feature in the interaction with the artefact and the manner in which these are to be interpreted need not be fixed in advance but can emerge from experience in the context.

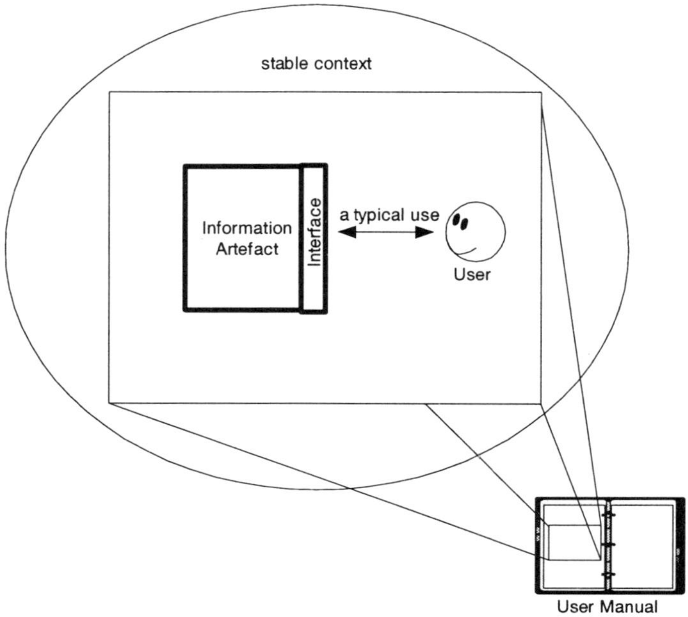

Fig. 1. Closed user-artefact interaction

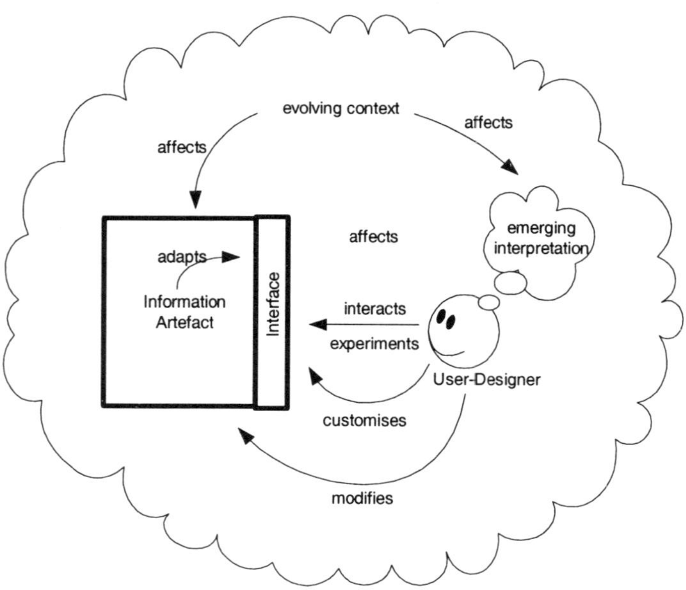

Fig. 2. Open user-artefact interaction

This open human-artefact interaction perspective is more in tune with a CT perspective. The use of an artefact, rather than being subject to a preconceived specification suited to a 'generic' user, is essentially hard to prescribe. It migrates in ways that cannot be anticipated, evolving with the users understanding and familiarity, and as the surrounding social and administrative context is adapted. The impact of this migration can be so radical as to embrace serendipitous patterns of interaction with the artefact that were originally uninterpreted.

This paper investigates a modelling technique for studying cognitive aspects of the use of an information artefact. This involves representing interaction with an information artefact by devising an interactive situation model (ISM) using principles and tools that have been developed by the Empirical Modelling (EM) research group at Warwick [14]. Our aim is to show that modelling with ISMs supplies a useful framework in which to examine cognitive aspects of user-artefact interaction, both open and closed.

2 Aspects of State in User-Artefact Interaction

The term 'information artefact' was introduced by Green and Blackwell in their study of CDs [8]. From the perspective of this paper, an IA is viewed as a construction whose state can be consulted or manipulated in such a way as to reflect the state of an external referent. Both non-interactive and interactive IAs fall within the scope of this definition, the distinction between these two kinds of artefact stemming from their capacity to undergo changes of state. For instance, a map is a non-interactive artefact whose referent is the geography of a region – though the map of itself undergoes no change of state, a state change is invoked when the map user points to a map location so as to refer to an external feature. In contrast, a digital watch – the primary information artefact used for illustrative purposes in this paper – is interactive, and makes state transitions in response to user actions.

For our purposes, the salient issues concerning an information artefact are that its use is situated, that it is perceived as having a state that can be to some degree changed in a deterministic manner by its user and that its function is prescribed by an abstract concept of 'appropriate use'. Appropriate use here refers to a norm of intended use, whereby the artefact serves to support particular activities, according to some standard conventions for interpretation. For a digital watch, the intended uses might include recording and potentially displaying the current time, fulfilling an alarm function, and serving as a stopwatch. In considering the use of an artefact, it is important to bear in mind that the user cannot necessarily recognise when interaction with the artefact conforms to this norm. For instance, to make appropriate use of the digital watch for telling the time you must know whether your watch is slow, whether it is currently British summer time, and which time zone applies to your current location.

Many aspects of state are relevant to the appropriate use of an IA. These aspects are typically specific to each instance of use, and have to be simultaneously apprehended by the user. These will be classified as follows:

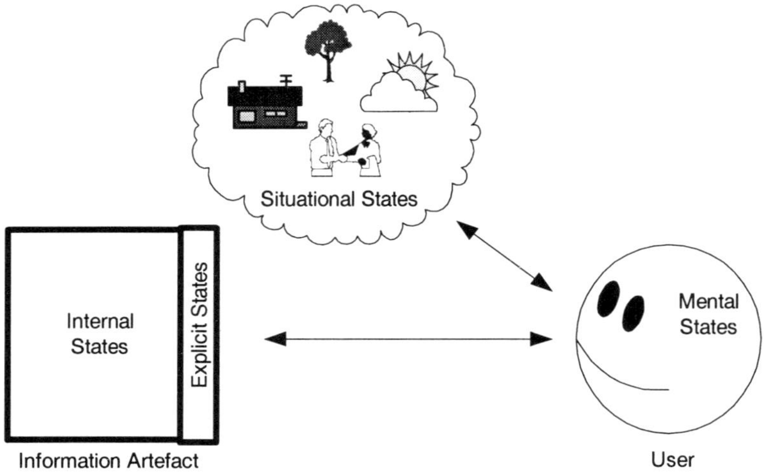

Fig. 3. SEMI aspects of state in user-artefact interaction

- explicit: the visible (or otherwise directly discernible) state of the artefact;
- situational: knowledge of the real-world context to which the artefact refers;
- mental: the state that we project upon the artefact when interpreting its current state and consulting expectations about its possible next state.

Aspects of the internal state of the artefact, of which the user is not in general aware, may also be relevant – these are certainly significant when the relationship between the artefact and its referent is perceived as inconsistent by the user. Such a perception may stem either from many sources, such as a misconception on the part of the user, a singular condition in the external situation, or a malfunction of the artefact.

These aspects of state can be elaborated with reference to a digital watch. *Explicit* state is what I can see merely by looking at the watch (simplified here by discounting other sensory channels, such as sounds the watch might emit, but without loss of generality). For instance, by looking at the watch display it may be impossible to tell whether it is in the 'display current time' mode, or shows the time at which the alarm is set. *Situational* aspects of state supply the norm for digital watch use. In appropriate use of the digital watch, knowing the actual time is significant. Knowing how to use the stopwatch function means having a fairly subtle understanding of external activities: such as 'running a race', 'lap time' etc. *Mental* state references complementary knowledge about state that has to be carried in the user's head to make sense of its behaviour. Though I may not be able to tell by looking at the watch whether it is in stopwatch mode or the 'display current time' mode, I may have reliable knowledge about this through recalling what abstract state transitions have been performed on the watch. For instance, I may know that it was displaying the current time, that I then pressed button X twice, and that this takes me into the stopwatch mode. The *internal* state is what someone testing or repairing my digital watch might consult – using special instruments to monitor the state of the digital circuitry etc whilst operating the buttons. The internal state of the watch is not usually accessible to the user, nor of concern to the user in so far as the watch operates reliably.

Activities that help to identify these aspects of state include:

- explicit state: take a snapshot, and show it to a third party who has not been engaged in interaction with the artefact;
- situational state: contrast playing with a digital watch that has already been set up for use and playing with a new watch, or contrast observing the watch in active use and experimenting with it in isolation;
- mental state: consider the knowledge of state that the user necessarily has to have to use the watch appropriately, but cannot be inferred from a snapshot of the current display.

The way in which these various aspects of state interact is highly complex. The context in which appropriate use of the artefact is set is teeming with empirically established assumptions. In general, the user is juggling with the relationships between all these aspects of state even as they try to put the watch to standard use. It can be difficult for the user to confirm that what is observed about the state of the artefact, what is simultaneously observed about the state of the world, what is inferred from knowledge of interaction with the artefact and what is presumed about integrity of the artefact and the user are all 'consistent'. Judging the consistency of such relationships between observations, presumptions and recollections of state is a dynamic empirical matter.

The cognitive demands of using an information artefact are shaped by the way in which the situational, explicit, mental and internal (SEMI) aspects of state are correlated in the mind of the user. The precise characteristics of this correlation differ from user to user, and will need to be determined by an empirical study of each individual user. Taken as a whole, the design of the IA is informed by a particular correlation between the SEMI aspects of state that corresponds to appropriate use of the artefact by a fully informed and committed user. Such a user is primed – for instance, by a user manual – about the idealised model of use with reference to each of these aspects of state. An account of idealised use of a digital watch will refer to situational state (e.g. "determine the current time"), explicit state (e.g. "when the alarm symbol is visible"), mental state (e.g. "recall that the watch can be in several different modes") and internal state (e.g. "when the display disappears, the battery needs replacing"), and to relationships between all four aspects.

The effective use of an information artefact requires experience of the artefact as well as familiarity with the user manual. Following standard practice in EM [14], the requisite experience can be the subject of an appropriately constructed ISM. Such an ISM aims to represent, in a manner that is both provisional and extensible, the way in which the SEMI aspects of state interact in use of the artefact. Because of the essential openness of the ISM, the ISM is at no point deemed to be a complete or perfect model, but can be readily refined and adapted to reflect different scenarios of use. In particular, an ISM that represents the designer's canonical model of use can serve as a 'seed' ISM from which a host of variants can be developed as needed. For instance, there will be variants to correspond to users with partial knowledge of the functions of the IA, perhaps with misconceptions about the correlation between SEMI aspects of state, and to correspond to different scenarios of use, both normal and exceptional.

A fuller account of the principles by which such ISMs can be created is the subject of the next section. Their application will be illustrated with reference to an ISM for the use of an actual digital watch.

3 ISMs and the Representation of SEMI Aspects of State

This section describes and illustrates the way in which an ISM to represent the use of an IA can be constructed. For simplicity, the discussion will focus on the construction of an ISM to represent the use of an actual digital watch, though the principles used are quite general, and have been applied in many different contexts [13,3,6,2].

3.1 ISMs as Construals

There is an intimate connection between an ISM to represent the use of an IA and the explanatory account that an expert, such as the designer of the IA, might give of its use. For instance, when the user first takes charge of the digital watch, they typically carry out a sequence of steps that involve consulting the SEMI aspects of state. They may change the internal state by inserting the battery, determine when the watch is in the `update_time` mode, consult the current time, set the time on the watch, then return the watch to `display_time` mode. In interpreting these actions in cognitive terms, we shall focus on the way in which SEMI aspects are correlated in the states that are visited, rather than on the sequence of steps as a recipe. For instance, whilst in the process of setting the time, the user may contemplate a state in which (in their view) the actual time is 12.30pm, the watch explicitly shows 12.28pm, the watch is in `update_time` mode, and there is a battery in it, so that the time kept by the watch is being updated.

Should the user make a mistake in setting the time (as when setting the watch to 12.30am rather than 12.30pm), the expert will typically be able to construe the error in similar terms: perhaps during the update the user thinks that the watch shows 12.28pm when in fact it shows 12.28am. The explanation for the user's interaction is framed with reference to observables that disclose the SEMI aspects of state, the user's expectations about the ways in which changes to these observables are interdependent, and the user's notion about what agency is operative. For instance, the user expects to be able to exercise control over the mode of the watch (agency), expects that if the mode is `display_time` the display will reflect the internal value of time (a dependency) as recorded and updated by the watch (agency), expects that on leaving the `update_time` mode to enter the `display_time` mode the internal time as kept by the watch will have been appropriately updated (agency).

The concept of *construing* the user activity that is informally introduced here is fundamental to the creation of an ISM. In studying the use of the watch it can be applied in many different ways. The expert construes the user's interaction with the watch, whilst the user simultaneously construes the states and responses of the watch. Different construals might be applied by the expert (respectively the user) to account for one and the same behaviour of the user (respectively the watch). In choosing to set the watch to 12.30am rather than 12.30pm, the user may be resetting the watch to reflect crossing the dateline, for instance, and the expert be mistaken about the current time (part of the situational state), rather than the user about the explicit state of the watch. If the watch keeps accurate time, it may seem appropriate to declare that the time on the watch depends on the current time, but this is a construal that would be confounded by taking the watch from one time zone to another. A watch that kept time by using the principle of the sundial, or exploited GPS to reset its time when moving between time zones would demand a different construal from a standard digital watch.

3.2 Developing an ISM for User-Artefact Interaction

Constructing an ISM involves identifying a family of observables and dependencies between them and finding ways to represent the current values of these observables using a suitable metaphor. In the modelling tools that we use to construct ISMs, families of observables and dependencies are represented by scripts comprising variables and definitions ('definitive scripts'). Where appropriate, the values of these variables are visually represented on the screen in geometric or iconic fashion. Within the modelling environment, there is an automatic mechanism to ensure that the values of variables are at all times consistent with their definitions, and with their associated visual representations (if any). The updating activity associated with this dependency maintenance is simpler than a general constraint satisfaction mechanism – it relies only upon propagating evaluation through an acyclic network of dependencies. The scope for distributing the ISM afforded by our modelling environment is an additional feature that has an essential role when we need to represent 'the same' observables as seen from the perspective of two different observers. In modelling the use of the digital watch, for instance, it is necessary to distinguish between the time as recorded by the watch and the time on the watch as *registered* by the user.

Constructing an ISM for the Digital Watch. Figures 4 and 5 together depict a distributed ISM to represent the use of an actual digital watch. Figure 4 represents those aspects of state that relate to the 'objective' state of the watch itself. In modelling an IA in isolation, the relevant observables typically refer to the explicit and internal aspects of state. When taking account of its use, these observables are complemented by others that refer to the mental and situational aspects of state. By way of illustration, the observables in Fig. 4 can be classified as associated with explicit and internal aspects of state. Sample observables to represent the explicit aspects of state refer to the digital display, the buttons and the alarm sound. Those associated with internal state include the time maintained by the watch and by its stopwatch subcomponent, the alarm settings that determine whether and when the alarm is triggered to go off, and the power level in the battery. Where the explicit state depends directly upon the internal state in normal operation of the watch, as when a bell icon appears on the watch face when the alarm is set, it is natural to conflate the corresponding internal and explicit observables. Though it seems pedantic to do otherwise, the distinction is significant in certain contexts. A watch engineer would always recognise the possibility that an explicit observable (such as the bell icon) could be inconsistent with the internal observable it is intended to reflect.

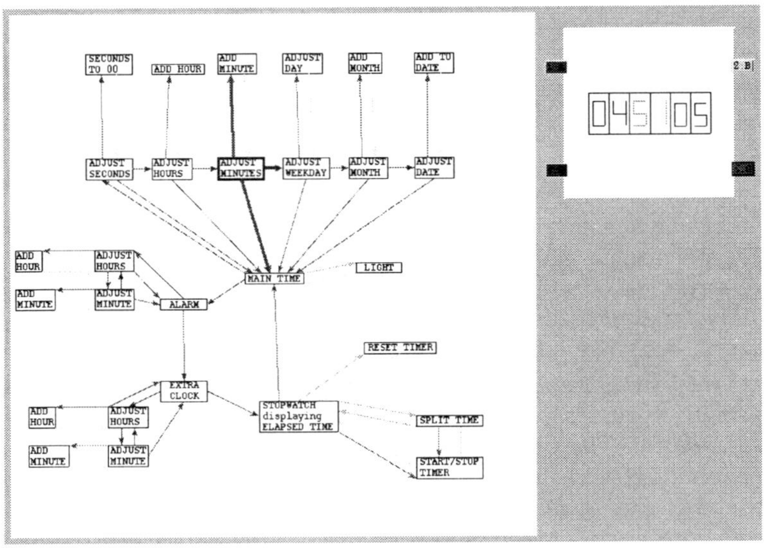

Fig. 4. Internal and explicit aspects of state in digital watch use

Changes of state within the digital watch, such as updating the time, are represented in the ISM by redefining the values of observables in the definitive script. Before an agent to perform this update automatically has been introduced into the ISM, it is possible for the modeller to emulate the state changes that a user of the watch might observe in a simple and direct manner by updating definitions manually. The epithets 'simple' and 'direct' here refer to the fact that a single redefinition will accomplish what is construed to be an indivisible change in the state of the watch. For instance, because the dependencies within the script faithfully reflect the dependencies between observables of the actual watch, a single redefinition to increment the internal time will automatically have the expected effect of ('simultaneously') updating the time as displayed in the `display_time` mode. The autonomous capacity of the watch to change state can be captured in the ISM by introducing agents that are primed to redefine the values of variables when certain preconditions are met. When using our modelling tools, these are represented via triggered procedures that are called whenever the values of specified variables are recomputed. Examples of such agents in the digital watch ISM include the mechanisms that update the internal time and that control the setting and sounding of the alarm.

A useful model of the digital watch has to include a representation for the internal state that is associated with its different modes. This aspect of the watch state is generally indicated on the display by an explicit observable in the form of a mnemonic, such as AL (for 'alarm mode'), that appears on the face. To be able to observe and manage the mode of the watch, it is necessary to be familiar with the mode-transition diagram. The nodes of this diagram are defined by the abstract modes of the watch and its edges by the transitions between these modes as specified by button presses. This diagram is the edge-coloured digraph to the left of the digital watch in Fig. 4. In interpreting Fig. 4 in its entirety, it is appropriate to consider the relationship between explicit and internal states of the watch as they are conceived by the watch designer, and communicated to the user via the user manual or by secondary notations [8] on the watch.

No observables to represent the situational aspects of state are included in the ISM shown in Fig. 4. Such observables would be an indispensable ingredient of the ISM if the function of the watch itself were to be dependent on its environment, as in the 'watch with automatic GPS reset' mentioned above.

Constructing an ISM for the Use of the Digital Watch. In modelling the use of the digital watch, the user's awareness of a whole range of SEMI aspects of state has to be taken into account. As will be explored in the next section, there is a sense in which creating an ISM to represent a user is an impossible task: it is at any rate a task that can never be completed, that can always benefit from additional empirical evidence, and that is confounded by the elusive and possibly ill-conceived notion of 'a typical user' (cf [5]). By its nature, an ISM is peculiarly well-suited to this difficult and obscure role: it is of its essence incomplete, acquires its significance through interaction, and has no formal functional or behavioural specification. It is equally apparent that the designer of a digital watch *does* have an notion of 'appropriate use' of the watch by an idealised user in mind. Such a user can reasonably be taken to be committed to using the digital watch in strict accordance with the designer's canonical model of use. To fulfil this role, it is not necessary for a user to be familiar with the entire functionality of the watch, but only with that part that relates to the specific uses of the watch that are to be exercised.

The observables and agency that define the internal and explicit aspects of the watch state (as represented in Fig. 4) are objective in nature, and at some level of abstraction reflect the designer's or the engineer's conception of the watch. The observables that represent the mental aspect of the state as perceived by a user are more controversial, and potentially more subjective. The role of these observables is to reflect the distinction between the current mode of the watch (as might be established by an electronics engineer through reference to its internal state), and the mode of the watch that the user presumes the watch to be in. The metaphor for the mental aspects of state that best suits an actual user's conception of the watch will be a matter for empirical determination, but for the idealised user, the abstract mode-transition diagram conceived by the watch designer can supply the appropriate framework. To this end, the partial – but so far as it goes perfect – knowledge of the watch use is depicted in Fig. 5 by the highlighted subset of the complete mode-transition diagram. This can be interpreted as representing the part of the watch functionality with which the user has become familiar.

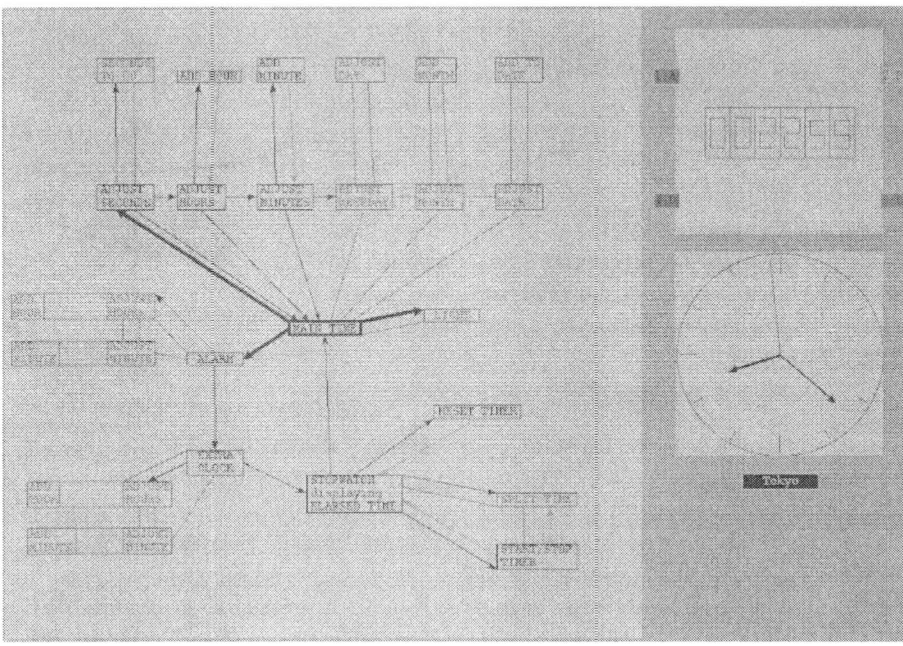

Fig. 5. Situational, explicit, and mental aspects of state in digital watch use

The situational aspects of state associated with use of the watch vary according to which particular mode of use applies. In so far as the standard timekeeping function of the watch is a persistent concern of the user, the current time and location are always part of the user context. In the ISM, this is reflected by the presence of the analogue clock to record the current local time in Fig. 5. Other situational observables become significant when specific user activities involving the watch are being studied, as when the stopwatch is being used to record the finishing times of two runners in a race (as discussed in Sect. 4 below).

3.3 The Identity of an ISM

The concept of importing new observables into the ISM according to what use is being made of the IA raises some fundamental issues about the integrity and identity of the ISM. An important and essential distinction between the ISM and a more conventional computer-based model is that it is inappropriate to identify the ISM with any particular fixed selection of observables or patterns of state transition. The ISM can only be explored state by state, and it is a matter of interpretation as to whether any particular transition should be seen as 'changing the ISM'. It is clear that many state transitions – such as the incremental changing of the time in normal operation – are to be viewed as changing the state of the watch rather than substituting a new watch. Other transitions, such as adding another button, are hard to interpret as anything other than changing the watch. It is also clear that there may be – and indeed always will be – observables of the actual watch yet to be taken into account in the ISM that might usefully be introduced. For instance, because of power considerations, the display might become fainter when the alarm is sounding. In general, any attempt to fix the identity of the ISM by declaring the specific states and state transitions it can undergo undermines the semantic role it serves for the modeller. The meaning of the ISM is experimentally mediated, and the modeller always has discretion over the interpretation of state transitions, whether they are associated with introducing new observables or giving different values to existing observables.

It is in this spirit that the ISM depicted in Figs. 4 and 5 can be regarded as reflecting the designer's construal of the digital watch. It represents the package that the designer consciously and explicitly offers when handing over the watch and its manual to the user. The cognitive processes of a user who experiments with the watch without first consulting the manual, the possible consequences of malfunction of the watch, and the arcane purposes to which the watch can actually be put (such as serving as the function of a protractor or a paperweight, for instance), are issues peripheral to the designer's remit. The rich variety of adaptations of the basic ISM that are accessible to the modeller can serve to represent this penumbra of actual rather than idealised interactions of the watch, as will be illustrated in the following section.

4 Illustrating the Use of the ISM

An ISM does not only itself serve as a construal – it can also be construed. First and foremost, the ISM is to be construed *as a construal*, but where the interaction and situation are appropriate, an ISM can be interpreted as a conventional computer program or as an IA for which specific user activities have been identified. The choice of interpretation adopted depends upon whether the interaction with the ISM is construed as open or closed user-artefact interaction.

As the ISM depicted in Figs. 4 and 5 illustrates, the way in which the modeller construes an ISM is both flexible and highly significant. The visualisation associated with the mode-transition diagram in Fig. 4 is to be interpreted as representing the mode of the watch as it is determined by its internal state. As it appears in Fig. 5, what is essentially the same visualisation refers to the state that the user projects on to the watch – it represents the mode the user thinks the watch is in. The choice of construal determines the dependency relationships recorded in the script and the kind of agency that can be exercised over them. In Fig. 4, the only change of state in the mode-transition diagram that is to be expected in normal operation of the watch results from a change to the internal mode of the watch initiated by a button press. Other changes of state in this diagram have to be interpreted as more radical in nature. For instance, illumination of the watch display that was not accompanied by the appropriate change of internal state might be construed as a watch malfunction, or the addition of a new node to the diagram construed as a redesign of the watch. In Fig. 5, changes of state associated with the mode-transition diagram are less constrained: it might indicate that the user had come to a new conclusion about the current mode of the watch, or had mastered a new aspect of its functionality.

There are numerous motivations for construing an ISM such as that in Figs. 4 and 5 in different ways. Different scenarios for use can be represented by a wide variety of modes of observation and agency. There is an entire agenda associated with teaching the use of the digital watch, and another with communicating about specific designs and design principles. From the perspective of stopwatch design alone, many further issues could be addressed. A school PE teacher, a sprinter and a long-distance runner all have different requirements: the teacher can operate the stopwatch whilst stationary, but the runners must use it whilst in motion; the sprinter needs to start and stop the watch in a way that does not interfere with their action; the long-distance runner would appreciate a watch that displays the time together with current heart rate.

Adapting the ISM to suit the whole range of useful construals involves reconfiguring the dependencies that link existing observables, the agency that can be applied to them and the way that they are distributed for observation. It can also mean introducing new observables together with dependency relationships to integrate them into the existing state. Making due allowance for the limitations of our current modelling tools, the ease with which such adaptations can be carried out is determined by the quality of the ISM as a construal. In particular, our ISM of the use of the watch in Figs. 4 and 5 is easy to adapt to new purposes to the extent that (at the appropriate level of abstraction) the ISM is a faithful reflection of how the watch works, and the purposes to which we want to direct the redesign or re-use of the watch are compatible with the way it works. Some specific adaptations of the ISM, with hints as to their possible application, will serve to illustrate this theme. They also show how the balance between situational, mental, explicit and internal aspects shifts according to whether the interaction with the ISM is more appropriately construed as open or closed.

There are many ways in which our distributed modelling environment can be used to study interaction with the digital watch. Different scenarios can be set up by distributing sections of the definitive scripts in Figs. 4 and 5 to mediate the actions and observations of demonstrators, observers and learners. For instance, button actions demonstrated by one user could be communicated to models on other users' screens. Alternatively, one agent in the network could be configured to monitor and record button presses automatically in the role of a passive observer so that learners' responses could be analysed later without their direct involvement. Learning with a computer based artefact can be less costly than interaction with a real world artefact since only a computer-based representation of the artefact under study needs to be distributed. Other possible ways of using the digital watch artefact in an educational environment are discussed in [13].

The ISM for the digital watch depicted in Fig. 4 is extensible. We can add new functionality to the watch very simply by including a few definitions. The functionality of watch model was derived from a real watch but was originally modelled with some features omitted in order to show how they could be incorporated at a later time. In this case, a 'second clock' feature that enables the user to keep track of the time in two different time zones simultaneously was left out and subsequently added 'on the fly' by introducing a short supplementary script. As another example, the watch – as designed – demonstrated viscosity [8] when the time was incremented beyond the target setting in the `update_time` mode. A small auxiliary script was sufficient both to remedy this problem and to make the necessary modifications to the mode-transition model associated with the internal state of the watch in Fig. 4. Different uses of the watch can likewise be introduced through adding observables to the situational state. To this end, the simple animated line drawing to represent two runners competing in a race shown in Fig. 6 can be added to the display. The watch user can then demonstrate how the stopwatch functionality of the digital watch can be used to record the finishing times of both the runners.

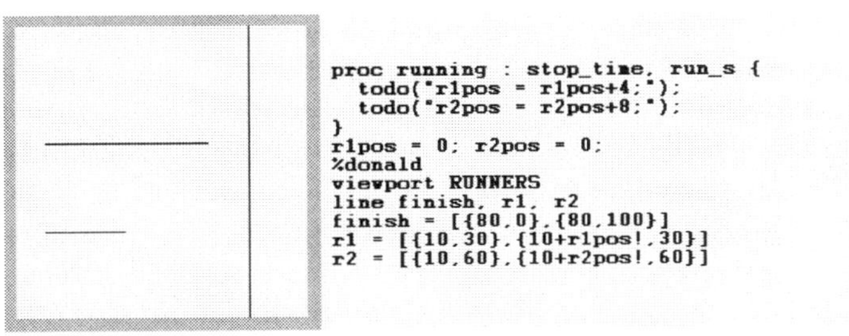

Fig. 6. Situational observables – timing two runners

The user-artefact interactions that neighbour on normal use include situations where environmental or perceptual obstacles interfere with the standard processes of observation. As a simple example, consider trying to determine the time from a digital display that is partially obscured by an item of furniture – for instance, as in observing a clock whilst lying in bed. A period of consistent and careful observation is typically needed before we can work out what the current time is, based on the partial displays of digits we can see, our knowledge of the pattern that governs the changes to these digits and our contextual knowledge of the approximate time of day that we believe it to be. The subtlety of observation in this scenario is compounded when we consider that the sleepy observer is liable to pass in and out of consciousness. One representative from the six sets comprising three simple redefinitions needed to transform the display appropriately is listed in Fig. 7. The boolean values in this listing could be replaced by predicates to take account of (e.g.) how the position of the observer affected clock visibility.

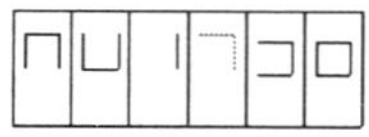

```
%donald
within led1 {
on3 = false
on5 = false
on7 = false
}
```

Fig. 7. A partially obscured digital display

5 ISMs, Cognitive Dimensions, and Invisible Computing

The construction of ISMs and the study of CDs share a common agenda with respect to understanding cognitive aspects of information artefact use. Both approaches aim to complement approaches based on aesthetic concerns [9] or counting user actions [4] by addressing interaction in conceptual terms, but they are quite different in character. As the previous section illustrates, user-artefact interaction can take such diverse and subtle forms that an empirical study of actual uses throws up far more information than it is feasible to document. CDs address this problem by abstracting from the specific experience of a user-artefact encounter, proposing general activities and issues to target in analysis. The application of ISMs involves creating an artefact that can implicitly offer a representation for this experiential knowledge.

CDs and ISMs put their primary emphasis on different kinds of user-artefact interaction. CDs focus on user activities of an established artefact, ISMs on modelling that is conceptually prior to the identification of the mode of use (if indeed there is to be any such identification). There is a useful parallel to be drawn with conventional programming – CDs are analogous to techniques for program comprehension, evaluation and testing, whilst ISMs are oriented towards the identification of program requirements. The development of IAs or programs from an ISM is an empirical activity that involves the identification of stable patterns of behaviour (cf. the empirical development of a manufacturing process in [6]). The analysis of SEMI aspects of state exemplified in Figs. 4 and 5 is relevant to the study of the artefact throughout this development, and converges to a view of IA-use similar to that described by Norman [10], in which the designer's model and the user's model are mediated via the system. CDs also offer higher-level abstractions to assist the analysis and comprehension of user-artefact interaction: some of these are associated with higher-order dependencies that could be introduced into EM, subject to placing them in their appropriate experiential context (cf. the incorporation of assertions about program state into an ISM in [3]).

The deconstruction of the user activities involved in creating an ISM helps to expose issues that relate to CDs. The ISM represented in Figs. 4 and 5 supplies a useful environment in which to explore the CDs of the watch. For instance, both the viscosity associated with decrementing the time whilst setting the watch, and the remedy associated with adding a decrement button, are reflected explicitly in the ISM. Many issues of hidden dependency are connected with the relationship between different aspects of state, and between explicit and internal aspects in particular. Creating ISMs can be a useful vehicle for demonstrating CDs and communicating about them. In exploring CDs, there are some advantages in being able to navigate the state space more freely than the actual information artefact itself allows.

Whether these points of contact between the use of ISMs and CDs are significant depends crucially upon the nature of the user-artefact interaction under consideration. If the CD analysis is directed at closed user-artefact interaction there is no particular advantage to be gained from the ISM modelling approach – indeed, there are optimisations to be made by constructing an OO model of the IA. In this case, the flexibility that the ISM affords primarily relates to issues of redesign of no specific relevance to CDs. There is more potential for interesting interaction between CDs and ISMs where the user-artefact interaction is open, or the CDs analysis is targeting users with different motivations and degrees of understanding. In this case, we can better exploit an ISM to construct models of use neighbouring on canonical use in the 'space of sense' [1].

Norman's vision of the future of computer technology [11] embraces information appliances each expertly engineered for its precisely specified and documented use and cooperating to support complex human activities. Odlysko [12] identifies problems of compatibility and intercommunication as major obstacles to the realisation of this vision in the short-term, and relates this to the trade-off between flexibility and ease of use. These discussions are framed from a perspective of closed user-artefact interaction, where ease of use is associated with 'delivering specific functionality in a way that is self-evident to the user' and flexibility is to be interpreted as 'offering more general functionality'.

The concept of an ISM as a vehicle for open user-artefact interaction relates to Norman's *invisible computer* culture in two respects. On the one hand, it suggests a framework to assist the identification of requirements and subsequent development of compatible and communicating information appliances. On the other, it points to a complementary vision of an alternative culture based on open user-artefact interaction. In this scenario, users will be educated – as in learning a natural language – to create their own personalised individual information artefacts for self-expression. In some respects, the significance of these artefacts will remain as private and subjective as a written document can be. With the will to understand each other, and through effort and cognitive demands similar to those we make when communicating in natural language, it will be possible for users to configure these private information artefacts to allow communication.

6 Conclusion

New technologies are changing the character of human-artefact interaction. They compel us both to confront and to establish more intimate relationships between human cognition and technology than were conceivable in the past. To this end, it is essential to give more support to an open user-artefact interaction perspective. The use of ISMs to model SEMI aspects of state is a promising direction for future research on this theme.

References

1. Beynon, W. M. "Liberating the computer arts". First International Conference on Digital and Academic Liberty of Information, Aizu, March 2001, to appear.
2. Beynon, W. M., Chen, Y-C., Hseu, H. W., Maad, S., Rasmequan, S., Roe, C., Russ, S. B., Rungrattanaubol, J., Ward, A., Wong, A. "The computer as instrument". In these proceedings.
3. Beynon, W. M., Rungrattanaubol, J., Sinclair, J., "Formal specification from an observation-oriented perspective", Journal of Universal Computer Science, Vol 6(4), pp407-421, 2000.
4. Card, S., Moran, T., Newell, A., "The psychology of human-computer interaction", Erlbaum, Hillsdale, 1983.
5. Cooper, A. "The Inmates are running the Asylum". Macmillan Computer Publishing, Indiana, 1999.
6. Evans, M., Beynon, W. M., Fischer, C. N., "Empirical Modelling for the logistics of rework in the manufacturing process" Proc. COBEM 2001, to appear.
7. Green, T. R. G., "Cognitive dimensions of notations". In People and Computers V, Sutcliffe, A., Macaulay, L., (eds) Cambridge University Press : Cambridge., pp443–460, 1989.
8. Green, T. R. G., Blackwell, A. F. "Design for usability using Cognitive Dimensions". Tutorial presented at the BCS conference HCI'98, 1998.
9. Neilsen, J., Molich, R., "Heuristic evaluation of user interfaces", Proceedings of ACM CHI'90 Conference, pp249–255, 1990.
10. Norman, D. A., "The Design of Everyday Things", The MIT press, 1998.
11. Norman, D. A., "The invisible computer", The MIT press, 1999.
12. Odlysko, A., "The visible problems of the invisible computer : A skeptical look at information appliances", In First Monday Vol.4, No.9 – September 6th 1999. Online at `http://firstmonday.org/issues/issue4_9/odlyzko/index.html`.
13. Roe, C., Beynon, W. M., Fischer, C. N. "Empirical Modelling for the conceptual design and use of engineering products". In Vakilzadian, H. (ed) Proc. International Conference on Simulation and Multimedia in Engineering Education, WMC'01, 2001.
14. The Empirical Modelling website at `http://www.dcs.warwick.ac.uk/modelling/`

Mediated Faces

Judith Donath

MIT Media Lab

Abstract. Incorporating faces into mediated discussions is a complex design problem. The face conveys social and personal identity; it reports fleeting changes of emotion and the cumulative effects of often repeated expressions. The face both expresses and betrays: it shows what the person wishes to convey – and much more. We are highly attuned to recognizing and interpreting faces (though these interpretations are very subjective). Incorporating faces into mediated environments can be quite desirable: it helps the participants gain a stronger sense of their community and can potentially provide finely nuanced expression. Yet there are significant disadvantages and difficulties. The immediate identifying markers revealed by the face, e.g. race, gender, age, are not necessarily the initial information one wants to have of others in an ideal society. And much can be lost in the path from user's thought to input device to output rendering. This essay discusses key social, cognitive and technical issues involved in incorporating faces in mediated communication.

1 Introduction

The face is essential in real world social interactions: we read character and expression in the face, we recognize people by their face, the face indicates where one's attention lies. Yet the face is mostly absent from online interactions – and this is in part why many people find cyberspace to be only a pale substitute for real world contact.

Today's fast graphics cards and high bandwidth connections have eliminated many of technical barriers to making the virtual world as fully visaged as the real world. Yet the problem goes beyond perfecting geometric models of facial structure, for there are complex social and cognitive aspects to how the face is used in communication that cannot be directly transplanted to a mediated environment. Furthermore, the desirability of faces cannot be assumed for all interfaces – some online communities have thrived because of the absence of faces and their immediate revelation of race, gender, age and identity.

Bringing the face to the interface requires radically reinventing the notion of personal appearance, while remaining grounded in the cognitive and cultural meanings of the familiar face. It requires analyzing applications to understand what aspect of the face they need to convey – personal identity? level of attentiveness? emotional expression? – and finding intuitive ways both to input and express this information. In some cases, the best interface is as realistic as possible, in others it has no face at all, while others may be best served by a synthetically rendered image that selectively conveys social information.

M. Beynon, C.L. Nehaniv, and K. Dautenhahn (Eds.): CT 2001, LNAI 2117, pp. 373–390, 2001.

Faces are used in many ways in computer interfaces, representing both people and machines. This paper focuses on the role of the face in computer-mediated human interactions in which the face represents a particular individual, communicating with other people in a real-time, online discussion.

Unlike much of the research in computer-mediated communication, we do not assume that the ultimate goal is to recreate reality as faithfully as possible. The computer makes it possible to go "beyond being there"[21] – to create environments that have features and abilities beyond what is possible in the ordinary everyday world. We can create environments in which the face shows expression, but does not reveal the user's identity; we can create worlds in which traces of the user's history of actions are sketched into the lines of the face. Yet introducing faces into a mediated communication system must be done carefully, for the face is replete with social cues and subtle signals; a poorly designed facial interface sends unintended, inaccurate messages, doing more harm than good.

2 Why Use Faces in Mediated Human to Human Communication?

There are many reasons to uses faces in mediated communication. The face is very cognitively rich and holds great fascination for us. Even newborn babies, a few hours old, will gaze longer at a face-like image than at a random array [24]. An environment filled with faces can be endlessly interesting to observe. People-watching is a perennially favorite pastime, in which we scan the surrounding scene for familiar faces and imagine the identity of the individual behind a stranger's visage [48]. An online environment populated with "people" with faces may seem more sociable, friendly, intriguing than a textual or purely abstract space.

Faces convey important social information about who you are and what you are thinking. We are cognitively wired to recognize and remember faces and your individual identity is uniquely represented by your face. The face also conveys social identity, with features that indicate basic categories such as age and gender as well as those that are associated with particular personality types. The face conveys emotional state and intent, displaying a wide range of expressions, from puzzlement to terror, fury to delight.

The faces helps to moderate and choreograph conversations. We use gaze to indicate attentiveness, to direct our remarks at an individual, to hold and yield the floor. Facial expressions soften our words, expressing humor, regret, etc. The face is very important in conveying responses, to show understanding, agreement, etc.

People behave more "socially", that is, more politely and with greater restraint, when interacting with a face. Sproull et al. [39] found that people responded quite differently to questions posed by computers when they were presented as text or as facial displays. For instance, when asked questions about themselves via text they answered with little embellishment but when queried by a facial display they attempted to present themselves in the best possible light.

Some of these reasons for using faces in mediated communication are advantages only in certain circumstances. The "social" responses that Sproull et al. detected can make a friendly discussion forum more sociable, but may be detrimental at other times. Isaacs and Tang [22] noted that non-facial interfaces could be more efficient, since the participants attended to the problems at hand, rather than to the time-consuming rituals of greetings and small-talk that ordinary politeness requires; Sproull and Kiesler [38] found that hierarchical distinctions were flattened in text-only discussions – it is plausible (though untested) that such distinctions would regain prominence in a mediated environment populated with visible faces (the desirability of maintaining or flattening such distinctions is context dependent).

The face allows us to characterize people at a glance. In the real world, the first things one learns about another are such social categories as age, gender and race, for the cues for these categories are embodied in the face. In an ideal world, would that necessarily be one's first impression? The online world has been touted as a place where one is identified first by one's words and ideas, free from the stereotypes imposed by such categorization; online spaces in which one's face is visible afford no such freedom. There is no simple metric for measuring the desirability of conveying this information, with numerous factors such as the purpose of the forum and the background of the participants affecting the evaluation. What we can do is understand fully what social cues the face does convey and use that knowledge to help determine where a facial display is appropriate.

Including faces in the interface is very difficult to do well. This is to a large extent due to the fact that the face is so expressive, so subtle, so filled with meaning. We ascribe character to and read emotion in any face, especially a realistically rendered one. There is no truly "neutral" face. A face in the interface is replete with social messages, but a poorly designed one will send many unintended ones.

In real world social situations we are constantly adjusting our face to do the appropriate thing – to hide or show our feelings and to gaze (or not) in the proper direction. We expect the same from mediated faces and when they elide a particular social protocol, we read an unintended message in the absence of a required expression or the accidental invoking of an inappropriate one. Making the "right" expression is extremely complex, for it is not a single motion, but a precisely timed choreography of multiple movements: a smile that flashes briefly conveys a different message than a smile that lingers.

One of the goals of this paper is to better understand what the fundamental limits are using mediated faces. Can the problems with mediated faces sending unintended messages be ameliorated with better input sensors and better renderings? Are the aspects of the face's social role that cannot be transferred to the mediated world? We will address these questions by first looking more closely at what social information the face conveys and then examining the technologies through which we bring these features to the mediated world.

3 What Does the Face Convey?

Almost every aspect of the face provides some sort of social cue and we are very adept at perceiving minute details of its configuration[1]. Knowing how to act toward someone and what to expect from them is fundamental to social interaction, and this knowledge depends upon being able to distinguish men from women, expressions of anger from those of joy, children from adults, friends from strangers – information that we read in the face. Our societal structures and mores have developed with the assumption that this face-conveyed information is available as the context for interaction.

The face conveys information through its structure, its dynamics, and its decorations[49]. The structural qualities include the overall head shape, the size and placement of the eyes and other features, the lines and texture of the skin, the color and quantity of scalp and facial hair. From these, viewers assess personality and make classifications such as race, gender and age. The dynamic qualities include gaze direction, pupil dilation, blushing, smiling, squinting, and frowning. From these, viewers read emotional expression and attention. Decorations include eyeglasses, cosmetics and hairstyle from which viewers read cultural cues, ranging from large scale group membership to subtleties of class distinctions and subcultural membership. There is also considerable interplay in how these qualities convey social cues. Haircuts affect the assessment of age, cultural mores modify the production and interpretation of emotional expressions, gender determination based on structural cues impacts the cultural interpretation of fabricated elements such as very short hair or lipstick. Recognition is primarily structural, though many times one will not recognize an acquaintance who has grown a beard or is shown in a photograph with an uncharacteristic expression.

The face conveys four major types of social information: individual identity, social identity, expression, and gaze. (This is not an all-inclusive list, for there are important functions that fall outside the scope of this paper, such as, as any lip-reader knows, displaying the words one is saying). These types may seem unbalanced: social identity is a broad conglomeration of all sorts information about one's gender, genetics, and geniality, whereas gaze is really a means by which the faces conveys information (such as conversational turn openings and attention). Yet this division is useful for thinking about mediated interactions, for addressing these communicative functions independently brings a great deal of flexibility and creative possibilities to the design of the interface.

3.1 Individual Identity

We are very adept at recognizing people. We recognize them at a distance, from various viewpoints, with different expressions and as they change with age [49]. We can

[1] Our ability to distinguish minute differences among faces is so acute that Chernoff proposed taking advantage of this ability to do multivariate statistical visualization with faces as the graphical representation: "Chernoff faces" map data to facial features as nose length, eye tilt, head shape, etc. [6]. The resulting faces may look happy, sad, surprised or pained - but the underlying data is independent of the interpreted social meaning of the face.

find a familiar face in a crowd with remarkable speed, especially considering how complex this task is: one's mental construct of the sought face is compared to each of the visible faces, all of which are quite similar in overall structure and are seen from different angles, in a range of lighting conditions, and feature different expression. There is strong evidence for specific neurological bases for recognizing faces. For example, injury to a particular area of the brain (the occipitotemporal section of the central visual system) leaves people with their vision intact, but nearly unable to recognize faces, a condition known as prosopagnosia[7]. Indeed, our notion of personal identity is based on our recognition of people by their face. To be faceless is to be, according to the Oxford English Dictionary, "anonymous, characterless, without identity."

In today's online, text-based worlds, facelessness is the norm and the extent to which participants are identified or left anonymous is a design feature of the various environments. Both anonymous and named forums exist and flourish, though each produces a different tone and is suited for a different purpose [10]. Anonymous or pseudonymous spaces provide an arena for exploring alternate personas and a safe haven for discussing highly sensitive subjects; they are also more likely to devolve into an endless exchange of flames or spam. Named forums bring the weight of one's real world reputation to the online world; in general, people behave in them more as they would in real life.

Online forums in which the participants' real faces are featured – as in, for example, a videoconference – are essentially named environments. Much of the discussion about the desirability of video as a medium focuses on issues such as bandwidth requirements and the common gaze problem (discussed below). The fact that it makes the forum into a public sphere in which everyone is seen and known needs to also be kept in mind, for it has a deep effect on the mores of the space.

3.2 Social Identity and Character

We recognize people not only as individuals, but also as types. Based on the cues we see in the face we quickly categorize people according to gender, ethnicity and age and make judgements about their character and personality.

These classifications tell us how to act toward the other, what behaviors to expect from them, how to interpret their words and actions. In many languages, it is difficult to construct a grammatically (or at least culturally) correct sentence without knowing the other's age, gender or relative social status. Such distinctions are also the basis of prejudice, with significant biases are found even among people who consciously decry race or gender based stereotypes[2]. More subtle but perhaps even more pervasive biases can be found in character judgements made on the basis of facial structure, e.g. a person with a baby-ish facial structure (large eyes, small nose, large forehead, small chin) will be judged to be more child-like in nature - trusting, naive, kind, weak [49]. This, and many other character judgements based on the face derive from "over generalization effects". According to Zebrowitz [49], we have very strong responses to cues for important attributes such as health, age, anger etc., so strong that they get over generalized to people whose faces merely resemble those with that attribute or emotion.

Cyberspace (the text version) has been touted as an ideal realm where the visual absence of these cues means that people are known and judged by their words, rather than by their gender, race, or attractiveness. Yet it is not simply a matter of text=good, face-based classification=bad. The cues we gather from the face are basic to much of our established social interactions, and many people find that they need to "put a face to a name" to go beyond a certain level of familiarity or comfort. Furthermore, simply eliminating the face does not eliminate the underlying cultural differences.

The distinction between structural, dynamic and decorative facial features is especially useful when thinking about mediated faces, for not only do these features serve different social purposes, they may also be electively and separately implemented. For instance, the decorative features – glasses, hairstyle, makeup, etc. – reflect one's choices and circumstances. This can be re-created in the decoration of online self-representations and indeed graphical MUDs and games such as the popular *Asheron's Call* feature avatars whose appearance derives from both the player's taste (today I wish to appear as a purple alligator) and role (but because I have not registered I may only choose between being a yellow or green smiley-face). While such simplistic decorations are far from the subtle social messages we communicate via our personal decorations in the real world, the potential certainly exists for these online decorations to become increasingly sophisticated as the mediated world evolves [41][40].

The dynamic features are also separable: there are motion capture facial animation programs that track the dynamic facial movements of a live actor and use them to animate a synthetic face [14][42]. The synthesized face can be that of the original actor (a technique used to achieve low bit-rate transmission of facial expressions [16]) or of any appropriately modelled face. While such techniques are used primarily to convey expression independently of other features, it is important to note that more information about social identity may be imparted this way than one might think: people can detect cues for age and gender in the dynamics of the face alone, as has been demonstrated with point-light experiments in which key points of the face are marked with dots and the rest is made invisible so that observers see only the moving dots [49].

The structural features are the most problematic in terms of stereotyping. It is the use of genetically determined features such as bone structure and skin color to assess someone's personality, morality, intelligence, etc. that raises the biggest concerns about unfair bias based on facial features. Cyberspace (the text version) has been touted as an ideal world in which such prejudice is eliminated because the initial cues by which such stereotypes are made are invisible. From this viewpoint, an interface that brings one's real face into cyberspace destroys this utopia, reintroducing the mundane world's bias-inducing cues. In practice the situation is more complex. For instance, gender differences permeate our use of language, and men and women are socialized to use apologetics, imperatives, etc. quite differently. Hiding one's gender online requires more than simply declaring oneself to be of the other gender: one must adapt one's entire tone and wording to the often subtle mores of the other. Thus, gender that is hidden online can be uncovered by writing style, albeit more slowly than such identification is made in the face to face world [10]. Furthermore, a lack of cues as to social identity does not lead to people thinking of each other as ciphers; rather, categorization still occurs, but with a high likelihood of error - an error which can have further consequences. For instance, if I mistakenly assume someone is a man who is actually a woman, and "he" uses locutions that would seem ordinary if spoken by a woman but coming from a man seem very passive and accommodating, I not only see him as a man, but as a particular type of man, timid and sensitive. Thus we see that while removing the face from the interface does remove some immediate social categorization cues, it does not eliminate such categorization entirely, and the ambiguity that ensues introduces new social problems.

3.3 Expression

One of the most important - and most controversial - communicative aspects of the face is its ability to convey emotion. We see someone smiling and know they are happy, we see someone frowning and know they are angry – or are they? Perhaps the smile was forced, a deliberate attempt to appear happy while feeling quite the opposite, and perhaps the frown indicates deep concentration, not anger at all. Although we are surrounded by expressive faces, there is still considerable controversy about how they communicate and what they really reveal.

Debate surrounds questions about whether the face reveals emotions subconsciously or whether it is primarily a source of intentional communication. Debate surrounds questions of whether our interpretation of emotions as revealed by face is innate, and thus cross-cultural, or learned, and thus subject to cultural variation [13][35]. Debate even exists about what emotions are [18] and whether they even exist or if they are a non-scientific construct, cobbling together disparate features ranging from physiological state to intent [17].

The most prevalent conceptualization of the relationship between the face and emotions is what Russell and Fernández-Dols call the Facial Expression Program [35], which has roots in Darwin's writings about the face [8] and is elucidated in the work of Izard [23], Ekman and others. The key ideas in this model are that there are a number of basic, universal emotions (7 is an often cited number: anger, contempt, disgust, fear, happiness, sadness and surprise), that the face reveals one's internal emotional state, though one may attempt to hide or distort this expressive view and that observers of the face generally are able to correctly read the underlying emotion from the facial expression [35]. Ekman's work has been quite influential in the computer graphics field, and this conceptualization of the relationship between emotions and facial expression underlies much research in facial animation (e.g. [47]).

In the context of designing face-based interfaces for mediated communication systems the debate about emotional expression vs. the communication of intent is especially relevant. Ekman's work emphasizes the expressive, often subconscious, revelatory side of facial expressions - indeed, one major branch of his research is the study of deception and the involuntary cues in facial expression and gesture that reveal that one is lying [12]. From this perspective, the advantage to the face is for the receiver, who may gain a truer sense of the other's intent by the involuntary cues revealed by the face (as well as gesture, tone of voice, etc.) than from the more deliberately controlled words. This model is rejected by Fridlund, who claims that the face's communicative functions must be to the advantage of the face's owner for if expression revealed information to the advantage of the receiver and the disadvantage of the owner, it would be evolutionarily untenable [17].

As a design problem, the issue becomes one of control – is the facial display controlled deliberately by the user or is it driven by other measurements of the user's affective state? If the display is the user's actual face (e.g. video) then the question is moot, it is the face, which may be displaying affective state or intentionality or both, but the system does not change this. If, however, the expressions on the facial display are driven by something else, the decision about what that something is important. To take two extremes, a very deliberate facial expression model is implemented when the face is controlled by pressing a button ("The Mood command opens a cascading menu from which you can select the facial expression of your avatar. Alternatively you can change the mood of your avatar by pressing one of the function keys listed in the cascading menu or use the mood-buttons in the toolbar."), as opposed to one in which the face's expression was driven by affective data gathered from sensors measuring blood pressure, heart rate, breathing rate, and galvanic skin response – bodily reactions that provide cues about one's affective state [32].

How universal vs. subjective is the interpretation of facial expression is also controversial. Even the smile, which seems to be the most universally recognized and agreed upon expressions, is used quite differently in different cultures. When it is appropriate to smile, for how long, etc. is culturally dependent. Much of the meaning we read in an expression has to do with minute timings and motions – what makes a smile seem like a smirk?

Context is also essential for understanding facial expression. Fernández-Dols and Carroll [15]caution that most studies of facial expression have been carried out without taking context into consideration, referring not just to broad cultural contexts, but the ubiquitous immediate context of any interaction. They point out that facial expressions carry multiple meanings and that the observer uses contextual information to interpret them. This is an important feature to keep in mind in understanding mediated faces, for mediated discussions occur in complex, bifurcated settings, where each participant is simultaneously present in an immediate and a mediated context. The smile I perceive may be one you directed at me – or it may have been triggered by an event in your space which I am not privy to. Such mixing of contexts occurs in real life too, for one's thoughts, as well as one's surroundings, constitute a context: "What are you smiling about?" "Oh nothing, I was just remembering something..." But in a mediated situation, with its multiple contexts, the observation of expressions triggered by and intended for other contexts may be a common occurrence.

3.4 Gaze

Gaze – where one is looking – is an important channel of social information[1][4][22][44]. We are quite adept at perceiving gaze direction (aided by the strong contrast between the eye's white cornea and colored iris) and use it, along with other contextual information, to infer other people's state of mind. Gaze is used in conversation, to determine whether someone is turning the floor over to another or is thinking about what to say next. Gaze is used to disambiguate language: I'm talking to "you",

you're welcome to "that". Gaze is both input and output: we look at something or someone because we are interested in them and our interest is revealed by the visible direction of our gaze.

The rules that govern how gaze is used in communication are complex and culturally dependent. Studies of gaze in conversation (see, for instance [1] or [22]) show an intricate ballet of words, gestures, and eye-movements that taken together are used to negotiate turn-taking, establish social control, reflect levels of intimacy, and indicate understanding and attention [4]. Research on gaze often focuses on its role as an indicator of attention. Yet in social communication, gaze has many functions – and averted eyes may not be an indication of averted attention. In a typical conversation, the speaker looks at the listeners to monitor their level of agreement and understanding, to direct an utterance at particular individuals, to command attention or persuade. The speaker may look away from the listeners in order to concentrate on a complex cognitive task, such as thinking about what to say next, or from embarrassment or discomfort (typically, speakers look at the listeners about 30-40% of the time [1]). Listeners look at the speaker more (about 60-70% of the time) and gaze directed at the speaker may signal agreement or it be an attempt to gain a turn. The listener's averted gaze may indicate very close concentration – or complete lack of attention. Furthermore, the length of time it is socially comfortable for two people to look at each other depends on their relationship: strangers look at each other more briefly and less frequently than acquaintances do, and prolonged mutual gaze is a sign of romance and intimacy [1].

There have been numerous attempts to bring gaze to computer mediated conversations. The problem – to show where each person is looking – is deceptively simple, but remains imperfectly solved. Some interfaces, such as many avatar-based graphical chats and current multi-party videoconferencing systems, simply ignore the problem, leaving the avatars to gaze off in random directions and the videoconference participants to appear in separate windows, each appearing to look intently at a spot just beyond the viewer's shoulder. Some interfaces take a very simplistic approach to gaze, using it to broadly indicate attention (e.g. [9]) but ignoring the myriad other social cues gaze provides. Some interfaces do attempt to recreate meaningful gaze in a mediated environment, but these quickly become immense and baroque systems: Hydra[37], a relatively simple system, requires $n*(n-1)$ cameras and monitors (where n is the number of participants) and Lanier describes an immersive approach [26] that uses numerous cameras, fast processors and more bandwidth than is available even at high-speed research hubs to facilitate a casual conversation in not-quite-real time.

Bringing gaze to the mediated world is difficult because gaze bridges the space between people – and the people in a mediated conversation are not in the same space. Addressing this problems requires not only developing a means for the participants to signal meaningful gaze patterns but creating a common, virtual space for them to gaze across.

Addressing this problem means finding some way to create a common, virtual space, as well as finding a way for the participants to control their gaze, whether algorithmically (as in [46]) or by detecting where they are actually looking (as in [26]).

With videoconferencing, the basic problem is that no common space is shared by the participants. With a two person system, the camera can (more or less) function as a stand-in for the one's conversational partner: when one looks at the camera, it will appear as if one were looking at the other person. The camera must be appropriately located; ideally, it is coincident with the video image of the other's eyes – and challenges are generated by both the opacity of video screens and the mobility of people's heads. Once there are more than two participants, the problem becomes far more difficult, for a single camera cannot stand-in for more than one person.

With avatar systems, the problem is that the user must somehow convey where he would like his avatar to be depicted gazing. Here, the act of indicating gaze is separated from the process of looking; the challenge is to motivate the user to provide this attention indicating information.

The face is highly expressive and informative, but it is not a quantitative graph. Almost everything it conveys is somewhat ambiguous and subjective, open to a range of interpretations and strongly colored by the observer's context. I may find a particular person's face to seem very warm and friendly, with a touch of mischievous humor – and much of that interpretation may be because of a strong resemblance of that person's structural features to those of a friend of mine, whose personality I then ascribe to the new acquaintance. Even something as seemingly objective as gaze is subjectively interpreted. If you are looking at me from a video window and you appear to glance over my shoulder, I may instinctively interpret this as meaning your attention is drawn to the activity occurring behind me, rather than to the activity in your own space beyond the camera.

4 Ways of Bringing the Face to the Interface

Once one decides to create a mediated social environment that includes faces, there are many ways of bringing the face to the interface. The face may be a photographic likeness of the person it represents, or it may be a cartoon visage, conveying expressions but not identity. The face may be still or in motion, and its actions may be controlled by the user's deliberate input or by autonomous algorithms. Each of these design decisions has an impact on the technological requirements and complexity of the system and significantly changes the social dynamics of the interface.

Bringing the face to the interface is a difficult problem and all of today's systems are steps towards achieving an ultimate goal, with many more steps yet to go.

For many researchers, the ultimate goal is to achieve verisimilitude, to make the mediated encounter as much like the experience of actually being in the same place as possible. Most work in video-based conferencing shares this goal, especially research in computationally sophisticated approaches such tele-immersion [26], in which multiple distant participants interact in a common virtual space. Some of the problems in this domain, such as today's poor image quality and lag, can be solved through increased bandwidth and computational power. Yet there are still immense challenges here; in particular, the need to create a common virtual space for the interaction while simultaneously depicting the subtle expressive shifts of the participants.

Yet verisimilitude is not the only goal. Hollan and Stornetta [21] termed reproducing reality as "being there" and urged designers to go "beyond being there", to develop new forms of mediated interaction that enable people to communicate in unprecedented ways that aim at being "better than reality". For example, we may wish to have an interface that uses an expressive face with gaze to provide the sense of immediacy, presence, and the floor control that we get in real life, but which does not reveal the user's identity. We may wish to have faces that change expression in response to the user's deliberate commands or, conversely, in direct response to the user's affective state as analyzed by various sensors. We may wish to have faces that function as a visualization of one's interaction history, an online (and hopefully benign) version of Wilde's *Picture of Dorian Gray.* Or faces that start as blank ciphers and slowly reveal identity cues as acquaintances grow closer. Some of these possible interfaces are relatively simple to implement, others are even more difficult than attempting verisimilitude. And they present a further design challenge, which is to know which, out of the universe of possible designs, are the useful, intriguing, intuitive designs.

4.1 Video and the Quest for Verisimilitude

Video technology makes it possible to transmit one's image across a network, to be displayed at a distant location. Video has the advantage of letting one's natural face be the mediated face. A slight smile, a fleeting frown, raised brows – expressive nuances are transmitted directly. Video reveals personal and social identity: you appear as your recognizable self.

Video can make people self-conscious. In real life, we speak, act, gesture without seeing ourselves; videoconferences often feature a window showing you how you appear to others. Also, online discussions may be recorded. The combination of appearing as oneself and seeing oneself in a possibly archived discussion can greatly constrain one's behavior. The desirability of this restraint depends on the purpose of the forum; it is neither inherently good or bad.

Contemporary videoconferencing technology has one camera per participant and each participant's image and audio is transmitted to all the others. The quality of the transmission is often poor, due to limited bandwidth. As we discuss the advantages and drawbacks of video as a conversational interface, we will attempt to separate problems that are solvable with increased computational power and faster networks from those that are inherent in the medium.

Video reveals identity, but it is not the same as being there. Studies indicate that although the face's identity cues are transmitted via video, something is lost in the process. Rocco [34] observed that people often need an initial face to face meeting to establish the trust needed to communicate well online, whether using text or video. This may be primarily due to the poor quality of today's video channel, which loses and distorts social cues by introducing delays and rendering gaze off axis. For instance, given limited bandwidth, it is known that given limited bandwidth, reducing audio lag is most important and that eliminating motion lag is more important than reproducing spatial detail [31], yet many social cues, such as subtle expressions, may be lost without this detail. The timing delays that do exist are jarring and can give a distorted sense of the other's responsiveness, interest, etc. While the delays may be measurably slight, they are perceptually significant, potentially creating a quite misleading (and generally not terribly flattering) impression of the other, an impression that might be interpreted as awkward, unfriendly, shifty, etc. - but is purely an artefact of the technology.

Video does improve social interactions, as compared with audio-only conferencing. Isaacs and Tang's research comparing collaboration via videoconferencing with audio conferencing and with face to face meetings has many interesting observations about the social role of the mediated face [22]. They found the greatest advantage of video to be making the interactions more subtle, natural and easier. They point out that while it may not make a group of people do a task more quickly (the sort of metric that has often been used to measure the usefulness of the video channel), it provides an important channel for social messages. For instance, it helps to convey one's level of understanding and agreement: people nod their heads to indicate they are following an argument, and may lift their eyebrows to show doubt, tilt their heads to indicate skepticism or frown to indicate confusion. Video is useful in managing pauses: one can see whether the other person is struggling to find the right phrase or has been interrupted by another activity. Video, they said, "adds or improves the ability to show understanding, forecast responses, give non-verbal information, enhance verbal descriptions, manage pauses and express attitudes... Simply put, the video interactions were markedly richer, subtler and easier than the telephone interactions."

Yet video also has some inherent drawbacks. Isaacs and Tang [22] enumerated a number of videoconferencing weaknesses, noting that it was "difficult or impossible for participants to: manage turn-taking, control the floor through body position and eye gaze, notice motion through peripheral vision, have side conversations, point at things in each other's space or manipulate real-world objects." These drawbacks arise because the participants do not share a common space.

Isaac's and Tang found these problems even in two person videoconferences. A key problem is gaze awareness: if I look at your image, I am not looking at the camera and the image you see appears to be gazing elsewhere. While this can be addressed with clever use of half-silvered mirrors and integrated camera, the gaze does not match our real world expectations. Indeed, being close may be worse, for once the awareness of the camera is lost, we attribute any oddity of gaze behavior to intent, rather than to the technology.

These problems are exacerbated once there are more than two participants. With two people, it is theoretically possible for the camera's to transmit from at least an approximately correct point of view; with more, it is not, at least not without more cameras. There have been a number of experimental designs made to address this problem. These fall into two categories: one can use multiple cameras and displays to extend the one-to-one videoconference model (e.g. Hydra [37]) or one can use a combination of three-D modelling and head-tracking gear to create a video driven synthetic space (e.g. tele-immersion [26]).

With the former approach, multiple cameras and displays are placed throughout one's space. Each participant is seen in his or her individual monitor and the setup is replicated at each site. For instance, a camera/monitor setup can be placed in each seat at a conference table, with each camera facing the one live person in the room. The video from the camera associated with your image at every node needs to be sent to you, as it then shows that person from the correct angle, as if you were looking at them from your seat. If implemented correctly, this method allows multiple participants to indicate attention by looking at each other and to share a common space, at least to the extent that the physical environment is replicated at each site. This approach requires multiple installations and (N)*(N-1) cameras and monitors. It provides little flexibility (e.g. one cannot leave one's seat to chat quietly with another person[2]). In the reduced case of N=2 participants, it is indistinguishable from one on one video conferencing, and thus shares the aforementioned advantages and disadvantages.

The latter approach attempts to create an environment that seamlessly blends the local and the remote in a common virtual space. Multiple video cameras capture the actions of each participant and using location information from various sensors and a considerable amount of computational power, each participant is mapped into a common virtual world. Such a system is far from implementation today and Lanier's estimates of the computational and network requirements for even minimally acceptable levels of detail put it at least 10 years in the future [26]. Furthermore, the quantities of gear required – cameras, head-tracker, eye-trackers, etc. – make the experience far from the seamless de-spatialization of daily experience that is the goal.

Ten years – or even twenty or fifty years – is a long time off, but it is not forever. We can assume that something like a seamless tele-immersive environment will one day exist, realistic enough to be just like being there. We will then have mediated environments in which the face, with all its expressive and revelatory powers, exists much as it does in daily life. We turn now to considering approaches to the mediated face that go beyond being there.

4.2 Avatars and the Quest for Expression

There are numerous and varied ways of bringing faces to the interface that do not attempt to fully imitate real life. There are simple graphical avatars and intelligently animated agents. There are video windows in virtual space and sensor-driven cartoons. A simple photograph replicates the user's appearance, but does not convey dynamically changing expression and gaze. A cartoon avatar may have a fictional visage while deriving its expression from an analysis of the user's speech.

There are a number of reasons why one would want to use a synthetic face. First, it supports interaction among large numbers of people in a common virtual space. The difficulty with video-based systems is integrating a number of separate spaces into a common environment; once one is no longer trying to bring in disparate real world elements, the common space problem disappears. Second, it allows for communication without necessarily conveying identity. Text-based online discussions support the full spectrum of identity presentation, from authenticated veracity to absolute anonymity: synthetic images can provide the same range within a graphical context (a synthetic

---

2 An interesting solution to this problem is Paulos and Canny's work on personal tele-embodiment using remote controlled mobile robotic devices that incorporate two-way video communication [30].

image may be entirely fictional or it can be derived from photographic and range data of the real person).

The goal with many systems is to bring the expressive qualities of the face to a virtual world; the challenge is sensing and producing expression in a socially meaningful way. Such systems are still at the very early stages of development. Commonly used avatar programs have only the most primitive style of expressive input (and output): expression buttons and keyboard shortcuts that let the user change the avatar's face to sport a smile, frown, etc. [19].

While these systems are simple, I will argue here that simplicity alone is not a problem, nor is complexity always desirable. Rather, the key is a balance between the information provided and the message that is sent. If minimal information is provided, a minimal message should be sent. The problem with many face-based interfaces is that they are sending too complex a message upon the receipt of too little data. The face is so highly expressive, and we are so adept at reading (and reading into) it, that any level of detail in its rendering is likely to provoke the interpretation of various social messages; if these messages are unintentional, the face is arguably hindering communication more than it is helping.

One solution is to stick with very simple faces. The ubiquitous "emoticons" – typed symbols that resemble sideways faces, e.g. the smile :-) the frown :-< and the wink ;-) – are extremely simple, yet function quite well at helping to communicate expressive information that clarifies the sender's intention. E-mail is notorious for generating anger due to miscommunication of irony, sympathy etc. Emoticons can make it clear that a statement is meant in jest, or that a writer is deploring, rather than celebrating, the incident they are reporting. Essentially new forms of punctuations, emoticons spread quickly because they were intuitive as well as needed. Their reference to familiar iconic facial expression makes them immediately accessible to readers[3].

Creating an avatar that is even somewhat reminiscent of a human being brings into play numerous requirements about its behavior. For instance, if I use a plain circle as the user's representation (see [45] for an example), I can move this circles across the screen by sliding it, and the movement seems perfectly reasonable. If I decide to use a more human-like representation and create an avatar with legs, then sliding it across the screen seems awkward – the avatar appears passive and inert. The legs make me want to have it walk, and to do so, one may either have the user painstakingly render each step, or have an automatic walking algorithm. The hand rendered one, far from being more expressively communicative, puts an onerous burden on the user, who must expend so much attention getting the avatar to put one foot in front of the other, that he or she has little time left over for actually communicating with others. So, one equips the avatar with automated walking algorithms. A simple interface might ask the user for a destination and would take care of getting the avatar there. Now, a behavior such as walking has some social information in it: we read moods, such as whether one is buoyant or dejected, from gait, as well as characteristics ranging from athleticism to

[3] Although cultural differences occur even here. Japanese emoticons differ from Western ones. For instance, in Japan, women are not supposed to show their teeth when smiling, as is depicted in the female emoticon smile (.) And the second most popular icon is the cold sweat (;), with no clear Western equivalent [33]

sexual attractiveness By providing the avatar with legs we then require it to walk, and walking is inherently expressive. All that the user has indicated is an endpoint, but via the avatar, has communicated much more.

The same is true of the face. Once there is a representational avatar, it requires behaviors and behaviors are expressive, introducing the big question of whether it is expressing what the person behind it wishes to express.

An interesting example is provided by Vilhjálmsson and Cassell's *BodyChat* [46]. Here, humanoid avatars in a chat environment are provided with automated social actions. The user indicates to the system the social actions he or she would like to perform and the avatar then performs a series of visible actions that communicate this intention. For instance, to indicate a desire to break away from a conversation, the user puts a "/" at the beginning of a sentence; the avatar then accompanies those words with a diverted gaze. If the other person responds with a similarly prefixed sentence, the conversation ends with a mutual farewell; if not, the conversation continues, until both parties produce leave-taking sentences. While the developers of *BodyChat* have addressed the whole body problem of avatar physical behavior, their approach – and the issues it raises – can be considered primarily in the realm of the face.

A key issue this highlights is communicative competence. The social signals that I send when I greet someone or take leave are not simply informative actions, but also displays of communicative competence. Let's compare the input and the output in this situation. In the real world, I decide I've had enough of the conversation - perhaps I am bored, perhaps I am late for another appointment, perhaps I sense that the other person needs to go and I don't want to detain them, perhaps a combination of all three. In each of these cases, the gestures I make to indicate leave-taking may be quite different – I may look around for a distraction, I may glance at my watch, or I may look directly at the other person as I take my leave. Each of these conveys a different message and each also expresses a different level of politeness and competence. If I am leaving because I sense the impatience of the other, the impression I convey will be quite different if I look down at my shoes, mumble goodbye and flee, or if I graciously and warmly shake hands, say some pleasant farewells, and go. My actions upon taking leave are modified by both my immediate motivations and my underlying social knowledge and style. As a participant in a conversation, I gather a lot of information from the leave-taking behaviors, only one bit of which is that the other intends to leave. I also get a sense of the leave-taker's reasons for leaving, level of concern for my feelings, social sophistication, etc. In the *BodyChat* system, the user conveys only that one bit - the forward slash that says "I intend to leave". The systems expands it into a more complex performance, designed to draw upon our social knowledge – a performance that the receiver interprets as the sender's intent. The problem is, much of that performance has nothing to do with anything that the sender intends. Is it better to have unintentional cues than none at all? The answer depends on the context - it is again a design decision. Vilhjálmsson and Cassell state that their research goals include pushing the limits of autonomous avatar behavior "to see how far we can take the autonomous behavior before the user no longer feels in control". Understanding these limits is an important contribution to understanding how to integrate the face into mediated communications.

There are numerous other approaches to creating mediated faces. Some use as their input the user's writing [28][29] or speech [11] to derive expression and drive the animation. Like *Body Chat* these systems all introduce some unintentional expressivity, for they are all translation systems, transforming their input into a model of the user's inner state or intentionality and then representing that state via an animation. Perhaps, as Neal Stephenson suggests in his novel *Snowcrash*[40], future expressivity will come in our choice of autonomous behavior avatar modules, much as we express ourselves via clothing today.

Systems that use video images or other measurements of the face to animate facial models ([5][14]) are interesting, for they do no such translation. Here, although the rendered face may be completely fictional (or photorealistic - such systems can thus run the gamut from anonymous to identified), its expressions, whether deliberate or subconscious, are derived directly from the user's face; it is the facial expressions themselves that are re-presented, not an implicit state.

5 Conclusion

The key problem in bringing the face to a mediated environment is to balance input and output. In our real world face, there are millions of "inputs" controlling the highly nuanced features, from the genes that determine the basic facial structure to the nerves and muscles that control the lips, eyes, and eyebrows. In the virtual world, the control structure is much coarser. We must understand what is the communicative ability of the system we create, and match the face to it. The face is an extraordinarily rich communication channel and a detailed face conveys a vast amount of subtle information, whether we wish for it to do so or not.

References

1. Argyle, M. and Cook, M.: Gaze and Mutual Gaze. Cambridge University Press, Cambridge (1976)
2. Aronson, E. The Social Animal. Freeman, NY (1988)
3. Ayatsuka, Y., Matsushita, N., Rekimoto, J.: ChatScape: a Visual Informal Communication Tool in Communities. In: CHI 2001 Extended Abstracts (2001) 327-328
4. Bruce, V. & Young, A.: In the eye of the beholder: The science of face perception. Oxford University Press, Oxford UK. (1998)
5. Burford D. and Blake, E.: Real-time facial animation for avatars in collaborative virtual environments. In: South African Telecommunications Networks and Applications Conference '99, (1999) 178-183
6. Chernoff H.: The use of faces to represent points in k-dimensional space graphically. In: Journal of American Statistic Association, Vol. 68 (1973) 331-368
7. Choissier, B.: Face Blind! http://www.choisser.com/faceblind/
8. Darwin, C. and Ekman, P. (ed.): The Expression of the Emotions in Man and Animals. Oxford University Press, Oxford UK (1998)

9. Donath, J.: The illustrated conversation. In: Multimedia Tools and Applications, Vol 1 (1974) 79-88.
10. Donath, J.: Identity and deception in the virtual community. In: Kollock, P. and Smith, M. (eds.): Communities in Cyberspace. Routledge, UK (1998)
11. P.Eisert, S. Chaudhuri and B. Girod.: Speech Driven Synthesis of Talking Head Sequences. In: 3D Image Analysis and Synthesis, Erlangen (1997) pp. 51-56
12. Ekman, P.: Telling Lies: Clues to Deceit in the Marketplace, Politics, and Marriage. New York: W. W. Norton. (1992)
13. Ekman, P.: Should we call it expression or communication?. In Innovations in Social Science Research, Vol. 10, No. 4 (1997) pp 333-344.
14. Essa, I, Basu, S. Darrell, T. Pentland, A.: Modeling, Tracking and Interactive Animation of Faces and Heads using Input from Video. In: Proceedings of Computer Animation '96 Conference, Geneva, Switzerland, IEEE Computer Society Press (1996)
15. Fernández-Dols, J. M and Carroll, J.M.: Context and Meaning. In: Russell, J.A, Fernández-Dols, J. M. (eds.): The Psychology of Facial Expression. University of Cambridge Press, Cambridge, UK (1997)
16. Forchheimer R. and Fahlander, O.: Low Bit-rate Coding Through Animation. In: Proceedings of Picture Coding Symposium. (March 1983) 113-114
17. Fridlund, A.J.: The new ethology of human facial expression. In: Russell, J.A, Fernández-Dols, J. M. (eds.): The Psychology of Facial Expression. University of Cambridge Press, Cambridge, UK (1997)
18. Frijda, N.H., Tcherkassof, A.: Facial expressions as modes of action readiness. In: Russell, J.A, Fernández-Dols, J. M. (eds.): The Psychology of Facial Expression. University of Cambridge Press, Cambridge, UK (1997)
19. Fujitsu Systems. New World Radio Manual. http://www.vzmembers.com/help/vz/communicate.html (1999)
20. Herring, S.: Gender differences in computer-mediated communication. Miami: American Library Association (1994)
21. Hollan, Jim, and Stornetta, Scott. Beyond Being There. In Proceedings of CHI '92
22. Isaacs, E. & Tang, J.: What Video Can and Can't Do for Collaboration: A Case Study. In: Multimedia Systems, 2, (1994) 63-73.
23. Izard, C.E.: Emotions and facial expressions: A perspective from Differential Emotions Theory. In: Russell, J.A, Fernández-Dols, J. M. (eds.): The Psychology of Facial Expression. University of Cambridge Press, Cambridge, UK (1997)
24. Johnson, M., Dziurawiec, S., Ellis, H. and Morton, J.: Newborns' preferential tracking of face-like stimuli and its subsequent decline. Cognition. 40. (1991)1-19.
25. Kunda, Z.: Social Cognition: Making Sense of People. Cambridge, MA: MIT Press. (1999)
26. Lanier, J.: Virtually there. In: Scientific American, April 2001. pp 66-76 (2001)
27. Nakanishi, H., Yoshida, C., Nishimura, T. and Ishida, T.: FreeWalk: Supporting Casual Meetings in a Network. In: Proceedings of ACM Conference on Computer Supported Cooperative Work CSCW'96, (1996) 308-314
28. Nass, C., Steuer, J. and Tauber, E.: Computers are Social Actors. In: Proceedings for Chi '94., 72-78 (1994)
29. Ostermann, J., Beutnagel, M. Fischer, A., Wang, Y.: Integration of Talking Heads and Text-to-Speech Synthesizers for Visual TTS. In: Proceedings of the International Conference on Speech and Language Processing, Sydney, Australia (1998)
30. Paulos, E. and Canny, J.: Designing Personal Tele-embodiment. In: IEEE International Conference on Robotics and Automation. (1998)

31. Pearson, D. E., and Robinson, J. A.: Visual Communication at Very Low Data Rates. In: Proceedings of the IEEE, vol. 4, (April 1985) 795-812
32. Picard, R. Affective Computing. MIT Press, Cambridge, MA (1997)
33. Pollack, A.: Happy in the East (--) or Smiling :-) in the West. In: The New York Times, (Aug. 12, 1996) Section D page 5
34. Rocco, E.: Trust breaks down in electronic contexts but can be repaired by some initial face-to-face contact. In: Proceedings of CHI '98. (1998) 496-502.
35. Russell, J.A, Fernández-Dols, J. M.: What does a facial expression mean? In: Russell, J.A, Fernández-Dols, J. M. (eds.): The Psychology of Facial Expression. University of Cambridge Press, Cambridge, UK (1997)
36. Scheirer, J., Fernandez, J. and Picard, R.: Expression Glasses: A Wearable Device for Facial Expression Recognition. In: Proceedings of CHI '99, Pittsburgh, PA (1999)
37. Sellen, A., Buxton, W. & Arnott, J.: Using spatial cues to improve videoconferencing. In: Proceedings of CHI '92 (1992) 651-652
38. Sproull, L. and Kiesler, S. Connections. Cambridge: MIT Press (1990)
39. Sproull, L, Subramani, R., Walker, J. Kiesler, S. and Waters, K.: When the interface is a face.In: Human Computer Interaction, Vol. 11: (1996) 97-124
40. Stephenson, N. Snow Crash. Bantam, New York (1991)
41. Suler, J.. The psychology of avatars and graphical space. In: The Psychology of Cyberspace, www.rider.edu/users/suler/psycyber/psycyber.html (1999)
42. Terzopoulos D.and Waters, K.: Analysis and synthesis of facial image sequences using physical and anatomical models. In: PAMI, 15(6) (1993) 569--579
43. Valente, S. and Dugelay, J.-L.: Face tracking and Realistic Animations for Telecommunicant Clones. In: IEEE Multimedia Magazine, February (2000)
44. Vertegaal, R.:. The GAZE Groupware System: Mediating Joint Attention in Multiparty Communication and Collaboration. In: Proceedings of CHI '99. Pittsburgh, PA. (1999)
45. Viégas, F. and Donath, J.: Chat circles. In: Proceeding of the CHI '99 conference on Human factors in computing systems, (1999) 9 - 16
46. Vilhjálmsson, H.H.and Cassell, J.: BodyChat: autonomous communicative behaviors in avatars. In: Proceedings of the second international conference on Autonomous agents. Minneapolis, MN USA. (1998) 269-276
47. Waters, K.: A Muscle Model for Animating Three-Dimensional Facial Expression. In: ACM Computer Graphics, Volume 21, Number 4, (July 1987)
48. Whyte, W.: City. Doubleday, New York. (1988)
49. Zebrowitz, L.: Reading Faces. Westview Press, Boulder, CO. (1997)

Implementing Configurable Information Systems: A Combined Social Science and Cognitive Science Approach

Corin Gurr and Gillian Hardstone

IRC for Dependability of Computer Based Systems (DIRC)
University of Edinburgh, 2 Buccleuch Place, Edinburgh EH8 9LW, UK
{C.Gurr, G.Hardstone}@ed.ac.uk
http://www.dirc.org.uk

Abstract. This paper outlines an interdisciplinary approach to tackling the issues of integrating medical information systems into existing healthcare environments where high dependability is a significant requirement. It focuses on the knowledge of system users (domain practitioners) and designers, and the potential use of diagrammatic representations of that knowledge during the implementation process in order to support communication between the two groups, and to serve as tools in assisting system reconfiguration to user requirements during implementation.

1 Introduction and Background

This paper outlines an interdisciplinary approach to tackling the issues of integrating medical information systems into existing healthcare environments where high dependability is a significant requirement. It focuses on the knowledge of system users (domain practitioners) and designers, and the potential use of diagrammatic representations of that knowledge during the implementation process in order to support communication between the two groups, and to serve as tools in assisting system reconfiguration to user requirements during implementation.

Integration of new technological systems into an existing organisational environment requires a clear understanding of technology as intrinsically social [14], rather than as predominantly technical, but with social aspects. This makes it easier to unravel some of the implications of implementing a technology in a particular environment, including changes in processes; shifts in power relations, responsibility, authority and access to information; and how these factors interact.

Knowledge is an important aspect of technology [14], and thus a key issue in system design and implementation. It is intrinsically social, both in terms of its substantive content (what is known) and its cognitive content (how it is known). New systems need to interface with users' existing knowledge of the domain(s) in which they operate, and the activities (practice) that need to be performed within the domain space in an organisational context. Designers also need knowledge about users' domains of knowledge and the specific context in which they

M. Beynon, C.L. Nehaniv, and K. Dautenhahn (Eds.): CT 2001, LNAI 2117, pp. 391–404, 2001.

put that knowledge into practice in order to communicate effectively during elicitation and requirements analysis. The relation between designers and users is critical when implementation involves extensive system reconfiguration [6] to user needs, as is usually the case with Hospital Information Systems (HIS).

The use of diagrammatic representations is common throughout engineering and design practice [8]. Adopting a social science-based approach to knowledge and practice in organisations, we intend to work from ethnographic descriptions of knowledge and practice in a specific empirical healthcare setting, as informed by taxonomies of knowledge [5,11,20], through to a representation of that knowledge in diagrammatic forms, in order to facilitate communication between users and designers. We anticipate that our methods will result in the development of a useful tool for systems implementation.

2 Knowledge and Practice in Organisations

A significant issue for the design and implementation of IT systems which are intended to support the business or operating processes of complex organisations, is how the deployment of these systems actually changes those processes, intentionally or otherwise. Integration of technological systems into an existing organisational environment requires a clear and visible understanding of the potential ramifications of the technology in that particular environment: which processes or practices will or could change; how responsibility, authority and access to information may change and how these three factors interact.

One approach to the issues outlined above is to consider knowledge as a key factor in computer system design and implementation, particularly when a new system is being designed to replace an existing system. System designers need to know about users' domains of knowledge: the content of that knowledge, how it is structured, and how it is used. To acquire this knowledge, they need to communicate effectively with potential system users to elicit and analyse requirements. Most importantly, the new system needs to interface with users' existing knowledge of the domain(s) in which they operate, and the activities that need to be performed within the domain space. Users also need to understand how to use the new system in an organisational context once it has been designed. The relation between designers and users is critically important if the implementation process involves extensive customisation or reconfiguration of a basic system to user needs, when design and innovation continue during the system's operation within the user organisation, as is often the case with Hospital Information Systems (HIS).

But how to understand and capture the complexity of organisational knowledge and practice during the system design and configuration process? And how to convey that knowledge between designers and users? Exploring who knows what in the domain space, and what they do with their knowledge is a useful point of entry into this area. A sociologically-influenced approach to domain knowledge is proposed.

Knowledge can be seen as an inherently social process, in terms of its cognitive and substantive content, distribution and mobilisation for practice [12]. There is an existing body of work, primarily in the sociology of science and technology, that deals with different ways of categorizing knowledge from a social science perspective. For example, Vincenti [20] has analysed the substantive components of domain knowledge, relating them to the knowledge-generating activities that create them, in the domain of aeronautical design engineering (see Table 1). This framework thus appears domain-specific, but can readily be adapted to other domains [11]. Although Vincenti's work does not specifically address the social nature of knowledge, it implies a division of labour within a given domain: different people will be carrying out various activities (such as research, experimental work or operation), and the distribution of the substantive content of knowledge will therefore vary accordingly, and be unequal. Such considerations clearly have implications for the implementation of a HIS, where there are many and varied occupational groups of users, including administration staff and clinicians from a range of domains and sub-domains.

Table 1. Categories of substantive (aeronautical design engineering) knowledge, and knowledge-generating activities [20]

Categories / **Activities**	Fundamental design concepts	Criteria and specifications	Theoretical tools	Quantitative data	Practical considerations	Design instrumentalities
Transfer from science			X	X		
Invention	X	X	X	X		X
Theoretical research	X	X	X	X		X
Experimental research	X					
Design practice		X			X	X
Production				X	X	X
Direct trial (including operation)	X	X	X	X	X	X

The cognitive content of knowledge is no less socially shaped and distributed. One taxonomy that captures these aspects of cognition is that developed by Fleck [5,7] (see Table 2). It goes beyond the conventional distinction between tacit and explicit knowledge, and may be useful in a systems design context, because it carries considerable explanatory power about social relations and context. For example, meta-knowledge about a domain is likely to be shared by most people working at a site, or forming part of the same department or occupational group. Although formal knowledge is often highly valued (just one reason for the status of clinicians) and rewarded, in a workplace context, it is often not

Table 2. Components and contexts of knowledge (After Fleck [7])

Components of knowledge

Cognitive components	**Description**	**Acquired through**	**Embodied in**
Formal knowledge	Theories, formulae; often in written or diagrammatic form	Formal education	Codified theories
Informal knowledge	Rules of thumb, tricks of the trade	Interaction within a specific milieu	Verbal interaction
Contingent knowledge	Widely distributed, seemingly trivial information, context-specific	On the spot learning	The specific context
Tacit knowledge	Rooted in practice and experience	Apprenticeship and training	People
Instrumentalities	Embodied in the use of tools or instruments	Demonstration and practice	Use of tools
Meta-knowledge	General cultural and philosophical assumptions; values, goals; may be specific to organisation, domain, occupational group, etc	Socialisation	The organisation

Contexts of knowledge

Context	
Domains	More or less well-defined 'parts of the world' to which a particular body of knowledge applies
Situations	Assemblies of components, domains, people and other elements (or 'human and non-human carriers of knowledge' [Hardstone1998]) present at any particular instant of expert activity (or 'knowledge mobilisation' [12])
Milieus	The immediate environments in which expertise is exercised; comprising sets of situations occurring regularly at particular locations, e.g. laboratories, operating theatres, offices, etc.

the most important or useful for everyday practice. For example, contingent, locally-specific knowledge is usually extremely important during the implementation of configurational technologies, such as HIS, but often undervalued.

Each individual, as a carrier of knowledge, can know the same thing or concept simultaneously in different ways (as different cognitive components). Thus practitioners may have formal knowledge about an aspect of their domain, but this will be internalized and amplified through experience, practice and local conditions to create informal, tacit and contingent knowledge. Hence the relative importance of each cognitive component to carriers changes over time and space. Dealing with tacit knowledge is perhaps not such a problem (for system designers, for instance) after all, as other non-formal knowledge components may be at least partially articulable.

By combining taxonomies of knowledge [5,11,20] that relate to the cognitive content of knowledge [5,11] and the substantive content of domain knowledge in practice [11,20], extended to include social and organisational knowledge [11], the knowledge related to a particular domain or task within that domain may be conceptualised as distributed across a grid, each square of which tells us something about the social nature of that knowledge. An example is shown in Table 3.

Table 3. The substantive and cognitive content of knowledge: Grid for analysis (of a particular activity, domain, situation or milieu – for example, bed management)

Substantive / **Cognitive**	Fundamental concepts (what it is; how it works; 'normal')	Criteria and specifications (quantitative goals)	Theoretical tools (maths methods; intellectual concepts)	Quantitative data (descriptive and prescriptive)	Practical considerations (incl. judgement)	Design instrumentalities (knowledge about procedures)
Formal knowledge						
Informal knowledge						
Contingent knowledge						
Tacit knowledge						
Instrumentalities						
Meta-knowledge						

To operationalise these concepts in a domain context, we can conduct grid and then gap analyses based on taxonomies of knowledge [5,11,20], identifying the knowledge being mobilised and its distribution, and charting the networks of people and objects involved in particular tasks or activities. The outputs constitute useful analytical tools, particularly when translated into diagrammatic representations. We can compare the old and new systems, identifying differences and problem areas from a knowledge perspective. This information can be fed back iteratively to designers and users, using clear and accessible representations where appropriate. These serve as communication artefacts or 'translators' [11, 12,20] to support communication between domains.

Domain knowledge is varied in content and unevenly distributed within the socio-cognitive structures [5] of technological systems, with significant overlaps between carriers. This distribution is shaped both by structural and by more contingent social factors, which can be described and analysed in sociological terms. It can also be mapped onto the squares of the grid described above, and a gap analysis conducted to discover whether unpopulated squares are either irrelevant or problematic, and which squares contain knowledge crucial to specific tasks or activities. By looking at how knowledge is put into practice in context within an organisation, it is possible to discover both how knowledge is distributed, and how it is mobilised for specific activities, including those supported by computer systems. The mobilisation of knowledge is almost always collective, occurring through the formation of temporary networks of human and non-human carriers. Over time, with repetition, some of these networks and mobilisation processes become institutionalised in communities and routines, independent of the specific individuals involved. These networks and groupings can be described and represented diagrammatically.

To operationalise these concepts in a systems design and configuration context, we can chart the diverse networks and communities of carriers assembled for particular tasks or activities, whether routine or *ad hoc*. We can combine this with use of the grid to identify the kinds of knowledge that are being mobilised and how they are distributed. The outputs (grids and network charts) should provide a useful analytical tool, particularly when combined with, or translated into diagrammatic representations. By conducting such an analysis of both the old and the new systems, we can compare the socio-technical system that is being replaced with the new system; identifying where the differences lie, and where problem areas might be, from a knowledge perspective. This information can be fed back iteratively to both designers and users, using clear and accessible diagrammatic representations where appropriate, as described below, to assist communication and discussion.

Studies touching on the mobilisation of knowledge by carriers from more than one domain [11,12,20], such as between system designers and domain users, suggest that some of the people and also the artefacts involved need to be able to operate in the networks of both domains, acting as 'translators' (especially in the light of domain-specific languages) for problem-solving to occur. The use of diagrammatic representations in the proposed empirical context may provide one means of translating between domains.

We propose to combine the above with previous research addressing the issue of communication concerning the design, assessment and deployment of complex, highly dependable computer-based systems, where that communication must take place across technical and non-technical boundaries. In that context, as here, knowledge concerning the (evolving) design, and the impact of changes to it, is distributed across a broad range of stakeholders representing multiple technical and non-technical disciplines, who hold diverse needs and goals. Our previous research has extensively studied the role of differing forms of representation, particularly diagrammatic, in facilitating the communication of knowledge in this context.

3 Representing Knowledge

The use of diagrammatic representations is common throughout engineering and design practice [8]. Previous research has compared diagrammatic and textual forms of representations from both semantic and cognitive perspectives [10]. This work will inform our design of diagrammatic languages and notations to capture the implications of our social analyses in accessible forms, facilitating the communication of knowledge during the design and deployment of complex, highly dependable computer-based systems across technical and non-technical domain boundaries. Thus we aim not only to study the potential impact of the proposed technological system, but also to make this impact visible and accessible to a broad range of stakeholders through the use of appropriately designed representations.

3.1 Designing Effective Diagrammatic Representations of Knowledge

Diagrams are popular, as many people find them more readily "accessible" than other forms of representation. Diagrams are also effective at presenting "the big picture"; that is, diagrams can typically contain far more visible structure than any text-based representation and this structure can be used to reflect the structure of whatever it is that the diagram represents. Diagrams are thus particularly popular and effective in design, where they are typically most effective at presenting high level overviews of entire systems, in which the relationships and interactions between components is highly visible, and thus more readily accessible.

An illustrative example of the significance of a well chosen representation in facilitating communication across technical boundaries for highly dependable systems is the HAZOPS (HAZard and OPerability Studies) hazard analysis technique [1]. HAZOPS is a technique that originated in the chemical industries which involves engineers and experts from a broad range of technical disciplines holding a series of "structured brainstorming" sessions to identify and assess the potential hazards of a proposed design. A typical HAZOPS is oriented around a diagram of the proposed design. For chemical plants, schematics of the physical plant layout (piping and instrumentation diagrams) are used. The HAZOPS team examine in turn each component depicted on the diagram and consider the hazards and likelihood of failures or deviations from its intended function. Typically each team member will have access to that information on the proposed design which is relevant to their field of expertise. Thus the team is able to bring a great breadth of experience and data to the analysis yet, by coordinating the analysis around a common focus (the diagram), individual team members need not be concerned with information beyond their own area of expertise. Furthermore, the diagram used in a HAZOPS typically represents the proposed design at a general enough level to be clearly understood by all team members, regardless of technical discipline and expertise, while still being sufficiently detailed to make an analysis based upon it worthwhile. The diagram thus plays the role of a *communication artifact*, an entity which guides and supports communication concerning the system under analysis.

An effective diagram is typically taken to be one that is "well matched" to what it represents. This is to say, that the logical and spatio-visual properties of structures inherent to the diagram are chosen so as to have some very direct correspondence with the structures that they represent in the semantic domain; and in particular that they are chosen so as to support desired reasoning tasks by making certain inferences *immediate* and *obvious*. A more detailed exploration of this issue, including a formalisation of the concept of well matched, is in [10].

In this section we present guidelines for both the design of effective diagrammatic languages, and the design of specific diagrams within such languages. These guidelines draw upon results from visual language theory, cognitive science, empirical psychology and graphic design. Integrating results from such diverse fields is a non-trivial task, which is here approached through a decomposition of the study of issues of effectiveness in diagrammatic languages according to analogous understandings of (written and spoken) natural languages. We present an overview of this study next.

3.2 Exploring Diagrammatic "Matching"

The study of natural languages is typically separated into the following categories: phonetics and phonology; morphology; syntax; semantics; pragmatics; and discourse. With the obvious exception of the first, the study of analogous categories in diagrammatic languages is at the same time both highly revealing of differences and similarities between the two forms of representation; and also provides a structure in which to explore the alternative means by which a diagram may capture meaning. Separating the study of diagrammatic languages into these categories permits us firstly to lay out the various means by which the structure inherent to diagrammatic morphologies and syntax may directly capture structure in the semantic domain; and secondly to consider how further pragmatic usage may convey meaning in diagrams. Such a study is undertaken in [10], which extends earlier work of [9] in decomposing the variety of issues pertaining to effectiveness in diagrams. This section presents an overview of this exploration, focusing on the alignment of syntactic features of diagrams to their semantics.

Morphology concerns the shape of symbols. The shape of a particular alphabetic character cannot convey much variation in meaning; an 'a' is an 'a' regardless of its font or whether or not it is bold or italicised. By contrast, the basic vocabulary elements in some diagrammatic language may include shapes such as circles, ellipses, squares, arcs and arrows, all of differing sizes and colours. These objects often fall naturally into a hierarchy which can constrain the syntax and, furthermore, inform the semantics of the system. This hierarchy may be directly exploited by the semantics of symbols so as to reflect the depicted domain.

A number of studies such as [3,17] have attempted to categorise diagrammatic morphology, Horn [13] reviews these and proposes a unified categorisation (for

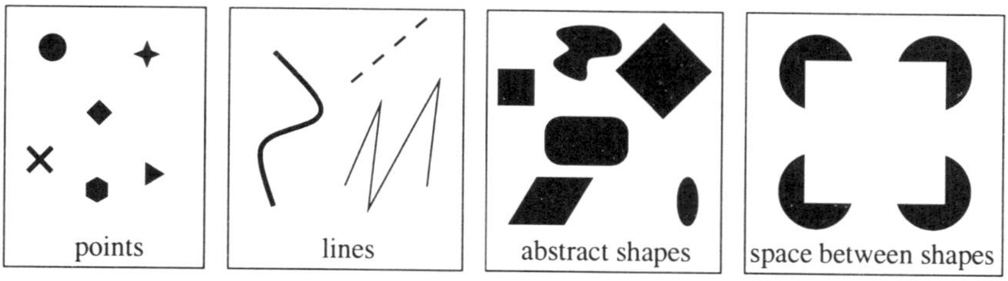

Fig. 1. Morphology of Shapes (Horn'98)

generic representations) whose most general categories are: words; shapes; and images. Here we focus on shapes, which Horn subdivides into: points; lines; abstract shapes; and "white space" between shapes – although we do not consider this latter here.

The category of abstract shapes, and potentially that of shaped points, may be further subdivided. For example, regular shapes may be divided into "smooth" and "angled" as determined by their corners. Such sub-categories may be further divided, leading to a type-hierarchy of shapes which may be directly exploited by the semantics of symbols so as to reflect the depicted domain. For example, consider a map on which cities are represented as (shaped) points. A categorisation of points divided into smoothed and angled could be exploited by a corresponding categorisation in th semantic domain with, say, smoothed points (circles, ellipses, etc) representing capital cities and angled points (triangles, squares, etc) representing non-capital cities. The division of smoothed and angled points into further sub-categories could similarly correspond to further sub-categorisations of capital and non-capital cities. Note however that there is no unique *canonical* hierarchy of shapes.

In addition to a morphological partial typing, symbols may be further categorised through graphical properties such as size, colour, texture, shading and orientation. For example, the meaning of symbols represented by circles may be refined by distinguishing between large and small, and different coloured circles. Thus, again, part of the structure in the semantic domain is directly captured by morphological or syntactic features.[1] The properties of graphical symbols we consider here – again modifying those suggested in [13] – are: value (e.g. grey-scale shading); orientation; texture (e.g. patterns); colour; and size. These are applied to points, lines and shapes as in Table 4.

In addition to exploiting the structure of the morphology of diagrammatic symbols, we may also exploit the structure and properties inherent to diagrammatic syntactic relations in ensuring that a diagram is well matched to its meaning. For example, the use of *inclusion* or *overlap* to represent semantic relationships which share logical properties with these syntactic relations. A promising exploration of the properties of various syntactic diagrammatic relations (primar-

[1] Note that textual tokens may also display such properties in a slightly more limited sense, such as font, italics, etc.

Table 4. Properties of primitives (2)

	Value	Orientation	Texture	Colour	Size
Point		min		X	
Line	lim	X		X	X
Shape	X	X	X	X	X

ily of relations between pairs of diagrammatic objects) is given by von Klopp Lemon and von Klopp Lemon [21], who define the logical characteristics of 12 properties and examine their presence or absence in around 65 syntactic diagrammatic relations.

Finally, in linguistic theories of human communication, developed initially for written text or spoken dialogues, theories of pragmatics seek to explain how conventions and patterns of language use carry information over and above the literal truth value of sentences. Pragmatics, thus, helps to bridge the gap between truth conditions and "real" meaning – that is, between what is *said* and what is *meant*. This concept applies equally well to diagrams. Indeed, there is a recent history of work which draws parallels between pragmatic phenomena which occur in natural language, and for which there are established theories, and phenomena occurring in visual languages – see [15] for a review of these.

3.3 Guidelines for Diagram Language Design

Our guidelines for diagrammatic language design are as follows:

1. identify the fundamental semantic concepts and any structuring which exists over these. Match this to the morphological structure of graphical primitives;
2. identify features and properties of these semantic concepts and match to properties of the chosen symbols and graphical syntactic features;
3. identify properties of semantic relationships between objects and match these to syntactic relations.

However, this matching must be in the context of consideration of the tasks which the potential diagrams are intended to support. These tasks should indicate the key features, and the syntax should be chosen so as to achieve maximum salience of these. This desire will also inform decisions when there is a choice of equivalent syntactic matches for some desired semantic feature.

Note that as certain graphical properties and syntactic relations may interfere, often a balance or trade-off is required when selecting the most appropriate syntactic match for some semantic aspect. Experience in graphic design (e.g [18, 19]) suggests a rule of thumb that *task* concerns outweigh *semantic* concerns; that is – where a trade-off is required, the preference should be whichever option supports greater salience of task-specific features.

Typically, for any non-trivial semantic domain and intended tasks, not all information may be captured directly through diagram syntax. Consequently the use of *labelling languages* for labels which may potentially contain significant semantic information is necessary for most practical diagrammatic languages. However, in an effort to increase the expressiveness, the unprincipled use of sophisticated labelling languages can perturb the directness of a diagrammatic language. Examples of languages which are diagrammatic at core, but have had their expressiveness enhanced through sophisticated labelling languages until any benefit to readers interpretation of the "diagrammatic aspects" is negated, are legion. This is a substantive and open issue which is beyond the scope of this paper, and so we merely issue the warning: treat labels with care.

Finally, the construction of any specific diagram must also ensure that any non-semantic aspects are normalised as far as possible, as random or careless use of colour or layout, for example, can lead to unwanted mis- or over-interpretation by the reader.

3.4 An Example of "Well-Matched" Diagrams

One practical application of the guidelines proposed above appears in a study by Oberlander *et al* [16] of differing cognitive styles in users of the computer-based logic teaching tool Hyperproof [2]. A language was devised for [16] which provided the reader with a salient and accessible representation of the significant differences in the use of Hyperproof by the two groups, named "DetHi" and "DetLo".

Examination of this semantic domain suggested that a simple node-and-link representation, where nodes represented Hyperproof rules (user commands), and directed links represented the "next rule used" relationship, captured the key concepts. The features seen as most necessary for presentation to the reader were the frequencies both of rule use and of transitions between specific pairs of rules. The preferred matching of these features to properties of boxes and arrows, as indicated by Table 4, was the use of *size* to represent frequency in each case. Thus the relative size of nodes directly corresponded to the relative frequency of rule use. Following the above guidelines, lines were restricted to being one of five discrete sizes, with increasing size indicating increasing frequencies. Thus each specific line width represented a range of frequencies relative to the issuing node, with frequencies of 10% and lower not being represented. Absolute transition frequencies are therefore represented by accompanying textual labels. The resulting diagrams are repeated here in Figs. 2 and 3.

The final consideration for the construction of these two specific diagrams in the devised language concerned the use of layout. The tasks for which the diagrams were to be put were of two kinds: the identification of patterns in a single diagram; and the identification of characteristic differences between two diagrams. Layout had a mild impact on the former task, suggesting that as far as possible the layout should place connected nodes in spatial proximity. Layout had a greater impact on tasks of the latter kind, suggesting that to facilitate comparisons firstly the layout of nodes in the two diagrams should be as similar as possible; and secondly that where size (area) of a node varied between the two diagrams, this variance should take place along a single dimension wherever

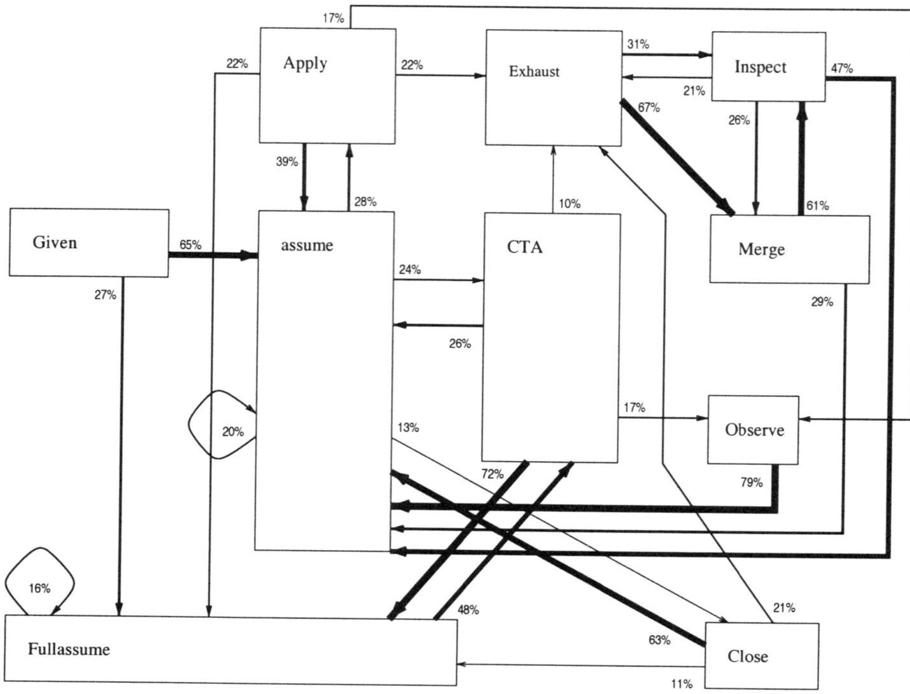

Fig. 2. Transition network for DetHi behaviour on indeterminate questions

possible (in accordance with the relative perceptual salience of comparison along identical uni-dimensional scale versus area, as indicated in empirical psychological studies such as [4]). One final point of note is that Hyperproof's `Close` rule was never used by DetLo subjects. Following the guideline that *task* concerns outweigh *semantic* concerns, the pragmatic decision was made that the `Close` node should be represented in Fig. 3 (rather than being of zero size). However, to indicate that this node categorically differed from all other nodes in that diagram, its bounding line was represented with a lesser value (i.e. a dashed line).

The effectiveness of this diagrammatic language for the required tasks should be readily apparent to the reader. Note, for example: the characteristic differences between the use of the `Observe` rule by DetHi and DetLo subjects; patterns of rule use such as `Merge-assume` by DetHi subjects which are completely absent in DetLo subjects; and the generally more "structured" use of rule-pairs by DetHi subjects – indicated by the greater number of thick lines, and fewer lines overall, in Fig. 3.

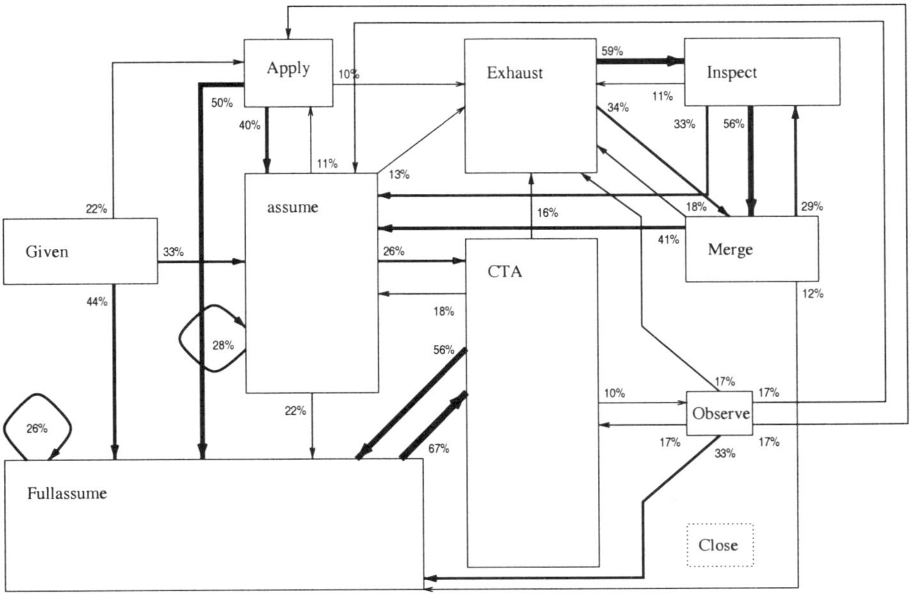

Fig. 3. Transition network for DetLo behaviour on indeterminate questions. Note that `Close` is not visited at all

4 Summary

Our initial application and evaluation of this work is in the domain of Healthcare Informatics. We will be working with a large NHS hospital, which is in the process of designing and implementing a new computer-based Health Information System (HIS). As various modules of the HIS are implemented, we will compare pre- and post-HIS working practices. Using the methods outlined above, we will provide feedback between users and designers throughout the design and implementation process.

The organizational structure of a hospital is typically one of great complexity and the needs and knowledge of system users are significantly diverse. In combination with the expectation that the proposed system will be subject to substantial local configuration for different medical and administrative departments, it is clear that the integration of this system into the existing hospital environment offers a fruitful opportunity for us to evaluate the efficacy of both our representations and our overall analytical approach to this task.

References

1. Chemical Industries Association. *A Guide to Hazard and Operability Studies.* 1992.
2. J Barwise and J Etchemendy. *Hyperproof.* CSLI Publications, 1994.

3. J Bertin. *Semiology of graphics: Diagrams, networks and maps.* University of Wisconsin Press, Madison, WI, 1983.
4. W S Cleveland. *The elements of graphing data.* Wadsworth, Pacific Grove, CA, 1985.
5. J Fleck. Innofusion or diffusation? the nature of technological development in robotics. PICT Working Paper Series 4, University of Edinburgh, 1988.
6. J Fleck. Configuration: Crystallising contingency. *International Journal on Human Factors in Manufacturing*, 1992.
7. J Fleck. Expertise: Knowledge, power and tradeability. In Williams et al, editor, *Exploring expertise: Issues and perspectives.* Macmillan, 1998.
8. C Gurr. Knowledge engineering in the communication of information for safety critical systems. *The Knowledge Engineering Review*, 12(3):249–270, 1997.
9. C Gurr, J Lee, and K Stenning. Theories of diagrammatic reasoning: distinguishing component problems. *Mind and Machines*, 8(4):533–557, December 1998.
10. C A Gurr. Effective diagrammatic communication: Syntactic, semantic and pragmatic issues. *Journal of Visual Languages and Computing*, 10(4):317–342, August 1999.
11. G Hardstone. *Robbie Burns' moustache: print knowledge and practice.* PhD thesis, University of Edinburgh, 1996.
12. G Hardstone. You'll figure it out between you: Problem-solving with the web-8. In Williams et al, editor, *Exploring expertise: Issues and perspectives.* Macmillan, 1998.
13. R E Horn. *Visual Language: Global Communication for the 21st Century.* MacroVU Press, Bainbridge Island, WA, 1998.
14. D A MacKenzie and J Wajcman, editors. *The social shaping of technology.* Open University Press, Buckingham, 2nd edition, 1999.
15. J Oberlander. Grice for graphics: pragmatic implicature in network diagrams. *Information design journal*, 8(2):163–179, 1996.
16. J Oberlander, P Monaghan, R Cox, K Stenning, and R Tobin. Unnatural language processing: An empirical study of multimodal proof styles. *Journal of Logic Language and Information*, 8:363–384, 1999.
17. F Saint-Martin. *Semiotics of visual language.* Indiana University Press, Bloomington, IN, 1987.
18. E R Tufte. *The Visual Display of Quantitative Information.* Graphics Press, Cheshire CT, 1983.
19. E R Tufte. *Envisioning Information.* Graphics Press, Cheshire, CT, 1990.
20. W G Vincenti. *What engineers know and how they know it: Analytical studies from aeronautical history.* John Hopkins University Press, Baltimore, MD, 1990.
21. A von Klopp Lemon and O von Klopp Lemon. Constraint matching for diagram design: Qualitative visual languages. In *Diagrams 2000: Theory and Application of Diagrams*, LNAI 1889, pages 74–88, Berlin, 2000. Springer.

Interdisciplinary Engineering of Interstate E-Government Solutions

Reinhard Riedl

Department of Computer Science, University of Zurich, Winterthurerstr. 190
CH-8057 Zurich, Switzerland

Abstract. We present a generic, inter-organizational approach to e-government for all, which relies on the structural engineering of distributed, administrative services. Citizens are enabled to initiate and control the secure exchange of trustworthy personal information about them. Our focus is on administrative services for migrating citizens in the European Union, but our system architecture generalizes to any co-operation of authorities exchanging personal data, and it guarantees that the strict European data protection principles are respected.

1 Introduction

A2C[1] E-government shall provide digital access to administrative services. So far, when a European citizen moves from one state to another, she has to spend a considerable amount of time on administrative tasks. First of all, she has to find information about administrative requirements at her new living place, such as how to register her new residence, or how to enroll her children in school and how to order the various different refuse collection services and so on. Then she usually has to contact authorities in her old and her new living place, she has to obtain personal documents from the authorities in her old living place and she has to deliver these documents to the authorities in her new living place. Thus, she can apply for various different public services, which means providing correctly filled in paper forms at the right place with the right additional documents at hand.

All this is extremely tedious. However, what makes these tasks so extremely difficult are the strong differences in civil services and public culture encountered in Europe. For example, while registration of residence is mandatory in Germany (where it has to follow strict rules in the general case) or in Italy (where civil servants have a lot of freedom in exception handling) there is nothing like registration of residence in the UK. Instead, in the UK citizens will use their power bills to provide evidence of their living place. Furthermore, while in Italy the traditional concept of family has a lot of importance, in the Netherlands homosexual marriage is possible and there is no legal distinction between father and mother: both are addressed as parent by the law. Nevertheless, a child may lose

[1] authority-to-citizen. i.e. the authority provides services for the citizen

M. Beynon, C.L. Nehaniv, and K. Dautenhahn (Eds.): CT 2001, LNAI 2117, pp. 405–420, 2001.

its legal father, when its parents move from the UK to the Netherlands as the Netherlands do not always accept British certificates of male parentship.

It may also come as a surprise to an Irish woman moving to Italy that from a certain administrative perspective, military bases and families are similar concepts in Italy. And it might surprise her profoundly that any of her grown up children might announce herself as the head of her family, which will be accepted by the Italian anagraf office. Thus interstate e-government in Europe faces the hard problem that cultural expectations of European citizens with respect to citizen-to-authority interaction differ strongly all over Europe. As long as no migration takes place this does not cause problems, but the European Commission intends to support mobility of European citizens and cultural exchange for various reasons.

Future e-government will have to address the problem of cultural heterogeneity. Compare also [1]. However, if e-government should serve all citizens, it must not assume cultural and physical skills as they are normal for a society, but it has to take the skills gap in society into consideration as well as physical handicaps. In particular, the latter is a main issue in the UK, where it is not admissible to provide 'first' solutions for non-handicapped people and to save money on extra solutions for handicapped people, if such solutions are possible. Differences in digital skills include positive and negative deviations, that is a lack of skills and a surplus of skills. Both have to be addressed in the system design as a better knowledge of IT systems often creates a more critical attitude and a reduced readiness to accept new technology, which has to be dealt with. Thus, cultural heterogeneity is 'enriched' with a somewhat orthogonal social heterogeneity of the users.

Further, a closer look at existing IT infrastructure in European municipalities reveals that the implementations of the principles of data protection differ strongly for different countries. While, cum grano salis, in Germany any exchange of personal data among authorities without consent of the person represented by the data or the authorization to do that signed by a judge is strictly forbidden, in Denmark there is a central database, which can be accessed by all public administrations. Social investigations indicate, however, that the Danish people accept this system only for historic reasons, while it would be impossible to implement a new centralized system in Denmark right now. Moreover, while in the UK the local authorities are very concerned about privacy and want to offer anonymization to their citizens, in Belgium authorities intend to install DNA databases. How such databases could be realized without violating the European regulations concerning biometric devices is still unclear. We face a completely heterogeneous situation with respect to data protection, too. That situation is strongly shaped by legacies and local traditions, and it partially contradicts data protection rules on the European level.

Winning public acceptance for European interstate e-government requires that local traditions be respected to a certain extent, which rules out imperialist standardizations. This paper is based on our work in a European interstate e-government project, which is funded by the Information Society Technology

R&D programme of the European Commission and which is affiliated with the Smartcard track of the e-Europe initiative. In the project, we have analyzed registration procedures for a new living place in seven European cities: Antwerp in Belgium, Belfast in Northern Ireland, Cologne in Germany, Grosseto in Italy The Hague in the Netherlands, Naestved in Denmark, and Newcastle-upon-Tyne in England. Theses analyses have exhibited severe differences on all levels: ontologies, laws and guidelines, administrative processes, legacy systems, and user expectations. Nevertheless, we managed to draw up a solution, which respects the differences and at the same time enables citizens to copy personal information from one organization to another. (Clearly, the underlying transfer of data has to include a translation from one ontology to another.) And it turned out that this is enough to provide basic interstate e-government services such as the registration of living place for a foreign European citizen.

Lots of discussions indicated that there is a ranking of priorities for abstract user requirements as follows: 1. trust and confidence, 2. & 3. access and usability, 4. benefits of the service Successful solutions will have to address all four types of requirements. Judgments for one and the same technical solution will differ significantly. The meaning and the relevance of trust and confidence depend both on the role of the user – e.g. citizen or civil servant – and on her cultural background. Note that trust an confidence differs from security as it describes the user acceptance of security solutions.

Access requirements vary with respect to the preferences for end-user devices, but universal mobile access may be considered as a global requirement, and technical problems arising from the use of different end-user devices is easier to cope with than the so-called 'soft' problems depicted above. (Still, universal access poses demanding requirements for the security design of an e-government application.) On the contrary, usability is strongly related with expectations and process comprehension, and thus design for usability has to deal with cultural diversity and the digital divide. Usability for civil servants requires built-in flexibility for exception handling, while usability for citizens requires a useful guidance and lots of self-explanatory help functionality. All three classes of users need to be able to trace legally relevant activities as well as inter-organizational work-flows.

The sociological analysts of our partners in the cited project have revealed that the main problem for migration citizens are due to the necessary interaction with private enterprises, in particular with banks. Thus future A2C e-government will have to design new digital forms of private-public partnership similar to the partnership responsible for the Fin-ID, which is a digital identity card which allows access to banking services. However, since the type of service provided seems to be less important than trust and confidence, it makes sense first to gain experience with e-government projects focusing on public services before we are rising to the challenge of interstate private-public partnership.

2 Support for Migrating Citizens

The support of migrating citizens is the main goal of the solution, which was developed and prototypically implemented in the project cited above. Citizens shall no longer have to carry personal documents in paper form with them in order to obtain certain administrative services. They shall not have to bother with finding out, which documents are needed in a foreign country. They shall not have to bother with obtaining these documents from various authorities and delivering them to various other authorities. Instead, they will be able to use a secure multi-application Smartcard, namely a JavaCard, for the access of governmental services at electronic kiosks with touch-screen interfaces. Those electronic kiosks will serve as contact points realizing so-called one-stop government.

Upon the request of the citizen, a kiosk contacts a local e-government application running a selected administrative service, and the application sends a work-flow description to the kiosk encoded as an XML-file. This enables the kiosk to guide the citizen through the administrative process. It informs her, which documents from which authorities are needed, and whether these documents are already stored on her JavaCard or in a virtual memory, which can be accessed securely with her JavaCard. In case that the documents are not available they have to be created ad hoc. In case that they are available on the JavaCard or its virtual extension, i.e. some virtual memory, which can be accessed solely by the Javacard, the citizen can choose either to use these documents or to request their new creation.

When a new document has to be created, the application server creates a request, which is time-stamped by the kiosk, and the kiosk asks the citizen to digitally sign the request for the new document with the JavaCard. Upon the consent of the citizen the time stamped request is handed to the JavaCard, and the citizen confirms her will to sign the request by pressing a finger on the fingerprint sensor, which is integrated into the Card.[2] The Card verifies the identity of its user and upon a positive result it signs the request and it hands it back to the kiosk, which hands it over to the application. The application in turn sends the request to a remote document service, which checks for the authenticity of the signing person and for her authorization to obtain the requested document, and then the document service creates the document, signs it, and encrypts it with the public key of the destination specified in the request. Authorization is verified against a list for service access rights held locally by the remote document service and the authenticity of the request is verified based on the validity of the digital signature provided by the JavaCard. The document service accepts only a restricted set of destinations, which are in our case the JavaCard and/or the application in charge of the administrative work-flow.

Once the local e-government application has received a personal document from the JavaCard or some remote document service, it proceeds similarly as the remote document service. It checks the relevance of the document and its authenticity based on the identity of the signing authority and its digital signa-

[2] Such an integration was not implemented in the cited project.

ture. In addition, possibly the local application requests further validations from validating agents, and data from different documents are checked with respect to consistency. An example of a remote validation agent is the police in Italy, which has to verify the citizen's statement on her new living place. The citizen may also be asked to verify the information contained in a document and to add her own signature, before the document is processed by the local e-government application, as this further increases the relevance of a document from a legal and administrative perspective. Once all documents and verifications needed are obtained, the local application presents the status of the administrative process to a civil servant for confirmation. Upon the confirmation of the civil servant, the administrative service is completed, data are written to the corresponding legacy system, and the citizen is informed about the completion of the service by e-mail and/or by surface mail. The main point from the perspective of digital identity and of organizational security is that the JavaCard speaks in effigy of the citizen with the local and the remote e-government applications and that it checks the authenticity of its user, every time she does so. Thus the user can access services transparently with respect to her current location. Since we provide asynchronous communication among e-government applications, the system provides additional time transparency. In the language of data privacy and data protection concepts, the citizen can create and control a context for the exchange of her personal data among different authorities. This complies with European data protection standards and it guarantees nearly optimal security.

The realization of the generic work-flow depicted above concept depends on the particular requirements for an administrative process. Apart from providing access to mandatory services, the kiosk further provides information on the optional municipal services and it offers recommendations on useful private services. The citizen is guided through the process depending on her chosen profile, which will be void, if she does not want any personalized guidance. In the future, further situated support will be offered for both civil servants and citizens based on the concepts of communities of practice, and communities of citizens, respectively. Of course, the application logic does not attempt to handle all exceptional cases automatically, but a human exception handling is supported in case that due to differences in administrative ontologies the appropriate information cannot be supplied automatically.

The electronic documents exchanged are XML- documents, according to an intermediary XML representation scheme for European citizens and items subject administrative considerations, such as cars, for example. The authority providing a document is responsible for the translation of data in its own ontology to a sub-representation in XML, and on the other side, the authority using a document is responsible for the translation from the representation in the intermediary XML-format to its local data scheme. That procedure avoids critical data accesses through an abstract tier. Moreover, in order to provide full further translatability into local ontologies, XML attributes are enriched with relevance attributes related to the source of information. Further, multiple versions of one and the same attribute with implication relations are used. However, in both cases it is not yet clear whether that type of intermediary representation scheme really provides optimal support for the relevance management by the receiver of a document.

The documents are created in two steps, first the e-document service provided by the e-government application translates data stored in the local legacy system into XML-files following a local XML scheme, and second this XML-file is translated into an XML-document following the intermediary representation scheme. When these documents are used for the supply of the administrative service, the same two steps are performed in reverse order, whereby the main processing is done with local XML-files and the translation to the local ontology is done only when data are finally written to the legacy system.

All documents shipped through the system are time-stamped and signed by the provider of the information. The signature assures the correctness of the document with respect to its explicitly stated context and the time-stamp of the document. The system guarantees the authenticity of the origin of information. Possible validation agents annotate meta-information on its relevance to the document, while the local e-government application decides on the actual relevance of the document with respect to the administrative service requested by the citizen. Thus a complete natural value chain for information management in virtual enterprises is implemented by the system.

Originally, we had intended to use the JavaCards as carriers for digital documents only, but it quickly turned out that this would not provide the functionality needed for A2C e-government. As a result of our research, we have exchanged the original approach in the project to create documents by handling states with the current approach to create states by shipping information. Although this might look like a rather philosophical issue, it is a major research result of the project. The original type of solution would have created a lot of problems for the organizational implementation and it would have provided significantly less useful services for the citizens than the current solution because of its restricted flexibility. The delicacy of the problem lies in the high costs for consistency management, which are avoided with the new system design. In fact, the new solution is a complete digital realization of traditional government, which is based on naturally, distributed and scattered states. Moreover, European laws on data protection are stringently respected and a complete control of all shipping of data representing information about her is given to the citizen.

It might appear that the application logic and the document handling described are purely technological artifacts and that they do not consider user requirements. However, the contrary is true. A comparison with conventional ideas and our original approach shows that the reflecting of functional user needs and constraints have essentially shaped the design of the system architecture on all levels, from the value chains and business process down to the integration of the IT-infrastructure. Many problems do not appear in our solution, which would have seriously constrained the usage of the original system. For example the power of attorney problem and the family concepts, can now be smoothly solved on the basis of the technically feasible delegation of access rights for services avoiding problems with family cards et cetera.

The critical decision taken in the project was to think back and to remember the universal cultural tradition of secure and trustworthy information exchange based on sealed documents. This enabled us to develop a conceptual decision, which reflects user requirements much better than the lists of requirements, which users provided in interviews, and which severely contradicted each other. We observed strong imperialist attempts by the various municipalities from different nations involved in the project and we only managed to settle the ever the same disputes at project meetings when we provided a universal structural concept for the technological solutions. In fact, once experts started to understand the technological issues on a structural level, they were much more open for compromise than they had been during the non-technological discussion of user requirements. Although people could not agree on user requirements they did agree to accept the generic customizable technological solution depicted above.

Originally, we had intended to perform an interdisciplinary, distributed interface-based engineering with a customary, industrial project management. This basically failed for two reasons. First, our industrial project management was not used to make strategic decisions requiring a detailed understanding of technical and cultural issues. And second, the understanding of the defined interfaces of the non-technological and the technological partners was incompatible. For example, there have been endless discussions on the meaning of evolutionary prototyping, which blocked the project for various months, while on the other hand, the project management applied management procedures known as waterfall model. Quite remarkably, the blockade was run, when the technological partners presented an economic model for e-government and a philosophical discussion of the cultural heterogeneity. In particular, the involved municipalities started to trust in the project as a whole then.

Although we aware that our prototypical solutions only addresses at best 80% of the problems and that we are far from having a solutions which could be used in real business, we think that our application scenario and our experiences collected in this project can be used as a reference case for the general problem of engineering large-scale inter-cultural e-business solutions and of holistic approaches to system design and to business technology. According to our experience, firstly, in many cases thinking back in history is mandatory for successful system design, and secondly, analytic structural thinking is a conditio sine qua non for project success, when highly heterogeneous user requirements rule out a clear cut logical solution.

3 Main Engineering Issues

In the following we shall discuss the importance of structural thinking for the design process and the importance of context management for the information exchange. Further, we shall shortly discuss user-interfaces and project management.

3.1 Structural Thinking

At the time of writing this paper, there are lots of local e-government project all over the world, and we are observing much less ambitious projects fail than the one discussed in this project. Local heterogeneity in large cities may cause project failure when it is not tackled appropriately. Classical process modeling as it is applied for e-business projects (with average success rates varying between 10 and 70 per cent depending on the size of the project, its degree of innovation a,nd the project team) is not good enough for complex e-government projects.

There are various essential differences between e-government projects and other e-business projects. In general, apart from exceptional cases, it is never possible to fully specify processes, as in any organization or company there is a human networking orthogonal to specified processes and sensible to the sympathy or antipathy among employees. Good e-business process models support self-organizing activities arising from human-networking and they nurture and support the cooperation in communities of expert workers. However, in e-government further considerations have to be taken into account, which require additional flexibility of processes. Governmental processes are designed to resolve contradicting interests and laws in a way which is accepted by the people. Management decisions in a private enterprise are usually not discussed by a public audience, while decisions taken by an authority are discussed in press and severe criticism might even force the resignation of the responsible member of the government.

In the course of decades well grounded decision processes have thus evolved as a consequence of public, political, and internal monitoring. These processes are capable of resolving exceptional cases which are either not covered by law and experience, or which are subject to contradicting laws and guidelines. That exception handling requires both a lot of freedom to violate process specifications and a subtle understanding when and how such violations are admissible. The latter relies on tacit knowledge developed through experience, thinking, observation, good advice by more experienced colleagues, and intuition. Process specifications and digital implementations must not hinder that flexible exception handling. That is only possible, if process designers have either succeeded in capturing the available tacit knowledge on exception handling, or they have implicitly understood where to put and how to design the boundary between specified processes and exception management. Fully specified processes are available for full digitalization and automation, while non-predetermined human decision processes, should only be supported and monitored by the information technology, but systems designers should not attempt to control that exception handling. It is thus a structural understanding of the cookable and the non-cookable, which is essential to the building of complex e-government applications.

The importance of tailoring technological solutions to preserve cultural diversity instead of trying to confront such diversity has been emphasized above, where we also have accounted, how major interface design issues can be avoided by a lower level of software engineering to implement an 'inter-cultural translation approach' to inter-governmental procedures. However, every pursuit of goals corrupts, and absolute pursuit corrupts absolutely. There are no inter-cultural solutions which do not effect the legacy processes. Successful management of a highly heterogeneous system requires convergence. Digitalization creates change and it will always be based on some compromise between optimization and conservative preservation of traditional business intelligence. In the optimal case a digital solution will facilitate procedures which did not take place behavior because they caused too much efforts before. However, even then, some people will perceive that as a violation of tradition and thus they may oppose it. Thus awareness of the need to violate tradition is a key success factor for an e-government project. That awareness has to be disseminated by the project management in a way which convinces the civil servants concerned that the change taking place results from essentially the same type of compromise which is characteristic for their everyday business. Putting it differently, engineering of e-government systems requires trust and confidence of the users that there needs have been considered in the design process. Of course, the same is true for business re-engineering in industry, but in e-government that trust and confidence is even more critical for project success than in industry.

A conceptual understanding of the interplay between digitalization, change, and user acceptance is a must for successful project management. Spectacular failures teach us what may happen without that understanding. The application of computers exists in an environment formed by human culture. In our e-government project we have tried to explore in practical way the impact of part of that culture – in particular societies definition of identity and privacy – and how computerization impacts this issue. This has caused more questions than answers (concerning lots of details not discussed in this paper) but it has provided us with a better understand of the risks which a commercial interstate e-government project would face.

Privacy is threatened by the use of contemporary Internet technology and therefore, constraints on its use were defined in the European data protection regulations. Our architectural design has demonstrated that sticking to these data protection principles does not prevent the digital facilitation of administrative services for European citizens. On the contrary, they provide guidelines for the design of a trust and confidence technology, whose usage is not restricted to e-government only. It equally fits with business models for dynamic virtual enterprises and for flexible strategic cooperation in supply-chains. That type of generic system architecture will not emerge from faithful process modeling, but it requires structural thinking and the analysis of underlying more basic concepts, such as identity and privacy. Clearly, further research work is needed on these issues.

In our prototypical e-government system the JavaCard speaks in effigy of the citizen with the application, while servers at the remote authority providing personal documents speak in effigy of the remote authority. This creates challenging psychological, social, cultural, and legal problems for the system engineering, which we could not fully solve yet. Part of the solution achieved so far is the straightforward concept of a double authentication performed by the JavaCard and described above, but various questions are still open. Whether servers have an identity which enables them to sign documents, or whether servers can speak automatically in effigy of an authority are still subject to controversial debate, as are the admissible forms of storage of biometric data. The bottom line of these problems is the challenge of (re)defining identity and the definition of the act of signing some document, which depend on cultural tradition. The main challenge of interstate e-government is the invention of generic solutions, which can be adapted to the (moving) state of discussion.

3.2 Context Management

Interstate e-government implies exchange of information between authorities using different ontologies. It implies exchange of data between work-flows implementing incompatible business processes, which are realized by incompatible legacy systems using incompatible data schemes. Therefore, the information technology must provide more functionality than a pure transfer of data. It must facilitate the communication of the context, which the data refer to. This is a necessary condition, although it is not a sufficient one. Further, in Europe it must take care of privacy issues.

One-to-one mappings between the various administrative ontologies in use might fulfill these requirements. Unfortunately, they are not really feasible for large scale systems since they scale with n^2. Hereby, in European interstate e-government n counts administrative regions rather than states in Europe. Virtual exchange spaces with a universal format for the representation of personal data are an alternative, which meets the requirements of highly heterogeneous systems in a better way. On the one hand, they define both the formats for exchange and the scheme for the representation of information, namely a namespace, plus affiliated information delivery services. On the other hand they delegate the responsibility for the creation and for the interpretation of documents to the sender of documents, and their receiver, respectively.

More precisely, such a virtual exchange space is characterized by a representation scheme (for the representation of information context and for the representation of information contents), a document scheme (capturing the representation of the content, the context of information generation, the validation of context and of the authentication, and the admissible usage), and by an authentication scheme (which distinguishes between qualitatively different forms of authentication). Further, the services are characterized by an authorization policy for the access to private data and by a usage policy defining which kind of usage is admissible for information represented by a document.

For scenarios similar to our project, there is a basic equation and various basic rules. The basic equation reads *"Relevance = correctness with respect to a well-specified context + translatability with respect to a well-defined scope + authenticity of the origin + actuality with respect to the usage context + (possibly confirmation by validation agents)"*.

This equation defines a virtual value chain for information exchange and thus it provides an economic model, which may be embedded into economic models for inter-organizational e-government and which generalizes to information exchange in any kind of virtual cooperations. Providing information quality, or relevance, respectively, with respect to one of the attributes appearing as summands in the equation may be considered as a value creating activity.

The equation can be implemented with a virtual information exchange space, which supports the exchange of data plus context definitions embedded into signed documents. This virtual exchange space may be considered as a boundary object [5] between different communities of practice, namely the various local authorities involved in the work-flow created by the supply of an e-government service to a citizen. Documents are sent to that boundary object or they are received from the boundary object, whereby it is important to understand that no true exchange of information is necessarily implied and that transparency in sense of computer science is the underlying architectural concept. Thus, the intermediary representation format of the virtual information exchange space is not a true ontology, but it some form of shared metaphor, whereby all sharing partners accept that the language terms used refer to different concepts and affordances.

The use of such a boundary objects as described above comes close to some form of standardization and it can nurture the convergence of processes in the system. At bootstrapping time, the political intention to rely on that boundary object is critical for its success, as the exchange of information will not run smoothly and in some cases, the meaning assigned to representation symbols will differ. However, since the virtual exchange spaces eases the burden of information exchange and since it virtually centralizes activities, it may lead to a consolidation of experience and thus a convergence of ontologies and processes can emerge. This is a far reaching concept, which applies not only to interstate e-government, but also to other forms of inter-organizational cooperation and to cooperation in interdisciplinary teams. As most project teams in e-business are interdisciplinary teams, the concept of virtual information exchange spaces could also be applied for project management in general e-business. We shall come back to this issue for e-government projects later.

Virtual information exchange spaces further facilitate the honoring of privacy issues. European data protection guidelines emphasize the concept of purpose/context. Collection and storage of personal data is only admissible if the purpose is legitimate, if the data handling procedures are appropriate for that purpose, and if lawful facts of permission have been established for that purpose. In this case the person represented by the data has to be fully informed about the scope of data storage and processing and about the identity of the person responsible for it. A change of purpose ends permission and starts a new process, whose lawfulness has to be proven. Reformulating that principle, processing and storage of data refers to a context, which must be well-specified, and which defines the scope and requirements for data handling. If documents contain a specification of the usage purpose or context, this will not prevent violations of privacy, but it supports the lawful handling of personal and private data.

The practical implementation of the relevance equation is rather straightforward. Relevance is always understood with respect to a well-defined usage context. The receiver of data can decide on it, based on the knowledge of the content which is represented by them and its relationship with the usage context, plus the knowledge of the trustworthiness of the data. That trustworthiness can be deduced from the the knowledge of the data generation process and the identity of the person or instance responsible for it. Transitivity of relevance decisions, or re-usability, respectively, is achieved by the creation of a digital document, which describes the context, which contains the information as data plus semantic annotations, and which is confirmed by a digital signature. If a digital document is received then its relevance statement may interpreted as a confirmation of the correctness of its contents, whose value depends on the trustworthiness of the signing person or instance. Translatability is never universal, and never total. It is achievable for a well-defined scope only, with some fuzziness at its boundaries. Clear procedures for human exception handling of this fuzzy domain are required. In many application scenarios, translatability is achievable with XML schemes as indicated above. Authenticity can be guaranteed by the validation of signatures. Actuality does not relate to states at the present time, but is a concept for dealing with information on states in the past. In most administrative processes rules for dealing with documents of a certain age are given and thus actuality can be calculated from the time-stamp on the document.

Validations take the form of a statement on a statement and they can be realized with signed document containers containing the original document and a statement on it. Alternatively, they can be realized with signed dynamic properties of document objects.

The usage of documents instead of data is central for relevance management in a distributed system, where data consistency is never fully achievable and good approximations are expensive. Shortcuts using shared or virtually shared data instead of sharing documents are not admissible, except for the case that the variable is itself a shared document, and no update functionality is available, which does not destroy the signature of the document. The latter implies that documents can only be created, copied or erased, and access rights for these services have to be defined. One realization of such shortcuts is document storage on the JavaCard] or in a virtual card extension, both of which may play the role of document caches.

The basic principle for the management of access to documents or to services providing digital documents is given by the rule that documents can be digitally requested if and only if the requesting citizen has the right to obtain the document according to data protection rules and administrative practice. A citizen has to provide digital proof of her right as she would have to do for non-digital requests, the only difference being that the physical authentication check is performed by a biometric device rather than by visual, human inspection of her and her documents.

If access rights belong to a group of people, or if a service may be requested by any individual a member of a group of people, rather than by its official delegate only, any member should be enabled to access the services or the data without the support of other members and with her own access device. This rules out the concept of a family card which speaks in effigy of a family rather than in effigy of an individual member of a family. Further the delegation of rights should be supported wherever this is facility is offered by the law. Hereby, delegation services have to be complemented with proper revocation services. Finally, it should be noted that data protection does not only apply to explicitly delivered data, but it equally applies to implicitly created data, which can be traced by the information technology. A clear storage and access policy for trace data is needed for any implementation of information exchange in e-government and e-commerce.

3.3 User Interfaces

So far we have discussed a prototypical underpinning technological solution for interstate e-government, but we have not yet mentioned two critical design problems: the interaction between the citizen and the e-government application, which is of major importance for user acceptance, and the interaction between the citizen and her JavaCard, which could be attacked by malicious code which uses the kiosk as a Trojan horse. Kiosks provide graphical user interfaces for the interaction between the citizen and the e-government application. This requires that the trustworthiness of the kiosk has to be guaranteed and user interfaces have to be designed in such a way that they can support users who are unfamiliar with the technology and the required administrative activities as well as users with physical handicaps.

The trustworthiness of kiosks may be achieved with certified, TPE-like kiosks, but the solution to be chosen will finally be decided by commercial issues, and it is difficult to predict which type of solution will succeed. We shall not elaborate on these issues here. The support for all users is a very challenging task, since different users have incompatible needs. Its requirements will be politically defined. A full discussion of the technical side of the whole issue is beyond the scope of this paper. We shall only list various ideas and questions here, as we would like to emphasis that further work is urgently needed.

The following discussion refers to the application scenario for the interstate e-government project described above, but mutatis mutandis it generalists to all e-business scenarios with inter-organizational information exchange and smartcard-based service access. Apart from the specific needs of a particular e-government service, the graphical user interface for citizens must provide seven basic functional components: context management, process management, form editing, document signing, help, history, and card and cache management. The user must be enabled to communicate her situated context to the system. This can be achieved with the definition of a personal and/or situational profile, which can either be done ad hoc or by the selection of a profile from a set of profiles stored on the card. Based on such a profile and a simple expert-system the system can offer personalized and situated guidance. Although such context awareness is a desired quality of user guidance, context transparency and freedom to choose a zero context profile has to be provided by the system, too, because otherwise the citizen would be forced to accept good recommendations without any possibility for re-negotiation. In addition to rule-based user guidance, the system should offer experience sharing facilities for citizens, but so far it is rather unclear how such a sharing could be implemented.

The process management facility of the GUI should enable the user to view the whole process, the state of her work-flow instance 'within' the whole process, and the set of actions she can take and the moves she can make from this state. A switching between unfinished process and a later take up should be supported. The form editing component has to enable citizens to view, fill out, and update forms defined by the work-flow specification for the e-government service. The document signing component is that interface component which is responsible for all communication between the kiosk and the system concerning the digital signing of document by the card. The help component should provide enough multi-lingual information on context and process management and on legal implications. The history function must provide access to traces of own activities for citizens. And the card and cache management component must provide a file service for the card.

For civil servants there are three types of interfaces: an instance editor, a process editor, and a supervisor interface. The instance editor must enable the civil servant to monitor the processing states of work-flow instances and to edit these instances. In addition, it must enable the civil servant to relate collected data sets with original documents and it must support a flexible exception handling, which is of particular importance for interstate e-government, where exceptions count as the norm. The process editor should enable the civil servant to edit and disseminate work-flow specifications and the rules for user guidance. And the supervisor GUI must support supervision of process handling and the administration of user accounts.

There are still lots of open problems, the more since we lack experience with user interfaces as they are depicted here. There is a critical trade-off between simplicity and universal usability, which is not well understood yet. On the one hand implementations of the sketched concept are nearly too complex for public usage. On the other hand any of the features can easily be argued for. Interdisciplinary cooperation would be needed for the engineering of the user interfaces, but it is hard to achieve because it requires that people involved are jointly aware of legal, psychological, and technical issues. In particular, although the integration of users in the design process is crucial, DSDM process models [2] integrating users are difficult to manage, because the functionality provided behind the graphical user interface is rather complex and difficult to understand for users.

3.4 Project Management

In the project described in this paper, the interdisciplinary cooperation disastrously failed, and still, it was the a conditio sine qua non, that we managed to draw up an architecture which facilitates cross-national, digital administrative services.

If we would not have had the possibility to work with experts from many different disciplines, we would not have designed and implemented the solution described in this paper. Nevertheless, it is pretty clear that we would have achieved much deeper results if the interdisciplinary cooperation would have really worked. It essentially failed, first of all because we had chosen an inappropriate process model and an inappropriate time scale. The process model did not reflect the high risk nature of the project, and the time scale was only 18 months, that is due to serious stuffing problems it was effectively around 15-16 months. As a result no convergence of ideas about the project could be achieved in the first half of the project, which created a lot of pressure and blocked various necessary activities throughout the whole project.

There are various options for project plans which can be followed for distributed interdisciplinary R&D-projects. The original idea in our project was a waterfall model with deliverables for knowledge transfer and with part time contributions by the partners depending on the phase of the project. This idea would have implied that delays of deliverables would have had direct impact on the project and the delivery of useless deliveries would have lead to complete project failure. Fortunately, the idea was rejected early in the project planning.

The actual project plan then understood deliverables as implementations of interfaces. It defined separate goals for all partners and it assumed parallel work throughout the whole project. That cooperation failed in so far as no interface worked in the way it was supposed to do. Nevertheless, many participants considered the project as a success. The project management used the project for training their staff and it was happy with how things proceeded. The municipalities had entered the project to find out what JavaCards are good for and they found out in the project that interstate e-government applications are feasible. Therefore, they applauded the fact that the project results provided an architectural concept, which they believe, could eventually work in practice. Most of them considered that achievement as practically relevant and politically usable, while only a minority of them complained about the fact that the implementation progress suffered a delay of various months (due to interface problems and staffing problems).

Possible alternatives would be the evolutionary delivery process and the incremental delivery process ([2]). In the project the word evolutionary prototyping was used a lot, but implications for the project plans were never discussed. Further the word rapid prototyping was used, but it was considered that this would only concern the technical partner and thus the general project management did not adapt its process model. We concluded that rather than one project, a multidisciplinary series of smaller projects should be carried out on the same topic, which are coupled by regular workshops without depending directly on the success of the other projects. The overall project management should be restricted to administrative tasks and the support of the convergence of the projects. It should focus on IT issues and on convergence of knowledge. (Compare [3]). Each single projects then could be led by an expert in the area and it could follow the the exploratory process models ([2]). Furthermore, partners from other projects should participate as junior partners, although that participation should not be critical for success.

Boundary objects could be used for convergence activities, but care has to be taken that their role is accepted by all partners. In our project, visual prototypes were introduced as boundary objects, but then they emerged as graphical user interfaces. This would have fit for a project with a predictable solution, but it blocked resources, which would have been needed for the prototyping of the document service infrastructure. A more detailed analysis of the management and engineering processes and the problems encountered thereby will be given in [4]. It may be considered as a rather special cynicism of the project that we managed to develop a prototypical context management for inter-organizational work-flows, but we failed to do the same for the project itself.

4 Conclusion and Outlook

Interstate A2C e-government could work in principle, but we have to perform more basic research on its risks and we have to master business technology for interdisciplinary engineering, before we may start to implement large scale solutions. We have learned a lot about the implications of digital identity, but social acceptance is still difficult to predict. The tailoring of our interstate e-government application framework for local needs (in order to preserve cultural diversity instead of trying to confront such diversity) will be essential in order to make a real step forward in e-government and to develop an open system. Our investigations on user requirements in a half a dozen European countries seriously indicate that the risk associated with this ambitious goal can be handled successfully. We have avoided various major interface design issues by a lower level of software engineering to implement an 'inter-cultural translation approach' to inter-governmental procedures, but there are still a lot of open design issues left. Right now, we are at the outside of digital society, where human identity will be complemented with its digital representation owning identical rights.

References

1. A. M. Oostveen, P.v. Besselaar, The Complexity of Concepts in International E-Government, to appear in Proceedings of 1st International IFIP-Conference on E-Commerce, E-Business, and E-Government, Zurich 2001
2. M. Ould, Managing Software Quality and Business Risk, Wiley 1999
3. Pawlowski S. D. Pawlowski, S.D. Robey, A. Raven, Supporting Shared Information Systems: Boundary Objects, Communities, and Brokering, Proc. ICIS 2001
4. R. Riedl, Limitations of Interstate E-Government and Interdisciplinary Engineering, to appear Proc. of the DEXA 2001 Workshop 'On the Way to E-Government'
5. E. Wenger, Communities of Practice: Learning, Meaning, and Identity, Cambridge University Press, 1998

Work, Workspace, and the Workspace Portal

Richard Brophy[1] and Will Venters[2]

[1]Active Intranet plc, Vanderbilt Court, 5 Victoria Avenue,
Harrogate, HG1 1EQ, UK,
rbrophy@ActiveIntranet.com
[2]Gemisis, University of Salford, Technology House, Lissadel Street,
Salford, M6 6AP, UK,
Will.Venters@gemisis.co.uk

Abstract. The workspace portals aim to provide the knowledge worker with access to all the information they need. This paper identifies social and technological requirements that are required to make the technology inclusive rather than exclusive. In discussing these requirements problems are identified, issues explored and solutions are suggested. This paper discusses how these issues are being addressed within the commercial context of the research and development activity of a software company.

1 Introduction

The aim of workspace portals is the reduction in user breakdown by providing a single integrated source of all the information they are likely to need in their work. Internet based technology is being used to achieve this. However the "social" issues of software need also to be addressed. There is a need to support software solutions with management and change programmes. It is easy to believe that improvements in software will make workspace portals easier to use and to be effectively introduced.

2 Work and the Workspace

What is the nature of modern work? Over the last fifty years computers and organisations have evolved together. New computer technology has allowed organisations to change, changes in organisational practice have led to changes in computing technology. Computer technology has also provided support for the bureaucratic (e.g. expense systems) and mechanistic functions (e.g. production control systems) of organisations. While computers are highly successful at supporting the bureaucracy of work [12], modern knowledge work is often not so "bureaucratic". Peter Drucker [14] is credited with coining the notion of the knowledge worker — a group of workers whose importance is growing with the emergence of a globalised, post-industrial

M. Beynon, C.L. Nehaniv, and K. Dautenhahn (Eds.): CT 2001, LNAI 2117, pp. 421-431, 2001.

economy in which knowledge displaces capital as the motor of competitive performance. Such a focus on knowledge furthermore requires changes in the structure, culture, and management styles employed within organisations [36]. Our mechanistic approach to information systems has been highly successful in Tailoristic problems (problems that can be easily "procedurised") however the problems faced by knowledge-workers are somewhat different.

Computer technology has also replaced many of the tools of work, the word-processor replaced the typewriter, the e-mail system replaced the office memo. Knowledge-workers employ these tools to deal with the complexity of work. Yet these generic tools have not been developed with specific tasks in mind. Organisations employ the same tools to perform differing tasks. One organisation may use e-mail to simply share official memos, others may use it to arrange all aspects of work, and to socialise. Due to the complexity of understanding the tasks knowledge-workers perform, the computer industry been forced to focus primarily on providing simple tools which may help.

Alfred Marshell, a forefather of neo-classical economics, was first to state explicitly the importance of knowledge within economic affairs: "Capital consists in a great part of knowledge and organisation ... knowledge is our most powerful engine of production" [28]. There is a need to understand the ways in which IS/IT might support knowledge work [5].

2.1 Technology to Support Knowledge Work

The typical characteristics of knowledge work have been described as follows:

- Variety and uncertainty in inputs and outputs;
- Unstructured and individualised work rules and routines;
- Lack of separation among process, outputs and inputs;
- Lack of measures;
- Worker autonomy;
- High variability in performance across individuals and time [13].

While definitions such as this have been criticised for failing to cover the different forms of knowledge work undertaken [36], they do succeed in highlighting the overall complexity of such work. This research focuses on:

1. The factors needed to support knowledge work, within the context of a particular technology: the workspace portal.
2. The factors which affect this technologies effective use.
3. The factors in installing such technology into the working practice of organisations.

In developing workspace portal technology to support the knowledge worker the following issues must be addressed:

Reuse issues:

- Knowledge workers are often faced with information overload
- The knowledge worker must decide information relevance

- Business transactions are increasing in speed

Association issues:

- Collaboration is core to knowledge work
- Cultural and social issues affect every aspect of work.
- In addition to formal work hierarchy, knowledge workers often interact through other structures, for example communities of practice [39].

These must be addressed by the technology to support the knowledge worker. The workspace portal is one proposed technological solution to the problem.

3 Workspace Portals

There is a need to provide a complete and integrated software solution, which builds on the advantages of standard tools to provide a complete solution to the problems of modern knowledge work. Such a solution is the workspace portal.

> "A workspace portal is a single coherent integrated portal that presents its users with all the information they need to carry out their jobs." [38]

The workspace portal can be compared to the Butler group's definition of a 3rd generation enterprise portal technology. Butler's definition of a 1st generation enterprise portal is the aggregation of internal and external information, with basic search. This generation of enterprise portal already exists. The 2nd generation of enterprise portal has an easy integration of externally hosted applications as well as personalisation and advanced search. This generation of enterprise portal is arriving.

3.1 Features of Workspace Portal

The concept of the workspace does not bring with it any new technologies. What it proposes is a synergy of existing technologies that should be greater than the sum of its parts. These parts are:

- Personalisation – of the data and of the presentation.
- Search and navigation – of information within, and external to, the organisation.
- Push – allows users to be notified when indicators move outside of an acceptable threshold.
- Collaboration and GroupWare – is the management of relationships.
- Task automation and workflow – is the automatic notification when a task needs to be performed.
- Applications – are accessible by users from one place.
- Infrastructure – must be scaleable, secure and available when needed.
- Integration – of disparate sources of information from ancient legacy systems data to the newest document.

3.2 What the Workspace Portal Aims to Achieve

In order to provide such a "coherent integrated portal", there is a need to efficiently support the work individuals perform. Such support should be tailored to the specific needs of the individual, and support the actions such individuals collectively perform. Within organisations individuals have a personal perspective environment, the complexity of which they manage. Everyone in the organisation will have developed differing approaches to dealing with the variety of this environment. We suggest that workspace portals should support the boundary of the individual's perception of their environment, aiding the individual perform tasks. We hypothesise that the aim of such portal technology is two fold: Firstly, to attenuate the variety supplied from the environment, ensuring that the information and knowledge presented to the individual is relevant to the variety faced. Secondly, to manage the amplification of variety of the individual, that is increasing the individual's options in acting upon the environment and making changes. Essentially we can view the individual as part of a cybernetic system, in which we must either amplify the variety of responses, or attenuate the variety of information provided [4,15]. We hypothesise that workspace portals may provide the structures, which are needed to involve all employees in the creation of possibilities and responses to external [environmental] disturbance [16]. Raul Espejo [15] states that if employees can operate in effective interdependent task loops which enable their organisational units to absorb disturbances [variety] then centralised strategic plans and information systems are redundant. Such organisations would have naturally distributed information and planning solutions built into their organisational structure. We suggest that workspace portals can enable such naturally distributed solutions.

Martin Heidegger proposes the notion of readiness-to-hand in which objects are not inherent in the individual's world, but arise only in an event of *breaking down* in which they become present at hand (discussed in [41]). Upon the breakdown of the tool – an unreadiness to hand, the tool will identify its existence within the domain of the user. An aim of software development is thus to understand and manage such breakdown. The aim of a workspace portal is to provide an interface between information and the individual; its existence should only support this process. By integrating the many informational tools into a workspace portal, it should be possible to reduce the probability of breakdown occurring through unfamiliar interfaces, and having to switch between screens, potentially reducing unreadiness-to-hand. Such integration may, however, increase the possibility of breakdown occurring by mismatch in expectation and action, that is what users expect to be possible and that which is possible. By careful analysis of the purposeful activity carried out by human actions within a company it is possible to improve this "fit" and thus improve the acceptance and use of the tool [8].

Within his latest book – "The Invisible Computer" [29], Don Norman also discusses the need to make tools that are based around the activities people are engaged in. He advocates the need for software targeted at the activities individuals undertake rather than homogenous general-purpose software packages. Yet a workspace portal is essentially such a piece of software. There can thus be identified a need to provide such software with the ability to "adapt" to the activities users engage in. Such intelligence may be provided by the use of intelligent agent software.

It should be noted that workspace portals, while aiming to provide all the information required to carry out their job, would not be the only source of information within the organisation. Discussions around the "Water cooler" have been identified as key to effectively sharing knowledge [6]. While the aim of creating workspace portals is to provide all the information an individual may require, this does not preclude the use of substitute solutions [31]. The solution should also appreciate and support alternative solutions. Since collaboration is seen as desirable, it should be encouraged by the solution.

3.3 Workspace Portals Impact

Workspace portals are not neutral artefacts within the modern work environment. They will affect and be affected by the work environment. As Shoshana Zuboff states: "It is a mistake to regard the new generation of information and communication technologies as neutral tools that can merely be grafted onto existing work systems" [42]. However, in developing workspace portals, third party software houses only have full control over the technological issues. As identified previously, such technology should adapt to the human activity, and to the organisational activity. In considering such issues we draw on the work of theorists such as Christopher Alexander [3], Wanda Orlikowski [31]; [33]; [32] and Claudio Cibborra [9]; [10]; [11]. Alexander [3] talks about the social aspects of software artefacts and in designing social software we need to come to terms with social responsibility. [7] make the point that if agents 'imitate or replicate human actions' then this introduces into software development 'both moral and social-institutional questions.'

The first environment a workspace must address is the company itself. For software artefacts to become a true workspace it must address company cultural issues as well as the work activity. Company culture can be defined by company 'Legends' [17] or "that is how it is done around here" [18]. Company cultures have stereotypes that can be used for classification. Different authors have different stereotypes. It must be recognised that any sort of classification is an approximation. No company will exactly be one stereotype, but will be amalgams of two or more stereotypes.

In providing new functionality to allow users to complete their work, the portal will change the way tasks are undertaken, thus changing the nature of the work. By providing a single interface for activity, these portals will integrate how tasks are performed. In developing software solutions it is easy to focus on the individualistic nature of the interaction. When designing a word-processor we only consider the needs of the individual "at the other side of the screen". As workspace portals are collaborative, the conceptual model users will have of the solution will include other users, with the portal mediating between them. In using the telephone we are unaware of the hardware, and become focused on the conversation. It is as though the other person is "within the telephone". In the development of GroupWare technology, of which workspace portal technology is an example, we must also consider the social nature of Human Computer Interaction (HCI), ensuring that in collaborative acts, the user is unaware of the workspace portal, and only aware of the collaborative act.

4 The Technology of Workspace Portals

In terms of technology, workspace portals could be an integration of intelligent agents, XML, GroupWare technologies and information systems.

4.1 Intelligent Agents

The role of the personal or interface agents that Maes [23] suggested has been questioned [30] and the experience of users using software artefacts like MS 'clipit' have coloured people's opinion. They seem to be a software artefact that you either love or hate, but not tolerate. An interface agent sits between the end user and the computer, watching, learning and helping when required. [30] suggest that personal information agents have an unarguable advantage in that they are software artefacts that help manage the explosive growth of information. One author has called this explosion 'infoglut' [24]. A factor, [30] cite, in the growing value of such agents is the growth of E-Commerce sites. Where manually searching the web is unproductive or at best inefficient. The creation of a tool that reduces the time spent searching has a definite, observable advantage to the knowledge worker. The problem, as [24] points out, with existing search engines is that they do not search in context.

Rather than providing passive "tools" which act upon user input, software becomes active looking for "relevant" material or action. It is the understanding of "relevance" which is difficult. Nicholas Negroponte talks of the "digital sister-in-law" who watches many films, and recommends those which she believes Nicholas will like [27]. Her understanding of Nicholas's taste allows such recommendations. Capturing factors such as "Taste" electronically would prove exceedingly difficult.

While software such as intelligent agents may aid the task of developing the workspace portal, the human and the digital are significantly and usefully distinct [7]. [23] suggested that agents would radically change the style of HCI. Intelligent agents are a powerful resource, however as Anthony Giddens argues "the humanizing of technology is likely to increase the introduction of moral issues into the now largely 'instrumental'" [19]. To use intelligent agents to make selection decisions on relevance is to provide the digital machine with choice on the user behalf.

4.2 XML

Another element in the development of workspace portals will be the use of XML (eXtensible Mark-up Language) as it allows the separation between content and delivery. Through the use of XML the workspace portal application can use a single item of information in increasingly sophisticated ways to support purposeful activity. An XML document can be easily applied to many different business applications. If a generic application, the workspace portal, is to effectively support the varied work styles and information needs of users it will require degrees of personalisation. The use of XML enriches such personalisation. Personalisation can be seen as tailorability and "personalised knowledge". Traditionally tailorability is defined (by [37]) as 'a feature of interactive software that allows the change of certain aspects of the software in order to meet different user characteristics and requirements'. We extend this to include personalisation of the information provided through the software as well as personalisation of the software itself. Therefore XSL (eXtensible Stylesheet Language) deals with tailorability and XML deals with the personalised knowledge.

4.3 Collaboration

One point that should be noted is that people tend to work collaboratively whenever possible. The surprising thing is that they still manage to do this when the tools they have actively places barriers in this process. [26] found that 'users distribute domain expertise by directly editing each others spreadsheets and by sharing templates'. It has been noted that local experts tend to create spreadsheets or word templates. These are then used by the group, and adapted to fit the task in hand. Could Martin Heidegger's notion of level of "breakdown" for a tool (also known as "throwness" [40]) be extended to include the breakdown of communities through the collaborative use of tools? Can the actions of one member of the community "breakdown" the readiness-to-hand of others? [25] point out that they see tailoring as indirect long-term collaboration between developers and users, mediated by a computer system and positioned within the 'different time, different place' domain of collaboration. They suggest that multiple representations and application units will support collaboration between these two groups. In addition [7] noted that increased personalisation may reduce the areas of common interest; if each user receives different material it may be harder to converse around common information.

4.4 Human Computer Interaction

Lessons learned from the creation of character, graphical and browser interfaces need to now be applied to the design of workspace portals. The growing sophistication of end users means that they will not tolerate software that does not fulfil their interface and usability requirements as well as functional and data requirements. [35] have suggested a framework of different requirement types. The advent of the internet has shifted power away from suppliers to customers. This shift will lead to a growing demand, from an increasingly sophisticated end user base, for software that is at least easy to use and preferably fulfils more usability and user interface requirements.

We suggest the workspace portals should include aspects of learning, both about the tool, and about the work. Aspects of Performance Support should be incorporated, in which help is provided regarding the task individuals are performing at a particular time.

5 Factors Outside the Technology

In introducing workspace portals it must be realised that lengthy and expensive analysis is unlikely to take place. Many organisations are now prepared to take risks in introducing potentially inappropriate "off the shelf" technology at low cost, rather than invest large sums in systems analysis and in-house development. It is thus left to the purchasing organisation to tailor and introduce the technology within the daily work practice.

In developing workspace technology we cannot discount the need to embed aspects of systems "analysis" activity within the application, ensuring that implementers within the purchasing organisation may effectively tailor the product to work practice. As such, workspace portals should be designed as toolkits which provide flexibility to the purchaser, enabling the purchaser to develop a business specific solution.

6 Our Activity

In order to identify the factors proposed in this paper, we are researching the way users operate the ActiveIntranet [2] product at present, and how they would make use of a workspace portal product in the future. [22] suggest that knowledge elicitation and requirements elicitation can learn from each other and that the two schools should merge. If the vision of the workspace portal is to be delivered there needs to be a merging of requirements engineering, knowledge engineering and task analysis. Requirements engineering for the functionality, knowledge engineering for the knowledge and task analysis for the task, all are needed to deliver a workspace portal.

Workspace portal development will need techniques from many fields of computer science such as requirements engineering, knowledge engineering and HCI. This diversity and the scale of the workspace portal mean broad brush techniques need to be used in gaining requirements. Highly individual techniques may work for the individual in a specific instance. This may not be scalable to a workspace solution because of the cost of using such techniques. Changing a person's tool has an effect on the task at hand, and vice versa. The design of the tools to help in the creation of the workspace portal will need to understand the relationship between task and function. In addition how can the requirements of a diverse user group be represented and modelled to give insight and applicability to the development of an interface? Kaindl [21] uses scenarios to model the domain, which are linked in to achieving goals by using functions. Using goals, scenarios and functions in his framework enables the acquisition of different requirements in a pragmatic manor.

It is proposed that there is a need to understand the methods by which software developers may effectively analyse the information processes of users, and map these effectively within the software. We propose the use of story telling in achieving this. [21] brings together in his combined model, a lattice framework, scenarios, goals and functions. We are using the combined model with a tagged extension, already documented in [34] as part of the TCS Scheme with Sheffield Hallam University, to try and deliver this.

Requirement acquisition has taken place using focus groups with present users. Focus groups have been described as 'a kind of group interview' [20]. Therefore they have been used to gain a broad spectrum of user feedback on the existing product. These focus groups have the following aims:

- Delivery of scenarios that are case specific and therefore domain dependant.
- Customer archetypes generated by customers. These are personality types of knowledge workers.

- Customer roles generated by customer. These are different roles the knowledge worker will take when using different tools of the product.

Further focus groups will be used to elicit requirements from a broader spectrum of users. These roles, archetypes and scenarios will then be used in designing the workspace portal technologies.

7 Conclusions

ActiveIntranet and Gemisis are exploring these issues within real world companies. Workspace portals aim to empower the knowledge worker by providing the information they require. Technological issues are only part of the problem, as software developers, we should also be mindful of the social issues. Emphasis should be made on the problem of effectively integrating workspace portals within the workspace, software should support not hinder this process.

Acknowledgements. ActiveIntranet plc develops the components of a Workspace Portal, collectively branded K*OS. K*OS enables companies to achieve sustainable profitability through the effective discovery, delivery and use of corporate knowledge.

Gemisis, based at the University of Salford, is a multi-million pound collaboration of corporate institutions and the public and private sectors. The award-winning project develops and delivers advanced online applications in the fields of business, education, health and the local community.

References

1. Ackoff, R.: *Redesigning the future*. John Wiley & Sons, New York (1974).
2. ActiveIntranet: http://www.activeintranet.com , ActiveIntranet plc. 20/03/01
3. Alexander C.: *The origins of pattern theory, The future of the theory and the generation of a living world*. IEEE Software September/October (1999) 71-82
4. Ashby W. R.: *An introduction to cybernetics*. London, Methuen & Co Ltd. (1956)
5. Bacon, C. J. and B. Fitzgerald: *The Field of IST: a Name, a Framework,and a Central Focus*, ESRC. (1999)
6. Brown, J. and E. Gray: *The People are the Company*. Fast Company: 78. (1995)
7. Brown, J. S. and P. Duguid: *The Social Life of Information*. Boston, Massachusetts, Harvard Business School Press. (2000)
8. Checkland, P. and S. Holwell: *Information, Systems and Information Systems*. Chichester, John Wiley & Sons. (1998)
9. Ciborra, C., Ed. : *Groupware & Teamwork - Invisible aid or technical hinderance*. Chichester, John Wiley & Sons Ltd. (1996)
10. Ciborra, C. U. and G. Patriotta: *Groupware and teamwork in R&D: limits to learning and innovation*. R & D Management 28(1) (1998) 43-52.
11. Ciborra, C.: *From Control to Drift: The dynamics of Corporate Information Infrastructure*. Oxford, Oxford University Press. (2000)
12. Dahlbom B and Mathiassen L: *Computers in Context: The Philosophy and Practice of Systems Design*. Oxford, NCC Blackwell. (1993)

13. Davenport, T., D. DeLong, et al.: *Successful knowledge management projects*. Sloan management review 39(2) (1998) 43-59.
14. Drucker, P.: *The Age of Discontinuity: Guidelines for our changing society*. New York, Harper & Row. (1969)
15. Espejo, R.: *Giving Requisite variety to strategy and Information systems*. Systems Science, Plenum Press. (1993)
16. Espejo, R. and R. Harnden, Eds.: *The Viable System Model*. Chichester, John Wiley & Sons. (1989)
17. Handy C.: *Inside Organisations, 21 Ideas for managers*. BBC Books. (1990)
18. French L.F. and Bell C.H.: *Organisations Development, Behavioral Science Interventions for Organisation Improvement*. Third Ed. Prentice Hall (1984).
19. Giddens A.: *The consequences of modernity*. Stanford CA, The Raymond Fred West Lectures, Stanford University Press. (1990)
20. Gorguen J.A. & Linde C.: *Techniques for Requirements Elicitation*, Proceedings from the International Symposium Requirements Engineering, IEEE, (1993)
21. Kaindl H: *Combining Goals and Functional Requirements in a Scenario-based Design Process*. People and Computers XIII, Hilary Johnson, Laurence Nigay and Chris Roast (eds), Springer-Verlag. (1998). 101-121
22. Loucopoulos P. & Karakostus V: *System Requirements Engineering*, McGraw Hill. (1995)
23. Maes, P.: *Agents that Reduce Work and Information Overload*, Communications of the ACM 37 (7), (1994). 31-40.
24. McKay, B.: *Search Containment, A knowledge index that works like a Holograph. Generates smart searches to Tame Infoglut – by leveraging how we learn*, at press (2001).
25. Morch A.I. & Mehandjiev ND: *Tailoring as Collaboration: The Mediating Role of Multiple Representations and Application Units*. In Computer Supported Cooperative Work 9. Kluwer Academic Publishers, (2000) 75-100
26. Nardi BA., Miller JR.: *Twinkling lights and nested loops: Distributed problem solving & spreadsheet development*. In Ed Saul Greenberg: Computer Supported Cooperative Work & Groupware. Academic Press Ltd, London, (1991)
27. Negroponte, N.: *Being Digital*. London, Hodder & Stoughton. (1995)
28. Nonaka, I. and H. Takeuchi: *The knowledge-creating company: how Japanese companies create the dynamics of innovation*. New York, Oxford University Press. (1995)
29. Norman, D.: *The invisible computer: why good products can fail, the personal computer is so complex and information appliances are the solution*. Cambridge Massachusetts, MIT Press. (1999)
30. Nwana, HS. & Ndumu DT: *A Perspective on Software Agents Research*. http://agents.umbc.edu/introduction/hn-dn-ker99.html 15.2.01
31. Orlikowski, W.: *The duality of technology: Rethinking the concept of technology in organisations*. Organisational Science 3(3): (1992) 398-426.
32. Orlikowski, W. Ciborra C.: *Evolving with Notes: Organisational Change around Groupware Technology: Groupware and Teamwork*. John Wiley & Sons Lts: (1996) 23-59.
33. Orlikowski, W. J. and Gash D. C.: *Technological Frames: Making Sense of Information Technology in Organizations*. ACM Transactions on Information Systems 12(2): (1994) 669-702.
34. Roast C. Brophy R. Bowdin A.: *Finding the Common Ground for Legacy Interfaces*. In Kobsa, A & Stephanidis, C. (eds.): User Interfaces for All., GMD (1998). 103-130.
35. Ryan M. & Sutcliffe A.: *Analysing Requirements to Inform Design*. In Johnson H. Nigay L. & Roast C.(eds.): People and Computers XIII, Proceedings of HCI '98, (1998). 139-157
36. Scarbrough, H. Currie, W. Galliers, B: *The management of knowledge workers. Rethinking Management Information Systems*. Oxford, Oxford University Press: (1999). 474-496.
37. Totter A., Stary C.: *Tailorability and Usability Engineering: A Roadmap to Convergence*. In Kobsa, A & Stephanidis, C. (eds.): User Interfaces for All., GMD (1998). 175-191
38. Wells D, Sheina A, Harris-Jones C.: *Enterprise Portals: New Strategies for Information Delivery*. Ovum Ltd. (2000).

39. Wenger, E.: *Communities of practice: Learning, meaning and identity*. Cambridge, Cambridge University Press. (1998).
40. Winograd, T.: *Bringing design to software*. New York, ACM Press. (1999).
41. Winograd, T. and Flores, F.: *Understanding computers and cognition*. Norwood,NJ, Ablex. (1986).
42. Zuboff, S.: *In the Age of the Smart Machine*. New York, Basic Books. Inc. (1988)

Experimental Politics: Ways of Virtual Worldmaking

Max Borders and Doug Bryan

Center for Strategic Technology Research (CSTaR)
Accenture, 3773 Willow Road, Northbrook, IL 60062 USA
{r.m.borders,douglas.bryan}@accenture.com

Abstract. We think that Massively Multi-user Online Role-Playing Games (MMORPGs) will soon evolve into Online Societies of political and economic interest. Studying them will require a methodology that balances observation, theory, and application. This paper outlines such a methodology. First we examine the nature of theory itself (meta-theory), including holism (a meta-theory about systematic interconnections between beliefs) and reflective equilibrium (a meta-theory on the recursive nature of theory and information). Next we determine starting points for our studies. We appeal to the production of "spontaneous orders," orders that are not designed but rather emerge from the behavior of agents in a simulation, and which proceed from simple, normative, non-teleological rules. Lastly we examine the character of those rules. We look to where morality and rational choice converge (Gauthier's minimax relative concession) and derive guidelines for designing rules.

1 Introduction

Like many discussions about technology, this essay is based on a prediction. The prediction is this: Online Societies will soon appear in Massively Multi-user Online Role-Playing Games (MMORPGs). Today millions of people play MMORPGs. Ultima Online from Electronic Arts has 150,000 paying subscribers, EverQuest from Sony has 350,000, Asheron's Call from Microsoft has 250,000, and Lineage from NCSoft has over 8 million registered players. It's common for 80,000 people to simultaneously take part in one of these games, and the average player sends 20 hours per week in game.

We begin with a set of assumptions about Online Societies. Then we'll move to issues of more theoretical interest: a sketch of a three-stage methodology for applied political economy, which seeks to utilize Online Societies as large-scale simulations. Stage I consists of metatheoretical considerations, as well as the manner in which theoretical investigations may proceed from these. Stage II involves the process of selecting "starting points" that will determine the presence and nature of rule sets for designing Online Societies (and which will serve as those starting points). Finally, Stage III fleshes out a range of appropriate guidelines in determining the character of those rule sets both in light of conclusions reached in Stages I and II, and due to further theoretical considerations.

M. Beynon, C.L. Nehaniv, and K. Dautenhahn (Eds.): CT 2001, LNAI 2117, pp. 432-441, 2001.

Let us, for the sake of discussion, accept the following:

- MMORPGs will evolve into immersive worlds with elaborate civic structures—*Online Societies.*
- Just as with online experiments in other fields, researchers will (cheaply) be able to harvest online societies for the interests of social anthropology and political economy.[1]
- Player-characters (PCs)[2] will contribute to the construction and administration of Online Societies.
- A methodology will be crucial to the enterprise of harvesting Online Societies for study.

It is with this last assumption that we will begin our discussion.

2 Metatheoretical Considerations (Stage I Methodology)

Ways of Virtual Worldmaking. The first methodological stage for the application of virtual worlds (V-worlds) to moral-political investigations is the constructivist stage. In this stage, the designer comes to grips with the very nature of her activity as a creator of worlds, especially as it relates to her particular conceptual schemata and theoretical assumptions. Nelson Goodman wrote [1]:

> Since the fact that there are many world-versions is hardly debatable, and the question how many if any worlds-in-themselves there are is virtually empty, in what non-trivial sense are there, as Cassirer and like-minded pluralists insist, many worlds? Just this we think: that many different world-versions are of independent interest and importance, without any requirement or presumption of reducibility to a single base.

Goodman understood that a network of symbols (linguistic, pictorial, and otherwise) make up or "construct" our conceptual schemata, hence constructivism. Since the demise of verificationism, theorists like Goodman, Quine, Sellars and Davidson have ushered in a paradigm dominated by more holistic theories of knowledge and truth. Holism, simply said, is the idea that our beliefs, whether scientific or everyday, are intimately connected within a network of other beliefs, the sum total of which is a gestalt encompassed by the term "theory." Sometimes our experiences cause disruptions in our theories, and we are thus often forced to revise them in light of new experiences. The revisions can be total and utter a la Copernicus, or they can be mere adjustments closer to the periphery of our theorization schemata. In any case, the content of these makes up what Goodman would call world-versions, and the process

[1] For example, economists are studying online behaviors; see, Rosenblatt, Joel: "Moving past rats: more economists study behavior in online experiments," *The Wall Street Journal*, 2 October 2000; and, Varian, Hal: "Economic scene: online users as laboratory rats," *The New York Times on the Web*, 16 November 2000.

[2] A PC is an amalgamation of a player and a character. For example, a player who is an accountant may play a character that is a warrior. The PC is part accountant and part warrior.

of systemization "worldmaking." Some of these worlds conflict, in other instances a plurality of worlds comes to bear. None of these holistic meta-theorists would argue that the "world-out-there" is simply up for grabs or up to anyone's arbitrary fantasizing. Rather, rigorous testing of statements in the entire corpus of theorization affords one a pragmatic middle ground between what Catherine Elgin calls the "absolute and the arbitrary."[3] So, theories leave room for useful conventions such as expression, modeling and metaphor, which often enjoy an inconspicuous presence even in our most rigorously scientific investigations. An understanding of theory can render judgments of fact, fiction, and falsity depending on the system employed.

What lessons can we take from this intimacy between linguistic systems and theory as it relates to V-worldmaking? Three things, at least:

1. This is the way science works, not to mention simulations and models.
2. The experience of V-worlds qua worlds is not reducible to bits and bytes, any more than the actual world is (conceptually) reducible to physics.
3. Holistic methodology preemptively avoids charges suggesting that the setting of parameters for any given V-world involves a vicious circularity. The background baggage of a theorist does not always taint his test of a theory. And though the online is not the actual, attention to relevant similarities can yield something instructive. For if we have learned anything since logical positivism, we have learned that beliefs are tested within background theories and there is no escaping the linguistic network inside which we operate.

V-worldmaking is, therefore, a highly constructivist enterprise, and is an extension of a holistic theorization style.

Possible Worldmaking and Reflective Equilibrium. V-worlds are, in many respects, the amalgam of various kinds of linguistic systems, which straddle a vaguely drawn line between what is real and imagined. V-worlds can thus be viewed as a quasi-reification of possible worlds. In some respects, V-worldmaking is a harvesting-ground for those possible worlds.

The process of V-worldmaking can also be said to involve the *reflective equilibrium* [2, pp. 17–8, 42–4, 507–8]. Once an investigator has a mostly complete set of principles in his moral-political theory, his V-worldmaking will proceed in relation to those principles. Further, noting relevant differences and similarities between actual and online societies, and incorporating edification garnered from these into the web of theorization, is part and parcel to a methodology involving the process of reflective equilibrium.

The species of reflective equilibrium applied while V-worldmaking will simultaneously address both the sentiment that the philosophical tradition is often too lofty and meta-contextual to be directly applicable, and the charge that subsequent "testing" of so-called lofty claims against some existing (actual world) status quo might involve a frightening alteration of that status quo—bringing, perhaps, serious harm to it. (Testing a theory of political economy on Argentina might worsen Argentineans' condition.) In V-worldmaking, we have the luxury of "coming back to reality," understanding—with humility—that we will have to sacrifice enough of the actual in our V-worldmaking to risk missing important areas of difference. That is why a cardinal

[3] Elgin, Catherine Z.: *Between the Absolute and the Arbitrary*. Cornell Univ. Press (1997).

corollary to this methodology must involve not only a refined sense of the difference between models and the phenomena for which they stand (and attempt to simulate), but also a refined sense of how the methodology can be productive when we possess a better understanding of what modeling is.

The Methodology of Modeling. V-worldmaking is, in our view, a form of large-scale modeling. Black tells us that a model, as a general type of metaphor, can bring two separate domains into a cognitive relation that utilizes the language of one domain as a lens for seeing the other [3, p. 236]:

> ... the implications, suggestions, and supporting values entwined with the literal use of the metaphorical expression enable us to see a new subject matter in a new way. The extended meanings that result, the relations between initially disparate realms created, can neither be antecedently predicted nor subsequently paraphrased in prose ... Metaphorical thought is a distinctive mode of achieving insight...

Black goes on to explicate two types of models: analogue and theoretical. The former is generally executed as a construction, while the latter is a description. An analogue model is an object, system, or process (realized in a new medium) designed to reproduce the "web of relationships" of an original. An analogue model will display certain point-by-point correspondences between its relations and those of the original, for "every incidence of a relation in the original must be echoed by a corresponding incidence of a correlated relation in the analogue model" [3, p. 222]. Black stresses that analogue models furnish the theorist with plausible hypotheses, not proofs. Indeed, this is what we can expect from our V-world simulations.

Now, if we conjoin the concept of a model that displays this web of corresponding relationships, with the conception of the theoretical model, we should be able to establish both the value of models/simulations to various sorts of investigation, as well as formulate a methodological framework for V-worldmaking. According to Black, theoretical models function (paraphrasing briefly) by (1) identifying a field that has some established regularities that need to be examined, (2) describing entities (regularities) in a secondary field, (3) establishing rules of correlation from the secondary field to the original field, and (4) using assumptions in the secondary field, together with the rules of correlation, to create inferences in the original field. These inferences may then be independently checked against data in the original field. If we can accept that V-worldmaking functions as a fusion of analogical and theoretical models by virtue of its possessing both the aspects of the former (form, function, and structure, embodied in the construction) and of the latter (linguistic, abstract, and theory-laden, embodied in the description), then we can move forward through the recognition that the simulations of V-worldmaking are grand-scale models valuable to applied social science and politico-economic theorization.

This kind of research, once it has taken root, will require a balance between the rigors of the traditional observation-based sciences, and the abstract and creative faculties of theoretical disciplines, between causality and plausibility.[4] In the next sections we will see why the challenges will require such a balance of research styles.

---

[4] Multi-agent simulation is perhaps the simulation field with the most experience in "plausible" social behaviors. We expect to draw lessons from that field as we proceed. See, for ex-

3 V-Worldmaking and Political Economy (Stage II Methodology)

The second stage of our V-worldmaking methodology consists of the choice of starting points. The first important question one has to ask when V-worldmaking is: What are we trying to learn? Of course, as a testing ground for some hypotheses, the V-world is a model that can be viewed in a number of ways—either as proceeding through time, or as a dataset within some time-slice, or at a certain point in time. But prior to this, the V-world will have a starting point from which the simulation proceeds. At the starting point, the designer is already at a large theoretical crossroad. She must decide if she will observe the natural orders that supervene on the behaviors of PCs, or if she will set out some initial rules in order to observe the phenomena of a macro-political superstructure within which the PCs must function (i.e., a massive political experiment). This is a theoretical decision one must be conscious of, for one starting point may deliver insights of more anthropological or sociological interest, while the other may tend to political economy. We should also hint at the idea that this crossroad is not completely without a middle way. We address this later. For the moment, let us deal with the two basic alternatives: proceed from a state-of-nature, and proceed from a predetermined rule set.

State-of-Nature as Starting Point. Though it is logically possible for anarchy to arise from a State of Nature, people tend to devise their own rules of behavior, which generally ground political superstructures. Nozick did (arguably) a good job of showing how agents in a State of Nature will likely enter into arrangements that eventually create some sort of state [4].[5] In any case, it is likely that PCs will form their own rules of conduct in the absence of "top down" rules provided by a designer. These rules, and resulting political structures, can vary greatly. Hence beginning from states of nature will yield diverse plausible hypothesis of anthropological interest.

The Predetermined Rule Set. Before discussing the latter starting point for V-worldmaking (a predetermined rule set), we should provide some preliminary justification for such an approach. It is "spontaneous order" that will be the leitmotif, and though it is a concept most often seen in biology, it gets its most robust articulation for political economy in Hayek. First, he makes a distinction between *kosmos* and *taxis* [5]. *Kosmos* is order that is self-generated, "spontaneous," while *taxis* is an artificially planned order. In short, Hayek believes cosmic order is superior to others insofar as it is a more pragmatic alternative to top-down designs marked by so much historical failure. Since self-generating order comes about as a result of the agents (PCs) "adapting to circumstances which directly affect only some of them, and which in their totality need not be known to anyone, it may extend to circumstances so com-

---

ample, Epstein, Joshua M. and Axtell, Robert L.: *Growing Artificial Societies: Social Science from the Bottom Up*. MIT Press (1996); Axelrod, Robert: *The Complexity of Cooperation: Agent-Based Models of Competition and Collaboration*, Princeton Univ. Press (1997); Axtell, Robert L. and Epstein, Joshua M.: "Coordination in transient social networks: an agent-based model of the timing of retirement," in *Behavioral Dimensions of Retirement Economics*, Henry J. Aaron (ed.), Brookings Press (1999).

[5] Also Axelrod experimentally showed the robustness of cooperation. See footnote 4.

plex that no mind can comprehend them all" [5, p. 41]. The vision is one of an efficient, yet organic, social order.

Hayek tells us that self-interest is not enough for spontaneous order: "For the resulting order to be beneficial people must also observe some conventional rules, that is, rules which do not simply follow from their desires and their insight into relations of cause and effect, but which are normative and tell them what they ought to or ought not to do" [5, p. 45]. So, it seems Hayek thinks that a moral-political rule set is an indispensable part of generating (and maintaining) spontaneous societal order; indeed, conforming to these normative rule sets makes such order possible. But Hayek is also aware that a state may seek to provide services that supersede the maintaining of the spontaneous order, and his account turns out to be an instrumentalist justification for the *rule of law* in liberal societies. It is at this point that Hayek thinks the legitimacy of the rule set comes into question. That is, the rule set requires justification beyond the existence of the order. This is why he makes yet a further distinction between (rules of) governments (a.k.a. law) and (rules of) organizations. Organizational rules differ relative to the role of each member, and are interpreted in light of the purpose of an organization. By contrast, the rules governing a spontaneous order should be purpose-independent and hold equally for all members. For "this means that the general rules of law that a spontaneous order rests on aim at an abstract order which is not known or foreseen by anyone" [5, p. 49]. For analogous reasons, the designer of an online society should *ceteris paribus* appeal to rules of government rather than rules of organizations, but rules of organizations will likely result as an emergent property. Thus organizational rules will be left almost entirely to the PCs.

Hayek warns that it is mistaken to think that contemporary society can be deliberately planned.[6] The complexity of civilization is self-generating. We preserve such complexity by enforcing and *improving* the simple rules that form and maintain spontaneous orders—not by directing agents toward a specific order. As Richard Epstein puts it (with respect to law) "the *minimum* condition for calling any rule complex is that it creates public regulatory obstacles to the achievement of some private objective" [6, p. 27]. So, the rule set should be simple, impartial, and non-end-specific. The resulting orders will turn out to be a species of pluralism, as each agent pursues her own unique aim.

To reiterate this idea, we know simple rules often lead to highly complex emergent systems. A designer cannot achieve the degree of complexity inherent in such systems. The more a designer attempts to regulate the activities of the PCs, the more likely the simulation is to fail in practice, and the less likely the methodology will yield fruitful insights.[7]

---

6 MMORPG designers have come to realize this as well. A founding designer of Ultima Online put it as, "You cannot keep pace with players ... there's a lot more collective IQ on their side" (Koster, Raph: "The rules of massively multiplayer world design" http://www.legendmud.org/raph/gaming/lawsindex.html) and an Ultima Online research fellow simply as, "central planning doesn't work" (Simpson, Z.B.: "The in-game economics of Ultima Online," presented at the Computer Game Developer's Conference, San Jose, CA, March 2000. http://www.totempole.net/uoecon/uoecon.html).

7 Investigations involving strange or complex rule sets may be of interest, but will be difficult to apply to political economy. Allowing such rule sets to fail spectacularly may yield insights analogous to physics' atom smashers.

Degree of Organic Unity. A parallel guideline for gauging the value of an initial rule set comes from a "degree of organic unity." Nozick introduces the concept in a discussion of value, which he garners from aesthetics. Something that has organic unity "unif[ies] diverse and apparently unrelated (or not-so-tightly related) material" [7]. For example, degree of organic unity can gauge a scientific theory, insofar as the theory is able to explain seemingly unrelated phenomena (e.g., Newton's laws explained both planetary motion and ballistics). Such examples have considerable illustrative power, and lend themselves in many respects to our earlier discussion of models.

According to Nozick, we will find valuable kinds of stasis wherever we find organic unity. With diversity being a critical aspect of V-worlds, we are left with a form of weak pluralism, since Nozick describes stronger pluralism as the recognition of diverse values with the belief that they may *not* be unified [7, p. 447]. Both monism (unity without diversity) and strong pluralism (diversity without unity) deny that stasis (value)[8] lies in that which is organically unified. Diversity constrains unity and vice versa. As Nozick points out, "a far-flung system of voluntary cooperation unifies diverse parts in an intricate structure of changing equilibria, and also unifies these parts in a way that takes account of their degree of organic unity" [7, p. 421]. Thus, designers can provide a super-structure or framework, which satisfies both the criterion of unity among the various agents (e.g., order, peaceful co-existence) and the criterion of appreciating diverse ends (ends which emerge, *pace* Hayek, from the more basic behavioral orderings of individual PCs). We can reiterate the idea of an organic order by speaking of the degree of autonomy available to an agent for utility maximization that does not upset the order. In the next and final section we examine this autonomy in detail.

4 Rules and Rationality (Stage III Methodology)

Gauthier and Decision Theory: Contractarianism as a Middle Way. This section may seem like a radical step for some readers. Indeed, it may smack of a highly prescriptive leap from an otherwise less loaded methodology. However, like most dichotomies, the aforementioned "crossroad" for our V-worldmaking seems to suggest a middle way—one between observing phenomena in the V-worlds that proceed from a virtual State of Nature (on virtual *latafundia*) and establishing a predetermined rule set that is a proto-codification of some body politic or planned economy. If an initial rule set is to apply with the hope of universal appeal—or at least, in some sense, exhibits the character of universality—we must assume the rule set satisfies a number of conditions. First, it seems reasonable to assume that the rule set be amenable to the rationality of PCs, where rationality is defined simply as the propensity to maximize utility (benefit, advantage) to oneself in relation to one's considered preferences. Such a condition takes into account certain assumptions about agents in some political economy and encourages both economic and cooperative activity. Second, the rule set

[8] In using Nozick's theory of value as the highest degree of organic unity, we would defer to the meta-ethics of Quine who said, "science, thanks to its links with observation, retains some title to a correspondence theory of truth; but a coherence theory is evidently the lot of ethics." (Quine, W.V.: *Theories and Things*, Columbia Univ. Press (1969), p. 59)

should allow for the maximization of expected benefit to PCs, but seek to eliminate free-ridership and parasitism. That is, eliminate relative disparities between a PC's contribution and gain. This condition is meant to prevent PCs from taking advantage of others. Third, the rule set should constrain harm. That a PC not be harmed arbitrarily seems fundamental to cooperative arrangements.[9]

In deriving a clearer articulation of such conditions, we think the best place to look is contemporary contractarianism. David Gauthier, in his celebrated *Morals by Agreement,* details a neo-Hobbesian approach to moral-political theory [8]. Simply said, the work is intended to show that certain types of ethical behaviors (constrained maximization activity) are amenable to the interests of agents in a community. In doing so, Gauthier wants to provide a justificatory framework for moral behavior—arrived at not through appeal to some grand set of moral truths, but derived through appeal to the rationality of agents [9, p. 23]:

> The key idea in many situations, if each person chooses what, given the choices of the others, would maximize her expected utility, then the outcome will be mutually disadvantageous in comparison with some alternative—everyone could do better. Equilibrium, which obtains when each person's action is a best response to the others' actions, is incompatible with (Pareto-) optimality, which obtains when no one could do better without someone else doing worse. Given the ubiquity of such situations, each person can see the benefit, to herself, of participating with her fellows in practices requiring each to refrain from the direct endeavor to maximize her own utility, when such mutual restraint is mutually advantageous. No one, of course, can have reason to accept any unilateral constraint on her maximizing behavior; each benefits from, and only from, the constraint accepted by her fellows. But if one benefits more from a constraint on others than one loses by being constrained oneself, one may have reason to accept a practice requiring everyone, including oneself, to exhibit such a constraint.

Many would argue that this approach, in seeking to form the very basis of morality, goes too far in relation to a plethora of competing theories. But from a methodological perspective, this approach provides a framework that is amenable both to self-interested behavior (rationality of PCs), and to the fundamental structures of interaction. Use of such rules will tend to produce both spontaneously generated orders and organic unities, so it is in our view the best way to establish starting points for V-worldmaking. Gauthier's Minimax Relative Concession (MRC) exemplifies this approach [8]:

> [MRC is meant to express] a measure of each person's stake in a [hypothetical] bargain—the difference between the least he might accept in place of no agreement, and the most he might accept in place of being excluded by others from agreement. And we shall argue that the equal rationality of the bargainers leads to a requirement that the greatest concession, measured as a proportion of the conceder's stake, be as small as possible.

[9] That doesn't mean the PCs won't harm, for this is often part of play, but that non-harm as a criterion is packaged into the rule set.

Once the basic structure of interaction for PCs is established with appeal to this kind of pre-moral structure, more complex political structures will emerge around them.

Getting Started: Rules of Thumb for MRC Rule Sets. In terms of simulation, an articulation of the ways large-scale maximization behaviors are governed by MRC might be too difficult a task to tackle here, and is an issue for future study. However, guidelines of the sort that—at least in a cursory way—lend themselves to MRC are already available to us. These, we think, do a good job of approximating MRC for the express purpose of giving designers a place to start. They are:

- Efficiency: Strive for Pareto-optimality. The thoroughly political responsibilities of externalities and non-Pareto-optimal conditions are dealt with in a manner that minimizes the divergence between private and public costs. Rule sets (and the public works that might flow from them) will—under the efficiency approach—always strive to create some sort of incentive structure for an agent's behavior.
- Simplicity: When one is in doubt about the choice of rule sets, "choose the simpler of two alternatives," says Richard Epstein [6, p. 33]. Epstein calls a rule complex if it creates obstacles to private objectives. Efficient orders are more likely to result from simple rule sets, as PCs pursue private objectives, if for no other reason than simple sets may yield a diverse collection of orders (efficient and otherwise).
- New Lockean Proviso (NLP):[10] Worsening another person's condition is prohibited unless one does so, through interaction with that person, to avoid worsening one's own. This kind of guideline is meant to delineate the sphere of an agent's activity relative to others. The NLP introduces a simple rights structure to the rule set, and ensures a mostly bottom-up orientation of the rule set.

The priority of the guidelines is then:

1. Defer to efficiency until violations of the NLP are clearly identified.
2. Give primacy to NLP in all matters where efficiency and NLP are at odds. This keeps the character of the rule set closer to that of MRC, and wards off the temptation to lapse into utilitarianism.
3. Defer to simplicity.

To sum up, the aim of this paper is to propose, with some justification, a tripartite methodology for using Online Societies as simulation in applied political economy. The first stage set the context of theoretical investigations *ueberhaupt,* while simultaneously justifying the very multidisciplinary and recursive nature of the enterprise. The second stage showed how certain kinds of rule sets could produce spontaneous orders, exhibiting organic unity, in actual and online societies. The third stage culminated in a distilled set of guidelines for the production of rule sets. The guidelines appeal to more elaborate theoretical considerations, namely, the convergence of moral theory and rational choice theory.

[10] Nozick and Gauthier both discuss the Lockean Proviso. The NLP as it is written here is Gauthier's final formulation of Nozick's "non-harm" interpretation of Locke.

Acknowledgements. We are grateful to Lucian Hughes, Accenture director of technology research—Palo Alto, for introducing us to the potential of MMORPGs. We also thank Mei Chuah, Kelly Dempski, and Ed Gottsman for many fruitful discussions during the development of our ideas.

References

1. Goodman, Nelson: *Ways of Worldmaking*. Harvester Press (1978).
2. Rawls, John: *A Theory of Justice*. Harvard Univ. Press (1971).
3. Black, Max: *Models and Metaphors: Studies in Language and Philosophy*. Cornell Univ. Press (1962).
4. Nozick, Robert: *Anarchy, State and Utopia*. Basic Books (1977).
5. Hayek, Friedrich A.: *Law, Legislation and Liberty*. Univ. of Chicago Press (1973).
6. Epstein, Richard A.: *Simple Rules for Complex World*. Harvard Univ. Press (1995).
7. Nozick, Robert: *Philosophical Explanations*. Belknap Press (1981).
8. Gauthier, David: *Morals by Agreement*. Oxford Univ. Press (1986).
9. Gauthier, David: "Why Contractarianism?" in, *Contractarianism and Rational Choice: Essays on David Gauthier's Moral's by Agreement*. Peter Vallentyne (ed.). Cambridge Univ. Press (1991).

Human Identity in the Age of Software Agents

John Pickering

Psychology Department
Warwick University, UK

Abstract. The human psychological condition arises in a strange loop. The mind is created within a cultural envelope which is itself created by the mind. Human identity is aquired as the cultural envelop is assimilated, a process in which other human agents are crucial. With the advent of information technology non-human agents have entered the loop. This paper discusses how this may affect our sense of who we are.

1 Introduction

This paper is about the cultural and psychological framework surrounding the creation of software agents. It suggests that psychological theories concerning the creation of self awareness through social interaction may be relevant to how agents are designed and used. It also examines some ways in which software agents might affect the way people interact socially and how this may influence how people think about themselves. It is hoped that it will help those actually creating software agents to see their work against a background of cultural change, particularly changes in the relationship between technology and society.

2 Constructive Postmodernism

Contemporary computing technology is relatively undirected. Compared with, say, the technology of power generation, or the technology of transport systems, information technology is remarkably open as to what it is actually for. The notable increase in the power and affordability of computers over recent times has meant that we have available more resources than we actually know what to do with. As a number of analysts have noted, the development of cognitive technology is not limited by our computers but by our imaginations. To stimulate our imaginations we will look here at how the creation of software agents relates to recent changes in psychology which in turn reflect the present cultural condition surrounding science and technology.

The present cultural condition is one in which runaway globalised technology, especially information technology is making into commodities things as basic as social contact and knowledge (Giddens, 1999). Media technology has created, the 'Global Village' foreseen by Marshal McLuhan. Within this community cultural blending and the synthesis of new meaning, as Vaclav Havel has put it is one of the central features of the postmodern condition (Havel, 1995). Another feature is a tendency to substitute simulation for reality, and then to leave reality behind altogether (Baudrillard, 1993).

M. Beynon, C.L. Nehaniv, and K. Dautenhahn (Eds.): CT 2001, LNAI 2117, pp. 442-451, 2001.

These aspects of the postmodern cultural condition are clearly relevant to creation of software agents.

Now 'postmodern' and related terms are all too often overstretched and misused. Postmodernist writings can seem merely to be the relentless deconstruction of beliefs and practices giving the impression that nothing can be said about anything. As a result they are too readily dismissed, occasionally with reason, as superficial jargon with little to interest those with a real job to do. When computer scientists and knowledge engineers encounter phrases like 'the metaphysics of the code', 'dissolving discipline boundaries' and 'ironic mimesis', they can easily be forgiven for reaching for their calculators in the belief that such talk is not for them.

But nothing could be further from the truth. Postmodernism also has a strongly constructive side (Griffin, 1988; Jencks, 1992). It helps to understand how systems of knowledge and practice, including science and technology fit into a wider patterns of cultural change. It is the effort to recognise how what we do and what we believe express the values of the age. David Harvey notes that information technology is central to what he has called "The condition of postmodernity" that surrounds and affects all our activities (Harvey, 1990). For one example, the mobility of information the principle factor behind the speculative world of 'voodoo economics' (Harvey, 1990, page 335). It powered the exponential jump in secondary financial activity, such as trading in futures and derivatives, that between 1960 and 1990 has created what one commentator has called the 'cultural logic of late capitalism' (Jameson, 1984).

Agent technology is intimately bound up with this logic. Market analysis, share trading and personal finance management are major areas of interest to those creating software agents. A brief survey of some major centres for software agent research in the US, such as the MediaLab at MIT and the Software Agents Group at Carnegie Mellon University, shows that a majority of the projects have to do with trading activities, to do with the management of information relevant to making economic decisions or scheduling work, as shown by the title of a paper in the area: "Agents that reduce work and information overload" (Maes, 1994).

This sense of overload reflects the way in which information and media technology have helped to create a recombinant culture in which images, sounds, logos and concepts are actively re-cycled, re-interpreted and new identity is continually synthesised. People can find it difficult to get their bearings and to work out their relationship to their surroundings. As a result, and as another central aspect of the postmodern experience, our ideas about human identity are also changing, calling into question the assumption, that seems so natural, of "... a Self as a stable, continuing entity apart from its own descriptions of being." (Zweig, 1996, page 145). For those who are creating software agents, this re-definition of the individual and of selfhood, what Anderson has called "The Pluralistic Person, The Disappearing Self", is likely to be a fruitful source of ideas (Anderson, 1996).

Thus, looking at this wider cultural perspective suggests new ways to think about what selfhood actually is. In psychology too, again as a reflection of the postmodern turn, the very notion of a self-like agent is being revised and extended. The self is a construct referring to a centre of agency and experience that exists only in relation to other selves. Taking this relational view, if software agents even approximate to the appearance of another self, then this will have some influence on the human self who encounters them. Agents will be significant in creating our sense of ourselves to the extent that it becomes possible to have a relationship with them that simulates the relationships we have with other people. Don Norman, a psychologist interested in the technological enhancement of human abilities, suggests the problems to be solved before software agents can be really natural and useful : "... are social, not technical. How will intelligent agents interact with people and perhaps more importantly how will people think about agents?" (Norman, 1994, page 68).

3 Interacting with Agents

There is no simple answer to this question at the moment, but it seems likely that to be effective, software agents will need to reflect some of the psychology of those who will use them. This is all the more true when we look at this issue as a matter of development. That is, can agents be built so that people will find it natural to engage with and hence to learn about them? In turn, can agents be given the ability, or the appearance of it, to learn about people?

Some enthusiasts who see agent technology as revolutionary appear to believe this will happen: "... agents will be intelligent enough to observe us and learn our habits and preferences as well as our personal demographics to serve us better" (Horberg, 1995, page 3). Others are more cautious, noting that the area is particularly prone to media hype. In a recent survey of software agent research the author expressed her belief that: "... agents can have an enormous effect, but that this will appear in everyday products as an *evolutionary* process rather than as a *revolutionary* one ...." (Nwana, 1999, page 140).

It is already clear that agent technology will develop as an evolutionary process. Biological models for software agents are common (e.g. Marrow & Ghanea-Hercock, 2000). Psychological models are also found but here there is a significant mutuality between recent developments in psychology and agent technology. Contemporary psychology is radically revising its ideas about cognition, the self and its development. New models have appeared based on dynamic systems theory (for a review see e.g. Port & Van Gelder, 1995). Earlier metaphors for the self, that of an *ab initio* centre of awareness for example, have been replaced by that of a relational network (Fogel et al. 1997). The process of development is the growth of that network and the dynamic internal change in it as it encounters external conditions (Dent-Read & Zukow-Golding, 1997). Sometimes the network metaphor is replaced by that of the internal society of cognitive agents (e.g. Minsky, 1985; Brooks, 1991) but the fundamental picture is similar: the unity of the self is actually a product of an underlying diversity of dynamic relations. Models of cognition too have moved on from the days of classic artificial intelligence. Cognition is now seen as embodied in active organisms and not as internal symbol manipulation (e.g. Clark, 1999; Wheeler & Clark 1999).

This goes hand in hand with the mutuality of biology, psychology and agent technology noted above. A recent critique of classic cognitive science has as one of its aims " ... to dispel the notion that the mind can be understood in the absence of biology." (Edelman, 1992, page 211). Thus just as information technologists are beginning to use biological metaphors to create self-like agents, psychology is turning to agent-like biological metaphors for the self and for cognition. Artificial life has replaced classic artificial intelligence as the principal psychological metaphor for mental activity. Artificial life is the attempt to "capture the logical form of life in artifacts"

(Langton, 1995). The 'life' in question is not only the life forms of which we know as the product of evolution, but also new life forms that will be the product of the imaginations and techniques of their creators. It has not been lost on commentators that at the same time as this project gathers momentum, the sequencing of the human genome has been heralded as a major triumph of genetic engineering.

This ironic exchange of identities in which mechanisms become organisms and organisms become mechanisms is precisely the re-negotiation of identity that is one of the hallmarks of postmodern culture. As software agents help with the information overload our media-saturated society produces, so the jobs in call centres and similar places of work require people to take the role of highly task-specific agents. This, of course, is not radically different from the role constraints that have been part of industrialisation for centuries. Information technology encroaches on intellectual work in the way that mechanical technology encroached on the domain of physical work. What is new however is that machines now participating in those activities through which people derive and sustain their identities.

The creation of software agents is a new phase of a long term cultural shift in which human activity is replaced by that of machines. However, what is distinctive about this phase is that more than ever before, the machines involved bring human autonomy into question. They are a specialised form of information processing technology in which agency, choice and psychological life, or at least a semblance of it, is created outside the human sphere. Again, sharing in the postmodern re-visioning of individual identity, it is quite possible that this will change how people think of human autonomy and how they exercise it in a social world shaped and mediated by information technology.

An enduring image of information technology in the past few decades is that of a potent force for social and economic good. It is represented as something that will open up our social and cultural lives, promoting transparency in administration, participation in government and autonomy in the design and control of the civic spaces, both real and virtual. There is a vigorous growth in computer mediated social life as people buy, talk, meet, share and plan what to do in virtual spaces accessed through screens and keyboards. The media celebrates the internet as a new electronic agora, in which software agents and human beings will learn to live together.

Such enthusiasm for this vision of our computer-enhanced lives usually comes from a highly visible and technologically sophisticated minority. It tends to conceal the more negative experiences and views of the less technologically adept majority. Here we are more likely find feelings of information overload, of useless gadgetry and of manipulation by media technology corporations. There are marked age differences here, the young tending to be enthusiasts, the old to be technophobic, something which tracks the passage of technological innovations from being novel and challenging to being a natural part of our lives.

This is the naturalisation of technology was noted by Heidegger (Heidegger, 1977). He observed that as our skills with a tool become well practiced, the tool itself disappears from consciousness, with important consequences that reach beyond the actual use of the tool to our experience of the world in general. This assimilation is central to understanding of how people learn to use information systems (see e.g. Winograd & Flores, 1987), and software agents will be no exception. We already see how a social life organised with the help of software agents is becoming natural to younger & younger age groups. Any parent will know how their children's activities are planned in an open fluid exchange of voice and text messages mediated by wearable computing. Communication service providers offer sophisticated message processing through which a lot of social negotiation can be done off line.

Now an important strand in the development of software agents is the attempt to make them able to adapt to individuals and to recurrent patterns in their working environment (Maes, 1995). If this can be done to any significant extent then as human individuals grow up in naturalised interaction with such agents, important parts of our experience of identity will be affected. Agents will be perceived as individuals and human individuals may well begin to share some of their identity with them. This sharing will also be perceived by others as an extension of their personalities, much as we presently accept the way in which people configure their email or voicemail answering systems. More importantly though, it may be perceived as such by the individuals themselves. If a person grows up sharing control of their lives with autonomous *doppelgangers* with whom they have a history of interaction, it will have a significant impact on how that person thinks of their own sphere of responsibility and how they are to exercise it.

4 Cultural Assimilation

The blending of people with machines has been a science fiction commonplace for decades. However, science fiction is far more likely than it was to become technological fact, given the point made above that what limits information technology innovation is imagination rather than computing power. It does not, in fact, require that much imagination to see the significance of agents with elementary psychological powers mediating the integration of machine intelligence with human activities. This is already creating skilled practices in which humans and machines interact socially. This technologised social interaction will mimic but will also shape human social interaction. The process of sociocultural learning in which human identity is formed will be changed as a result. Here, again, it may help those who are creating software agents to familiarise themselves with the way psychologists and sociologist have described the sociolocultural context in which human selfhood is created.

Rogoff offers a three level model of how individuals assimilate sociocultural practices (Rogoff, 1995). The most explicit level is apprenticeship, being what happens when learners who know they are learning participate with teachers who know they are teaching to develop specific skills and knowledge. At the next level there is guided participation. Here, explicit instruction is not involved. Nonetheless individuals learn though taking part with others in collective activities that leave a residue of skills and knowledge. The highest level is participatory appropriation in which individuals make for themselves a style and a unique set of practices which are the means to achieve goals they have set themselves.

The important point to note here is that Rogoff only considers the 'others' involved here to be human beings. In time, and through a process of cultural evolution, software agents could also fulfill this role. This will be all the more probable if they are able to learn and share learning. Thus social interaction with machine intelligence is becoming more a realistic possibility. It is not important here whether machine intelligence is considered to be "the same as" human intelligence. Nor is it important to decide whether agents that learn and evolve are in fact autonomous. Whether the agency of artifacts comes from human design or whether it emerges as the artifact develops, what is important is that artifacts can become agent like. Simulation is enough. The possibility of interacting with artifacts as if they were agents will mean that they are treated as such.

As software agents begin to mediate social relations during human development, so the human experience of those relations will change. As they become involved in skilled human practice as participants rather than mere tools it will influence the way in which human beings experience themselves and others. The growing network of relations which constitute the self will spread to include relations with non-human agencies. In the longer perspective of human social evolution this is not new. It is after all a natural feature of a life in which people interact with animals. What is new is that we can now see that it will in time be possible to create non-human agencies that are surrogates for real human actors. The social role of software agents, although it is very small at present will grow rapidly. Depending on the rate and extent of this growth there will be a corresponding force for change in human experience that will need to be understood and managed if it is not to cause damage.

Looking at the broader cultural context will helps us to learn how to live with the forces unleashed information technology, just as we learned lived with the forces of heat engine technology. Here, Lewis Mumford's remarkably far-seeing works, especially *Technics and Civilisation* are very useful, since they explore the role of technology in cultural evolution (Mumford, 1940, 1968). He distinguished what he called the eotechnic, paleotechnic and neotechnic phases to cover, respectively, tool making, power generation and information handling. These phases do not just chart the development of techniques. They track the incorporation of technology into the human condition. Two principal observations that Mumford makes about this incorporation are relevant here. First, it makes possible radically new conditions of human association. Second, technology is not merely the means by which relatively unchanging human needs are met but is the means by which new ones are created. It helps create a system from which emerge fundamentally new forms of human activity and the values these activities express.

5 Agents and Social Relations

Human-agent interaction is already creating new values and new sensitivities, especially in the young. It is already an important part of the way people live in technologically sophisticated cultures. Social spaces will gradually transform under the impact of networked communications and in response to the skills and practices that people develop to use them. Seen in the long perspective of cultural evolution, this transformation is part of the mutual evolution of people and machines, an aspect of the postmodern blending of identities. This blending is prefigured in the developments in psychology noted above and in recent changes in evolutionary theories. These developments are giving a greater role to action, both in development and in evolution. Recent accounts of how the brain develops suggest that that it is more influenced by the environment than was previously thought (Edelman, 2000). New models of the interaction between learned and innate factors in development are being developed and advances in evolutionary theory also point to a more significant role for social interactions in human evolution (Elman et al., 1996; Deacon, 1997).

These models may well be useful to those developing software agents able to participate in human social interaction. As they do it will certainly influence human practices and values, as Mumford pointed out. For example, it may become relatively unimportant to distinguish between technologised human agents and humanised technological artifacts, especially for the young. Children who grow up amid sophisticated software agents are very likely to develop habits and values that will influence how they communicate with each other and with people in general.

This reflects how machines are now more than ever involved in human social practice and this involvement is bound to accelerate. Human relations will be technologised to the extent that such artifacts are able participate as agents in social interaction rather then merely to mediate it. The encounter with these artifacts will occur earlier and earlier in human development. They will thereby take part in the sociocultural learning by which skilled practices, and the values they express, are transmitted. The attribution of human like agency to artifacts will change the image of both machines and of human beings.

As social relations are increasingly mediated by machines the practices that are thereby created will transmit technologised values and sensitivities. This will influence how human beings communicate with each other and how they think of themselves. For some observers of software agent research this creates the dangerous possibility that we will devalue the person (Lanier, 1995). From the earliest days of artificial intelligence, it was apparent that ethical issues arose in the creation of systems to mediate human interaction in areas like education and therapy. Here, it was objected, was something that one human being does for another and not something that could be handed over to a machine (e.g. Weizenbaum, 1976).

These concerns are even more relevant now. We are becoming aware that as well as being an adjunct to human practices, digital technology will create attitudes, tastes and modes of social interaction. These will shape the cultural conditions within which people develop their abilities to live together. These conditions now include agent like artifacts with which human beings will need to co-exist. Given the destructiveness of the geopolitical forces unleashed by technology, a better understanding of the impact this will have on human social relations is urgently needed.

Technology creates the cultural envelop within which human beings exist and, perhaps more significantly, develop. The boundary between the natural the artificial is now more problematic and more contested than ever before. As technology becomes more organic and social, so the organic is technologised and increasingly brought within the sphere of human social concern and action. These developments remind us that the human condition is not exclusive a function of what is in the head or even in the genes. It is inextricably entwined in a system of relations that comprise the biological order of the body, the cultural order outside the body and the mutually evolved processes that integrate them into a seamless whole.

6 Agents Are Simulacra

An analysis of these processes were part of the extraordinarily prescient writings of Walter Benjamin (Benjamin, 1979). He recognised how fundamentally technology alters the cultural order and with it human sensibilities. Benjamin's unfinished *Arcades Project* was about the production of human consciousness within a commodified and technologised envelop (Buck-Morss, 1989). His views reflected the dark conditions of the 1930's in Europe. He saw that the massive overproduction unleashed by mechanical technology was leading culture away from authentic being and towards alienated violence. In his *The Work of Art in the Age of Mechanical Reproduction* is Benjamin discusses the detachment of artistic work from domains of traditional aesthetic and cultural values as part of this alienation. It is interesting to note that the means for this detachment is the mobility duplication of images, what Benjamin called the domain of the simulacrum.

This is precisely the mobility that digital technology has now unleashed on an even greater scale. It underlies the recombinant culture of the postmodern era (Winner, 1996) A software agent is a simulacrum of a more active sort. It is suggested that such agents may become able, perhaps through learning, to simulate natural sociability as closely as makes little difference (Donath, 1997; Mitchell, 1995). Such simulation creates the condition that Baudrillard calls 'hyperreality' (Baudrillard, 1993, 1983). Baudrillard points out a transition to a political economy of simulation mediated by information technology. Simulation here moves beyond mere imitation, that is, the fabrication of a copy of something real. The significant step is that simulation becomes the real: "Simulation is no longer that of a territory, a referential being or a substance. It is the generation by models of a real without origin or reality: a hyperreal." (Baudrillard, 1983, See the section entitled 'The Map Precedes the Territory').

Software agents are a social form of hyper-reality. Baudrillard's view of the destructive and alienating character of hype-reality runs counter to the early and optimistic humanism of techno-enthusiasts like Lewis Mumford. They depicted technology as a stage in a progression towards productive human association, very much in keeping with the benign visions of Donath and of Mitchell, who present agent technology as an extension of human co-operation by digital means. Looked at from a global perspective, however, technology does not seem quite so wholesome (Giddens, 1999). For Benjamin, Baudrillard and Harvey too, the drift into virtualisation offers a darker and more dystopian image of postmodern culture.

This is the transition from value to sign, from production to reproduction and from copies to simulacra. It is a continuation of Benjamin's central insight into the effects that technology would have on human sensibilities and self-consciousness. What he foresaw was the significance of the simulacrum, the multiple and mobile sign that now, in the shape of software agents, is coming to participate in essential parts of human social interaction. For some (e.g. Lanier, 1995) this is alienating. For others it is to be celebrated as part of a cultural trajectory towards the transhuman condition (e.g.Gray, 1995; see also: www.cyborgmanifesto.org).

7 Conclusion

The purpose of this paper has been to sketch of some broader ideas that may be of interest to cognitive technologists involved in the creation of software agents. This work is often represented by those who do it, and those who fund it, as the means to a future of enhanced technological power used for human good. But the future arrives earlier than it used to, giving us less time to prepare. We are learning that we often have not been able to anticipate the full effect of technological innovations. Baudrillard alerts us to the danger of a drift into hyper-reality, a process of virtualisation in which simulacra become the reality of social and political life. Software agents are simulacra with attitude, and we need to be aware of their wider cultural effects. Walter Benjamin, concerned at the technologising of human consciousness, foresaw the power of the simulacrum and that violence results when society cannot contain the forces created by technology. The social force of agent technology needs to be understood as a matter of urgency.

These warnings need to be set against the media hype of information technology. The virtual social space opened up by agents and networks is often presented as a virtuous social space in which democratic interaction between autonomous citizens and benign software agents will be the means to more openness and justice in the global community. But this presentation is a representation, a simulacrum that is in danger of becoming reality. We need to be aware that agent technology may also harmfully degrade how people value themselves and treat each other.

References

1. Anderson, W. (1996)(Ed.) *The Fontana Postmodernism Reader.* Fontana: London.
2. Baudrillard, J. (1983) *Simulations.* Tran.s P. Foss. New York, Semiotext(e).
3. Baudrillard, J. (1993) *Symbolic Exchange & Death.* Sage, London.
4. Benjamin, W. (1979) The Work of Art in the Age of Mechanical Reproduction. In *Illuminations*, translated by Zohn, H., pages 219 - 253. Fontana, London.
5. Brooks (1991) Intelligence Without Representation. *Artificial Intelligence*, 47: 139 - 160.
6. Buck-Morss, S. (1989) *The Dialectics Of Seeing, Walter Benjamin And The Arcades Project.* London, MIT Press, c1989
7. Clark, A. (1999) Towards an embodied cognitive science? *Trends in Cognitive Science*, 3(9): 345 - 351.
8. Deacon, T. (1997) *The Symbolic Species.* Penguin:London.
9. Dent-Read, C. & Zukow-Golding, P. (1997) *Evolving explanations of development.* Washington, DC : American Psychological Association.
10. Donath, J. (1997) *Inhabiting the Virtual City: The Design of Social Environments for Electronic Communities.* Doctoral thesis presented Februrary 1997 to the Program of Media Arts and Sciences, School of Architecture & Planning, MIT. See: judith.www.media.mit.edu/Thesis/
11. Edelman, G. & Tononi, G. (2000) *Consciousness: how matter becomes imagination.* Penguin: London.
12. Edelman, G. (1992) *Bright Air, Brilliant Fire.* Penguin: London.
13. Elman, J. et al. (1996) *Rethinkng Innateness: A Connectionist Perspective On Development.* London: MIT Press.
14. Fogel, A., Lyra, M. & Valsiner, J.. (Ed.s) (1997) *Dynamics and Indeterminism in Developmental and Social Processes.* New Jersey, Erlbaum.
15. Giddens, A. (1999) *Runaway World: how globalisation is reshaping our lives*, London: Profile Books.
16. Gray, C. (1995) *The Cyborg Handbook.* Routledge, London.
17. Griffin, D. R. (1988) (Ed.) *The Reenchantment of Science: Postmodern Proposals*, State University of New York Press, Albany, NY. p 173.
18. Harvey, D. (1990) *The Condition of Postmodernity.* Blackwell, Oxford.
19. Havel, V. (1995) Self Transcendence. In *Resurgence*, No. 169, March, pages 12 - 14. Also in Anderson, W. (1996)(Ed.) *The Fontana Postmodernism Reader.* Fontana: London.
20. Heidegger, M. (1977) *The Question Concerning Technology.* (Trans. Lovitt, W.) New York: Harper Row.

21. Horberg, J. (1995) Talk to my agent: software agents in virtual reality. *Computer Mediated Communications*, 2(2): 3 - 11.
22. Jameson, F. (1984) Postmodernism, or the cultural logic of late capitalism. *New Left Review*, 146:53 - 93.
23. Jencks, C. (1992) (Ed.) *The Postmodern Reader*. Academy Editions, London.
24. Langton, C. (Ed.) (1995) *Artificial Life: an Overview*. MIT Press, London.
25. Lanier, J. (1995) Agents of Alienation. *Journal of Consciousness Studies*, Vol. 2 (1): 76 - 81.
26. Maes, P. (1994) Agents that reduce work and information overload. *Communications of the ACM*, 37(7): 31 - 40.
27. Maes, P. (1995) Modeling Adaptive Autonomous Agents. In *Artificial Life: An Overview*, edited by Langton, C., MIT Press, London.
28. Marrow, P & Ghanea-Hercock, R. (2000) Mobile software agents. *BT Technology Journal*, 18()4): 129 - 139.
29. Minsky, M. (1985) *The Society of Mind*. New York: Simon & Schuster.
30. Mitchell, W. (1995) *City of Bits*. MIT Press, Cambridge.
31. Mumford, L. (1940) *The Culture of Cities*. London, Secker and Warburg.
32. Mumford, L. (1968) *The Future of Technics and Civilisation*. Freedom Press, London.
33. Norman, D. (1994) How might people interact with agents? *Communications of the ACM*, 37(7):68 - 76.
34. Nwana, H. (1999) A perspective on software agent research. *Knowledge Engineering Review*, 14(2): 125 - 142.
35. Port, R. & Van Gelder, T. (1995) (Ed.s) *Mind as Motion*. MIT Press: London.
36. Rogoff, B. (1995) Observing Sociocultural Activity on Three Planes. In *Sociocultural Studies of Mind*, edited by Wertsch, J. et al., Cambridge University Press, Cambridge, UK.
37. Weizenbaum, J. (1976) *Computer power and human reason : from judgment to calculation*. - Freeman, San Francisco.
38. Wheeler, M. & Clark, A. (1999) Genic Representation: Reconciling Content and Causal Complexity. *The British Journal for the Philosophy of Science*, 50: 103-135.
39. Winner, L (1995) Who Will Be In Cyberspace? *The Information Society*. Vol. 12, No. 1, pages 63 - 72.
40. Winograd, T. & Flores, F. (1987) *Understanding Computers and Cognition*. New York: Addison-Wesley.
41. Zweig. G. (1996) The Death of the Self in the Postmodern World. In Anderson, W. (1996)(Ed.) *The Fontana Postmodernism Reader*. Fontana: London.

Tracing for the Ideal Hunting Dog: Effects of Development and Use of Information System on Community Knowledge

Anna-Liisa Syrjänen

University of Oulu, Department of Information Processing Science
Linnanmaa, P.O. Box 3000, FIN-90401 University of Oulu, Finland
anna-liisa.syrjanen@oulu.fi

Abstract. The paper highlights the effects of an IT-based tool on people's everyday activities when sharing a common interest. The basic aspect is practical knowledge, which will be approached from an individual viewpoint. It describes how the people concerned enriched their knowledge of their shared interest by shaping their concepts and how they made good progress with the help of an information system of their own making. The research shows that an information system can assist its users' thinking in many ways, and can thus be one of the crucial factors in the knowledge-enriching process. The paper emphasizes the weight of practical knowledge within this research topic, which is not so widely known and has been insufficiently examined as yet, in spite of the increasing popularity of IT-based tools in every sector of human life.

1 Introduction

Computer systems are a pervasive feature of current life, and they certainly influence what we think about the world around us. Evidently having a greater or lesser impact on people's every-day work and recreations within all social environments [1]. In spite of the rapid growth of computers and information systems, however, there are relatively few descriptions of their influence on people's activities or their ability to maintain the sense of collective experience which is essential to any organization, community or individual person and which help them to understand both their achievements and success and their woes. This is not only a matter of know-how and the processing of information into suitable and necessary knowledge, but also a matter of gaining a time perspective and understanding how and why the world as it is now is based on how it has been before [2]. In this aspect, we have a good reason to study in detail human activities that have a long culture behind them and incorporate experiences of a certain domain through the knowledge resources that have been acquired, i.e., how individual persons process information into knowledge by shaping their own concepts, developing their own meanings and solving practical problems.

The purpose of this paper is to present an example of this, with some features that makes the case illustrative and informative. It concerns the evolution of a hunting dog, the Karelian Bear Dog (KBD), from the 1980s into the new millennium supported by an IT-based breeding system. The focus is on the role of the information

M. Beynon, C.L. Nehaniv, and K. Dautenhahn (Eds.): CT 2001, LNAI 2117, pp. 452-462, 2001.

system with a concept shaping process. The case example is a very convenient one, as the subject is unambiguous and small enough, and can thus be quite explicitly defined and handled within clear limits. The dog's evolution and the history of the breeding information system are still fresh in the memory of the persons concerned and are sufficiently well documented. The case is also of personal interest to the author, who has been an insider of the KBD community for over twenty years.

As a research subject, the case offers many challenges on account of the multifarious nature of an undertaking such as the breeding of hunting dogs. For example, the most important characteristic of a hunting dog, its hunting instinct, is problematical in many ways. If you are looking at a dog, a great deal of its genetic makeup, and especially its hunting instinct, will be invisible to the human eye, unlike its external appearance, for instance. The only way to single out its hunting instinct is by observing the dog's behaviour and character. It is thus difficult to measure this feature, but evidently much more difficult to understand how such an instinctive and mental feature can be bred at all? Moreover, the breeding of *this particular dog* has encountered many other hardships, too. Soon after breeding was started, the Winter War of 1939 broke out and Finland lost precisely the part of Karelia from which the basic dog population had been collected, and this nearly destroyed the breed. The people concerned were forced to continue the breeding work with only a handful of the basic dogs. Another obvious drawback was that there was little genetic information on the hunting instinct, but also little information on the hunting dog breeding in general [3],[4]. All the above cumulative factors created difficulties in formulating a concept of how the desired hunting instinct and other qualities could be achieved in the breed, especially, when they were dealing with a living creature, which was then (and still is) "such a valuable part of the Finnish cultural history"[5]. Another interesting feature of this case was that, like all the breeding activities as such, the work of developing the computer-based breeding information system was carried out voluntarily by the FSC's hunting dog enthusiasts (mostly non-IT-professionals) over the last twelve years, and not for business purposes, but unpaid, for the sake of the dogs and practical hunting needs[6].

Hence, the research emphasizes *practical aspects* [7] of knowledge, by highlighting the subject through the individuals' thinking as mediated by the tools used, and by explaining how the people processed the information that gave them a better understanding of their special shared interest. To highlight *knowledge enrichment* within this special context, the subject will be examined through empirical research (thematic interviews with six persons [6] and historical materials, as the author was an insider for years), placing emphasis on the viewpoint of one group of people, the consultants as breeding experts. To understand the practical aspect of their knowledge and the concept-shaping process within this domain, it will be studied in detail by attempting to answer the question of *how an information system can assist the thinking of an individual user?* In this paper *community* is seen according Lesser et al [30] as "a composed of people who interact on a regular basis around a common set of issues, interest, or needs…participation is often voluntary". Correspondingly, community *knowledge* is seen here as something that can be revealed and shared with others, based on the ideas of Blum [23] that "data are the items given to the analyst, information is the data organised in some useful way and knowledge is the cumulative experience in manipulation, exchange, and creation of information". This paper will first describe the development of the system, including a definition of the concept and some short reviews, and then it will analyse the effects of the system on the users' thinking and end with conclusions and a discussion of the results.

2 Tracing Steps with Short Concept Shaping Reviews

The case concerns the work of enthusiasts of the Karelian Bear Dog (KBD), which is one of the aboriginal dog breeds in Finland. The breed organization is the Finnish Spitz Club (FSC), which is affiliated to the Finnish Kennel Club (FKC). KBD breeding was started in 1936, with the goal of "creating a sturdy dog which barks at big game"[8]. To guide the breeding activity, the FKC laid down rules for elk-hunting trials, which were approved in 1943 and the first trial was held in 1945 [3].

First Steps. In the late 1980s, before the computer-based information system was set up, the manual breeding system contained information on about 20.000 dogs, about 5000 dog shows and the results of about 4000 hunting trials, stored in various forms on paper [3],[4],[9]. The dog population embraced thousands of living dogs, and they had some top-level dogs (mostly males), but "the peak was a narrow one" and after fifty years of breeding attempts they were still suffering from a lack of good breeding dogs (especially females). Although all those involved had done their best "public opinion hardly believed in bear dogs as good elk hunting dogs on a more large scale". When the breeding consultant at that time retired, there were hardly any people willing to take on this responsibility. When a new consultant started in 1986, all the dog information stored on paper was available but no guidelines as to how the job should be done, because all the information on past experiences "had disappeared inside the head of his predecessor". Hence, the consultant decided to use his common sense: "if we want to have hunting dogs, then the breeding dogs should bark at game and their external appearance is less important". To understand the dog-breeding situation, he started to make his own tools, various lists, tables, and hierarchically or circularly ordered charts to which he added further information obtained from many scattered sources. This system functioned well enough for a while, but it was quite hard work, it took a lot of time, the manual tools were hard to maintain and the tables and charts became more limited. Also, although he found some good breeding combinations, the results were not as good as he and everybody else expected [6].

Review. The above description already shows that from the early days the goal of "creating a sturdy dog which barks at big game" has been quite a problem, because of a lack of shared methods for breeding the hunting instinct, although the people concerned knew what kinds of dogs they wanted [3]. Like "wicked problems" in general [10], it failed to yield any answers that could definitely be said to be "true" or "false", but rather partial solutions that were more or less "good" or "bad". In this respect, the consultant had to do what he could and use the available resources: hunting trial results, dog registrations and his "common sense". Hence the first two steps of *enriching* breeding *knowledge* were 1) *reducing the goal* to the breeding of a hunting dog by developing *the most essential feature* for the purpose of a hunting dog's being, the hunting instinct, and 2) *by defining the other known features of the dog only marginally* (cf. outward appearance, registered purity of breeding). The basic concept, a hunting dog, was framed tentatively, but it was still quite vague and the advisory work was laborious because of all the different and numerous paper sources [6].

Next Steps. The FSC started to develop its own computer database information system in 1987, and then "the main purpose was to help the annual publishing work". Once this part of the system worked, they developed a first application aimed directly at serving the needs of breeding guidance, a pedigree function. The KBD community acquired the new system in 1990, and it quite soon proved to be a very useful tool for planning new dog pairs. The consultant began to study dogs' pedigrees more closely together with the trial results, i.e. the records of dogs' behavior made by trained judges during elk-hunting tests, and this was easier now because "all the necessary information was in the same place and each dog could be compared with the whole set". As he had expected, the dogs that had been successful in hunting trials, mostly males, also had more successful offspring than the others. These observations convinced him that he "was on the right track", and this was easy to believe because it conformed with his common sense concept of heredity [6].

Review. In this case one can say that the system and the existing information were used to clarify a user's hunch by comparing a part with the whole. The next step in enriching breeding knowledge can be stated as 3) *using the available resources* (i.e. common sense, hunting trial results, registration documents) *in a different way from before.* The concept of a hunting dog received a certain bench mark, that at least one of the dog's parents should have passed a hunting trial, i.e. it should itself be a good hunting dog.

Epoch-Making Steps. By comparing breeders who had bred more dogs, the consultant became involved in another very thorny problem: that some breeders who had used good male hunting dogs as sires for years and had managed to breed some good dogs before were not being successful any longer. A solution came suddenly with a program called Breed, developed by an enthusiast of the Finnish Spitz Dog who was worried about the detrimental effect of inbreeding on the incidence of health problems. In 1993, when the program was adopted into KBD's system, "nobody believed that it could be useful for bear dogs", as most breeders used a method called "linear inbreeding", i.e., the best breeding results can be achieved if the dogs are relatives. Among the breeders quite a general conception was that this method was the accepted view of the genetics of domestic animals then, thus recommended by the consultant [4], but also used for dog breeding because "we knew of no better method at that time". Hence some of the breeders believed that the closer the dogs were related, the better the pair. After comparing degrees of inbreeding with hunting trial results he came up with an epoch-making discovery: that the dogs with high degrees of inbreeding were not as successful as those with a lower degree and that the first instances of inherited eye disease were to be found among them. Arguments against linear inbreeding were presented at annual meetings and published in the club's magazines, and eventually most breeders gave it up when "their eyes opened to see its negative influences on the whole dog population", and the outcome was that breeders obtained more and better dogs than before [6].

Review. The consultant evidently had no idea at the time about the detrimental effects of inbreeding on the dog's hunting instinct, and the notion went against the common concept of breeding methods that all those concerned had believed in for over fifty years. This step can be described as the developing of a tool for 4) *broadening the selection base by the identification of risks* (i.e. degrees of inbreeding) and 5) *uniting the new pieces of evidence* and *testing new combinations* in further hunting trials. The concept of a hunting dog gained a new solid basis, a wide gene pool of parents instead of a very narrow gene pool, and the combining of this with earlier findings improved the practical understanding of how to breed hunting dogs.

Final Steps. By the mid-1990's the community had acquired more consultants, and it was easier now because "the whole system can be given to a new person, and all the information and tools needed are in a small package, not only inside old men's heads". The consultants were assisted by a tool called "male recommendation", introduced in 1995 that automated the process of selecting a male dog by using pre-determined quality conditions. Hence, every selection is now unique and is determined by the characteristics of the particular female dog in relation to information, which is essential only within this unique situation. The tool gives alternatives, but the consultant chooses three to five male candidates and sends his recommendations to the breeder, who makes the final choice. Thus, the advisory work had become easier, but the consultants were still puzzled about why not all the litters of good hunting dogs with low degrees of inbreeding were equally good. In 1996, they started to look for an answer by developing a system based on the common sense assumptions that "the younger the dog, the less it has learned" and "a puppy inherits its characteristics from both parents, not only from its sire". These ideas were verified using a tool called "breeding points". By combining all these and earlier observations, consultants could more closely argue what combination of dogs provided the best background for the forthcoming litter [6]. KBD breeders can now get litters in which several dogs have a good hunting instinct and can take part in hunting trials at a much younger age than earlier dogs, e.g. seven dogs out of a litter of nine have passed the hunting trial by the age of two years [11].

Review. As one can see, the system was again used to clarify hunches and vague ideas regarding a solution, and the system convinced the consultants by offering them concrete evidence. If we remember what a difficult problem this tracing of the breeding features for a good hunting dog had been earlier, these final steps were much easier, because the people concerned had already realized how to use the search instrument, the information system. Thus the subsequent steps were 6) *tracing the origins of the features* (breeding points), combining them and testing again, 7) *automating the selection process* by using pre-determined quality conditions based on tested results (male recommendation), and finally, for the present, 8) *choosing the dogs with the most intense hunting instinct* (on the basis of their behavior in hunting trials) in relation to all the relevant features and the particular female, in accordance with the idea of *quality perfection*. The quality level of the whole dog population can be scaled up only by using breeding dogs with a *higher than average hunting instinct intensity*, and hence the aim should always be to plan the *best possible* dog combination with regard to the basic features of breeding dogs. The achievement of this quality in practice, however, demands that both the consultants and their clients should not alone share the same meanings but also have the same concept of reliability and relevance of the knowledge as it is perceived [12]. In summary, the current concept of how to breed good hunting dogs is that breeding dogs should have a wide gene pool, that both the sire and the dam must be tested and shown to be good hunting dogs themselves, and that the younger the parents were when they passed their first hunting trials, the better chances they have for passing their hunting characteristics on to their offspring. In view of the fact that KBD registrations have remained at the same level for over the past ten years but more good hunting dogs have been available than before [11], one can recognize that the new KBD breeding system has been quite successful in promoting breeding activity.

3 Analysis of the System's Effects on Users' Thinking

As the above account suggests, the breeding consultant's advisory work before the new system was introduced was quite laborious, involving the use of numerous paper sources of different kinds. To manage amongst the chaos of papers, the consultant made his own tools, which proved to be cumbersome to use and maintain, so that carrying on with the work was hard and almost impossible. We are therefore entitled to ask how the new information system has assisted the work and thinking of the individual user in the context of achieving his intention of breeding good hunting dogs? The first effect can be described as *shaping of thought – as a whole and with a sense of time*. The biggest advantage at first was that more dynamic, easily maintained, remodeled tools became available, so that processing, searching for, ordering and linking of information were considerably more effective. This allowed the user to form larger thought-entities and different kinds of bodies of thought and to supplement and reshape them when needed. As one interviewee said "a whole dog population was at my fingertips and I had a fair chance of going as far back as necessary, to the roots of the first dogs if I had to ". This could also help a user to acquire a "sense of time", i.e., why the world is like it is now and how it is now [2], [13]. As one person argued: "I can clearly see what this breeder has tried to do and why his situation is as it is now [6].

Secondly, the system was mostly used to again a *better understanding of causality*, i.e. reasoning in the sense of logical problem-solving [14], by comparing, verifying and combining information. One of the greatest advantages was that the system gave the user an excellent instrument for ordering and reordering the information that was needed when he had to formulate his intentions. Using these functions it was easier to shift or form new perspectives by looking at what characters particular entities had, how similar or different they were, etc. When totally different kinds of entities can be ordered hierarchically, linearly, chronologically or in some other appropriate way within the same contextual dimension, this will assisted a person's thinking by helping him to shape links and relationships between his thoughts and to expose gaps in order to form a better understanding of the factors affecting the subject concerned [15], [16], [17].

Thirdly, a user and the system were partners in quite a suitable symbiosis of human and machine interaction, thus the system could *promote the narrative side of human thinking* [18], [19],[20]. When the system is taking care of information processing and ordering, it can structure a contextual knowledge space [21] managed by a human agent, doing all the work in which it excels: counting, performing routines and most of the simplest tasks, canceling and reprocessing, storing and maintaining information, etc. Hence it can assist people's thinking, imagination, sketching, planning, decision-making, etc. - all the things which humans are good at [22],[23]. In this case the symbiosis worked, because the user knew exactly what all the marks and letters meant which were used to describe concepts of his special domain, the knowledge that constructed the reality, which he was trying to improve [21]. As the interviewees said: "the practice - participation in hunting with dogs or hunting trials - supplies the second part of the machine. You can get much more out of a machine if you know what is needed in practice." In that sense the enthusiast of hunting dogs had found a suitable way of reducing their practical knowledge to concepts that were appropriate for use in computing. Hence the information system could support the narrative mode of the users' cognition [20]. Every symbol, such as AVO1, KVA, XXX, ibd 5%, etc. told the user long and eventful stories, and he could always look for more information from the system to analyze it in greater detail. Based on his practical experience, he could conclude how believable and valid the story mediated by the symbol was [12].

The fourth class of effects can be described as *support for experimental thinking* (know-how) [15]. One of the best features of any information system is that it makes routines more effective than when performed by a human, by storing, processing and organizing information based on what the human has already thought [22]. By this means, the system helped its users to recall their earlier thoughts and to continue to evolve and redirect their thinking. If, in addition, the system contains not only information and results relevant to the users' own earlier decisions but also to those of others, as this system did, it can provide feedback that is essential for supporting new decisions. This meant that everybody could judge how good the decisions had been and learn from them [2]. As one person said: "It was easy to see what pairs of dogs each person had planned and how successful they had been". Moreover, the system could help in the planning and decisions by offering new alternatives and perspectives that the user had never thought about: "I had to admit that the machine was wiser than I was. It showed me alternatives that I didn't know of before." Hence, as Blum [23] states: "computing provides ... a mechanism that moves us from the plan-and-build paradigm to the build-and-evolve paradigm." As another of the interviewees said: "when I plan a new dog combination, I also try to think how this can help the future of the whole population". Thus, if the system shows you the whole entity, it also gives you a possibility to make your single decisions in relation to that entity, and thereby to create future innovations and goals [6],[21],[24].

Finally, the case description shows that there is quite a clear alternation between development work and use of the system in connection with problem solving and thinking processes. Obviously, when the users shifted the perspective of their dog breeding guidance, they did so by developing the information system, mediating the activity through instruments that were capable of adaptation [25],[26]. By combining two practices, dog breeding and IS development work, they arrived at a system that was advantageous for both [22], in addition to which their shared experience, i.e. the hunting trials, enabled the system developers to understand the users' context of operation without extensive or detailed additional requirements, while the users in turn easily understood the concepts of the system's interface [6],[12].

4 Conclusions

As an object of research, the current concept of the breeding of Karelian Bear Dogs in order to maximize their hunting instinct can be seen as a theory of breeding guidance ("theory-in-use" [27]). This is because certain factors have an influence over the propagation of the desired quality characteristics in the dog population and this process can be guided consciously. The process of knowledge enrichment took place through the development and adoption of a new information system. With the help of the system the persons concerned managed to find a way to guide breeding activity in accordance with the desired practical purpose, and the system promoted this by assisting the users' thinking in many ways. *In this case* the main effects can be formulated as follows. The information system assisted the users' thinking 1) by supporting the shaping of thought entities with a sense of time, 2) by giving a better understanding of causality, 3) by promoting the narrative side of human thinking, and 4) by supporting experimental thinking. Hence the users could shift perspectives or form new ones with the help of the system, and the system can today be seen as "an integral part of their practice" [28]. It can also be said in summary that by storing the results of earlier breeding decisions, the system encapsulates the evolution of this breed of dog, with all the accumulated knowledge regarding its breeding and the history of breeding activity, i.e. it memorizes the past, which implies that it can also give hints about the tacit knowledge possessed by the earlier people concerned [2].

We can also say that in spite of the good progress made in breeding guidance, the people concerned *do not think* that they know how the dogs' hunting instinct is inherited, nor do they know why the breeding system works (except that "it makes sense"). Furthermore, they do not think that breeding success depends only on consultants and the computer-based information system. On the contrary, *they do think* that there was no suitable genetic information available on this special subject (the heredity of the hunting instinct) and that no better breeding system exists that could be as useful under the circumstances affecting this breed. They also understand that success in breeding is a complex matter that is influenced by many factors, e.g. the *people involved and their readiness to co-operate*, which ensured that concerted action could evolve among the many participants in the KBD-community.

All in all, however, the KBD breeding consultants seem to have a very useful system, although there are still many open questions concerning the tricky problem that had to be faced. For instance, is it possible that they could had shifted their thinking perspectives in any case and would have found these factors influencing the breeding of hunting dogs without the information system? Moreover, would they have been capable of defining the system's functions in detail in 1987 as requirements to be implemented by an outsider if they had chosen to order a complete information system? In the light of the above arguments, this seems doubtful. They would obviously have made some kind of progress with breeding guidance without the new system, but it would have been considerably slower. The consultant would without doubt have had serious problems in identifying the highly detrimental effects of inbreeding on the hunting instinct, for instance, especially when he knew that the method was supported by the genetics of domestic animals and had been used for near a half a century. Hence it is most likely that he would not have found these defects and other features without the new system.

Also, the arguments stated here emphasize that the definition of an information system of precisely the given kind would had a matter of chance rather than a high probability in 1987. Some parts of the system would had been defined fairly easily, but it is less likely that the application for analyzing the degree of inbreeding would have been invented then, when nobody even believed in the subject seriously, especially when we recall that this and many of the other functions were found only after the system had been in use for years. As Blum [23] states "... once a product exits it changes our understanding of what should be done". Moreover, the role of an information system of one's own was seen to be totally different at that time. The scenario in the FSC 50th anniversary yearbook in 1987 states that: "after ten years we shall have effective computers ... to help maintain a register of members and their addresses, arrange the printing of documents, organize book-keeping and filing, collect information for breeding ... it is the duty of the Finnish Kennel Club to analyze and investigate these things in more detail ... we are not ready for all that yet" [4]. Finally, based on the above reviews, it is evident that the KBD consultants would not have shifted their thinking perspectives, and in part even reversed their opinions entirely, without powerful arguments, which would have been hard to find without the new system. It is easy to understand this in the light of their traditional practical culture [13], [24], [29]. They needed to see the hard evidence before they believed, as they traditionally do with their comrades and their dogs, for it has been written that "a man must be worthy of respect for the sake of his deeds and not for the sake of his words, and a black dog for barking at game and not for wagging its tail" [5].

Finally, the research leaves many open questions. For example, the consultants reported that they used the same information as before, but in a different way somehow. How? What features of this information helped them to solve the problems, and how did they convince the breeders that the transfer knowledge on new kinds of dog pairs into the field was possible? In spite of the success of these new thinkers, not all KBD breeders have abandoned their old ways of thinking and acting, even though in some cases this persistence seems to be to their disadvantage. Why? Also, there may be others who simply "don't care a damn" about breeding guidance or the common goals of the KBD-community. Obviously these attitudes may have something to do with the community's social system, but are there other reasons, too?

References

1. Brown, J.S. and P. Duguid (2000). *The Social Life of Information*. Harvard Business School Press, Boston, MA. p. 320.
2. Leonard, D. and S. Sensiper (1998).*The Role of Tacit Knowledge in Group Innovation*. California Management Review, vol.: 40 (3. Spring 1998): p. 112-132.
3. Perttola, J. (1998, original 1989). *The Story of the Karelian Bear Dog*. Suomen Pystykorvajärjestö - Finska Spetsklubben ry. Painokotka Oy 1998. p. 119. The FSC's permission to publish a photo: the Karelian Bear Dog named Usvan Murri, a Champion and Field Champion male owned by P. Karjalainen from Anjalankoski.
4. Simolinna, J. (1987). *Suomen Pystykorvajärjestö - Finska Spetsklubben r.y. 1938-1987 50 vuotta*. Pyhäjokiseudun kirjapaino oy, Oulainen 1987. p. 242.
5. Karelian Bear Dog Division (2001). *Karelian Bear Dog - Finnish breed for big game hunting*. Jalostusnumero, Pystykorva 1 B, 2001, vol.: 44. (1B): p. 72.
6. Syrjänen, A-L. (2000). *Thematic Interview of FSC people:* Audiotapes, lettering files.
7. Cheng, P.W. and K.J. Holyoak (1985). *Pragmatic Reasoning Schemas*. Cognitive Psychology, vol.: (17): p. 391-446.
8. Finnish Kennel Club (1996). *Official Breed Standard. FCI classification group 5. Section 2.4. Nordic hunting dogs. With working trial.KARELIAN BEAR DOG (Karjalankarhukoira)*.
9. Finnish Spitz Club (2001). *Karelian Bear Dog Database (not a public archives)*.
10. Rittle, H.W.J. and M.M. Webber (1973). *Dilemmas in a General Theory of Planning. (rp. 1984 as Planning Problems are Wicked Problems)*. Policy Science (rp. in Developments in Design Methodology, New York: John Wiley & Son), vol.: 4 p. 155-169.
11. Seluska, H. and S. Tiensuu (2000, 2001).*Spitzdog, annuals 1999, 2000 Karelian Bear Dogs*. Gummerrus Kirjapaino Oy, Saarijärvi. p. 253, 302.
12. Biggam, J. (2001). *Defining Knowledge: an Epistemological Foundation for Knowledge Management*, in *Proceedings of the Thirty-Fourth Annual Hawaii International Conference on System Science (HICSS-34), January 3-6, 2001*. Maui, Hawaii (CD-ROM): The Institute of Electrical and Electronics Engineering, Inc.
13. Virkkunen, J. and K. Kuutti (2000). *Understanding organizational learning by focusing on "activity systems"*. Accounting, Management and Information Technology. Elservier Science Ltd., vol.: 2000 (10): p. 291-319.
14. Gooding, D. (1990). *Mapping Experiment as a Learning Process: How the first Electromagnetic Motor Was Invented*. Science, Technology, and Human Value, vol.: 15 p. 165-201.
15. Robillard, P.N. (1999). *The role of knowledge in software development*. Communications of the ACM, vol.: 42 (1): p. 97-92.
16. Brown, J.S. and P. Duguid (1998). *Organizing Knowledge*. California Management Review, vol.: 40, No 3, Spring 1998 p. 90-111.
17. Simon, H. (1979). *Information processing models of cognition*. Ann. Rev. Psych., vol.: 30 (1979): p. 363-396.
18. Bruner, J.S. (1986). *Actual Minds, Possible World*. Cambridge, MA. Harvard University Press.
19. Bruner, J.S. (1990). *Acts of Meaning*. Cambridge, MA. Harvard University Press.
20. Boland, R.J. and R.V. Tenkasi (1995). *Perspective making and perspective taking in communities of knowing*. Organization Science, vol.: 6 (4): p. 350-372.

21. Krogh, G.v., K. Ichijo, and I. Nonaka (2000). *Enabling Knowledge Creation. How to Unlock the Mystery of Tacit Knowledge and Release the Power of Innovation.* Oxford University Press, Inc. p. 292.
22. Norman, D.A. (1993). *Things that Makes Us Smart.* Reading, MA: Addison-Wesley Publishing Co.
23. Blum, B.I. (1996). *Beyond programming. To a New Era of Design.* Oxford University Press 1996. p. 423.
24. Wenger, E. (1999). *Communities of Practice. Learning, Meaning, and Identity.* Learning in doing: Social, Cognitive, and Computational Perspectives. Cambridge University Press. p. 318.
25. Kuutti, K. (2000). *Community Knowledge.* Tutorial material of CSCW' 2000.
26. Virkkunen, J. (1999). *Community Knowledge as an Object of CSCW Research - An Activity Theoretical Interpretation,* in Workshop on Community Knowledge at ECSCW'99, Copenhagen, September 1999.
27. Culff, D. (1991). *Architecture: The Story of Practice.* Cambridge, MA: MIT Press.
28. Brown, J.S. and P. Duguid (1992). *Enacting Design for the Workplace.*, in *Usability: Turning Technologies into Tools,* P.S. Adler and T.A. Winograd, Editors. Oxford University Press. New York. p. 164-197.
29. Leont'ev, A.N. (1978). *Activity, consciousness and personality.* Englewood Cliffs, NJ: Prentice Hall.
30. Lesser, E. L., M. A. Fontaine, et al. (2000). Knowledge and Communities. Butterworth-Heinemann. p. 260.

Critique of Pure Technology

Ho Mun Chan and Barbara Gorayska

City University of Hong Kong
{sachm,csgoray}@cityu.edu.hk

Abstract. Every new technology fundamentally changes social and organizational structures. The danger is that technology, when applied with little thought, will *dictate* the changes and we may not like the results. This paper draws parallels between our critique of the dangers inherent in pure technology ('pure' in the sense of being free from all associations) and Kantian critique of pure reason. Two fundamental questions are posed: What are the limits of technological solutions to human related problems? (analogous to the question 'What are the limits of human thought and reason?') What are the preconditions under which people *can* make sense of the technological world? (analogous to the question 'What are the preconditions in which people can make sense of experience?'). We explore a phenomenon of unthinking application of pure technology with reference to (1) human inability to perceive the thresholds beyond which technological solutions no longer apply to the human related problems they were originally intended to solve and (2) human tendency for a decontextualized understanding of the technologies involved.

1 The Phenomenon of Pure Technology

Our lives have been *organized* around artifacts like the clock or the computer in ways that often make us forget they have been invented as instruments to be *used* by people. For example, the clock is essentially an instrument for measuring time which allows us to synchronize our activities. According to The New Encyclopaedia Britannica (Ready Reference, Micropaedia, 3: 392) its original purpose may have been to alert the sacristan in monasteries to toll a bell that called the monks to prayer. The first computer was originally a mechanical device intended for purely numerical operations. It is hard to believe that the inventors of the clock and the computer could in any way intend and foresee *a priori* that their operations will come to dictate how we live our lives. However, the original purely instrumental roles of the clock and the computer have long since ceased to be the most important ones. In these and numerous other examples, we have gone over the threshold at which the relation between a technology and our lives is reversed. Instead of interacting with the technologies serving us, over a period of time we have gradually adapted ourselves to those technologies, such that our lives are structured by their inherent operations.

In the world dominated by the artifact, our attention tends to be focused on what works, i.e. on techniques and short-term rewards. It is rarely focused on what does not

M. Beynon, C.L. Nehaniv, and K. Dautenhahn (Eds.): CT 2001, LNAI 2117, pp. 463-476, 2001.

work, i.e. on undesirable consequences of applying a technique for the participants engaged by the technique and any long-term costs of that engagement. It comes as no surprise that in technological societies human related problems in areas such as healthcare, education, or the environment, are often regarded as predominantly technological in nature and, consequently, people seek technical means to solve them. Technological solutions generate new human related problems for which new technological solutions are sought. As this process is recursive, spiraling, and accumulative, in the end often more serious human related problems are created than solved.

Technology tends to be applied without limit and whether it has been applied appropriately is often only revealed *a posteriori* through use. Seen from the perspective of the human, there is a danger in such an approach to technological progress, since generating needs for new technologies by an unthinking application of technology can be self-perpetuating. Coupled with a gradual reversal of the roles between tools or instruments and people who use them, an abundance of technologies in human habitats leads many to believe that we ourselves are technological by nature. As Habermas (1987) points out, this often results in maladapted people colonized by technology within their own lifeworld.

The analogy to Kant's critique of pure reason becomes apparent. Kant believed that we can only apply reason appropriately if we know its limits and that we can only know the limits of pure reason when we understand what reason is. By the same token, we will only know how to appropriately apply technology to solving human related problems if we know its limits and we can only know the limits of pure technology when we understand what technology is.

2 Factors That Affect Understanding of the Technological World

Kant's thesis which is of particular interest to investigations in Cognitive Technology is that things in the world affect us and that the mechanism and the objects of perception are co-dependent. Empirical, scientific knowledge and scientific investigation can only lead to knowledge of how things in the world, including ourselves, appear to us and not of what they are in themselves. All that we experience is thus "thoroughly conditioned by features which have their source in the human subject" (Strawson, 1987:406). *Mutatis mutandis*, this thesis applies to how we, as human subjects, understand the technological world. Every technology originates as an idea (a principle or an algorithm) in our situated minds (Gorayska & Mey, 1996). When those principles or algorithms are subsequently embodied as technologies in the material world, they become objects of experience; our understanding of them and of the dynamic, mutual adaptation that results must therefore be limited by the way we are, i.e. by the perceptive capabilities inherent in our own constitution. Below, we identify three important areas of such limitations: cognitive bias in estimating risks, overgeneralizing technologies intended for limited use, and disregard for the pragmatics of interpretation.

2.1 Cognitive Bias in Estimating Risk

Focus on a quick fix that technological solutions are able to provide, along with the human tendency to overlook or overestimate the long-term adverse consequences that a solution may bring about, result from different forms of our cognitive bias. Typical examples include:

a) If chances of undesirable consequences of tool use are small, we *ignore* the fact that they can accumulate in the long run; we continue to use that technology and bring about the undesirable consequence.
b) If the undesirable consequences concern the remote future, we tend to *overlook* their seriousness.
c) Sometimes we tend to *underestimate* the adversity of any consequence that goes beyond our imagination.
d) Sometimes we go to the other extreme to *overestimate* the adversity of any consequence that goes beyond our imagination.
e) Most of the times we are *overly optimistic* that remedial actions can promptly be taken when things go wrong.

Such forms of bias stem from the computational limitation of the brain and the complexity of the tasks involved in estimating long-term consequences of our actions. Technological hazards and colonization of the lifeworld by technology are often identified only *a posteriori* and too late. Real life examples abound. Cigarette smoking, drug addictions, wearing safety belts are well known instances of (a), (b), (c) and (e). A nice example of (d) is that we tend to overestimate, and therefore are often obsessed about, the danger of consuming cholesterol to an extent not always supported by the real risk factor.

An interesting difference of attitudes towards developments in genetic engineering shows up in current, heated arguments for cloning humans and against genetically modified food. This difference can be accounted for with reference to the ease with which we can bring in future, real life contexts for estimating the magnitude of impact ((b),(d)). The impact of cloning humans (Gibbs, 2001) is still perceived by many as relatively contained inasmuch as it is directed at those few who would stand to benefit and therefore of their free choice would wish to participate. Such individuals normally do not contemplate scenarios related to genetic determinism, e.g. what it would be like to bring up a baby that looks like a deceased, beloved husband but grows up to become a different person. Nor are considerations such as these a pressing, global, public concern. For obvious reasons, the impact of genetically modified food is different in this respect. The situation is akin to our current antismoking campaign that only reached its peak and began to enjoy a great deal of success after the negative effects of passive smoking had been brought to the attention of the general public. Pragmatically speaking, we are more likely to ignore the dangers of technology only if, and for as long as, we fail to perceive them as an immediate and direct threat to ourselves.

2.2 Overgeneralizing Technologies Intended for Limited Use

There is a popular myth that technology is almighty. Yet recent developments in complexity analysis show that many problems are so intractable that they cannot be solved by ideal machines in real time. Since these limits are intrinsic, we can only manage but not overcome them by doing ecological analysis (Chan 1996). Instead of aiming at designing almighty artifacts that can help us solve any possible problem in any possible situation (*vide* our attempts at developing general problem solvers in the early days of AI), we can only aim at constructing artifacts that can solve specific problems with which we need to deal in our natural and social environments. However, we often fail to note that such small scale artifacts only work in some specific domains and cannot work properly when they are used in some other contexts. What is worse is that we may allow our lives to be structured by the inherent operations of artifacts even when they fail to work properly. The overuse of microwave ovens in China is a case in point. When these were first introduced to China, people mistakenly began to use them for all kinds of Chinese cooking that they were not suitable for, e.g., steaming fish. Such was the fascination with the new technology that, when the steamed fish turned out dry, it was the inferior taste of fish that people settled for, rather than abandoning the microwave oven.

2.3 Disregard for the Pragmatics of Interpretation

These and other numerous examples (of context dependent attitudes to and beliefs about the merits of modern technology) point towards a parallel between language and technology which, if drawn, would enable us to understand properly the relationship between human life and the technologies we use. This parallel can be viewed in two distinct ways. We can look at language as a technology (Vygotsky, 1962; Gorayska and Mey, 1996). Conversely, we can look at technology as a language. In the former case, we will be highlighting the instrumental nature of language (i.e. a kind of mental tool). In the latter case, explored here, we will be highlighting the pragmatic nature of technology. Below, we discuss several points of contact.

2.3.1 Hermeneutics of Technology

To echo Wittgenstein (1953), the meaning of a symbol is not dictated entirely by its syntax and semantics but largely by the way and the context in which it is used. In the case of interpreting text, neither the text itself nor the author can dictate how it is going to be interpreted. Texts have lives in themselves not because their meanings are fixed but because their meanings keep on transforming across different epochs and different cultures. Texts become alive and unfold their histories via their interaction with human beings. In this process we become transformed, by being 'in-formed', as a result of interpreting and re-interpreting those texts. As we become so transformed, the language we employ in creating texts is itself evolving accordingly (Olson, 1994).

The relationship between technology and the user is akin to the one between the text and the reader. Although an artifact has an internal structure of design and a form perceived at the interface, neither the social and personal significance nor the subsequent development of a piece of technology can be solely derived from these. There is no unique and decontextualized answer to the question "What is this artifact for?" For example, not many people today would understand a crankshaft device commonly used around the ore and coal mines of the 17th Century unless they knew that it used the same principle as the lever system that propels the wheels of a steam locomotive (Haberland, 1996). Nor can the change in social significance or the subsequent development of technology be prevented. The popularity of video games, word processing, E-mail, Internet and so on was not anticipated when the computer was first invented. The meaning of a mobile phone to individuals and the society which they cohabit unexpectedly transformed itself when the phone's use spread to hectic cities like Hong Kong. It no longer means the same there as it did when it was being used in emergencies when travelling in remote areas of, say, Northern Europe.

2.3.2 Tripartite Relation of Pragmatics

The relationship between the speaker, language and the hearer can be shown to correspond to the relationship between the designer, technology and the user.

The mistake of regarding technology as "pure" arises from the misconception that the significance of a technology can be once and for all determined at the outset. This misconception also presupposes that the designer is the principal interpreter and source of the meaning of a piece of technology and that he or she, in spite of being a third party to subsequent interaction, *can* nevertheless deduce, *a priori*, the universality of its significance to all of us at all times. If this were indeed the case, users would be adapting themselves not only to the technology but also to the will of the designer. When this happens, as it often does, the inequality of power between those who design and produce and those who use has important social consequences. Since in such cases the use of a given technology is imposed on us by others, those who cannot adapt carry a social stigma and are seen as lagging behind. The development process of making technology serve human ends in a more natural way is hindered, and so is the creative ascription of meanings to the technology by the user.

Since human thinking is often biased, there is no strong reason to believe that the designer is intrinsically in a better position to foresee the long-term development of a particular technology and the potential hazards that it may bring about. On the contrary, many examples have shown that designers often fail to see either. Of all participants in technological progress, they seem to be least aware of their own human limits, in particular those limits that relate to foresight and sound, unbiased consequential thinking. Ian Wilmut – the scientist who cloned Dolly but has come out publicly against cloning humans – only tried to help farmers produce better sheep. His intention was not to help sheep to have genetically identical children (Eric Parens in Gibbs, op. cit.: 40).

Those who can and do foresee the long-term adverse consequences of the technologies they have designed may not always want to lay bare to the public all important details of the design due to a quest for power and the motive of protecting one's self interest. In the world dominated by the ideology of pure technology where innovation has to sell, market forces operate that make it difficult, if not impossible, for the consumer to overcome the cognitive bias, develop a more sound sense of ethics, and employ much more accurate strategies for consequential judgement. Although in principle we are all able to shape the development of technology by exercising choice, by being skillfully manipulated by marketing media (see also Clark, 1997) we are caught up in a perpetual myth that *more* (of the same) always means *better* (for ourselves and the society concerned). Such a market-generated failure is largely a result of our cognitive failure that leads to the uncritical acceptance of the ideology of pure technology and to our lives being structured by a reified technology.

3 The Pragmatics of Technology

3.1 Pragmatics of Scientific Explanation

The significance of contexts in furthering understanding permeates the pragmatics of (scientific) explanation. According to Van Fraassen (1988) an explanation gives an answer to a question about why this or that happens (which he calls the why-question). Such a question is always asked in the context where a) there is more than one possible description of what may have happened (which he calls a contrast-class) available to those who interpret an event and b) reasons which refer to the perceived causal structures in physical reality are provided that support each of those descriptions. An answer to the why-question enables people to select the right description of what happened because it gives a plausible reason for why what happens is relevant to the phenomenon in question. For example, when we consider the question "Why did Dr Warren help Mrs Smith to die?" outside the context in which it has been raised, this question is unclear with respect to its communicative intention. We may want to know why it is Dr Warren but not anyone or anything else that helped Mrs Smith, or why Dr Warren helped Mrs Smith to die but did not do something else, such as saving her, or why it is Mrs Smith but not anyone else who was helped to die. Yet, if we know the context in which the why-question has been raised, it is often also possible to know which information is being sought. For example, if one is interested to learn why it is Dr Warren but not anyone else who helped Mrs Smith to die, the contrast-class may consist of

> Dr Warren helped Mrs Smith to die, Dr Rise helped Mrs Smith to die, nurse Sally helped Mrs Smith, ... and so on

The statement that explains the act of helping Mrs Smith to die, say "Dr Warren is a person who strongly believes in euthanasia (but others do not)", is an answer to the why-question. It gives us the reason and enables us to select him as the person who helped Mrs Smith to die as the correct description of what happened.

According to Van Fraassen, the content of a why-question (Q) in a certain context is determined by three factors: <P, X, R>, where P is the topic of the question, X is the contrast-class of which P is a member, and R is the relation between an answer to the why-question A and (P, X). A is relevant to Q if A stands in a relation R to P and X such that P but not other members of X are true.

Concerned with the question of explanation, Van Fraassen recognizes the need for contexts and reasons for actions. But even though his reasoning is done in the framework of context, he does not explicitly address how we establish relevance from A to Q. To put the pragmatics of (scientific) explanation, and consequently the pragmatics of technology, on a more Kantian footing, however, we need to address this question, and we need to anchor decisions about relevance in our constitution as the perceiving, human subjects.

When we consider the above example in greater detail, it becomes apparent that establishing relevance is a recursive mental process. On the one hand, we have those who seek to explain what has happened. To them, knowing that Dr Warren strongly believes in euthanasia is relevant to the purpose at hand as it provides a motive which *to them* is the reason for Dr Warren's action. On the other hand, we have Dr Warren choosing to perform his act of mercy because, given his set of beliefs about euthanasia, helping Mrs Smith to die satisfies his motive. The act is therefore also relevant to *him* and fully justified.

In both cases, there is a specific intent entertained by some individual(s) participating in the event (this intent is often referred to as the goal that needs to be satisfied). Note that the goal of people seeking an explanation is different from Dr Warren's goal. This goal dependence in choosing what is relevant to what and for whom is reflected in the semantics of the natural language term 'relevant' which is taken to refer not only to the causal structures of physical reality but also to mental mechanisms for storing and processing goal knowledge (Gorayska & Lindsay, 1993, Lindsay & Gorayska, 1994).

In both cases, too, there is a set of beliefs about the world that the individuals concerned bring to bear at the time the relevance to the goal is being decided. This set contains the belief that euthanasia is desirable and that it is Dr Warren who holds this particular belief. This set of beliefs determines the relevance relation. Contrary to what Van Fraassen's account may imply, such contexts for interpretation are not givens (Sperber and Wilson, 1986) but comprise that which, at the time when decisions are made, *people* can readily perceive as true in the world around them. Therefore, the relevance they act upon is always subjective.

Establishing (subjective) relevance to action involves the conscious recognition of five factors: <G, E, P, A, M>, where E is an element of some action sequence P (e.g., whatever Dr Warren did to help Mrs Smith to die). G is (are) the goal(s) from the characteristics of which action sequences are inferred (e.g., an intent to help Mrs Smith to die). A stands for a person or some agent capable of implementing an action sequence P (e.g., Dr Warren), while M is a world model in which a goal G is found, in which an action sequence P is constructed, and in which A operates (e.g. a set of beliefs about desirability of euthanasia). The relevance equilibrium is maintained at the value T(rue) for any individual A operating in a world model M, entertaining (a) goal(s) G, and constructing or implementing action sequences P if and only if an instance of E is essential to P and P is sufficient to achieve G in M.

In terms of mental processing, this relational description captures the constraints on operations of internal cognitive mechanisms that integrate goals, actions, effective action sequences, agents, and world models into relevance schemata. Although the constraints remain fixed, the variable values are in constant flux. Modifications of any existing value result in a cognitive dissonance and lead to a necessary adjustment of the remaining values if the internal relevance equilibrium is to be maintained at value R=T. Difficulties arises due to the complexity of the branching factor. In our example, the relevance of Dr Warren's action to his conviction that euthanasia is desirable needs to be recognized before this conviction can be accepted as a relevant explanation. In reality, however, we have to deal not just with two but with multitudes of such endogenous or socially and culturally enjoined goals, multitudes of possible depicted models of the world, multitudes of possible effective actions and multitudes of interacting human agents capable of bringing about personal change and new social order. In support of the cognitive processes of self-reflection, integration, analysis and discrimination, therefore, the meta-processes that establish relevance need to be *cognitively fluid* across different domains and different problem spaces (term due to Mithen, 1996). Errors (such as cognitive bias or overgeneralizing, as discussed earlier) arise from faulty beliefs, incomplete knowledge of the world, or discrepancies of the processing power among individual cognitive mechanisms. As will be shown in Sect. 3.2, technology often plays a vital part in these types of (mal)adjustment.

The pragmatic factors that Van Fraassen identifies, as well as the importance of relevance in human action, apply also to the pragmatics of technological growth. When a technology is invented, it often has a single purpose and application. When its design is adopted, it brings to the foreground problems not previously anticipated or perceived. Up to a threshold point at which the technology turns into an unwarranted constraint on freedom, people tend to rethink the original purpose of its design. For a variety of reasons, new relevance is established and the technology may be modified to bring about further success. New problems are uncovered and further modifications happen. Hence the development of technology is a dynamic process. A parallel to the pragmatics of (scientific) explanation is given in Table 1.

Table 1. A correspondence between the pragmatics of scientific explanation and the pragmatics of technology

Scientific Explanation	**Technological Development**
Topic (purpose of explanation)	Original purpose of application
Contrast-class of descriptions	Contrast-class of possible applications
Reasons	Reasons
Relevance	Relevance

3.2 An Example: To Live or to Die, That's the Question

Some believe, rightly, that, if a technology exists, it will be used. At the same time it is also true that modern technology has advanced to the point where it forces us to make choices. The question of euthanasia may serve as an example (Chan 1991, 2001).

The Shorter Oxford English Dictionary defines *euthanasia* as either "a quiet and easy death" or "the means of procuring this", i.e., "the action of inducing a quiet and easy death" (C. T. Onions, ed., 1983, 1: 689). From the vantage point of the patient, euthanasia may be classified as voluntary, non-voluntary or involuntary. Voluntary euthanasia means that euthanasia is carried out with the patient's consent or at his/her request. Non-voluntary euthanasia takes place when the patient is without the capacity for self-determination. Involuntary euthanasia takes place when the patient objects to it. Euthanasia could be further classified as active or passive. Active euthanasia is one effected by taking measures (e.g., injecting a lethal substance) to cause the death, while passive euthanasia is effected by interrupting life sustaining measures or treatment.

Many terminal patients are severely disabled by their illness. Many fall into an indefinite coma and are deprived of the capacity of self-determination. Confronting such situations is a vast moral dilemma for the medical profession, for the patient (if he or she still possesses sufficient capacity for determination), and for the patient's family who have to decide whether or not the patient's life should be terminated.

Every member of the medical profession has sworn on the Hippocratic Oath to honor and prolong human life by administering only beneficial treatments, according to individual judgements and abilities, and to refrain from causing harm or hurt. Euthanasia is therefore perceived by many as incompatible with the mission of the medical profession which is entrusted to cure the sick, to help the sick in the course of their recovery and to eliminate or reduce their suffering. The code of medical ethics envisages the medical profession as being under no obligation to make value judgements on the life of the patient. Even if the medical professional is of the view that the patient's life is without value, he/she should never terminate the patient's life, nor should the medical professional procure and assist the patient to terminate his life (American Medical Association, Code of Medical Ethics, 1994; Callahan, 1992; Pellegrino, 1994).

This code of practice is currently a topic of heated debate. The advancement of medical science has intervened in the dying patient–doctor relationship, creating a discord in the relevance equilibrium. Significant adjustments now have to be made to the goals medical professionals set for the actions they consider effective in meeting those goals, and for the instruments they choose to interface with in support of their actions.

Death is the state when the life has gone. No living person can experience death, as death comes only when life evaporates. As Wittgenstein rightly said, "death is not an event in life" (Wittgenstein, 1961:72). People are not afraid of death itself, but of the dying process, especially the suffering accompanying it, and of dying without dignity. The dying process is inevitable. Therefore, it is important that death should be commensurate with the way we would wish to have lived (Dworkin, 1993). Technology does not always make dying more dignified.

Modern medical science has revealed many intermediate stages between life and death, which are now connected with the brain functions instead of with those of the heart. Such advancements have deepened our understanding of life and death with respect to the biological and physiological phenomena, which are now monitored by instruments of an unprecedented precision. This has released the medical profession from the need to be sensitive to the external bodily and behavioral symptoms and has necessitated the development of new relevance schemata, i.e., effective action sequences that interface not with the patient him/herself, but with the available medical equipment.

Unfortunately, this advance has not improved our understanding of life and death as a psychological or spiritual process. The undue reliance on a technology that monitors or sustains physiological states of the patient has increased the objectification of medical science and care and has further obscured the threshold between living and dying. Medical professionals, informed by their instruments, are now better capable of saving a patient from the brink of death (Scaneiderman & Jerker, 1995). However, by focusing their attention on sustaining biologically defined life, which they see as their mission or goal, on many occasions they fail to acknowledge, even to themselves, that in so doing they are only prolonging the dying process and not saving the patient in any sense.

At times when medical science and the resulting technology were less advanced, our cognitive limitations made it impossible to imagine that intermediate stages between life and death would emerge as a result of medical technology. Nor could we imagine the longer-term, adverse consequences of applying resuscitation or life supporting equipment. The inventions were welcome for the following reasons:

- The span of human life was in general shorter. (This had prevented doctors from recognizing certain problems, which now could be properly diagnosed as a consequence.)
- Saving life meant return to good health.
- The ability to prevent sudden death caused by acute illness, accidents and emergencies allowed medical professionals more room to cure the patient.
- The doctors' duty to fight death to the very end at all cost was fully justified, hence ethical in every respect.
- The curative model of medicine dominated the mind of medical professionals. (The objectification inherent in technology led to patients not being regarded as subjects to be cared for, but rather as mechanisms to be repaired.)
- In addition, patients were far less knowledgeable about medical issues. Medical paternalism was the dominant practice and the professionals had the greatest say in making medical decisions.

In the above world model, believed to be relevant to medicine, medical professionals became habituated to the use of technological equipment in all situations where it could be applied; also, in most cases the application was beneficial to the patient.

However, the excessive reliance on medical technology that only treats biological symptoms and diverts attention from the sociological or psychological factors causing them, impairs what we above have called 'cognitive fluidity' and causes a gradual transformation of our lives. Life expectancy has increased to the point where the terminal stage is often of very low quality, leading to many personal and social problems. Yet, a vast majority of medical practitioners, not fully alerted to the changes, still continue the old practice. Medical paternalism and objectification of medical science and care often bring physical and psychological suffering to those under treatment and their close relatives. There is little understanding that dying is a process that is shared (Chan, 1999). The subjective feelings of the patient and the family are often ignored. In numerous cases the application of sophisticated equipment has gone beyond the threshold of benefit and, consequently, the limits of the original intent (i.e., the goal of medical treatment as saving and prolonging life at all cost) has to be revised. New needs (hence new goals) have arisen: the need to understand emotion and psychological states at the time of death, the need for effective techniques to control pain, and the need for a range of hospice technologies that bring comfort to the dying. At the same time, the unthinking use of the current technology has become a major factor that has triggered the movement to legalize euthanasia.

It is not our intention in this paper to make a judgement on whether or not euthanasia should be legalized. In many situations, euthanasia would not be needed if we could restrict the excessive use of resuscitation and life supporting equipment. For this to happen, we need to adjust our world models in which relevance is sought. The following views need to be widely accepted:

1. Failure to provide resuscitation and refusing life supporting equipment to prolong the dying process do not constitute an act of killing the patient.
2. Patients have the right to refuse treatment. Respecting such a refusal by a competent patient does not mean killing the patient. It means respecting their self-determination.
3. The curative model of medicine needs to be gradually replaced by the care-based model of medicine.
4. Adequate palliative and hospice services are necessary to manage pain and provide terminal care for the dying patient.

This revised set of beliefs implies for new goals to be generated; A new effective medical practice will follow. Foregoing life-sustaining treatments will no longer be regarded as practicing euthanasia. Current goal conflicts may be resolved, as medical professionals will no longer be obliged by their Hippocratic Oath to provide such treatment. In some cases the Oath will oblige them to do just the opposite. As a result, resuscitation and life support will be confined to the productive use of prolonging a life of quality, so as to make room for those professionals not only to cure but also to provide care for the patient.

4 Concluding Remarks

We invent technology to make complex things simple. Medical technology supported by medical science provides us with increasingly more precise information about human biology and physiology. It is exactly our need for this type of precision—precision that would eliminate uncertainty and thus relieve us from having to make blind guesses and moral judgements—which appears to drive the quest for greater technological advance and technological support (see also Gorayska, Marsh & Mey, this volume). Paradoxically, this advance and precision often only add to the complexity of relevant side-effect factors that we have to deal with, and to moral dilemmas that such advances were meant to eliminate. For example, a more precise definition of the biological life and death and the availability of life-sustaining technologies have broadened our awareness of the extension of 'euthanasia' yet, at the same time, we have introduced a greater variety of ways that we must now choose among in order to cure instead of help to die. The same considerations apply to many other inventions in the modern, technological world.

Since we have already socially adjusted to accept the dominance of technologies over the way we live our lives, for us to regain autonomy in the use of tools, education and social change need to induce proper and well-informed behavioral change. What needs to happen is an increased awareness of the dangers inherent in the unthinking use of pure technology. This can be achieved through an open dialogue—an ongoing and collective deliberation of how our environment and our lives should be shaped by technical means and what kinds of risks we as the public can collectively bear. Such collective deliberations—which ultimately lead to collective responsibility in the face of adversity—require access of a wider public to details of the design and the already known evidence of potential short-term and long-term risks in adopting a given technology.

Again, the parallel with Kantian philosophy and the critique of pure technology becomes clear. Kant believed that pure reason had its own limits and that there were fundamental questions which could only be addressed by our practical reason and aesthetic judgement. The parallel in the case of pure technology is that we also need to acknowledge the limits of technology and admit the fact that human related problems in a technology-oriented society have to be dealt with not only at the level of cognitive, rational, technology, but also at the cognitive, emotive, psychological and spiritual levels. Most importantly, they will have to be dealt with at the ethical, social and political levels as well.

References

American Medical Association (1994). *Code of Medical Ethics*. Chicago: American Medical Association.

Callahan, D. (1992). "When Self-Determination Runs Amok", in Tom L. Beauchamp & LeRoy Walters (eds.) *Contemporary Issues in Bioethics*. Belmont, California: Wadsworth Publishing Company, 1994.

Chan, H. M. (1996). "Levels of Explanation: Complexity and Ecology". In Barbara Gorayska and Jacob L. Mey (eds.), *Cognitive Technology: In Search of a Humane Interface*, Amsterdam: Elsevier Science.

Chan, H. M. (1999). "Sharing Death and Dying", paper presented at the International Conference on Applied Ethics, organized by New Asia College, Faculty of Medicine and Department of Philosophy, Chinese University of Hong Kong, Dec 1999.

Chan, H. M. (2001). "Euthanasia or Terminal Care?", in R Z Qiu (ed.) *Bioethics: Asian Perspectives (Philosophy and Medicine Series),* Dordrecht, Netherlands: Kluwer Academic Publishers, 2001.

Clark, A. (1997). *Being There: Putting Brain, Body, and World Together Again.* Cambridige, MA: MIT Press.

Dworkin, R. (1993). *Life's Dominion: An Argument about Abortion, Euthanasia, and Individual Freedom.* New York: Alfred A. Knopf.

Gibbs, N. (2001). Human Cloning: Indecent Descent? *Time* 157(8), 36-41.

Gorayska, B. & R. Lindsay (1993). The Roots of Relevance. *Journal of Pragmatics* 19(4), 301-323.

Gorayska, B. & J. L. Mey (1996). "Of Minds and Men". In B. Gorayska and J. L. Mey (eds.), *Cognitive Technology: In Search of a Humane Interface*, 1-24. Amsterdam: Elsevier Science.

Haberland, H. (1996). "And Ye Shall Be As Machines"- Or Should Machines Be As Us? On the Modeling of Matter and Mind. In B. Gorayska and J. L. Mey, eds., *Cognitive Technology: In Search of a Humane Interface*, 89-98. Amsterdam: North Holland.

Habermas, J. (1987). *The Theory of Communicative Action*, (translated by Thomas McCarthy). London : Heinemann.

Kant, I. (1781). *Critique of Pure Reason*, (translated by Norman Kemp Smith), London: MacMillian, 1933.

Lindsay, R. & B. Gorayska (1994). *Towards a General Theory of Cognition.* Unpublished Manuscript.

Mithen, S. (1996). *The Prehistory of the Mind: A search for the origins of art, religion and science.* London: Orion Books Ltd.

Olson, D. R. (1994). *The World on Paper: The conceptual and cognitive implications of writing and reading*. Cambridge: Cambridge University Press.

Pellegrino, E. D. (1994). "Euthanasia as a Distortion of the Healing Relationship". In Tom L. Beauchamp & LeRoy Walters (eds.) *Contemporary Issues in Bioethics.* Belmont, California: Wadsworth Publishing Company, 1994.

Schneiderman, L. J. & N. Jecker, (1995). *Wrong Medicine.* Baltimore, London: The John Hopkins University Press.

Sperber, D. & D. Wilson, (1986). *Relevance.* Oxford: Blackwell.

Strawson, P. (1987). Entry on Kant. In Gregory, R.L., ed., *The Oxford Companion to the Mind*, 406-8. Oxford: OUP.

Van Fraassen, B. C. (1980). *The Scientific Image*. Oxford: Clarendon Press.

Vygotsky, L.S. (1962). *Thought and Language.* Translated by Gertrude Vakar. Cambridge, MA: MIT Press.

Wittgenstein, L. (1953). *Philosophical Investigations.* Oxford: Blackwell.

Wittgenstein, L. (1961). *Tractatus Logico- Philosophicus.* London: Routledge & Kegan Paul.

The Computer as Instrument

Meurig Beynon, Yih-Chang Ch'en, Hsing-Wen Hseu, Soha Maad, Suwanna Rasmequan, Chris Roe, Jaratsri Rungrattanaubol, Steve Russ, Ashley Ward, and Allan Wong

The Empirical Modelling Research Group, Department of Computer Science, University of Warwick, Coventry CV4 7AL, U.K.
http://www.dcs.warwick.ac.uk/modelling/

Abstract. A distinction is drawn and discussed between two modes of computer use: as a tool and as an instrument. The former is typical for the use of a conventional software product, the latter is more appropriate in volatile environments or where close integration of human and computer processes is desirable. An approach to modelling developed at Warwick and based upon the concepts of observable, dependency and agency has led to the construction of open-ended computer-based artefacts called 'interactive situation models' (ISMs).

The experience of constructing these ISMs, and the principles they embody, exemplify very closely the characterisation of instruments as 'maintaining a relationship between aspects of state'. The framework for modelling that we propose and report on here seems well-suited to account for the dual 'tool-instrument' use of computers. It is also sufficiently broad and fundamental to begin the deconstruction of human-computer interaction that is called for in any attempt to understand the implications of computer-based technology for human cognitive processes.

1 Introduction

Current frameworks for developing technological products reflect a limited conception of their role. In designing such a product, the emphasis is placed on what can be preconceived about its use, as expressed in its functional specification, its optimisation to meet specific functional needs, and the evaluation of its performance by predetermined metrics. This perspective on design is not sufficient to address the agenda of cognitive technology [13]; it takes too little account of the interaction between a technology, its users and its environment. For instance, it is well-recognised that developments in technology can be the result of uses of a product outside the scope of those envisaged by its designers.

Such considerations apply in particular to computer-based technologies. Standard software development methodologies begin by identifying the precise roles that the computer has to play (e.g. through the study of use cases [11]), and focus on designing programs to fulfil these roles as efficiently as possible. Because each use of the computer is tightly constrained by specifying such roles, the trend in designing business processes is to prescribe the interaction between human and computer agents exactly, and optimise their operation accordingly.

M. Beynon, C.L. Nehaniv, and K. Dautenhahn (Eds.): CT 2001, LNAI 2117, pp. 476–489, 2001.

In this respect, traditional software development favours the conception of the computer as a tool, developed specifically to serve a particular purpose.

In practice, business environments and technologies are volatile, and are liable to evolve in ways that subvert the intended preconceived processes. A major concern in modern software engineering is the need to develop software in such a way that it can be readily adapted to changes in its environment, and to the reengineering of business processes. A conception that is better suited to computer use, both in this context and with reference to the agenda of cognitive technology, is that of the computer as instrument.

Our paper will be in three main sections: the first elaborating on the distinction between the tool and instrument perspectives, and the issues concerning human interaction with artefacts it raises; the second outlining principles and tools for computer-based modelling that we have developed in order to address these issues; the third discussing some relevant case studies.

2 Instruments and Tools

The purpose of this section is to highlight key features of tool and instrument use that motivate the principles for computer-based modelling to be introduced and illustrated in Sects. 3 and 4.

2.1 What Is an Instrument?

The term 'instrument' is here being used to refer to a piece of technology that maintains a relationship between two aspects of state. This broad definition is intended to encompass scientific instruments – such as an ammeter, prosthetic devices (such as a pair of spectacles), and musical instruments. An ammeter maintains the position of a needle according to the current flowing in a circuit, a pair of spectacles maintains a relationship between an external scene and the image on the wearer's retina, and a musical instrument maintains a relationship between the emotional state of the performer and an aural effect. The informality of the references made to 'state' and 'maintaining relationships between state' in this characterisation is acknowledged; later sections of the paper will supply more context for their interpretation.

All three examples of instruments mentioned above have a characteristic feature in common: they establish a correspondence between states that is conceptually direct and immediate. A change in current moves the needle. A change in the external scene changes the image on the retina. A change in the performer's emotional state effects a change in the sound emitted by an instrument. A significant distinction between the three examples is the different roles that human agency plays in each case. No human intervention is needed to maintain the position of the needle on the ammeter. A pair of spectacles serves its function through cooperation between human and technology where the human element is typically unconscious. The most effective performance of the musical instrument demands great intensity of awareness and responsiveness in exercising human skills.

Our primary concern is with *interactive* instruments, where the role of the human in maintaining state resembles that of the performer of a musical instrument. Within the exceptionally broad framework of study to be invoked in this paper, other instances of instruments can be interpreted as derived from this most general case, in the sense that – for instance – the ammeter is the product of a sophisticated empirical process arising from human interactions with the world that involved an awareness and responsiveness of comparable subtlety. In what follows, the term 'instrument' will be used to refer to an interactive instrument.

The characterisation of an artefact as a tool or instrument is not to be interpreted as an either-or classification. The surgeon's scalpel can be (at one and the same time) both a tool to perform a function, and the subject of a performance quite as engaging and open to environmental influences as any musician's. The terms 'tool' and 'instrument' are to be regarded as interpretations put upon the use of an artefact. The dictionary definition of an instrument as 'a tool for delicate work' [12] suggests a similar association between the concept of an instrument and a particular quality of attention required for its use. Potentially the computer can serve as both tool and instrument, and both perspectives may be appropriate at one and the same time. The principal issue to be examined in this paper is: how can we complement our formal view of computation, which favours the computer as tool, to address the potential of the computer as an instrument?

2.2 Characteristics of Instruments

The distinction between an instrument and a tool is associated with particular characteristics of use. In practice, the emphasis when using instruments is on exercising personal skills, whilst the use of tools is typically associated with performing a specific function in an organised framework for interaction in which other human agents or observers are involved. Instruments and tools are respectively correlated in this fashion with subjective and objective interactions. For instance, where the pianist is engaged in a highly personal way with their performance, and judges its success in subjective terms, the mechanic wielding a spanner is generally taking a specific action following a well-defined procedure to attain a particular goal that can be objectively validated. The relationship between instruments and tools identified in this paper accounts for this subjective versus objective emphasis in terms of closely related, but more primitive, aspects of interaction with artefacts.

Both tool and instrument use are particular cases of interaction with artefacts. The very concept of identifying an artefact as a tool or as an instrument involves establishing some characteristic mode of interaction with it. The use of a hammer is appropriate to a context where the characteristic action is hitting a target object with the head of the hammer. A piano is normally used by striking the keys with the fingers. In practice, the potential interactions with an instrument are more open-ended in nature, but they are focused around a range of specific skills that can be evaluated by experienced exponents. In the case of the piano, examples of such skills might include the ability to play scales and arpeggios, to harmonise a melody, or to play pieces within a particular genre.

The standard activities associated with tools and instruments in this way – though very diverse in character – have this in common: they are all to some degree examples of ritualisable experience that can be reproduced by a suitably skilled agent. Recognising such ritualisable experience is not necessarily an objective matter – it is enough that the personal experience of the executant acquires a degree of consistency, and reflects authentic knowledge of their own capabilities, the qualities of the artefact and the essential context. It is in this spirit that – whatever the independent judgement of an experienced musician – the amateur pianist speaks of 'playing the Moonlight Sonata' and of 'not being able to play it with the cat on my lap'.

Both tools and instruments are rooted in the use of artefacts associated with activities that are sufficiently familiar, well-rehearsed and practised that they can be repeated and so can reliably carry us to specific goals; moreover, these activities may be sufficiently rich as to be valued in themselves, for the experience they offer in execution, and the promise of unexpected novel interest and delight. The distinction between tool and instrument perspectives is then a matter of emphasis. In tool-like use of an artefact, we are concerned with efficient and reliable progress towards specific goals (possibly sacrificing any concern for satisfying engagement in the activity). In instrument-like use of an artefact, we give greater priority to appreciation of the experience than to achievement of the goal. A balance of both perspectives is often appropriate, as – when playing chess – we want to win, but also want to explore interesting and novel scenarios, or – when playing music – we aim to play accurately, but aspire to emotional intensity.

The most significant characteristics of the use of an instrument rather than a tool can be illustrated with reference to musical performance. The performer experiences interaction with the instrument as a continuous engagement, where feedback from the instrument and the environment is involved. The outcome of the engagement between performer and technology is more than the accomplishment of a preconceived function. The performance will differ according to situation, and be open to influences (such as the acoustics of the hall, the response of the audience, the precise characteristics of the instrument, the mood of the performer) that are shaped through negotiation and evolve dynamically. The unpredictable manner in which these factors are reflected in the physical and mental state of the performer contrasts with the stereotyped and goal-oriented view of state that is expressed in the familiar proverb "for a man who has only a hammer, the whole world looks like a nail". There is also the possibility that a performance ventures beyond preconceived limits – there is scope for spontaneous action, experiment and improvisation.

3 Computer-Based Modelling for Instruments

This section discusses the extent to which Empirical Modelling (EM), an approach to modelling under development at the University of Warwick [20], provides a conceptual framework for studying the use of instruments and practical support for their construction using the computer. The essential concept behind EM is the analysis of experience in terms of agency, dependency and observation and its representation through the construction of computer-based 'interactive situation models' (ISMs) [14]. A number of special-purpose software tools have been developed to support the construction of ISMs, and a large number of such models created through student projects over the last 10–12 years. Experience gained from this modelling activity indicates strong points of connection between interaction with ISMs and interaction with instruments, as characterised above. In particular, the construction of an ISM is a situated activity that can develop in an open-ended fashion in response to the modeller's evolving focus of interest, and involves exploration and experiment.

3.1 Principles of ISM Development

The principles of ISM development will be illustrated using a simple exercise in modelling a traditional clock (see Fig. 1).

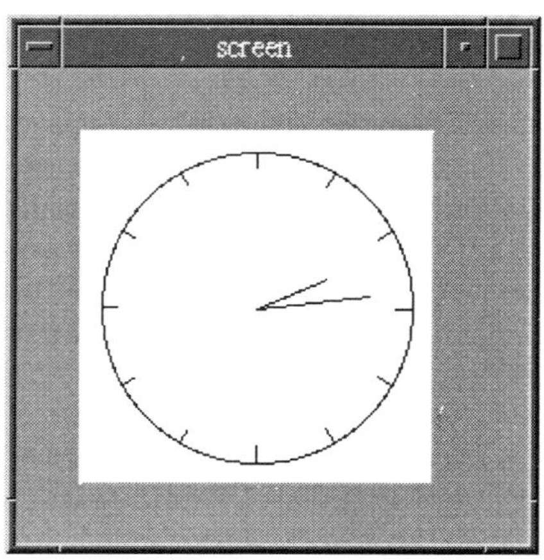

Fig. 1. A simple clock model

This illustration is quite unrepresentative of the scale of ISMs that have been built using EM tools, whose scripts may include several few thousand definitions, but it does indicate the nature of the incremental construction that is involved in creating and using such ISMs. The definitions in the script for this model include the following:

```
openshape clock
within clock {
  real sixthpi
  line eleven, ten, nine, eight, seven, six, five, ..., one
  line noon
  point centre
  real radius
  circle edge
  sixthpi = 0.523599
  radius = 150.0
  eleven = rot(noon, centre, -11 * sixthpi)
  ...
}
```

The variables in this script represent observables in the clock: the rim of the face, represented by the circle `clock/edge`, its centre `clock/centre` and the divisions `eleven`, `ten`, `nine` ... that indicate the hours. A complementary set of definitions represent the dependencies that link the positions of the hour and minute hands to the current time (represented by the variable `clock/t`).

```
within clock {
  line minHand, hourHand
  real minAngle, hourAngle
  real size_minHand, size_hourHand
  int t
  size_minHand, size_hourHand = 0.75, 0.5
  minAngle = (pi div 2.0) - float (t mod 60) * (pi div 30.0)
  hourAngle = (pi div 2.0) - float (t mod 720) * (pi div 360.0)
  minHand = [centre + {size_minHand*radius @ minAngle}, centre]
  hourHand = [centre + {size_hourHand*radius @ hourAngle}, centre]
  centre = {200, 200}
  ...
}
```

Notice how these are specified in such a way that both the position of the minute hand and the hour hand depend on the time via independent definitions. An alternative way to express this dependency that might more aptly describe the physical relationship between the hands of a mechanical clock would express the position of the minute hand as linked to the position of an internal mechanism, and derive the position of the hour hand by a definition representing the chain of cogs that might connect the hour hand to the minute hand.

```
within clock {
  minAngle = (pi div 2.0) - float (t mod 720) * (pi div 30.0)
  hourAngle = (pi div 2.0)  - ((pi div 2.0) - minAngle) div 12.0
  ...
}
```

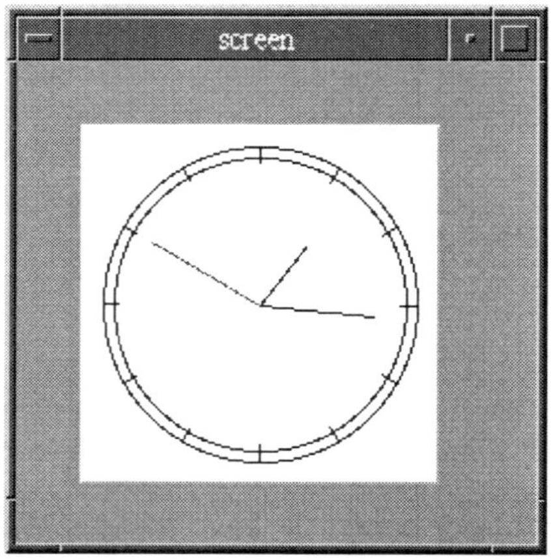

Fig. 2. Clock with details added

Whilst the current time `clock/t` is unspecified, the hands are omitted from the clock face. In specifying this time, the modeller can adopt many different viewpoints. For instance, they may act as if in the role of:

- a user, setting the clock to the current time;
- a designer, seeking to place the hands in a significant configuration;
- the clockmaker who connects the clock mechanism.

When defining the clock mechanism, a simple agent can be introduced to update the clock according to the real time. This is programmed to 'observe' the time on the computer system clock, and to increment the variable `clock/t` every minute.

There are many other instances of potential redefinitions that represent plausible actions on the part of different agents. These effect only very simple changes to the generated display, but nevertheless can correspond to rich thought processes and changes of perspective on the part of the modeller. In the role of a user, the modeller will consider such issues as starting and stopping the clock, or setting the time to reflect a new time zone. In the role of designer, the modeller may consider the appearance of the clock face, the possibility of changing the colour of the hands or adding a second hand (see Fig. 2). The modeller can also act in a role that is outside the scope of either the designer or the user, as when reconfiguring the display to a convenient size for demonstration, or adding physically unrealistic features to the clock. Other possibilities include simulating an exceptional event, such as occurs when the minute hand comes loose and hangs vertically. These modifications highlight two fundamental ideas behind EM:

- the construction and structure of scripts mirrors the way in which the modeller construes state-change to occur;
- the modeller's perspective on the script is subject to change from moment to moment, and involves internal human activity (relating to thought pro-

cesses, situation and agency) that is much richer and more complex than the external computer-based change.

In these respects, constructing an ISM differs from the mathematical approach to creating a model using a computer, where the normal practice is to decide the precise functionality of the model in advance, and to implement from a functional specification. Modelling activity in EM is closer in spirit to creative work in the arts, such as making a sculpture or composing a piece of music. The interaction between the artist's state of mind and the work they are creating is dynamic, and the meaning of the work of art is shaped as it is being developed, as in bricolage [9].

3.2 ISMs as Instruments

There are many ways in which experience of constructing ISMs can illuminate – and has informed – the characterisation of tools and instruments introduced in Sect. 2. To simplify the discussion, and to avoid technical detail, an ISM will be viewed at a rather high level of abstraction as comprising a definitive script that defines a conceptual state, a display interface made up of one or more screens that embodies some part of this state, together with a collection of agents, each with certain privileges to amend a definition in the script or add a new definition, subject to context and cue. These agents will in general include a variety of human interpreters, who might be in the role of users of the ISM or be one amongst several in a distributed team of modellers. The act of making a redefinition in the script may itself be embodied in an external interaction, such as the movement or an action of the mouse, through a control interface. Where the ISM is not distributed, so that all the state is localised in a single artefact, there is a conceptual role for a locally omnipotent interpreter of the ISM, who is privileged to modify the definitive script directly in whatever fashion they please.

One of the practical aspirations for Empirical Modelling is to develop software tools and/or a more general computer-based technology that can support this 'idealised' vision of an ISM and more. The idealisation reflects the illustrative models that we have constructed in practice, making allowance for the limitations of our current tools. It would clearly be appropriate to extend the concept of embodiment in respect of display and control to take account of more advanced technologies than a typical workstation supplies. For the purposes of this paper, such an extension is not essential, though it is relevant to the issue of using ISMs to construct tools and instruments of the degree of sophistication we are accustomed to see around us.

The characterisation of an instrument as 'maintaining a relationship between aspects of state' is vividly represented in working with ISMs. The concept of shaping the state-as-experienced of an ISM to correspond to that of an external referent is prominent in EM, and in itself characterises an ISM as an interactive instrument. Within an ISM, there are dependencies that maintain the relationship between different subscripts, such as the definitions that link the internal value of the time to the position of the hands, or that determine whether the alarm is ringing with reference to the current time, the alarm time and whether the alarm is set. The agency that is introduced into the clock linking the display to the current time illustrates another mechanism for maintaining relationships between aspects of state.

Analysing what is conceptually involved in the ISM as an instrument reveals the fundamental abstraction to be dependency between states in the physical world. Each such primitive dependency is associated with an experimental observation about how a change to one observable indivisibly effects changes to others. The ISM builds layer upon layer, each based on activities of an instrumental character: the implementation of the dependency maintainer in our interpreter, the compiler for the interpreter, the design of the workstation – at each level, engineered for the maintenance of relationships between state. The significance of such dependency is for the most part hidden from the modeller, but can be exposed – for instance – by substituting a computer too slow to implement an agent that updates in real-time, or to re-evaluate a definition within the lifetime of the modeller.

Viewed in this way, the ISM itself is a complex hierarchical organisation of agency and dependency. Subject to avoiding chains of interdependent definitions of pathological length, there is no practical need to deconstruct the dependencies expressed in definitions by taking the interpreter, the compiler and the hardware into account, but such a deconstruction is essential in order to appreciate the semantics of the ISM as an instrument. In particular, an ISM can refer to relationships between aspects of state embracing observables that are explicit in a definitive script and those in the external environment. It is for this reason that part of the definitive script for the clock can be interpreted as defining "the state of the screen display".

4 Computer-Based Instruments and Tools from a Cognitive Technology Perspective

The impact of technology upon our cognition is a central theme of Cognitive Technology (CT). Much thinking about computer use and technology necessarily tries to address this issue without taking full account of the complexity of the relationship between the experiences offered by the computer and the experiences of users: how these experiences depend on the physical and social context, on the personal characteristics of the user, and how they are liable to evolve. The concepts of ease-of-use [15] and of invisible computing [16] will no doubt play a significant practical role in exploiting computer-based technology, but – where CT is concerned – they are only one peripheral aspect of a much bigger agenda. The most satisfying activities – such as playing a musical instrument – are not generally easy, and though they eventually involve invisible interaction, they are learned through sometimes painful, sometimes rewarding engagement of mind, body and soul. To understand the use and implications of computer-based technology more fully, it is essential to undertake some deconstruction of human-computer interaction, exposing its empirical roots not only in human experience and technological performance, but also in its physical, social and administrative context. Exploring the potential for marrying human and computer activities through the use of interactive instruments provides an appropriate focus. A key objective is to be able to understand the dual tool-instrument perspectives within a single framework.

4.1 Paradigms for Computer-Based Instruments and Tools

The ISM can be seen as an archetype for interactive computer-based instruments. In its essential substance and nature, it is well oriented towards this role. A definitive script is an intricate net of observations about relationships between changes to observables – the product of a family of experiments. Within the script, each definition can be viewed as an instrument, maintaining a relationship between one feature of the state and another. Taken as a whole, the definitions in an ISM, each associated with an experimental context, form a tower of dependencies composed hierarchically in a manner resembling the network of experimental observations that validates a well-conceived engineering product. To construct such an ISM, the mind of a human interpreter must visit every composition of such dependencies, construing it with reference to the agency that is to exploit it. This is the justification for making interactive instruments (see Sect. 2) our primary concern.

Though each ISM has the same characteristic substance, its quality is crucially dependent upon two factors. The first is the way in which the dependencies in the ISM are assembled by the modeller: this relates to the structure of the ISM, empirically established by the modeller according to how they construe its intended behaviour with reference to observables, agency and dependency. The other is the experiential foundation supplied by the constituent experiments. In each case, the reliability with which a relationship between aspects of state can be maintained is an empirical matter. The delicacy of the human control over the instrument is one of these constituents of the experimentally shaped responses of the ISM: it is the basis for the ritualisable experience of the skilled performer.

Numerous ISMs demonstrate these principles practically in relation to modelling real-world phenomena. In that context, the modeller's construal refers most especially to how the phenomenon itself is explained. A simulation of the Clayton Tunnel railway disaster is one case study of this nature [17]. Other research, carried out by Cartwright in collaboration with Adzhiev and Pasko [2], has involved the development of a geometric instrument based on a definitive front-end to the HyperFun geometric modelling environment [3]. In this context, the application builder's construal is concerned with giving the user appropriate control over the geometry described by the geometric modeller.

In contrast, conventional programming paradigms are oriented towards tool-building by computer. The possible contexts of application of the program as a tool are determined by its specification, and the program code is an explicit account of the functions that the tool can perform. Procedural and declarative programming styles approach the characterisation of a tool by specifying its functions explicitly and implicitly respectively, as is indicated by their substance. A procedural program is a complex pattern of sequences of changes to values of observables (an explicit account of a process). A logical or functional specification is a complex aggregate of assertions about relationships between values of observables (the set of predictions of a theory).

4.2 Instruments and Tools in the EM Perspective

Reliability of experience is crucial to the successful development of tools, and to the subagendas of ease-of-use and invisibility in particular. Unlike ISMs, traditional computer programs, being optimised to serve particular functions and operate in specific situations, are constructed in ways that do not necessarily give any insight into the fashion in which the programmer construes the domain (though this is recognised to be highly relevant to the process of identifying a requirement). They are generally designed to exploit the computer's capacity for performing exceedingly complex state-change, and to make the role of the user as clearly defined and simple to enact as possible. These qualities derive from specifying and fashioning the context for the program execution tightly, in somewhat the same manner that a train runs along pre-engineered tracks. In software system development, the analogue of laying track is the identification and contrivance of reliable experience. Providing this essential foundation for software system applications was what first motivated Pi-Hwa Sun to introduce the concept of an ISM [1].

The use of ISMs to trace the activities involved in developing algorithms and processes in environments that initially support only unconstrained and unsystematic interaction is illustrated in two studies. Our study of heapsort [8] shows how an environment in which logical invariants of the algorithm appear as observables can be embedded into an environment similar to that a lecturer might use when introducing the algorithm on a blackboard. A second study illustrates how a manufacturing process and an associated rework process can be fashioned from primitive production and assembly style activities by building an ISM that combines process automation with the possibility of human intervention in managing non-routine rework [18].

The way in which tools are locked into their context of use accounts for their relative inflexibility. A traditional computer program can be versatile, in the sense that it can perform a compendium of diverse functions, like a Swiss Army knife, but it is constrained by the sharply prescribed user-computer boundary, and does not admit open and interactive re-interpretation in use. In contrast, an instrument such as an ISM invites the human interpreter to engage their imagination in whatever ways suit the situation. This potential for an eclectic projection of meanings that can be subjective and provisional onto an ISM is evident even in the simple clock illustration. The result is that re-use in EM is often associated with re-interpretation and a relatively seamless reworking. Indeed, several variations on clocks and digital watches deriving from a single ISM are featured in previous work: these include ISMs, including distributed ISMs, to represent a combined statechart and digital watch, for a chess clock, and for the explicit state and mental model of an actual digital watch [6], [7]. There is likewise an ISM associated with a family of OXO-like games [5].

The intimacy of instrument and mind is nowhere more apparent than in the ways in which instruments can migrate from the external domain of the technology so as to become invisible to the human interpreter. This is commonplace in everyday technology, as when the use of a lens as a subject in the study of optics leads to the development of spectacles. In our characterisation of the instrument as maintaining a relationship between aspects of state, this can be interpreted as merging one aspect of state with another, enlisting the instrument in the service of the model. In EM terms, this is directly interpretable with reference to partitioning definitive scripts in different ways and so reconfiguring the aspects of state whose relationship is the subject of attention. An instance of this migration from referent to model occurs whenever a fragment of script is first developed in isolation, then embedded into the ISM under construction. It is through such migration that this fragment becomes associated with one of the constituent experiments of the ISM.

EM supplies a useful framework in which to integrate the dual tool-instrument perspectives. Though they have an open, uncircumscribed functionality, our ISMs can be exercised as if they were designed for a specific purpose. In this role, ISMs are not as efficient as conventional programs optimised to this function, and in this sense they can be viewed as instruments for prototyping tools (see [19]). As described in [1], they can also be used to explore the contexts for reliable interaction that precede the specification of tools.

An ISM establishes an intimacy of human-computer association that is quite unlike a conventional program in character. From a CT perspective, the most important implication of this is the way that – like the spreadsheet [10] – it has the power to change the culture of use. In principle, the openness of the ISM allows the human agents to exploit the technology in what is characterised in [4] as an 'idealist' rather than a 'realist' frame of mind. Where the objective of the realist is to use technology to save effort and obtain results automatically, the idealist is primarily motivated by a concern to complete the task in a way that gives satisfaction and achieves results that are highly optimised to the particular situation. The first significant practical application of this concept was the use of the Temposcope [4] to timetable some 120 student project orals in March 2001. It is perhaps encouraging that the administrator who made use of this ISM for the first time this year made no comment on the quality of the software, but declared herself much happier about the resulting timetable than on previous occasions.

5 Conclusion

It remains to consider more closely the relevance to Cognitive Technology of the computer-based instrument culture associated with EM.

It is surely too much to expect that CT can predict or fully explain the complex interactions between technology, mind and society. It is difficult to imagine how any study could remove all controversy from issues such as the survival of the QWERTY keyboard, how certain musical instruments are forgotten whilst others have become the carriers of an entire musical tradition, or what social conventions are needed to sustain a language. That said, current accounts of technology are not well-suited for the discussion of such concerns, and EM provides an alternative perspective that gives much greater prominence to the empirical roots of knowledge. In particular, as a conceptual framework, EM can help us in studying the emergence of the ritualisable activities that support tools and instruments from our casual and serendipitous interaction with artefacts. As our discussion of the tool and instrument perspectives has demonstrated, the construction of ISMs can also be used to record and explore insights that are difficult to frame in language alone.

It is unclear to what extent CT is concerned with guiding the future development of technology. In so far as CT draws our attention to a complex evolutionary activity, there is a speculative analogy to be made with Darwinian evolution, and the developments – inconceivable to Darwin's contemporaries – that have eventually led to genetic engineering. Studies in CT can certainly guide us, when developing technologies, to anticipate *some* of the unfortunate implications for people and society that are currently unintended and unexpected and to promote technological developments that are more rewarding and potentially less dangerous in human terms. Somewhat paradoxically, the essential rationale for CT is that – no matter how technologies are developed – they will always evolve in ways that take us by surprise. In so far as CT is concerned with helping us to deal with the effects of this evolution, EM is of interest as an approach to developing computer-based technology that acknowledges that requirements change – indeed that there *is* no fixed requirement – and promises to deliver resources that are less prescriptive and integrate more effectively with human activities. Our ongoing research on the Temposcope [4] and Cartwright's research on applying dependency maintenance to interactive TV applications [2] is indicative of the potential here.

In our current state of knowledge, the principal agenda for CT is perhaps to expose and describe the phenomena that we observe in the interaction of technologies with people and societies. It is our belief that the EM approach of construing phenomena in terms of observables, dependency and agency, and embodying these construals in ISMs, is philosophically and practically well-suited for tackling this agenda, and can assist in understanding and developing instruments of mind.

References

1. Sun, P-H., Distributed Empirical Modelling and its Application to Software System Development, PhD thesis, University of Warwick, July 1999.
2. Cartwright, R. I., "Distributed shape modelling with EmpiricalHyperFun", First International Conference on Digital and Academic Liberty of Information, Aizu March 2001, to appear.
3. `http://www.hyperfun.org/`
4. Beynon, W. M., Ward, A., Maad, S., Wong, A., Rasmequan, S., Russ, S., "The Temposcope: a Computer Instrument for the Idealist Timetabler", Proceedings of the Third International Conference on the Practice and Theory of Automated Timetabling, Constance, Germany, August 16–18, 2000.
5. Beynon, W. M., Joy, M. S., "Computer Programming for Noughts-and-Crosses: New Frontiers", Proceedings of PPIG '94, Open University, 27–37, January 1994.
6. Fischer, C. N., Beynon, W. M., "Empirical Modelling of Products", International Conference on Simulation and Multimedia in Engineering Education, Phoenix, Arizona, January 7–11, 2001.
7. Roe, C., Beynon, W. M., Fischer, C. N., "Empirical Modelling for the conceptual design and use of products", International Conference on Simulation and Multimedia in Engineering Education, Phoenix, Arizona, January 7–11, 2001.
8. Beynon, W. M., Rungrattanaubol, J., Sinclair, J., "Formal Specification from an Observation-Oriented Perspective", Proceedings of the Fifteenth British Colloquium in Theoretical Computer Science, Keele University, April 1999.
9. Levi-Strauss, C., "The savage mind", University of Chicago Press, 1966.
10. Nardi, B. A., "A small matter of programming – Perspectives on end user computing", MIT Press, Cambridge, Mass, 1993.
11. Jacobson, I., "Object-oriented software engineering – A use case approach", ACM Press, Addison Wesley, 1992.
12. Concise Oxford Dictionary of Current English, 8th Edition, Clarendon 1990.
13. `http://www.cogtech.org`
14. Beynon, W. M., "Empirical Modelling and the Foundations of Artificial Intelligence", *Proceedings of CMAA'98*, Lecture Notes in AI 1562, Springer, pp322–364, 1999.
15. Roberts, D., Berry, D., Isensee, S., Mullaly, J., "Designing for the User with OVID: Bridging User Interface Design and Software Engineering", Macmillan Technical Publishing, 1998. `http://www.ibm.com/easy/`
16. Norman, D. A., "The Invisible Computer", The MIT Press, October 1999.
17. Beynon, W. M., Sun, P-H., "Computer-mediated communication: a Distributed Empirical Modelling perspective", Proceedings of CT'99, San Francisco, August 1999.
18. Evans, M., Beynon, W. M., Fischer, C., "Empirical Modelling for the logistics of rework in the manufacturing process" COBEM 2001.
19. Allderidge, J., Beynon, M., Cartwright, R., Yung, Y. P., "Enabling Technologies for Empirical Modelling in Graphics", Research Report CS-RR-329, Department of Computer Science, University of Warwick, Coventry, UK, July 1997.
20. `http://www.dcs.warwick.ac.uk/modelling/`

Computational Infrastructure for Experiments in Cognitive Leverage

Christopher Landauer and Kirstie L. Bellman

Aerospace Integration Science Center
The Aerospace Corporation, Mail Stop M6/214
P. O. Box 92957, Los Angeles, California 90009-2957, USA
{cal,bellman}@aero.org

Abstract. The purpose of this paper is to raise some hard and interesting questions about the new relationships possible between humans and their artifacts:

- What happens when we can have collaborative relationships with our responsive and knowledge-bearing artifacts?
- What happens when group minds are mediated through new types of computing system that can support new and subtle forms of interaction among thousands of imaginations?

The second purpose is to share our work on several enabling technologies that make it possible to experiment with these new types of relationships among humans and machines in new ways. We describe some of the new computing challenges that occur when we have more than one human interacting with the computing systems and with each other. Lastly, we raise some issues about remaining human and creating technology that we can not only live with but thrive with.

1 Introduction: Raising Questions

The primary purpose of this paper is to raise some hard and interesting questions about the new relationships possible between humans, their computational artifacts, and each other:

- What happens when we can have real collaborative relationships with our responsive and knowledge-bearing artifacts? That is, what happens when our artifacts are intelligent and interactive enough to become an "other" to us, one that we allow into our intimate psychological world of concepts?
- What happens when group minds are mediated through new types of "Constructed Complex System" that can support new and subtle forms of interaction among thousands of imaginations (a *Constructed Complex System* is a complex heterogeneous system, managed or monitored by computer programs)?

The new relationships are possible because of the incredible increase in computer system speed and capacity, amounting to a revolution in capability of our computational artifacts. These changes have led to some new approaches to the

M. Beynon, C.L. Nehaniv, and K. Dautenhahn (Eds.): CT 2001, LNAI 2117, pp. 490–519, 2001.

use of computer systems by humans [52] [14] [38], and a recognition that the context of use is extremely important in understanding how human behaviors are affected by their computing systems [15] [40] [43].

Part of the revolution is that certain cognitive skills will be de-emphasized (e.g., the oral tradition type of rote memorization), and new skills will come to the fore (e.g., rapid indexing of information like the meta-knowledge of a library). There will also be new types of social skills required.

Hence, in the second part of this paper, we share our work on several enabling technologies that make it possible to experiment with these new types of relationships among humans and machines in new ways:

- Wrappings provide the necessary highly flexible infrastructure that is explicit, interpretable, sharable, and reflective;
- Virtual Worlds provide a new kind of experimental testbed for the empirical modeling that we believe is needed;
- Computational semiotics studies the nature of symbol systems and their use in computational systems, which we believe will eventually be an essential component of any kind of cognitive embodiment;
- Conceptual Categories provide a relatively new kind of flexible knowledge representation that helps us approach the variety and multiplicity of meanings that computing systems will need to interact with humans.

These enabling technologies support a much more flexible definition and use of information and information services in computing systems than is usually available, and it is our contention that such flexibility is necessary for studying these questions. These problems are hard enough with one human. There are further new computing challenges that occur when we have more than one human interacting with the computing systems and with each other.

Finally, after the approaches are developed and experimentation is performed, we need to keep in mind that our criterion is to remain human and create technology that we can not only live with but thrive with.

In the rest of this introduction, we set the stage for our discussion of the issues we hope to study, the experiments that we expect to provide information about those issues, and the technologies that support those kinds of experiments. We start with a description of our attitude towards tools, and what it means when the tools become less physical and more conceptual.

1.1 Tools

Many tools have allowed us to leverage ourselves physically. A few, like books and writing, have allowed us to leverage ourselves cognitively, to a certain limited extent. Now we are seeing the beginnings of a revolution about how much we can leverage ourselves cognitively, and what happens emotionally and socially to us when we do so. The question to be addressed here is what happens when we leverage ourselves using this computational power, i.e., when we "embody" these new cognitive tools. By *embody*, we mean that the tools become extensions of our self-concept; we perceive them to a limited extent, but it is as if they become parts of ourselves. Experienced drivers, for example, will often describe the handling of a car as being embodied as they "hug the road" or "feel their way along" on a foggy night.

We have always embodied our tools and sensors, incorporating them into our body image at multiple size scales:

- Hammers, other hand tools, prosthetic limbs, eyeglasses, computer games (the tools and sensors do not need to be real);
- Vehicles, such as cars, trucks, buses, airplanes, wheelchairs;
- Very large vehicles and equipment, such as heavy earth-moving equipment (shovels, carriers, graders), shipyard and construction cranes;
- Microscopic tools and other waldos (electronic and mechanical prosthetics [41]), such as those for single cell manipulation under a microscope, radioactive and other hazardous materials handling, and tele-operated robots and surgical tools.

Humans are remarkably adept at adjusting to different "dynamic ranges" of activity (the *dynamic range* of an activity, a term borrowed from acoustics, refers to the interval between the smallest distinguishable effect and the largest interpretable effect), but this adaptability has its limits, and not understanding those limits leads to problems. When the engineering works in these cases, it is because the engineering has mapped appropriately from the modes and dynamic ranges of the external phenomenon into the modes and dynamic ranges that humans can handle (including speed, amount, and kinds of motion, rate, amplitude range, and variability of sensory inputs).

The tools we described above are physical amplifiers or transformers. We have entered a time when we can speculate that computer-based systems will become powerful and interesting enough to amplify us cognitively. Of course, at present most so-called "tools of thought" are only for informational amplification or transformation, which is almost entirely external (although some computer visualization and exploratory data analysis techniques do rely on human pattern recognition capabilities).

This is where we seem to be now in our computing system interface design concepts, working at developing the appropriate mapping models and activities for different application domains [37] [38]. This work is important, but it is really only a first step towards a larger question that we regard as much more important:

- How are we changed when we embody our new information amplification and transformation capabilities?
- How will those embodied capabilities change us (both individually, and in our various groups)?

The problem is to design systems so that not only are the static and dynamic ranges appropriate for a human being physically, but also cognitively, emotionally, and socially.

1.2 Issues

What happens when we really embody extra computational power? Because it is representational, does it also become cognitive power? If so, then new types of cognitive and social skills will have to be taught. Cognitive ones like indexing with meta-knowledge as a library does, and social ones for social brainwork ("group mind").

What happens when we can even encode the right context as we can in a Multi-User Virtual Environment (MUVE)? It is something like off-loading complexity by its embodiment into artificial worlds, which allows new partnerships and additionally complex roles and systems. These roles can include such things as partnerships within a group mind because we can explicitly track interactions of a more complex and subtle type. Lacking this, we would have the equivalent of an intellectual dog pile with no emergent organization instead of a group mind. This new level of complexity handling and explicit information is technically difficult, but it has its rewards: it allows us to orchestrate complex computational processes in ways that are not possible without such information about their internal behaviors, assumptions, and requirements. In fact that is an interesting image: some of our best geniuses have been conductors and composers keeping abreast of intricate musical roles and interactions in compositions. Computationally doing this would allow new kinds of evaluation criteria, symbolic and computational ones, instead of physical auditory "sounds good" ones. With these evaluation criteria also explicit, we can begin to study the evaluation processes themselves.

One (increasingly prevalent) use of this new type of computational embodiment is how we are off-loading memory and changing our memory requirements. We already no longer memorize information in the same way as we used to, since there are books (modern oral histories are few and far between, and modern oral historians are even fewer). We already allow them to extend our "personal" memories. In a way, any personal writing does that, too. Now all we need to do is to recall a reference and a meta-tag on the types of information contained in that resource, and maybe also a specific retrieval problem, to gain access to that additional information.

At a different scale, consider what happens when we embody an entire "data wall" or a whole computational environment. The idea is that this new cognitive power embodiment can be applied at any scale. It need not necessarily occur only for a small tool like a book or computer screen that helps some type of unseen operations inside the user's head, but rather can include something very large of which the user is only a part.

That is, up to now when we have discussed cognitive leverage, we have discussed it as if the enhancement is occurring inside a human's mind and revealed in cognitive behavior. One of the interesting aspects of VWs is the ability to project the contents of one's imagination and mind on to a Virtual World where one can now occupy a part of that world in an avatar and share the occupation of that world with other humans and artificial beings as agents. Simulation has always been a way to "act out" and project an artificial world of a limited sort as part of one's reasoning process. The difference here is that we can now occupy our simulations and hence experience them and reason about those experience in both formal and informal ways. Furthermore we can share that reasoning with others – again, both human and artificial others.

The result is a profoundly new element possible in our mental projections; the ability to have a relationships with our creations. Informally, authors have long spoken of feeling like they are living with their characters. The science fiction author Ray Bradbury (in a 1999 presentation at Border's Bookstore in Thousand Oaks, California) stated quite strongly that after a while he felt that he was discovering the personality of his characters and not inventing them. Apparently, Melville felt that way about his characters in Moby Dick. With responsive, animated agents and Virtual Worlds, we go far beyond this. We have already had the experience of being "surprised" by the discoveries of theorem provers and other analytic programs. If we add sophisticated computational capabilities, a wide knowledge base, the ability to retain memories of previous interactions, and a "personable" interface (e.g., good human language parsing, the appearance of friendly and helpful behavior, good graphics and sensory output displays), we are well on our way not only to be surprised by our helpful artificial companions, but increasingly dependent upon their behaviors and their reactions to our behavior in solving problems. At that point, we then go beyond embodiment (feeling that object as part of oneself) to relationship (feeling that there is another).

How would it effect users to have tools that are embodied by and for themselves, that not only enhance some capability (like a car) but also have some kind of relationship to us. In a way, we already embody some of our relationships: we often try to live up to the expectations of those others or enact their expectations. What then does that mean when the "other" is a computer program, especially if it is rigid, disfunctional, or valueless?

1.3 Experiments

Our claim is that we do not know enough yet to answer these questions, or even to decide whether or not these are all (or even most) of the right questions. We believe that we need to create a new kind of infrastructure to support the necessary experimentation on these (and other) questions. There are (at least) several levels of experimentation needed:

1. Information Amplification: With Wrappings and Virtual Worlds, described below, we can track, learn, and experiment with what is needed when. This is what is beginning to happen now.
2. Individuals: Human embodiment of these new capabilities will provide projected spaces (i.e., allow inner minds to be acted out, projected into the computing environment more easily).
3. Groups: When groups of humans embody these capabilities, we expect surprising new things to be possible, and our experimental platforms must support traceability and tracking of group behaviors.

Since we do not know enough to make the right choices yet, we need much more and much better empirical modeling. We need better domain models for the application domains in which our Constructed Complex Systems will operate, including both the dynamic and static models (i.e., spaces of variability and dynamic ranges within them), better modeling languages and notations, and better model development systems in which and with which to design, define, create, analyze, combine, and understand these models.

All of this model development needs a helpful infrastructure, for integration among disparate models, recording of appropriate interaction information, applying different kinds of analytic tools, and combining their results into understandings about the phenomena that are modeled. We also need experimental testbeds in which multiple models, multiple roles, diverse interactions, and data collection and analysis can take place.

The enabling technologies we discuss come to the fore in several ways here. First, Wrappings are an approach to creating and using explicit meta-knowledge in order to integrate and adapt large heterogeneous systems of resources. They provide an infrastructure for creating, utilizing, and tracking the meta-information (and how we are using it) for problems in an explicit and machine interpretable way, in order to manage these Constructed Complex Systems. Later on, we discuss the advantages of using Wrappings to support these new types of complex system by making the use of meta-knowledge explicit, machine interpretable, sharable, and reflective.

Second, we will need new ways of capturing and processing subtle interactions with and uses of our new types of responsive tools. Virtual Worlds (VWs) are a way of presenting, embodying, and distributing this explicit information by embodying the capabilities or information as an object (to be manipulated by agents or users), an agent (for collaboration or other interaction), or as the defining "physics" of the setting or place within the VW. VWs are just one way of apportioning this explicit information, but the information itself can still be stored in Wrapping Knowledge Bases and processed in the ways we describe for Wrapping-based systems. The objects and processes in the VWs become resources in the Wrapping system. This allows us to do several things: (1) the adaptive additions of new objects, interpreters and other processes; (2) the processing and reasoning about how things are being combined for different uses; and (3) the generation of rooms and spaces on the fly, given certain users and needs. This also allows us to move objects across VWs because the right semantics are made explicit.

Making information explicit is the first step towards making it sharable, which we also want, since we want to work on complicated group mind organizations of resources. One of the problems here is that making information explicit for humans is very different from making it explicit and interpretable for computing systems.

Computational Reflection is about processing information about the use of resources and the mapping between resources and problems. Part of the use of reflection in this context involves being able to reason explicitly about how one is using resources, how one is setting up problems, and the relationships of the problems, outcomes, and methods used. These outside views of the computational processes are extremely useful for explaining and understanding the experimental behaviors. For example, in user interface design, it is important for constructing effective interfaces to allow (or require) the system to explain itself [15], that is, to have an account of what it has done, why it was done, and what it might do next. Computational Reflection is essential to this process.

1.4 Structure of Rest of Paper

The structure of the rest of the paper is as follows: We start by describing, in Sect. 2, Wrappings as a flexible infrastructure for Constructed Complex Systems, that we expect to use for most of our system developments. Then in Sect. 3, we describe the use of Virtual Worlds (VWs) as a new kind of experimental testbed for these studies, for individuals and for "group mind", in which we regard the creative action of multiple minds as something different from and frequently outside any one of the participants. Environments like these Virtual Worlds are going to require us to deal with meaning.

In Sect. 4, we describe some of the more difficult connections between computer-based information and the meanings that make sense to humans, at the level of complex and flexible knowledge representation (conceptual categories), and at the level of basic symbol systems (studies in computational semiotics).

Finally, in Sect. 5, we worry about it all, and present some cautionary words about this enterprise.

2 Wrappings

We have argued that we need a powerful and flexible integration infrastructure for managing this new revolution, and we propose that Wrappings fit the bill, because they are more permissive, more flexible, semantically more powerful, and more supportive of formality in the analyses of the processes and products of integration than most other approaches.

In this section, we describe briefly what Wrappings are and how they work; there are many references for more details [25] [28]. Then we describe the capabilities that Wrappings provide in detail.

Our original motivation was very large space systems [6] [34], which combine hundreds of organizations, thousands of people, millions of lines of computer programs, and tens of thousands of component devices into a system that works. We started developing Wrappings about twelve years ago for system engineering of these systems, and discovered that they have much wider application, to any complex heterogeneous systems managed by computing systems, a class of systems that we call *Constructed Complex Systems* [21] [1] [28].

2.1 Wrapping Properties

We start by describing our notion of "Integration Science" [9]. Integration as a process is importantly an issue of defining suitable relationships among components under new contexts. Above all else, an "Integration Science" must have the formal basis and the techniques for dealing with the representation and processing of context information. Context takes the initial component to be integrated into a system and reinterprets, changes, and biases it, whatever its processing, use, or goal.

Most discussion of integration focuses on the results of integration: the integrated theory, formalism, language, program, system or technique. Integration is treated as a one-time process, to be completed and not considered further. We think that this choice is at least over-simplified and often simply wrong. Our research has concentrated on the processes and infrastructure of integration: the kinds of component resources that are to be integrated, the information services that they are expected to provide, the kinds of knowledge about how those resources interact, the kinds of processes that use the knowledge to perform the Intelligent User Support (IUS) functions [1], to Select, Adapt, Combine, Apply, and Explain resource use.

The resources that are expected to perform information services need to be described. To this end, we insist that proper integration needs explicit meta-knowledge to describe the component resources to be integrated, and that, moreover, integration needs both the meta-knowledge and processing algorithms that use it. The information alone is not enough, since we often want to interpret these descriptions in different ways for different purposes. The lack of explicit and accessible interpreters is the main deficiency of Prolog and spreadsheets as programming styles.

The Wrapping approach has two advantageous simplicities: (1) a simplifying uniformity of description, using the meta-knowledge organized into *Wrapping Knowledge Bases* (WKBs), and (2) a corresponding simplifying uniformity of processing that meta-knowledge using algorithms called *Problem Managers* (PMs), which are active integration processes that use the meta-knowledge to organize the system's computational resources in response to problems posed to it by users (who can be either computing systems or humans).

The Wrapping theory has four essential properties that underlie its simplicity and power:

1. ALL parts of a system architecture, at all levels of detail, are *resources* that provide an *information service*, including programs, data, user interfaces, infrastructure services, architecture and interconnection models, and everything else (implementors choose a level of detail below which they do not want to decompose the services into separately selectable units).
2. ALL activities in the system are *problem study*, (i.e., all activities *apply* a resource to a *posed problem* in a particular problem context), including computations, user interactions, information requests and announcements within the system, service or processing requests, and all other processing behavior (implementors choose a level of detail below which they do not want

to decompose the activities into separately selectable units). We therefore specifically separate the problem to be studied from the resources that might study it.

3. *Wrapping Knowledge Bases* (or WKBs) contain *Wrappings*, which are explicit machine-processable descriptions of all of the resources and how they can be applied to problems to support what we have called the *Intelligent User Support* (IUS) functions [1]:
   - *Selection* (which resources to apply to a problem),
   - *Assembly* (how to let them work together),
   - *Integration* (when and why they should work together),
   - *Adaptation* (how to adjust them to work on the problem), and
   - *Explanation* (why certain resources were or will be used).

   Wrappings contain much more than "how" to use a resource. They also provide information to help decide "when" it is appropriate, "why" it might be the right one for the problem, and "whether" it can be used in this current problem and context.
4. *Problem Managers* (PMs), including the *Study Managers* (SMs) and the *Coordination Manager* (CM), are algorithms that use the Wrapping descriptions to collect and select resources to apply to problems. Making these infrastructure resources also explicit is one key to the flexibility afforded in Wrapping systems. They use implicit invocation, both context and problem dependent, to choose and organize resources. The PMs are also resources, and they are also Wrapped.

The Wrapping information and processes form expert interfaces to all of the different ways to use resources in a heterogeneous system that are known to the system [21] [6] [34].

The most important algorithmic simplification is the Computational Reflection provided by treating the PMs as resources themselves: we explicitly make the entire system Computationally Reflective by considering these programs that process the Wrappings to be resources also, and Wrapping them, so that all of our integration support processes apply to themselves, too. The entire system is therefore Computationally Reflective [22] [19] [31], which means that it has a processable model of itself, that is, a complete model of its own behavior (to some level of detail), so it can analyze what it has been doing, what it is about to do, and what it is doing, to gain some perspective over and control of its activities (this is essential for good explanation, and useful for flexibility in processing) It is this ability of the system to analyze and modify its own behavior that provides some of the power and flexibility of resource use.

In summary, the infrastructure of such a flexible system needs to put pieces together, so it needs the right pieces (resources and models of their behavior), the right information about the pieces (Wrapping Knowledge Bases), and the right mechanisms to use the information (Study Manager, Coordination Manager, and other Problem Managers). Our Wrapping approach provides all of these features and more.

2.2 Wrapping Processes

The processes that use the Wrapping information are as important to us as the information itself, and are one of the main differences between our approach and most others in software engineering (even those called "Wrappers" or "Wrappings"): almost all of those approaches consider Wrappers to be bits of interface code that surround a program module, making it easier for outside programs to use. There is usually no mention of the processes used to construct that code. Our Wrappings are bits of explicit information that can be used to produce those bits of interface code, as and when they are needed. This requirement places some severe constraints on what we must have in the Wrappings and how it can be processed.

The Wrapping processes are active coordination processes that use the Wrappings for the Intelligent User Support functions [1], which generates the usual interface code on the fly from the meta-knowledge about how to use a resource. They also provide overviews via perspective and navigation tools, context maintenance functions, monitors, and other explicit infrastructure mapping activities. This section describes them briefly. Other Wrapping references have more information [25] [28].

The two main classes of Problem Managers are the Study Managers that coordinate the basic problem study process, and the Coordination Managers that drive the system. We describe these next.

Coordination Manager. The alternation between problem definition and problem study, and the determination of an appropriate context of study, is organized by the *Coordination Manager* (CM), which is a special resource that coordinates the Wrapping processes. The basic problem study sequence is monitored by a resource called the *Study Manager* (SM), which organizes problem solving into a sequence of basic steps that we believe represent a fundamental part of problem study and solution.

The default CM runs a sequence of steps that manages the overall system behavior (see Fig. 1):

Coordination Manager Steps
Find context : determine containing context from user or by invocation
indefinite loop:
- **Pose problem** : determine current problem and problem data
- **Study problem** : use SM to do something about the problem
- **Present result** : to user (problem poser)

Fig. 1. Default Coordination Manager (CM) Step Sequence

We explain each of these steps in turn. To "Find context" means to establish a context for problem study, possibly by requesting a selection from a user, but more often getting it explicitly or implicitly from the system invocation. It is the reference to resources that convert from that part of the system's invocation environment that is necessary for the system to represent to whatever internal context structures are used by the system. To "Pose problem" means to get a problem to study from the problem poser (a user or the system), which includes a problem name and some problem data, and to convert it into whatever kind of problem structure is used by the system (we expect this is mainly by parsing of some kind). To "Study problem" means to use an SM and the Wrappings to study the given problem in the given context, and to "Present results" means to tell the poser what happened.

Study Manager. The Study Managers (SMs) embody the central algorithm of our problem study strategy. There are several kinds of Study Managers; we only describe the simplest one. The purpose of any SM is to organize the resources that process the Wrappings.

The SM process begins with a problem poser, a problem defined by its name and some associated data, and the context in which the problem was originally posed. It assumes that it is given the context, problem poser, problem and associated data (usually by the CM). The default SM step sequence is as follows (see Fig. 2):

Study Manager Steps
Interpret problem :
- **Match resources** : get list of candidate resources
- **Resolve resources** : reduce list via negotiation, make some bindings
- **Select resource** : choose one resource to apply
- **Adapt resource** : finish parameter bindings, use defaults
- **Advise poser** : describe resource and bindings chosen

Apply resource : go do it
Assess results : evaluate the results

Fig. 2. Default Study Manager (SM) Step Sequence

We explain the steps in detail, for clarity. To "Match resources" is to find a set of resources that might apply to the current problem in the current context. It is intended to allow a superficial first pass through a possibly large collection of Wrapping Knowledge Bases. To "Resolve resources" is to eliminate those that do not apply. It is intended to allow negotiations between the posed problem and each Wrapping of a matched resource to determine whether or not it can in fact be applied to this problem in this context, and make some initial bindings of formal parameters of resources that still apply. To "Select resource" is simply to make a choice of which of the remaining candidate resources (if any) to use. To "Adapt resource" is to set it up for the current problem and problem context, including finishing all required bindings. To "Advise poser" is to notify the problem poser (who could be a user or another part of the system) what is about to happen, i.e., what resource was chosen and how it was set up to be applied. To "Apply resource" is to use the resource for its information service, which either does something, presents something, or makes some information or service available. To "Assess results" is to determine whether the application succeeded or failed, and to help decide what to do next.

SM Recursion. Up to this point in the description, the SM (by itself) is just a (very) simple type of planning algorithm that considers only one step at a time. The Computational Reflection that makes it a framework for something more comes from several additional design features. First, all of the Wrapping processes, including the CMs and SMs, are themselves Wrapped, as we mentioned before. Second, the processing is completely recursive: "Match resources" is itself a problem, and is studied using the same SM steps as we described above, as are "Resolve resources", "Select resource", and ALL of the other steps listed above for the SM and for the CM, that is, every step in their definitions is a posed problem. The simple form we described above is the default SM at the bottom of the recursion. Third, there are other SMs that have slightly more interesting algorithms (such as looping through all the candidate resources to find one that succeeds). These three properties mean that, for example, every new planning idea that applies to a particular problem domain (which information would be part of the context) can be written as a PM that is selectable according to context; it also means that every new mechanism we find for adaptation or every specialization we have for resource application can be implemented as a separate resource and selected at an appropriate time. It is this recursion that leads to the power of Wrapping, allowing basic problem study algorithms to be dynamically selected and applied according to the problem at hand and the context of its consideration.

The recursion in the SM immediately gives it a more robust and flexible strategy, since the resources that carry out the various steps of the processing can be selected and varied according to context. At every step, the SM has the choice of *posing* a new problem for that step or using a basic function that "bottoms out" the recursion. The choice is generally made to pose new problems, unless there would thereby be a circularity: same problem, same context (the definition of context is such that this condition is easy to check). The important point is that the SM is only our most basic mechanism for controlling these steps; more advanced versions of matching, selecting and so forth will be implemented by resources that are chosen like any others, using the same recursive steps.

The recurrence of posing and studying problems is managed by the CM. The poser (i.e., any of the resources applied by the SM to the "Pose problem" problem) reads expressions from somewhere, as determined by context, and the SM interprets them. The poser usually has a parser (different posers may have different parsers), which reads text and makes symbol structures, within a particular context defined by the "Find context" step of the CM.

The Wrapping approach provides a very straightforward place to perform studies of different kinds of resources needed for application system development. It also makes a very good approach to the problem of software infrastructure for large system integration (nothing requires all of the resources to be software).

2.3 Problem Posing

The separation we made above of problems from resources can be seen as a programming paradigm that applies to all programming and modeling notations: the *Problem Posing Interpretation.* [25] It is a different interpretation of notations that greatly facilitates our search for flexibility. It uses what we have called *Knowledge-Based Polymorphism* to map from problem specifications to the computational resources that will provide or coordinate the solution. Problem Posing can therefore be viewed as a new programming paradigm that changes the semantics, but not the syntax, of any programming or modeling language (in accordance with our wish to make the interpreters explicit and separate from the programs themselves). It can even interpret imperative programs in a declarative way.

In any programming or modeling language, whether imperative, functional, object-oriented, or relational, there is a notion of *information service providers* (e.g., functions to be called, state to change, messages to be fielded and acted upon, and assertions to be satisfied), and a corresponding notion of *information service requests* (the function calls, assignments and branches, messages to send, and assertions that cause those service providers to be used). In almost all of the languages, we connect the service requests to the service providers by using the same names, i.e., the connections are static and permanent, defined at program construction time [25].

The Problem Posing interpretation breaks this connection and generally moves it to run-time, recognizing that all of the language processors can tell the difference between the service provider and the service request. Then the language processors take the service requests and turn them into *posed problems* (hence the name), and use the Wrapping processes described earlier (or any other mapping process) to allow a context-dependent resource selection process to select an appropriate resource for the problem in the problem context. The selection process is guided by Knowledge Bases that define the resources, the kinds of problems they can address, and the specific requirements for applying the resource to the problem in the context. This is what we mean by "Knowledge-Based" Polymorphism. It allows a convenient and flexible mapping from problems to configurations of resources that can deal with them.

Programs written in this style "pose problems". They do not "call functions", "issue commands", "assert constraints", or "send messages". Program units are written as "resources" that can be applied to problems. They are not written as "functions", "modules", "clauses", or "objects" that do things.

Problem Posing also allows us to reuse legacy software with no changes at all, at the cost of writing a new compiler that interprets each function call, for example, not as a direct reference to a function name or address, but as a call to a new "Pose Problem" function, with the original function call as the specified problem and problem data. With this change from function calls to posed problems, the entire Wrapping infrastructure can be used. In particular, as the usage conditions for the legacy software change (which they always do), that information can be placed into the problem context, and used to divert the posed problems to new resources written with the new conditions in mind (only the timing characteristics will change, but those changes are frequently completely subsumed by using faster hardware). The gradual transition away from the legacy code is extremely important. Writing such a compiler is a well-understood process, and it is often worthwhile to do so.

3 Virtual Worlds

A human using a computer program of any kind is presented with a Virtual World, that is, an environment in which the user is allowed to perform some limited set of control actions, and is presented with some limited set of information displays. The number and variety of available control actions is almost always very small, and only occasionally determinable. The number and utility of information displays is almost always not enough. These appallingly limited worlds are so restrictive in their scope and so poor in their quality of interaction that using them proficiently becomes an exercise in excessive focus on certain details, and can only be performed successfully by a few people. This deficiency is the foundation of our interest in Virtual Worlds: to make them more interesting, more understandable, and more humane [43]. We need tools that not only enhance our cognitive abilities, but because we are now trying to understand new roles for ourselves with our inventions, we need to enhance our ability to reflect upon and monitor those activities.

In this section, we describe Virtual Worlds (VWs) as our proposed experimental testbed, and our approach to the necessary infrastructure for this type of experimentation. We describe what VWs are, some technical aspects of their construction, and what they allow us to do for our experiments. We sometimes use the term "Multi-User Virtual Environment" (MUVE) to emphasize that there are multiple users, which is a significant departure from most computing systems, and that requires a system to support human-to-human interactions. An older and still popular generic term for these programs is MUDs, for "Multi-User Domains".

3.1 MUVE Architecture

For simplicity in this discussion, we consider MUVEs that are derived from one of the simplest MUVE servers, called TinyMUD, and the oldest TinyMUD still active (and the oldest continuously active non-combat MUD of any kind), called DragonMUD [44] [48], since it has all the essential technical features we want. It is not too complex to extract and describe those features easily, and because it illustrates one of the most important aspects of MUDs: it is the writers and artists that define the culture (not the technical substrate), and it is the culture that makes a MUD viable or not.

The architecture of this kind of text-based MUVE is organized as a central server, to which remote clients connect across the Internet. The client-server interaction protocol is very simple; it allows the external users to send what we call input *interaction items* (i.e., typed text, button pushes, and other operations) to the server, and to receive what we call output *interaction items* (i.e., presented text, screen object motions, and other operations) from the server. The most frequently exchanged interaction item is the text string, sent from users to users as "talk", and uninterpreted by the server (this is the activity that some chat rooms get right).

MUVE Server Architecture. The architecture of the MUVE server has three distinct conceptual layers [7] [9] [27]:

- the Connectivity Layer,
- the Virtual World, and
- the Infrastructure Layer between them.

The Connectivity Layer is responsible for the transitions in both directions between users (i.e., humans or programs that use the MUVE) and the MUVE server program(s). We have explicitly set out the Connectivity Layer because it is responsible for the multi-user capabilities of a MUVE. It has four main functions:

- Connection Management,
- Command Order Arbitration,
- Distribution of Commands to Interpreters,
- Distribution of Results, Actions to Users, and
- Fair Scheduling.

The first function is to listen to the Internet on a certain port that is well-known so that other programs can find and connect to it. The second is to guarantee that each *interaction item* is treated as a unit, so that, for example, simultaneously arriving text is not overlapped. The third function is to distribute the interaction items to the appropriate resources for interpretation, and the fourth is redistribution of uninterpreted items and interpretation results using a kind of local multicast, based on place and virtual proximity. Finally, some MUVEs also try to enforce fair scheduling and load balancing.

MUVE Users. The users of a MUVE share a conceptual environment called the "Virtual World", which consists of a set of locations called "rooms", and interconnections between rooms called "exits", generally implemented as a database that holds or carries the world.

There is nothing in the definitions that requires MUVE users to be humans. There are many examples of software agents or "softbots" that run outside the

MUVE server, and connect to it using exactly the same protocols as human users do. They move around using exactly the same set of commands also.

It is sometimes hard for new users to tell at first when they are interacting with a robot. This is where it begins to become important to consider socially intelligent agents [4] [24] [45] [46] [5], so that the interactions can be more natural.

MUVEs and Wrappings. We have shown [9] [25] [28] how our Wrapping approach can be used in this kind of architecture to allow multiple construction languages, context-based interpretation of user commands, and "regional physics", that is, different interaction rules in different parts of the space. All of these capabilities come from using a specialized Study Manager to connect user interaction activities with computational resources in a context- and location-dependent way.

The resources in a Wrapping-Based MUVE are the places and connections; the objects, tools, and users; and the command interpreters and user interfaces (the agents are not resources, since they are external to the program, as are the human users; only the interfaces to them are resources).

The problems correspond to the commands: connect and disconnect; move, talk, and other actions; the building commands; and some miscellaneous commands directed at the server itself.

Humans in MUVEs. It turns out that there are many important non-computer technical aspects of MUD Virtual Worlds also, and these are somewhat beyond the scope of this paper, though we recognize that it is usually these non-technical features that determine whether a MUD thrives or disappears (the most important two are the aspect of being "well-written" as a story, and the behavior of characters towards others). DragonMUD is an example of this phenomenon, since it uses one of the simplest server programs (a TinyMUD), and yet one of the longest lived of the MUDs [44] [48].

The most important aspect of the MUVE for human users is the shared sense of "presence" [35]: the feeling that one is actually "in" the Virtual World in a fundamental way, and moreover, that one is "in" the Virtual World *with other humans.* Building computing systems that support this sense of presence places a great burden on our computationally realized semantics of place.

The second most important aspect is the notion of "place", that the interface presented by the system to the users engages our sense of place [16] [18]. VWs provide both of these properties and more [44] [14] [2].

3.2 VWs as Integration Places

Over the last nine years, we have been exploring the use of collaborative virtual environments called Virtual Worlds (VWs) as a new type of testbed for experimenting with ways of organizing and integrating diverse types of computational and human processes [35] [8] [9] [27]. In other papers, we have discussed a wide variety of applications to fields as diverse as education [7], software engineering [28], and mathematical research [8]. Here we would like to focus on some of the experiments that we are currently building in these environments. The idea of using the VW as a testbed for research was briefly described in several earlier papers; here the focus is specifically on strategies for dealing with the integration of different kinds of processes, both human and artificial. Integration is becoming one of the key barriers to better system modeling and development in complex and large systems. One can partly define "complexity" in a system by the number of different viewpoints, models, and analytic techniques required to represent and reason adequately about such a system [10] [9]. The problem is that different viewpoints and formal methods are not completely disjoint; rather they are often partially overlapping and incompatible in subtle ways in their assumptions, definitions, and results. The outcome of these interactions is that there is often no well-defined decision process for combining the results of these different levels and types of models. In the worst case, the layman and the professional can only guess, and often resort to an unjustified strategy of just "adding them all up". Below we describe briefly the integration problem and our approach to this classic problem using VWs.

Unlike a formal mathematical space, or even the usual kind of homogeneous simulation system, part of the strength of a VW is its ability to become the common meeting ground for a variety of different types of symbol systems and processing capabilities. These different symbol systems and processes occur in a variety of forms within the VW. Information and processing capabilities can be packaged or encapsulated as "agents" [12] [38], who often interact with human users in natural language, and can freely move and act within the VW in the same way as a human user; as an "object" within the VW that is manipulated by human and / or agent users; or as part of the "setting", e.g., the description, capabilities, contents, and "virtual physics" of one of the many "places" within a VW. The packaging of some process to become an object, an agent, or part of a setting in a VW hides, as does any good Application Program Interface (API), many details about the process and how that process works. The VW gives the appearance of a uniform world of objects, actors, and places, all acting within the common "physics" of a setting, and seen and heard in the same way. This is a reasonably successful and good strategy for integration [8] [9] [26] [28].

However, if one looks one level deeper, this common meeting ground is also the theoretical meeting ground for how communication will occur among different types of formal systems, computational systems, and even humans. It therefore makes a new hopeful tack for the very hard traditional problem of integrating different formal systems.

3.3 VWs as Testbeds

We have used VWs as testbeds for our own research on intelligence and autonomy for some time [22] [2] [29] [31], and more recently on emotions and self [4] [5]. These areas of research are full of problematic language (overuse and misuse of metaphors, for example), but they have important consequences. We need an environment in which we can operationalize some of these concepts and actually observe such mechanisms in use. One of the most important things we need is an environment in which we can explore these very difficult "mappings" between goals, agent capabilities, agent behaviors and interactions with the environment, and consequences or results in that environment. One of the most difficult issues has been that, heretofore, since we could not completely instrument the real world, we certainly could not capture all interactions between a system and its world. Now in Virtual Worlds we have an opportunity to do so. The disadvantage of course is that these worlds are not nearly as rich as real worlds. However, it is our experience that when one starts filling these worlds with information, objects, processes, and agents that move around the different rooms of these worlds and interact with each other, then the worlds are rich enough to be complex worlds. If one now adds humans behaving through a number of modalities in these worlds while interacting with these objects and agents, we have more than sufficient complexity to be interesting enough for the foreseeable future.

Especially important is that these worlds force us to describe explicitly what the salient features in the "world" are that will be noticed by the agent, what the actions are that will be performed in this world to cause what results, and so forth. This, to our mind, has been the missing partner in the usual concentration on building agents to interact with the world. Lastly, we have a simplification not enjoyed in the real world of having hard boundaries between the inside and outside of a system (although one can walk into an object in a Virtual World and have it become a setting [3]).

Equally important is that we need a testbed and a style of experimentation that allows us to build, observe, refine, and accumulate knowledge about our agent experiments and implementations. Computer scientists, unlike other scientific fields, do not have a good track record in building off of each other's implementations. Partly due to the wide variety of languages, machines, etc., and the problems of integration, each researcher tends to rebuild capabilities rather than using someone else's. Virtual Worlds hopefully will encourage a new attitude and process for conducting and observing each other's efforts and sharing experiments and implementations, since they have such a low entry cost, both conceptually and computationally.

3.4 Group Mind

One of the main promises of computing technology has always been that we will be able to work better, with computer-supported activities and resources. For example, the educational communities are fond of using the terms "anywhere, anytime, anybody" in regard to the learning process, by separating the knowledge from the instructor.

The idea of putting "all human knowledge" on the Internet, making it available to everyone, has become more prominent recently, with the success and wide availability of the World-Wide Web, but it is not a new idea [51] [13] [39], or even a particularly well-defined idea, since the most that we have ever done along these lines is to make descriptions of knowledge available for simple kinds of computing. Even if we think that it is desirable, it is still very hard. The main technical problem has been finding the right methods for indexing such a large body of information, but the more important problem is, more disturbingly, who will decide what is and is not knowledge (this is where the Web's diversity and decentralization of authority is both a blessing and a curse). This is a problem that has not been adequately addressed as yet.

One of the main recent promises of computing technology has been that we will be able to work together better, with computer-mediated distributed interactions. The ability to embody computational power allows us not only to reason about complex systems as individuals but allows us to participate in groups, and leads to the notion of "group mind", which is the result of many humans cooperating. There are many groups attempting to make "groupware" products to support this kind of interaction, from "collaboratories" through "computer-supported cooperative work". The common reaction to most of this work is that it doesn't.

We think that some of the problems that occur in these systems, when systems are designed for interactions among multiple people, are easier to avoid in a VW.

For this reason, we are building VWs as the experimental testbed, because they automatically allow multiple humans to interact in a common environment. The Web is a very powerful access medium, which became popular because of its low entry cost, but the Web is not enough for group mind, since it is only about shared artifacts.

We have argued that shared presence in addition to shared artifacts make a collaboration much more effective [35] [27].

Shared presence is a very powerful force, as can be seen even in the impoverished environment of chat rooms [49], but it takes much more power in a MUVE [17], even if it is just a TinyMUD (a particular kind of MUVE server that has been available for over ten years) [14] [48]. While it seems to some that the relative anonymity of virtual interactions is safer (and it is for some kinds of interaction), it is also true that the protective social conventions of face-to-face interactions in public and in private are not available in a virtual environment, so some of the "presumed privacy" and "protective coloration" available to humans in conversation is missing in VWs. This lack means that some virtual talk goes directly from one human to the other without a conventional protective filter.

This power is why we use a VW for our experiments. Even though a VW is in itself a shared artifact, constructed by its users (some have been built by thousands of users over multiple years), it supports shared presence very well: the users can share each other's presence in their artifacts.

Of course, if we are going to use a VW as an experimental platform, we need to understand what it means to study people working together. The first requirement is that there is "virtual work" for them to do [2], whether it is collaborative construction of the world or cooperation on some external project. The second requirement is that there are tools and other resources for them to use in that work. Finally, there must be measurement facilities embedded in the system, as invisibly as possible.

Here, VWs clearly have a tremendous advantage. Because a VW is a mediated environment, we can record all interactions between people and tools and each other. That is, we have a new ability to capture and analyze all human-to-human interactions, as well as the human-to-tool and tool-to-tool interactions. We still need to develop methods for analyzing them, but such information has never before been so readily accessible. This property is why VWs make a new kind of testbed, and why we have emphasized them so much in this paper.

4 Meanings

All of our cognitive helpers are limited by how well they track our meanings. Trying to deal with meanings is difficult. So far, the computing systems we describe make no use of meaning. At most, they transfer interaction items between users without attempting to interpret them. We think that they can do more than that. To do that, we need to consider the different kinds of assumptions that are common in computing system development, and that lead to the rigid and brittle systems that are (rightly) castigated [36], and see where we can get by changing the assumptions.

In this section, we describe two approaches to providing more computational assistance for processing meanings [32]. Neither is completely well-developed or guaranteed to work, but both show interesting properties. They address different kinds of assumptions that are common in computing system development.

The first is a new flexible knowledge representation technique based on some noted deficiencies in using sets to represent categories, and the second opens some possibilities for very different kinds of computer processing, based on the notion of a computational symbol system as an explicit point of control and flexibility in a system.

4.1 Conceptual Categories

We have advocated using a Wrapping-based infrastructure for Constructed Complex Systems, which uses knowledge representation techniques for the WKBs.

In this subsection, we describe a new method of knowledge representation that is based on the fact that humans create and use categories in a way that is very different from the usual computing style [20].

We have defined "conceptual categories" as a new mechanism for flexible knowledge representation [26] [30] [33]. The basic notion is a generalization of set theory in four directions: our collective objects have (1) indefinite boundaries and (2) indefinite elements; (3) the context is allowed to leak into the interpretation of the objects; and (4) there is a notion of multiplicity of structures that correspond respectively to considering the same object or class from different points of view, in different contexts. This notion allows us to model the modeling decisions explicitly, and to keep track of the modeling simplifications so we can relate them to each other.

It is important to get away from the use of sets as the only model for categories of knowledge, since they artificially limit the kinds of categories that can be considered [20]. Sets have both definite boundaries and specific elementary units. There are many models of "uncertainty" that generalize the first constraint: probability distributions, fuzzy sets, belief functions, rough sets, and many others. As near as we can tell, there is no appropriate model that generalizes the second constraint. This lack is mainly due to the nature of mathematics as our most concrete form of reasoning: the elementary units must be defined before we can start most mathematical reasoning processes. This constraint does not seem to be present in linguistic reasoning [50] [47]. We believe that this difference is significant, and we are using it as a way to approach the problem.

We can gather any linguistically expressible concept into a category, and then change the focus, from a domain in which the expression has meaning to a domain containing the expression. Categories have indistinct boundaries. As they become more the central focus, they become more precisely determined. This means that the computational structures we define depend on our state of knowledge or interest. The focus of attention determines what the categories seem to be for a computing system that uses them.

At the meta-level, we have terms (symbols and symbol constructs) that refer to categories, and we know some things about the categories, but not everything. We want to reason about the relationships among the terms, without having to resort to information about the categories that we have not expressed, and we want to determine what information is still needed from what we are trying to derive.

Elements of a category either have properties or perform actions. All actions are changes in symbol structures (e.g., many modern computers have a notion of input and output operations as reading and writing of specific addresses), and all are mirrored by changes in the "memory" of performing the action.

This conceptual data structure allows us to represent the modeling decisions that underlie different models, and to record and compare the corresponding viewpoints. We believe that this approach will bring more of the modeling process into the system, where the system might benefit from being able to treat it mathematically, or at least systematically. The hard part of the process is the domain modeling that identifies the assumptions, and the interacting linguistic framework that allows different kinds of assumptions to be compared.

This organization provides the system with a kind of intelligent ontology, in which the structure of knowledge is not only context- and problem-dependent, but also much more flexible than the usual structures, because it does not need to rely exclusively on a single logic for its reasoning capabilities. In particular, the system can maintain multiple viewpoints that are not necessarily consistent, and make appropriate selections according to context.

4.2 Computational Semiotics

Computational semiotics is the study of the use of symbol systems by computational systems [23]. Our studies of symbol systems are trying to find flexibilities at an even more basic level than the knowledge representations can provide.

We have described integration mechanisms that allow many different kinds of resources to interact with each other and with human and computational users in a very flexible environment. These enabling technologies provide a setting for our experiments, but they are still computer programs of the usual kind, reducible to and fully grounded in bits and bit-pattern interpreters. This well-foundedness is a barrier, since it prevents us from elaborating "underneath" the symbol system in use, in addition to the usual elaboration "above" it that we expect to use.

In this subsection, we show that there are some possibly unexpected limitations in the ways even humans can use them.

The "get-stuck" theorems are phenomena that are well-known in computing circles, at least operationally; they have been observed in "extensible" languages, large and growing knowledge bases, and inheritance hierarchies for large object-oriented systems: no matter what fixed finite set of elaboration mechanisms are used, no matter what finite set of initial structures are used, the constructed structures eventually become constrained too much by what is already there, and stagnate, or at least become extremely difficult to extend further. To change that phenomenon, which is a kind of creeping rigidity of partial success, we need construction mechanisms that can themselves be changed and elaborated. This is a key approach to avoiding the problem [29].

There are actually two kinds of "get stuck" theorems. We do not prove them here; details can be found elsewhere [23]. The first ones show that there is a limit on how finely a system can discriminate using a fixed symbolic system, and the second ones show that there is a limit on how finely a system can be elaborated using a fixed symbolic system, even if humans are doing the elaboration.

Our interpretation of the first kind of theorem is that even though we seem to be able to expand (tree-structured) hierarchies indefinitely (e.g., directory structures, ordinary ontologies, context-free grammars), they cannot express some of the important connections between different kinds of knowledge; it is the cross-connections that give us much richer expressive mechanisms (e.g., directed graphs, context-sensitive or even general phrase-structure grammars). We "get stuck" in our attempts to express important aspects of our complex environments.

In other words, our computing systems can discriminate more situations more finely by using larger and larger description structures, but the limited number of structures of each size means that the structure size grows very fast relative to the number of situations to be discriminated. There eventually comes a point at which the time and cost of description or processing is too large for any useful response to be computed in the time available. That point is a limitation on the expressive and computational mechanisms; they cannot do any more any faster.

To avoid this problem, which is the first "get-stuck" problem, we have to allow the expressive mechanisms to contain cross-connections. But they do not solve the problem for us. They lead to the second "get-stuck" problem.

The system needs more and more structures for finer and finer discrimination. However, having more structures implies making it more cumbersome to change: there are more interconnections, more obscure interconnections, more surprises, and more unforeseen implications. We eventually "get stuck" in our attempts to extend them. This is the creeping rigidity of partial success.

These expressive mechanisms, even when designed to be extensible, seem to get stuck by over-constraining the possible refinements, even when the refinements are organized and implemented by humans (in other words, this is not simply a Turing or Gödel theorem about what is or is not computable or decidable by algorithms). We think that if the system can retain the fluidity in the original structures, that is, if the system can replace the basic symbolic units and the corresponding construction mechanisms, this problem might go away, at least partially.

Another aspect of the theorem is that when the domain is widened (which we can see will reduce the density and would hope would change the results), there are many nodes that were previously part of the context that now need to be expressed explicitly, so that the density may in fact not decrease at all, and even if it does, it is one-time only, and the result of the theorem continue to hold. Either way, rigidity increases.

For example, consider Knowledge-Based Systems and ordinary computer programs. In each case, we start with a fixed set of entity types or type constructing mechanisms (all of the types are finitely constructed). The program can build new objects and relations. It can even build new types, but it cannot build new type constructing mechanisms. On the other hand, we, as developers, can add new mechanisms. We seem to "get stuck" either way.

However, we note that it is the assumption of a fixed symbol system that leads to the problem, so we are investigating systems that can change their own symbol systems [23] [29].

5 Building a Livable Technology

In the introduction to this paper, we claimed that new technologies such as Virtual Worlds and agent technology add a profoundly new element possibly to our mental projections; the ability to have relationships with our creations. Although many authors have always discussed the feeling of having a relationship with their characters, with responsive, animated agents and Virtual Worlds, we go far beyond this. We speculate that if we add sophisticated computational capabilities, a wide knowledge base, the ability to retain memories of previous interactions, and a "personable" interface (e.g., good human language parsing, the appearance of friendly helpful behavior, good graphics and sensory output displays), we are well on our way not only to being surprised by our helpful artificial companions, but increasingly dependent upon their behaviors and their reactions to our behavior in solving problems. At that point we then go beyond embodiment (feeling that object as part of oneself) to relationship (feeling that there is another).

The problem here is that relationships with others are supported by our communication systems, our cognitive and emotional capabilities, and our social and cultural experiences and behaviors. These are all areas that biological systems have evolved over millions of years. What happens when we create human-artificial systems where the "others" involved in an intimate relationship are odd and limited in inhumane, indeed, un-animal-like ways? We already know from the world of psychotherapy, that family systems can be impacted negatively when even one member of the family is "dysfunctional". Of course, dysfunctional is a funny word to apply to a set of computer programs, but then up to this decade the word "emotional" and "intelligent" for a robot would have only belonged in a science fiction story. The term "dysfunctional" isn't just an issue of a system not adequately performing its tasks (as it is within verification and validation for example). Rather, dysfunctional has the connotation that there are normative standards for behavior within a community of performers. And it has been known for a long time, that dysfunctional members of a group harm the group. Dysfunctional here can be in terms of emotional responses, personality, communication skills, activities, and social behavior. Some might argue that the rigidity or artificiality in the communication and social behaviors of a robot will never have a negative impact because in fact we will never think of it as an intimate other person but rather as perhaps an intimate other species. Maybe robots will become like domesticated companion animals to us.

Relationships always take us outside of the individual and place them within a context. This context can be in terms of the physical system, via both its physics and its ecology. When this physical system also includes conspecifics (members of the same species) then we start to enter the world of social and cultural systems. None of our biological modeling or concepts has ever dealt very well with how to handle the complicated subject of domesticated animals, that is, other species that we love. Although there have of course been numerous studies about the impact of domesticated animals (both ways), they fall into a peculiarly overlapping area in terms of our models of ecology versus cultures.

In a way, we do "embody" our relationships or to use some popular psychological jargon, we "internalize" the attitudes of others – and according to clinical experiences, live up to or enact the expectations of these others. Again the issue is what happens if this other is a computer program. Essentially we are discussing the impact of intimacy with our robots and agents on us as humans. This reminded us of some of Thomas Moore's interesting and painful essays on the problems of a computer-involved culture [42]. Why quote a theological philosophical essay, written by a theologian and psychotherapist, in a scientific paper? Because therapists are on the front lines of trying to keep people's human experiences healthy and whole, explicable and livable in a world that for some feels inhumane and overwhelmingly painful and colorless. Therapists and spiritual leaders are often, in our scientific culture, the only witnesses to people struggling with meaning and trying to make sense of their existence. If we want to create a technology that we can live with, we must include a serious discussion of their concerns and observations.

Although his answers are not always ours, we believe that we must examine seriously some of his fundamental questions. For example, his essays are concerned with how we structure meaning for ourselves within a highly scientific culture. For all the inadequacies of psychotherapy and its lack of rigorous scientific understandings and theory, nonetheless it retains its ability to treat explicitly and support characteristically human experiences. Hence,

> "In therapy, we never understood dreams completely, but we developed a closer relationship to the intimate inner world of the imagination – we glimpsed some of the narrative themes that were influencing life. More importantly, we translated daily experience into the language of dream to glimpse the strong imagination that was at work making meaning." (p. 73 of [42])

Another issue he deals with is the decline of individuality – of a healthy eccentricity. Again, one can see how our technologies such as Virtual Worlds can both allow thousands of humans to share for the first time creative inventions and images of their minds or become parts of pre-existing virtual games, where even hundreds share the same anonymous avatar and they learn to sublimate all their individuality into a limited creativity of competition.

The purpose of our research and development is to make our technologies work to enhance our unique imaginations and all of our personality, social and emotional needs. Because at the moment, it is the human biological system that is flexible and adaptive, we constantly invent creative ways to live with our inflexible creations. But our adaptations are not without strain and costs – some subtle. For example if typing is difficult, some quickly adjust and limit their language to type quick emails. But they have sacrificed some of their creativity, expressiveness and color to do so. That may not matter to some but what happens when the humans interactions - for Cyber-medicine or tele-education – become similarly abbreviated to accommodate a "spellchecker" level of machine-supported communication? There the loss of expressiveness could lead to serious treatment errors or to a degraded treatment due to subtle human interactions and needs for comfort and comfortable relationships. Because our new information technologies do not immediately mediate physical devices and hence physical effects on human users, we tend to believe our inadequate social and system engineering has less serious outcomes. But that simply is not the case. In some ways the immediate success of these new tools to support our activities is secondary to our ability to create tools that support the way we want to live with ourselves and with others.

That is, we need to balance the capabilities of our tools and our science to know and control our Virtual Worlds with our needs to cultivate a human culture that supports humans capable of thriving in the real world which will never be as controllable or as known as we humans would wish. As Moore poetically states it,

> "Our age is Promethean. Beneath our attempts to explore and analyze the whole of life is the wish to be immortal and all-knowing. The fire of the gods, which we have stolen, flickers in the glow of computer and television screens and blinds us in the brilliance of a rocket blasting off or a nuclear bomb exploding. We believe ourselves to be evolved, better than our ancestors and certainly more knowledgeable. We trust our motives are generally humanitarian, but it is becoming gradually clearer – at least felt if not understood – that the implied repression of passion and the closing off to mystery leave us vulnerable to madness and its acting out." (p. 115 of [42])

Are we capable of creating such systems? Unfortunately we know we are capable of creating systems that are harmful to us and to our environment. In fact, our record for how to intervene medically, ecologically, and socially has been a painful one of learning from our mistakes and recognizing repeatedly the consequences of side-effects, trade-offs among complex configurations of variables, and inadequate models of ourselves.

In this paper, we described our best strategy for developing the necessary testbeds so that we can take advantage of our new technologies. These testbeds must include the ability to observe ourselves in relationship to these new technologies. These testbeds are necessary for us to conduct explicit experimentations and to collect experiences with these new types of cognitive leverage. The use of testbeds also acknowledges explicitly that we do *not* know how to design all the uses of these new technologies or how to limit the use of these technologies to their most successful and appropriate applications. Ironically although Virtual Worlds may allow us to creatively enact mythic worlds, they can also serve to debunk the myths of our current technologies.

Because the rate of technology change has been so fast, there has been the harmful idea that their incorporation into our lives and work would follow correspondingly quickly. Although technologists have complained bitterly about the technology adoption problem, this conservatism has probably helped us more than we know by giving us more time to create more appropriate models of ourselves. Now we need to use technology to help speed up the collection of data that will lead to these better models of ourselves and our uses of technology. In other words, part of the goal of this paper is to harness technology to become part of the solution in creating the critically necessary models of our parts and our roles in human-machine systems. If we are not willing to sacrifice thousands of patients or students or others by simply seeing what happens (which we are not) then we need to make it clear to our communities of interest that we need real studies and not just implementations in cyberspace. Furthermore we must make it clear that these studies need to focus as much on the psychology and sociology of humans as on the mechanisms and implementation of devices. In a way we need to assert two of our most human characteristics: the power to reflect and reason from a social and emotional point of view, and the power to build systems that embody the values we promote.

References

1. Kirstie L. Bellman, "An Approach to Integrating and Creating Flexible Software Environments Supporting the Design of Complex Systems", pp. 1101-1105 in *Proc. WSC'91: The 1991 Winter Simulation Conf.*, 8-11 December 1991, Phoenix, Arizona (1991)
2. Kirstie L. Bellman, "Sharing Work, Experience, Interpretation, and maybe even Meanings Between Natural and Artificial Agents" (invited paper), pp. 4127-4132 (Vol. 5) in *Proc. SMC'97: The 1997 IEEE Int. Conf. on Systems, Man, and Cybernetics*, 12-15 October 1997, Orlando, Florida (1997)
3. Kirstie L. Bellman, "Towards a Theory of Virtual Worlds", pp. 17-21 in *Proc. VWsim'99: The 1999 Virtual Worlds and Simulation Conf.*, 18-20 January 1999, San Francisco, SCS (1999)
4. Kirstie L. Bellman, "Emotions: Meaningful Mappings Between the Individual and Its World" (invited paper), *Proc. Workshop on Emotions in Humans and Artifacts*, 13-14 August 1999, Vienna (1999)
5. Kirstie L. Bellman, "Developing a Concept of Self for Constructed Autonomous Systems", pp. 693-698, Vol. 2 in *Proc. EMCSR'2000: The 15th European Meeting on Cybernetics and Systems Research, Symposium on Autonomy Control: Lessons from the Emotional*, 25-28 April 2000, Vienna (April 2000)
6. Kirstie L. Bellman, April Gillam, Christopher Landauer, "Challenges for Conceptual Design Environments: The VEHICLES Experience", *Revue Internationale de CFAO et d'Infographie*, Hermes, Paris (September 1993)
7. Kirstie L. Bellman, Christopher Landauer, "Playing in the MUD: Virtual Worlds are Real Places", *Proc. ECAI'98: The 1998 European Conf. on Artificial Intelligence, Workshop w14 on Intelligent Virtual Environments*, 25 August 1998, Brighton, England, U.K. (1998)
8. Kirstie L. Bellman, Christopher Landauer, "Virtual Worlds as Meeting Places for Formal Systems", in *The 7th Bellman Continuum, Int. Workshop on Computation, Optimization and Control*, 24-25 May 1999, Santa Fe, NM (1999); (to appear) in *Applied Mathematics and Computation*, (May 2001, expected)
9. Kirstie L. Bellman, Christopher Landauer, "Integration Science is More Than Putting Pieces Together", in *Proc. 2000 IEEE Aerospace Conf. (CD)*, 18-25 March 2000, Big Sky, Montana (2000)
10. Richard Bellman, P. Brock, "On the concepts of a problem and problem-solving", *American Mathematical Monthly*, Vol. 67, pp. 119-134 (1960)
11. Jeffrey M. Bradshaw (ed.), *Software Agents*, AAAI Press (1997)
12. Jeffrey M. Bradshaw, "An Introduction to Software Agents", Chapter 1, pp. 3-46 in [11]
13. Vannevar Bush, "As We May Think", *The Atlantic Monthly*, Vol. 176, No. 1; pages 101-108 (July 1945)
14. Jen Clodius, "Computer-Mediated Interactions: Human Factors", (invited keynote presentation) *MUDshop II*, September 1995, San Diego, California (1995); at URL `http://www.dragonmud.org/people/jen/keynote.html` (Last checked 20 March 2001)
15. Paul Dourish, Annette Adler, Brian Cantwell Smith, "Organizing User Interfaces Around Reflective Accounts", in *Reflection'96 Symposium*, 21-23 April 1996, San Francisco, California (April 1996); also at URL `http://www.parc.xerox.com/csl/groups/sda/projects/reflection96/index.html` (last checked 6 May 2001)

16. Winifred Gallagher, *The Power of Place: How Our Surroundings Shape Our Thoughts, Emotions, and Actions*, Harper Perennial (1993)
17. Billie Hughes, "Educational MUDs: Issues and Challenges", (invited keynote presentation) *MUDshop II*, September 1995, San Diego, California (1995); at URL `http://www.pc.maricopa.edu/community/pueblo/writings/MudShopBillie.html` (Last checked 20 March 2001)
18. Edwin Hutchins, *Cognition in the Wild*, MIT (1995)
19. Catriona M. Kennedy, "Distributed Reflective Architectures for Adjustable Autonomy", in David Kortenkamp, Gregory Dorais, Karen L. Myers (eds.), *Proc. IJCAI-99 Workshop on Adjustable Autonomy Systems*, 1 August 1999, Stockholm, Sweden (1999)
20. George Lakoff, "Women, Fire, and Dangerous Things", U. Chicago Press (1987)
21. Christopher Landauer, "Wrapping Mathematical Tools", pp. 261-266 in *Proc. EMC'90: The 1990 SCS Eastern Multi-Conference*, 23-26 April 1990, Nashville, Tennessee, SCS (1990)
22. Christopher Landauer, Kirstie L. Bellman, "Computational Embodiment: Constructing Autonomous Software Systems", *Cybernetics and Systems*, Vol. 30, No. 2, pp. 131-168 (1999)
23. Christopher Landauer, Kirstie L. Bellman, "Situation Assessment via Computational Semiotics", pp. 712-717 in *Proc. ISAS'98: The 1998 Int. MultiDisciplinary Conf. on Intelligent Systems and Semiotics*, 14-17 September 1998, NIST, Gaithersburg, Maryland (1998)
24. Christopher Landauer, Kirstie L. Bellman, "Computational Embodiment: Agents as Constructed Complex Systems", Chapter 11, pp. 301-322 in Kerstin Dautenhahn (ed.), *Human Cognition and Social Agent Technology*, Benjamins (2000)
25. Christopher Landauer, Kirstie L. Bellman, "Generic Programming, Partial Evaluation, and a New Programming Paradigm", Chapter 8, pp. 108-154 in Gene McGuire (ed.), *Software Process Improvement*, Idea Group Publishing (1999)
26. Christopher Landauer, Kirstie L. Bellman, "New Architectures for Constructed Complex Systems", in *The 7th Bellman Continuum, Int. Workshop on Computation, Optimization and Control*, 24-25 May 1999, Santa Fe, NM (1999); (to appear) in *Applied Mathematics and Computation*, (May 2001, expected)
27. Christopher Landauer, Kirstie L. Bellman, "Virtual Web Worlds: Extending the Web for Collaboration", pp. 90-95 in *Proc. WETICE'99: Workshop on Web-based Infrastructures and Coordination Architectures for Collaborative Enterprises*, 16-18 June 1999, Stanford, California (1999)
28. Christopher Landauer, Kirstie L. Bellman, "Lessons Learned with Wrapping Systems", pp. 132-142 in *Proc. ICECCS'99: The 5th Int. Conf. on Engineering Complex Computing Systems*, 18-22 October 1999, Las Vegas, Nevada (1999)
29. Christopher Landauer, Kirstie L. Bellman, "Architectures for Embodied Intelligence", pp. 215-220 in *Proc. ANNIE'99: 1999 Artificial Neural Nets and Industrial Engineering, Special Track on Bizarre Systems*, 7-10 November 1999, St. Louis, Mo. (1999)
30. Christopher Landauer, Kirstie L. Bellman, "Relationships and Actions in Conceptual Categories", pp. 59-72 in G. Stumme (Ed.), *Working with Conceptual Structures – Contributions to ICCS 2000, Auxiliary Proc. ICCS'2000: Int. Conf. on Conceptual Structures*, 14-18 August 2000, Darmstadt, Shaker Verlag, Aachen (August 2000)

31. Christopher Landauer, Kirstie L. Bellman, "Reflective Infrastructure for Autonomous Systems", pp. 671-676, Vol. 2 in *Proc. EMCSR'2000: The 15th European Meeting on Cybernetics and Systems Research, Symposium on Autonomy Control: Lessons from the Emotional*, 25-28 April 2000, Vienna (April 2000)
32. Christopher Landauer, Kirstie L. Bellman, "Symbol Systems and Meanings in Virtual Worlds", *Proc. VWsim'01: The 2001 Virtual Worlds and Simulation Conf.*, 7-11 January 2001, Phoenix, SCS (2001)
33. Christopher Landauer, Kirstie L. Bellman, "Conceptual Modeling Systems: Active Knowledge Processes in Conceptual Categories", *Proceedings ICCS'2001: The 9th International Conference on Conceptual Structures*, 30 July-03 August 2001, Stanford (August 2001)
34. Christopher Landauer, Kirstie L. Bellman, April Gillam, "Software Infrastructure for System Engineering Support", *Proc. AAAI'93 Workshop on Artificial Intelligence for Software Engineering*, 12 July 1993, Washington, D.C. (1993)
35. Christopher Landauer, Valerie E. Polichar, "More than Shared Artifacts: Collaboration via Shared Presence in MUDs", pp. 182-189 in *Proc. WETICE'98: Workshop on Web-based Infrastructures for Collaborative Enterprises*, 17-19 June 1998, Stanford University, Palo Alto, California (1998)
36. Thomas K. Landauer, *The Trouble with Computers: Usefulness, Usability, and Productivity*, MIT (1995)
37. Brenda Laurel (ed.), *The Art of Human-Computer Interface Design*, Addison-Wesley (1990)
38. Brenda Laurel, "Interface Agents: Metaphors with Character", Chapter 4, pp. 67-77 in [11]
39. Ulrike Lechner, Beat Schmid, Salome Schmid-Isler, Katarina Stanoevska-Slabeva, *Structuring and Systemizing Knowledge on the Internet – Realizing the Encyclopedia concept on Internet*, Study, 1998, 01/98; at URL `http://www.netacademy.org/netacademy/publications.nsf/all_pk/1036`, January 1998 (Last checked 20 March 2001)
40. Maja J. Mataric', "Studying the Role of Embodiment in Cognition", pp. 457-470 in *Cybernetics and Systems*, special issue on *Epistemological Aspects of Embodied Artificial Intelligence*, Vol. 28, No. 6 (July 1997)
41. James W. Moore, Review of *Waldo and Magic, Inc.* by Robert A. Heinlein, URL `http://www.wegrokit.com/jmwami.htm` (last checked 6 May 2001)
42. Thomas Moore, *Original Self: Living with Paradox and Originality*, Harper Collins Publishers (2000)
43. Bonnie A. Nardi and Vicki L. O'Day, *Information Ecologies: Using Technology with Heart*, MIT (1999)
44. Mike O'Brien, "Playing in the MUD", Ask Mr. Protocol Column, *SUN Expert*, Vol. 3, No. 5, pp. 19-20, 23, 25-27 (May 1992)
45. Paolo Petta, "The Role of Emotions in a Tractable Architecture for Situated Cognizers", (invited paper), *Proc. Workshop on Emotions in Humans and Artifacts*, 13-14 August 1999, Vienna, Austria (1999)
46. Paolo Petta, Carlos-Pinto Ferreira, and Rodrigo Ventura, "Autonomy Control Software: Lessons from the Emotional", in Henry Hexmoor (ed.), *Proc. Agents'99/ACS'99: Workshop on Autonomy Control Software*, 1 May 1999, Seattle, Washington (1999)
47. John D. Ramage, John C. Bean, *Writing Arguments: A Rhetoric with Readings* (3rd Ed.), Allyn and Bacon (1995)

48. Reed Riner, Jen Clodius, "Simulating Future Histories", *Anthropology and Education Quarterly*, Vol. 26, No. 1, pp. 95-104 (Spring 1995); at URL
`http://www.dragonmud.org/people/jen/solsys.html`
(Last checked 20 March 2001)
49. John Schwartz, "A Terminal Obsession", Washington Post Style Section (27 March 1994); summary posted by Mich Kabay to RISKS digest 29 March 1994 (Vol. 15 Issue 71); at URL
`http://www.infowar.com/iwftp/risks/Risks-15/risks-15.71.txt`
(Last checked 20 March 2001), and at URL
`http://catless.ncl.ac.uk/Risks/15.71.html\#subj3`
(Last checked 20 March 2001)
50. Douglas N. Walton, *Informal Logic: A Handbook for Critical Argumentation*, Cambridge (1989)
51. H. G. Wells, "World Brain: The Idea of a Permanent World Encyclopaedia", Contribution to the new *Encyclopédie Française* (August 1937); also in H. G. Wells, *World Brain*, Doubleday, Doran, Garden City, NY (1938); also at URL `http://sherlock.berkeley.edu/wells/world_brain.html` (Last checked 17 March 2001), and at URL `http://art-bin.com/art/obrain.html` (Last checked 17 March 2001)
52. D. D. Woods, "Cognitive Technologies: The Design of Joint Human-Machine Cognitive Systems". *The AI Magazine*, pp. 86-91 (1987)

Author Index

Lecture Notes in Artificial Intelligence (LNAI)

Vol. 1952: M.C. Monard, J. Simão Sichman (Eds.), Advances in Artificial Intelligence. Proceedings, 2000. XV, 498 pages. 2000.

Vol. 1955: M. Parigot, A. Voronkov (Eds.), Logic for Programming and Automated Reasoning. Proceedings, 2000. XIII, 487 pages. 2000.

Vol. 1967: S. Arikawa, S. Morishita (Eds.), Discovery Science. Proceedings, 2000. XII, 332 pages. 2000.

Vol. 1968: H. Arimura, S. Jain, A. Sharma (Eds.), Algorithmic Learning Theory. Proceedings, 2000. XI, 335 pages. 2000.

Vol. 1972: A. Omicini, R. Tolksdorf, F. Zambonelli (Eds.), Engineering Societies in the Agents World. Proceedings, 2000. IX, 143 pages. 2000.

Vol. 1979: S. Moss, P. Davidsson (Eds.), Multi-Agent-Based Simulation. Proceedings, 2000. VIII, 267 pages. 2001.

Vol. 1986: C. Castelfranchi, Y. Lespérance (Eds.), Intelligent Agents VII. Proceedings, 2000. XVII, 357 pages. 2001.

Vol. 1991: F. Dignum, C. Sierra (Eds.), Agent Mediated Electronic Commerce. VIII, 241 pages. 2001.

Vol. 1994: J. Lind, Iterative Software Engineering for Multiagent Systems. XVII, 286 pages. 2001.

Vol. 2003: F. Dignum, U. Cortés (Eds.), Agent-Mediated Electronic Commerce III. XII, 193 pages. 2001.

Vol. 2007: J.F. Roddick, K. Hornsby (Eds.), Temporal, Spatial, and Spatio-Temporal Data Mining. Proceedings, 2000. VII, 165 pages. 2001.

Vol. 2014: M. Moortgat (Ed.), Logical Aspects of Computational Linguistics. Proceedings, 1998. X, 287 pages. 2001.

Vol. 2019: P. Stone, T. Balch, G. Kraetzschmar (Eds.), RoboCup 2000: Robot Soccer World Cup IV. XVII, 658 pages. 2001.

Vol. 2033: J. Liu, Y. Ye (Eds.), E-Commerce Agents. VI, 347 pages. 2001.

Vol. 2035: D. Cheung, G.J. Williams, Q. Li (Eds.), Advances in Knowledge Discovery and Data Mining – PAKDD 2001. Proceedings, 2001. XVIII, 596 pages. 2001.

Vol. 2036: S. Wermter, J. Austin, D. Willshaw (Eds.), Emergent Neural Computational Architectures Based on Neuroscience. X, 577 pages. 2001.

Vol. 2039: M. Schumacher, Objective Coordination in Multi-Agent System Engineering. XIV, 149 pages. 2001.

Vol. 2056: E. Stroulia, S. Matwin (Eds.), Advances in Artificial Intelligence. Proceedings, 2001. XII, 366 pages. 2001.

Vol. 2062: A. Nareyek, Constraint-Based Agents. XIV, 178 pages. 2001.

Vol. 2070: L. Monostori, J. Váncza, M. Ali (Eds.), Engineering of Intelligent Systems. Proceedings, 2001. XVIII, 951 pages. 2001.

Vol. 2080: D.W. Aha, I. Watson (Eds.), Case-Based Reasoning Research and Development. Proceedings, 2001. XII, 758 pages. 2001.

Vol. 2083: R. Goré, A. Leitsch, T. Nipkow (Eds.), Automated Reasoning. Proceedings, 2001. XV, 708 pages. 2001.

Vol. 2086: M. Luck, V. Mařík, O. Štěpánková, R. Trappl (Eds.), Multi-Agent Systems and Applications. Proceedings, 2001. X, 437 pages. 2001.

Vol. 2087: G. Kern-Isberner, Conditionals in Nonmonotonic Reasoning and Belief Revision. X, 190 pages. 2001.

Vol. 2099: P. de Groote, G. Morrill, C. Retoré (Eds.), Logical Aspects of Computational Linguistics. Proceedings, 2001. VIII, 311 pages. 2001.

Vol. 2100: R. Küsters, Non-Standard Inferences in Description Logics. X, 250 pages. 2001.

Vol. 2101: S. Quaglini, P. Barahona, S. Andreassen (Eds.), Artificial Intelligence in Medicine. Proceedings, 2001. XIV, 469 pages. 2001.

Vol. 2103: M. Hannebauer, J. Wendler, E. Pagello (Eds.), Balancing Reactivity and Social Deliberation in Multi-Agent Systems. VIII, 237 pages. 2001.

Vol. 2109: M. Bauer, P.J. Gmytrasiewicz, J. Vassileva (Eds.), User Modeling 2001. Proceedings, 2001. XIII, 318 pages. 2001.

Vol. 2111: D. Helmbold, B. Williamson (Eds.), Computational Learning Theory. Proceedings, 2001. IX, 631 pages. 2001.

Vol. 2116: V. Akman, P. Bouquet, R. Thomason, R.A. Young (Eds.), Modeling and Using Context. Proceedings, 2001. XII, 472 pages. 2001.

Vol. 2117: M. Beynon, C.L. Nehaniv, K. Dautenhahn (Eds.), Cognitive Technology: Instruments of Mind. Proceedings, 2001. XV, 522 pages. 2001.

Vol. 2120: H.S. Delugach, G. Stumme (Eds.), Conceptual Structures: Broadening the Base. Proceedings, 2001. X, 377 pages. 2001.

Vol. 2123: P. Perner (Ed.), Machine Learning and Data Mining in Pattern Recognition. Proceedings, 2001. XI, 363 pages. 2001.

Vol. 2132: S.-T. Yuan, M. Yokoo (Eds.), Intelligent Agents: Specification, Modeling, and Application. Proceedings, 2001. X, 237 pages. 2001.

Lecture Notes in Computer Science